Techniques and Concepts of High-Energy Physics IV

NATO ASI Series

Advanced Science Institutes Series

A series presenting the results of activities sponsored by the NATO Science Committee, which aims at the dissemination of advanced scientific and technological knowledge, with a view to strengthening links between scientific communities.

The series is published by an international board of publishers in conjunction with the NATO Scientific Affairs Division

A	**Life Sciences**	Plenum Publishing Corporation
B	**Physics**	New York and London
C	**Mathematical and Physical Sciences**	D. Reidel Publishing Company Dordrecht, Boston, and Lancaster
D	**Behavioral and Social Sciences**	Martinus Nijhoff Publishers
E	**Engineering and Materials Sciences**	The Hague, Boston, Dordrecht, and Lancaster
F	**Computer and Systems Sciences**	Springer-Verlag
G	**Ecological Sciences**	Berlin, Heidelberg, New York, London,
H	**Cell Biology**	Paris, and Tokyo

Series B: Physics

Techniques and Concepts of High-Energy Physics IV

Edited by

Thomas Ferbel

University of Rochester
Rochester, New York

Plenum Press
New York and London
Published in cooperation with NATO Scientific Affairs Division

Proceedings of a NATO Advanced Study Institute on
Techniques and Concepts of High-Energy Physics,
held June 19–30, 1986,
in St. Croix, U.S. Virgin Islands

Library of Congress Cataloging in Publication Data

NATO Advanced Study Institute on Techniques and Concepts of High-Energy
    Physics (4th: 1986: Christiansted, V. I.)
    Techniques and concepts of high-energy physics IV / edited by Thomas
Ferbel.
        p. cm.—(NATO ASI series. Series B, Physics; v. 164)
    "Proceedings of NATO Advanced Study Institute on Techniques and Con-
cepts of High-Energy Physics, held June 19–30, 1986, in St. Croix, U.S. Virgin
Islands"—T.p. verso.
    "Published in cooperation with NATO Scientific Affairs Division."
    Includes bibliographical references and index.
    ISBN 978-1-4684-5403-1   ISBN 978-1-4684-5401-7 (eBook)
    DOI 10.1007/978-1-4684-5401-7
    1. Particles (Nuclear physics)—Congresses. I. Ferbel, Thomas. II. North
Atlantic Treaty Organization. Scientific Affairs Division. III. Title. IV. Series.
QC793.N38   1986
539.7′21—dc19   87-21293
                                                                     CIP

© 1987 Plenum Press, New York
Softcover reprint of the hardcover 1st edition 1987
A Division of Plenum Publishing Corporation
233 Spring Street, New York, N.Y. 10013

PREFACE

The fourth Advanced Study Institute (ASI) on Techniques and Concepts of High Energy Physics was held once again at the Hotel on the Cay, in the scenic harbor of Christiansted, St. Croix, U.S. Virgin Islands. The ASI brought together a total of 67 participants, from 17 different countries. It was a great success, due to the dedication of the inspiring lecturers, the exceptional student body, and, of course, the beautiful setting.

The primary support for the meeting was again provided by the Scientific Affairs Division of NATO. The ASI was cosponsored by the U.S. Department of Energy, by Fermilab, by the National Science Foundation, and by the University of Rochester. A special contribution from the Oliver S. and Jennie R. Donaldson Charitable Trust provided an important degree of flexibility, as well as support for worthy students from developing nations.

As in the case of the previous ASI's, the scientific program was designed for advanced graduate students and recent PhD recipients in experimental particle physics. The present volume of lectures should complement the material published in the first three ASI's, and prove to be of value to a wider audience of physicists.

It is a pleasure to acknowledge the encouragement and support that I have continued to receive from colleagues and friends in organizing this meeting. I am indebted to the members of my Advisory Committee for their infinite patience and excellent advice. I am grateful to my distinguished lecturers for participating in the ASI. I thank John Bythel of the West Indies Lab for providing us with a description of the geology and marine life of St. Croix, and Albert Lang for talking him into it. I thank Melissa Franklin for organizing the student presentations. I also wish to thank Earle Fowler, Bernard Hildebrand and Bill Wallenmeyer for support from the Department of Energy, and David Berley for the assistance from the National Science Foundation. I thank Leon Lederman for providing me with access to the talents of Angela Gonzales at Fermilab. At Rochester, I am indebted to Judy Mack, Sal Spinnichia, and especially Connie Jones, for organizational assistance and typing. I owe thanks to Shirley and Ray Boudreau and to Hurchell Greenaway, the managers of the facilities at the Hotel on the Cay, for their and their staffs' hospitality. I wish to acknowledge the generosity of Chris Lirakis and of Mrs. Marjorie Atwood of the Donaldson Trust. Finally, I thank Dr. Craig Sinclair of NATO for his continuing cooperation and confidence.

T. Ferbel
Rochester, New York
May 1987

CONTENTS

OLD PHYSICS, NEW PHYSICS AND COLLIDERS

I. Hinchliffe

Lawrence Berkeley Laboratory
University of California
Berkeley, California 94720

Introduction

In these lectures, I shall discuss some topics in the standard model of strong and electroweak interactions and shall then point out some problems in this model and indicate how these problems are attacked in some theoretical models. I shall begin with a discussion of radiative corrections in the Glashow-Weinberg-Salam model,[1] stressing how these corrections may be measured at LEP and the SLC. I shall then discuss some features of QCD which are relevant to hadron colliders. This discussion will complement the lectures of Luigi DiLella,[2] who has shown impressive evidence from the CERN Sp$\bar{\text{p}}$S collider for the correctness of QCD. In my discussion of the unsolved problems of the standard model I shall not discuss supersymmetric theories since their theoretical aspects and phenomenological consequences are discussed by other lecturers at this school.[3] I shall however, discuss some aspects of models in which quarks and leptons are composite particles.

1. Testing the Weinberg-Salam model.

The Lagrangian describing the weak and electromagnetic interactions of the quarks and leptons is given by[1]

$$
\begin{aligned}
\mathcal{L} = &-\frac{1}{4}\,F^a_{\mu\nu}F_a^{\mu\nu} - \frac{1}{4}\,G_{\mu\nu}G^{\mu\nu} \\
&+ i\bar{\psi}_{Li}\gamma^\mu D_\mu \psi_{Li} \\
&+ \mu^2(\Phi^+\Phi) - \lambda(\Phi^+\Phi)^2 \\
&+ \lambda_{ek}\bar{\ell}_{L,k}\Phi e_{R,k} + \lambda_{ukl}\bar{q}_{L,k}\Phi^+ u_{R,l} + \lambda_{dkl}\bar{q}_{L,k}\Phi d_{R,l}
\end{aligned}
\tag{1.1}
$$

where

$$
F^a_{\mu\nu} = \partial_\mu W^a_\nu - \partial_\nu W^a_\mu + g_2 \epsilon^{abc}W^b_\mu W^c_\nu
$$

and

$$
G_{\mu\nu} = \partial_\mu B_\nu - \partial_\nu B_\mu
$$

are the field strength tensors for the three gauge bosons of $SU(2)_L$ (W_μ^a) and $U(1)$ (B_μ), which have coupling constants g_2 and g_1. The indices on the fermion fields are generation indices which take values in the range (1,2,3).

The left-handed fermions appear in $SU(2)_L$ doublets

$$\psi_{iL}: \quad \ell_L = \frac{1}{2}(1-\gamma_5)\begin{pmatrix} \nu \\ e \end{pmatrix}, \quad q_L = \frac{1}{2}(1-\gamma_5)\begin{pmatrix} u \\ d \end{pmatrix}$$

which have $U(1)$ charges -1, and 1/3. The right-handed fermions appear as $SU(2)_L$ singlets

$$\psi_{iR}: \quad e_R = \frac{1}{2}(1+\gamma_5)e, \quad u_R = \frac{1}{2}(1+\gamma_5)u, \quad d_R = \frac{1}{2}(1+\gamma_5)d$$

with $U(1)$ charges -2, 4/3 and $-2/3$, respectively. This pattern is, of course, replicated for the second and third generations which contain the μ and τ leptons and the strange, charm, top and bottom quarks. The Higgs doublet Φ has $U(1)$ charge -1. The covariant derivatives D_μ are given by

$$D_\mu = (\partial_\mu - ig_2 T^a W_\mu^a - ig_1\frac{y}{2}B_\mu).$$

Here y is the $U(1)$ charge of the representation on which D_μ acts. For an $SU(2)$ doublet $T^a = \tau^a/2$, where τ^a is a Pauli matrix, while for an $SU(2)$ singlet, $T = 0$.

This Lagrangian contains seventeen parameters. There are two gauge coupling constants g_2 and g_1 describing the interactions of the $SU(2)$ and $U(1)$ gauge theories. Two parameters μ and λ determine the Higgs mass and the interactions of the Higgs field with itself. The remaining parameters are the quark and lepton Yukawa couplings λ_i. Let us examine the spectrum of physical states in the model.

For $\mu^2 > 0$ the ground state of the theory is given when the Higgs field ϕ has a non-zero vacuum expectation value (VEV):

$$\langle\Phi\rangle = \begin{pmatrix} 0 \\ \frac{v}{\sqrt{2}} \end{pmatrix} \tag{1.2}$$

with $v = (\mu^2/\lambda)^{1/2}$. This non-zero VEV results in a mass for three of the four gauge bosons. The charged gauge bosons of $SU(2)_L$ have mass

$$M_W^2 = g_2^2\frac{v^2}{4}. \tag{1.3}$$

There is a massless neutral gauge boson, the photon,

$$A_\mu = \sin\theta_W W_\mu^3 - \sin\theta_W B_\mu$$

and a massive boson

$$Z_\mu = \cos\theta_W W_\mu^3 + \cos\theta_W B_\mu$$

with mass $M_Z^2 = v^2(g_1^2 + g_2^2)/4$. Here the weak mixing angle θ_W is given by

$$\tan\theta_W = \frac{g_1}{g_2}. \tag{1.4}$$

Table 1: Couplings of physical particles in the Weinberg-Salam model (unitary gauge). f is a fermion of charge Q_f and weak isospin T_3 ($= +1$ for u, c, t quarks and neutrinos, -1 otherwise).

$$W f \bar{f} \qquad \frac{g_2}{2\sqrt{2}} W_\mu \bar{f} \gamma^u (1 - \gamma_5) f$$

$$Z \bar{f} f \qquad \frac{e}{2 \sin \theta_W \cos \theta_W} Z^u \bar{f} \gamma^u (v_f - a_f \gamma_5) f$$

$$V_f = T_3 - 4Q \sin^2 \theta_W$$

$$A_f = T_3$$

$$HWW \qquad g_2 M_W H W_\mu^+ W_\mu^-$$

$$HZZ \qquad \frac{g_2}{2 \cos \theta_W} M_z Z_\mu Z_\mu$$

$$H f \bar{f} \qquad \frac{g_2 m_f}{2 M_W} H \bar{f} f$$

The electric charge of the electron is given by $e = g_2 \sin \theta_W$. The non-zero value of v results in lepton masses

$$m_e = \lambda_e \frac{v}{\sqrt{2}}. \tag{1.5}$$

The quark masses are more complicated since weak interactions allow transitions between different generations, i.e., $s \to u + W^-$. The Yukawa interactions of the up quarks can be chosen to be diagonal, i.e.,

$$\lambda_{u_{ij}} = \lambda_{u_i} \delta_{i,j}.$$

The masses of the charge 2/3 quarks are then given by

$$m_{u_i} = \lambda_{u_i} \frac{v}{\sqrt{2}}. \tag{1.6}$$

The down quark mass matrix contains seven parameters which are the masses of the d, s and b quarks and the four angles of the Kobayashi-Maskawa mixing matrix. The final parameter is the Higgs mass $m_H = \sqrt{2\lambda}\, m_W / g_2$. The theory has a large number of parameters but is able to describe a wealth of experimental data. The most important parameters are v, g_1 and g_2 which control the strength of weak and electromagnetic interactions. Most experimental tests of the model do not depend upon quark or lepton masses (or alternatively, on the quark and lepton Yukawa couplings) so that experimental success is more remarkable.

The W and Z bosons couple to quarks, leptons and the remaining physical Higgs boson H with interactions shown in Table 1.

I have so far discussed the model at the tree level, i.e., to lowest order in the coupling constants g_1 and g_2. Before discussing tests of the theory, it is worth noting the approximate size of the radiative corrections which can be expected. These corrections will depend upon the fine coupling constant $\alpha = e^2/(4\pi)$. In addition, tests will be made over a large range of momenta. Momentum transfers can be very small (for example, in Thompson scattering) or very high (for example, the production of a Z or W boson). The gauge interactions produce effects which depend logarithmically on these scales. Hence an order of magnitude estimate of radiative corrections will give $\alpha/\pi \log(M_W^2/m_e^2)$. This is of order 5%. Some experiments are already sensitive to corrections of this size; experiments at the Z^0 resonance performed at LEP[4] or the SLC[5] will be more sensitive so it is important to discuss radiative corrections in some detail.

I will begin with the radiative corrections in Quantum Electrodynamics. Consider the scattering of two charged particles of mass M, at momentum transfer Q. To lowest order in α, this scattering is described by the exchange of a single photon (Fig. 1a). If the theory contains a particle of mass m_e, the effect of this particle can appear at next order in perturbation theory via the graph of Fig. 1b. The relevant Feynman diagram is the one-loop correction to the photon self energy shown in Fig. 2. This graph is given by

$$\Pi_{\mu\nu}(Q^2) = -e^2 \int \frac{d^4k}{(2\pi)^4} \frac{\text{Tr}[\gamma^\mu(k + m_e)\gamma^\nu(k - Q + m_e)]}{(k^2 - m_e^2)((k - Q)^2 - m_e^2)}. \tag{1.7}$$

This integral is divergent; we can regulate it by performing the loop integral in n dimensions, i.e. by making the replacement[6]

$$\frac{d^4k}{(2\pi)^4} \rightarrow \frac{d^n k}{(2\pi)^n}.$$

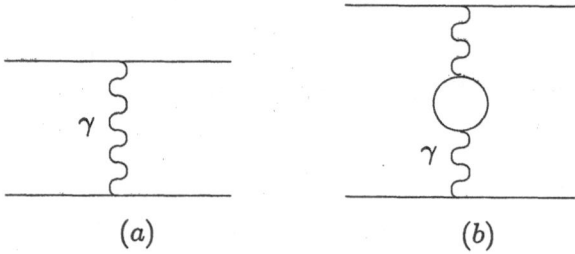

(a) (b)

Figure 1:   (a) Feynman diagram showing a contribution to the scattering of two charged particles in QED; (b) A higher order contribution.

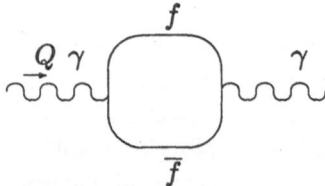

Figure 2:   A contribution to the photon self energy at one loop due to a charged fermion of mass m_e.

We then have

$$\Pi_{\mu\nu}(Q^2) = (Q_\mu Q_\nu - Q^2 g_{\mu\nu})\Pi(Q^2)$$

$$= (Q_\mu Q_\nu - Q^2 g_{\mu\nu})\frac{2i\alpha}{\pi}\left[\frac{2}{4-n} + \log(4\pi) - \gamma_E \right. \tag{1.8}$$

$$\left. - \int_0^1 z(1-z)\log\left(\frac{Q^2 z(1-z) - m_e^2}{\mu^2}\right)dz + O(n-4)\right].$$

Here γ_E is the Euler-Mascheroni constant ($\gamma_E = 0.577$) and $\alpha = e^2/4\pi$. I have expanded the result around $n = 4$ and not written the terms which vanish as $n \to 4$. The scale μ has been introduced since, in n dimensions, the interaction of an electron with a photon has the following form

$$e\mu^{(n-4)/2}\bar{\psi}\gamma^\mu\psi A_\mu. \tag{1.9}$$

The coupling constant e (or α) then remains dimensionless in n dimensions. μ is an arbitrary constant — physics cannot depend upon it.

The scattering of the heavy particles then has an amplitude of the following form

$$\frac{e^2}{Q^2}[1 + i\Pi(Q^2)]. \tag{1.10}$$

In order to make contact with physics, the divergence at $n = 4$ must be removed, i.e., the theory must be renormalized. Two renormalization schemes are often used:

(a) Minimal subtraction.[7] Here the term $\frac{2}{n-4}$ and the attendant constants γ_E and $\log 4\pi$ are thrown out. This amounts to defining a renormalized charge

$$e_R^2 = e^2 + \frac{e^2}{4\pi^2}\left[\frac{1}{(4-n)} + \frac{1}{2}log4\pi - \frac{\gamma_E}{2}\right]. \tag{1.11}$$

This scheme is very easy to use but it is unphysical; the resulting renormalized coupling constant is not directly related to any physical quantity. This definition is the one normally used in QCD.

(b) Define the renormalized charge so that as $Q^2 \to 0$, the scattering amplitude is e_R^2/Q^2.[8] Hence, $e_R^2 = e^2[1 + i\Pi(0)]$. This definition has the advantage that it can be related directly to a physical quantity, the scattering rate at small momentum transfer. It is the definition used in Quantum Electrodynamics; it corresponds to the value of $\alpha = 1/137$ measured from Thompson scattering.[*] This definition cannot be used in QCD since perturbation theory is not reliable as $Q^2 \to 0$.

In the limit of large Q^2, the scattering amplitude has the following form

$$\frac{\alpha_R}{Q^2}\left[1 + \frac{\alpha_R}{3\pi}\log(Q^2/m_e^2)\right]. \tag{1.12}$$

I have retained terms of order $\log Q^2/m_e^2$ only and have used the definition (b) of α. We can introduce a running coupling constant $\alpha(Q^2)$ by

$$\alpha(Q^2) = \alpha\left[1 + \frac{\alpha}{3\pi}\log(Q^2/m_e^2)\right]. \tag{1.13}$$

---

[*]The most accurate measurements of α come from the Josephson Junction.[9] Since the momentum transfers in this case are extremely small, the value obtained corresponds to the definition discussed here.

The scattering amplitude is then proportional to $\alpha(Q^2)/Q^2$.

In the standard model there are contributions to $\alpha(Q^2)$ from all charged particles. The only one whose mass is not known is the top quark. Each charged particle begins to contribute when $Q^2 > m_i^2$. The evolution of $\alpha(Q^2)$ is shown in Fig. 3. At $Q^2 = M_W^2$ [10]

$$\alpha_{em}(M_W^2) \sim 1/128. \qquad (1.14)$$

Having defined the electromagnetic coupling α_{em}, it now remains to specify the other parameters of the Weinberg-Salam model. Apart from the fermion masses and the Kobayashi-Maskawa angles there are three parameters. In the Lagrangian these are μ, λ and one of g_1 and g_2. (The other is fixed by α). At lowest order these parameters can be taken to be M_W, M_H and g_2. A renormalized mass is easy to define; it is simply the position of the pole in the particle's propagator which corresponds to its physical mass. g_2 remains to be defined.

At present the W mass is not well measured. Consequently, we shall not take it to be one of the fundamental parameters. The Higgs mass must be taken as one parameter. The two remaining ones will be directly related to two accurately measured physical quantities. The muon lifetime is extremely well measured. This can be used to extract the Fermi constant G_F.[11] We need one other quantity. I will take this to be the mass of the Z boson which will be well measured at the SLC or LEP in the near future. The fundamental parameters are therefore: α_{em}, obtained from the Josephson Junction; G_F, obtained from the muon lifetime; M_Z and the Higgs mass M_H which does not play a crucial role in the subsequent discussion. This procedure eliminates the need to define the coupling constant g_2.

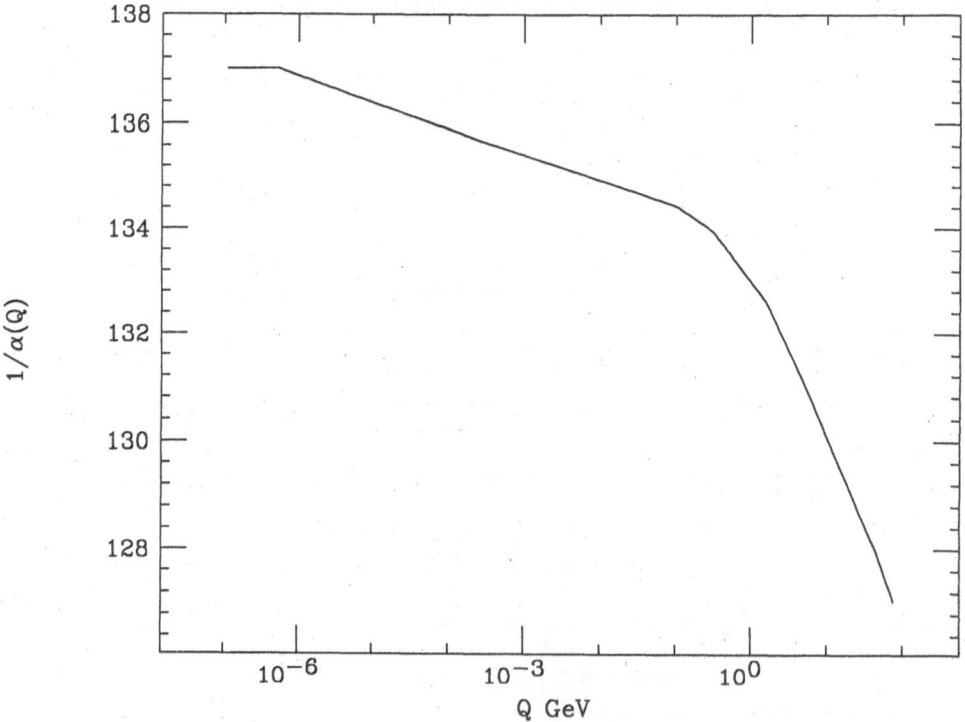

Figure 3: The behavior of $\alpha_{em}(Q)$ with Q, which shows the increase of $\alpha_{em}(Q)$ due to the known quarks and leptons.

In lowest order the Fermi constant G_F is related to the muon lifetime (τ_μ) by

$$\frac{1}{\tau_\mu} = \frac{G_F^2 m_\mu^3}{192\pi^3}\left[1 - \frac{8m_e^2}{m_\mu^2}\right].\tag{1.15}$$

It is traditional to include higher order QED corrections from the graphs of the type shown in Fig. 4. The right-hand side of Eq. (1.15) is modified by a factor[11]

$$1 + \frac{\alpha}{2\pi}\left[\frac{25}{4} - \pi^2\right]\left[1 + \frac{2\alpha}{3\pi}\log(m_e/m_\mu)\right].\tag{1.16}$$

Since α is known, the muon lifetime can be used to extract G_F. [†]

Once the Z mass is determined the W mass is predicted[12] to be

$$M_W = \frac{M_Z}{\sqrt{2}}\left[1 + \left[1 - \frac{4\pi\alpha}{\sqrt{2}M_W^2 G_F(1 - \Delta r)}\right]^{1/2}\right]^{1/2}.\tag{1.17}$$

Δr includes the effect of radiative corrections, for a top quark mass of 35 GeV and a Higgs mass of 100 GeV

$$\Delta r = 0.0696 \pm 0.0020.\tag{1.18}$$

There is some uncertainty in Δr. Apart from the unknown top and Higgs masses, the contribution of light quarks in the loops of Fig. 2 is uncertain. This contribution must be gotten from measurements of the cross-section for the process $e^+e^- \to$ hadrons since QCD corrections are not small and cannot be calculated when $Q^2 \leq 1$ GeV^2.

What has happened to the weak mixing angle θ_W? In the approach that I have used it is not a fundamental parameter. It can be defined by $\cos\theta_W \equiv M_W/M_Z$. The coupling constant g_2 can now be defined by $g_2 \equiv e/\tan\theta_W$. In lowest order when Δr is zero, g_2 is related to G_F in the usual way

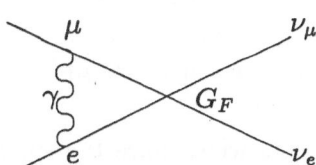

Figure 4:   Feynman diagram showing a QED correction to the muon lifetime.

Table 2: Values of M_Z (or $\sin^2 \theta_W$) extracted from various experiments.

Process	M_Z	$\sin^2 \theta_W$	Ref.
$e^+e^- \to \mu^+\mu^-$	84 ±4.8	0.17 ±0.02	16
$\nu p \to \nu p$ $\bar{\nu} p \to \bar{\nu} p$	91.8±2.8	0.23 ±0.02	17
ed asymmetry	93.3±2.1	0.220±0.014	18
$\nu_\mu e \to \nu_\mu e$	91.8±2.8	0.23 ±0.02	19
Parity violation in atoms	98.5±7.9	0.19 ±0.04	20
$\bar{\nu} N \to \mu X, \bar{\nu} X$ $\nu N \to \mu X, \nu X$	92.4±0.6	0.226±0.004	21

$$G_F = \frac{g_2^2}{4\sqrt{2}M_W^2}. \tag{1.19}$$

There is an alternative renormalization scheme to the one I have described. e and g_2 can be defined by minimal subtraction[13] (call these $\bar{e}$ and $\bar{g}_2$). The weak mixing angle is now defined by

$$\tan \overline{\theta_W} = \bar{e}/\bar{g}_2. \tag{1.20}$$

$\bar{\theta}_W$ and θ_W are related by $\sin \bar{\theta}_W = \sin \theta_W + 0.006$.[13] In view of the possible confusion I shall not use $\sin \theta_W$ in the subsequent discussion.

The present measurements of M_Z from UA1 and UA2 has a large error.[14] Table 2 shows the value of M_Z extracted from the analysis of a large set of low energy experiments. For example, the ratio of cross-sections

$$\frac{\sigma(\nu_\mu e \to \nu_\mu e)}{\sigma(\nu_\mu e \to \nu_e \mu)} = \frac{7M_Z^4 - 20M_Z^2 M_W^2 + 16M_W^4}{13M_Z^4 - 28M_Z^2 M_W^2 + 16M_W^4} \tag{1.21}$$

can be used to extract M_Z using the formula of Eq. (1.17) for M_W. This expression for the ratio of cross-sections is given in lowest order. The errors are too large for radiative corrections to be relevant.

The two types of experiments which have the smallest quoted errors are deep inelastic neutrino scattering[21] and the asymmetry in polarized electron deuterium scattering.[18] The ratio of the cross-sections for neutral and charged current deep inelastic neutrino scattering from a nucleon ‡ is given by

$$\frac{\sigma(\nu_\mu N \to \nu_\mu X)}{\sigma(\nu_\mu N \to \mu X)} = \frac{\frac{1}{2} - x + \frac{20x^2}{27} + \epsilon\left(\frac{1}{6} - \frac{x}{3} + \frac{20x^2}{27}\right)}{1 + \epsilon/3}. \tag{1.22}$$

In lowest order $x = 1 - M_W^2/M_Z^2$. ϵ is the ratio of the fraction of nucleon's momentum carried by antiquarks to that carried by quarks. Radiative corrections cause a small shift in x.[22] This shift depends upon the kinematics of the experiment and is of order 0.005. The value of $x = 0.226 \pm 0.004$ which is quoted has an error which is smaller than this higher order correction.

---

‡ N here refers to a target which consists of an equal mixture of protons and neutrons.

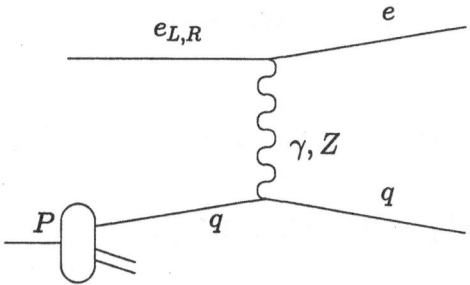

Figure 5: Diagram relevant for the scattering of polarized electrons from a nuclear target.

There are a number of uncertainties in the value of x extracted from deep inelastic scattering. Firstly, there are QCD corrections to the structure functions. A more important source of uncertainty is that due to the charm quark. There is a contribution to the charged current cross-section from the reaction $\nu + s \rightarrow \mu^- + c$. This rate is affected by a threshold factor which depends on the charm quark mass. The neutral current cross-section is affected to a lesser degree since the process $\nu + c \rightarrow \nu + c$ is inhibited due to the small number of charm quarks in the nucleon. If the charm quark mass is allowed to vary from 1.2 to 1.8 GeV, there is an uncertainty of order $\pm$ 0.005 in x.[21] Since the charm quark mass is not determined within this range, I am forced to conclude that this measurement is not sensitive to radiative corrections.

In the case of the scattering of a polarized electron from a deuteron,[18] one observes the interference between Z and photon exchange (see Fig. 5). The following asymmetry is predicted

$$\frac{\sigma_L - \sigma_R}{\sigma_L + \sigma_R} = Q^2 \left[a_1 + a_2 \left(\frac{1 - (1-y)^2}{1 + (1-y)^2} \right) \right],\tag{1.23}$$

where $\sigma_L (\sigma_R)$ is the cross-section for a left (right) handed electron and

$$a_1 = \frac{-G_F}{2\sqrt{2}\pi\alpha} \frac{9}{10} \left[1 - \frac{20x}{9} \right],$$

$$a_2 = \frac{-G_F}{2\sqrt{2}\pi\alpha} \frac{9}{10} [1 - 4x].\tag{1.24}$$

The kinematical variable y is the fractional energy loss of the electron: $y = \frac{E_1 - E_2}{E_1}$ where E_1 (E_2) is the incoming (outgoing) electron energy. In this case the higher order corrections computed for the kinematics of the SLAC ed scattering experiment are $\delta x = 0.005$.[13] Again this is comparable to the experimental error. It appears, therefore, that only the next generation of experiments will be able to see higher order corrections.

Table 3 shows the values of M_W and M_Z found by the UA1 and UA2 collaborations.[14] The predicted value of M_Z and M_W inferred from the results in Table 2 are shown for comparison. The agreement is remarkable.

Table 3: Values of M_W and M_Z measured by the UA1 and UA2 collaborations[14].

	UA1	UA2	Values[15] obtained using Table 2
$M_W (GeV)$	$83.5^{+1.1}_{-1.0} \pm 2.7$	$81.2 \pm 1.1 \pm 1.3$	81.4 ± 0.6 (79.8 without radiative corr.)
$M_Z (GeV)$	$93.0 \pm 1.4 \pm 3$	$92.5 \pm 1.3 \pm 1.5$	92.5 ± 0.5 (90.2 without radiative corr.)

As I have stressed, most of the radiative corrections are due to known quantities in the standard model such as the coupling of the electron to the gauge bosons. In principle, the accurate measurement of such radiative corrections can give information on the two main unknown parameters, namely the top quark and Higgs masses. In order to illustrate this, consider the effect of these particles upon the relationship between the W and Z masses (see Eq. (1.17)).

There are contributions to the W and Z self energies from the t and b quarks which are shown in Fig. 6. The evaluation of these graphs at zero external momentum gives[23]

$$\Pi_W^{\mu\nu} = -\frac{3ig^{\mu\nu}g_2^2}{32\pi^2} \left[m_t^2 \log m_t^2/\mu^2 + m_b^2 \log m_b^2/\mu^2 \right.$$
$$\left. + \frac{m_t^2 m_b^2}{m_t^2 - m_b^2} \log m_t^2/m_b^2 - \frac{1}{2}(m_t^2 + m_b^2) \right],$$

$$\Pi_Z^{\mu\nu} = \frac{3ig^{\mu\nu}g_2^2}{32\pi^2} \frac{M_{Z_{Lo}}^2}{M_{W_{Lo}}^2} (m_t^2 \log m_t^2/\mu^2 + m_b^2 \log m_b^2/\mu^2). \quad (1.25)$$

I have used dimensional regularization and dropped the terms proportional to $\frac{1}{n-4}$, γ_E and $\log 4\pi$. The quantities $M_{W_{Lo}}$ and $M_{Z_{Lo}}$ are the lowest order values for the W and Z masses. These contributions cause shifts in the W and Z masses

$$M_{W,Z}^Z = M_{W,Z,Lo}^Z \left[1 + \frac{\Pi_{W,Z}^{\mu\nu}}{g^{\mu\nu} M_{W,Z,Lo}^2} \right]. \quad (1.26)$$

These contributions then modify the relationship between the W and Z masses

$$M_W^2 = M_{W_{Lo}}^2 \left[1 + \frac{3G_F}{8\sqrt{2}\pi^2} \left[\frac{2m_t^2 m_b^2}{m_t^2 - m_b^2} \log\left(\frac{m_t^2}{m_b^2}\right) + m_t^2 + m_b^2 \right] \right]. \quad (1.27)$$

Notice that the μ dependence (and the terms proportional to $1/(n-4)$ had I written them) have canceled, i.e., this correction is finite and independent of the renormalization scheme. Notice that the effect of these corrections is to increase M_W as the t quark mass rises. Current constraints from low energy experiments and from the measured value the W mass imply $m_t \leq 320 GeV$.

There are also shifts in the W and Z masses which arise from radiative corrections involving the Higgs.[24] The Feynman diagrams are shown in Fig. 7. The graphs of Fig. 7c do not contribute to the Δr of Eq. (1.17), since these graphs correspond to

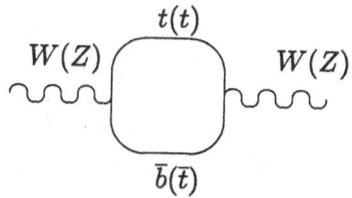

Figure 6: Feynman diagram showing contributions to the W and Z self energies from the t and b quarks.

a renormalization of the Higgs VEV, whose value does not affect M_W/M_Z. In the limit of large Higgs mass we have

$$M_W^2 = M_{W_{L0}}^2 \left(1 - \frac{3G_F}{8\sqrt{2}} M_Z^2 \left(1 - \frac{M_W^2}{M_Z^2}\right) \log(m_H^2/M_W^2)\right).$$ (1.28)

The dependence upon M_H is rather weak; M_W falls slowly as m_H is increased. Figure 8 shows the relationship between the W and Z masses for different values of top quark and Higgs masses.

The result that as m_t or m_H increases the radiative corrections increase, may seem to be contrary to intuition. Consider a theory with a particle of mass M. If this theory is probed with energies much less than M, then there is a general result known as the decoupling theorem which states that the effect of the heavy particle is proportional to $1/M^p$, where p is some positive number. As $M \to \infty$, the particle decouples from low energy physics.[25] Thus, for example, the effects of the τ lepton on $(g-2)$ of the muon are very small. This theorem is proved under the assumption that the coupling of the particle does not vary with M. This is the case for a heavy lepton in QED whose coupling to the photon, α_{em}, is independent of M.

In the case of the electroweak theory, the couplings cannot be held fixed as M is increased. Recall that the top quark mass is related to the W mass via

$$m_t = \sqrt{2}\frac{\lambda_t}{g_2} M_W$$ (1.29)

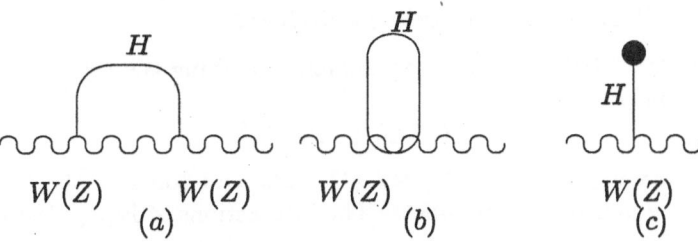

Figure 7: Feynman diagrams showing contributions to the W and Z self energies from the Higgs boson.

11

where λ_t is the Yukawa coupling of the t quark to the Higgs boson. If M_W is held fixed and m_t is increased, then λ_t must increase; the top quark interacts more strongly with the Higgs. Clearly, for a sufficiently large m_t, λ_t will be so large that perturbation theory ceases to be reliable. I shall return to this point later. Recall also that the Higgs mass is related to the W mass and the Higgs self interaction (λ) by

$$m_H^2 = 2\lambda M_W^2/g_2^2. \tag{1.30}$$

Again, if M_W is held fixed then a large m_H implies a large λ.

The graphs of Figs. 6 and 7 appear to depend only upon the gauge couplings g_1 and g_2 and not upon λ_t and λ. This is illusory as the following argument will demonstrate. Before the $SU(2) \times U(1)$ symmetry is broken, the theory contains four massless gauge bosons and four scalars (the components of the complex Higgs doublet Φ). After symmetry breaking, three of the gauge bosons acquire mass. In order to do so they must each gain an additional degree of freedom, their longitudinal polarization states. Three of the scalars supply those degrees of freedom. Hence, "the physical W and Z bosons have some Higgs in them". Couplings of physical W's and Z's are therefore sensitive to the Higgs Yukawa coupling.[§] Hence fermions and Higgs bosons of large mass in the standard model do not decouple and can affect the relationship between the W and Z masses.

At lowest order, there is a relationship (cf. Eqs. (1.3) and (1.4)) between g_1, g_2, M_W and M_Z, viz.,

$$\frac{M_W^2}{M_Z^2} = \frac{g_2^2}{g_1^2 + g_2^2}. \tag{1.31}$$

The form of this relationship is due to the breaking of $SU(2) \times U(1)$ via a Higgs doublet. In models with more complicated Higgs sectors (for example Higgs triplets) this relationship is lost. We can introduce an additional parameter ρ to take account of this possibility

$$\rho \equiv \frac{\sqrt{2}M_W}{M_Z \left(1 + \left(1 - \frac{4\pi\alpha}{\sqrt{2}G_F M_Z^2(1-\Delta r)}\right)^{1/2}\right)^{1/2}} \tag{1.32}$$

so that $\rho \equiv 1$ in the standard model. (Note that $\Delta r = 0$ in lowest order; in higher orders it should take the value predicted in the minimal model.) If the low energy data listed in Table 2 is analyzed with the parameter ρ, one gets[15]

$$\rho = 1.006 \pm 0.008. \tag{1.33}$$

Since ρ is consistent with 1 it is reasonable to ask if this implies that the Higgs sector of the $SU(2)_L \times U(1)$ model is severely constrained.

The part of the $SU(2)_L \times U(1)$ Lagrangian describing the self interaction of the Higgs fields, namely

$$\mu^2(\Phi^+\Phi) + \lambda(\Phi^+\Phi)^4 \tag{1.34}$$

has a larger symmetry than $SU(2)_L \times U(1)$. One can consider the complex Higgs doublet as having four real components. The interactions of Eq. (1.34) are invariant

---

[§]This argument can be seen clearly by writing the theory in 't Hooft-Feynman gauge.

with respect to rotations among these components, i.e., there is an $O(4) = SU(2) \times SU(2)$ symmetry. When the $SU(2)_L$ symmetry is broken as one component of the Higgs doublet gets a non-zero VEV, the $O(4)$ symmetry is broken to $O(3) = SU(2)$. It is this $SU(2)$, known as custodial $SU(2)$,[26] which insures that $\rho = 1$. It does this because the resulting mass for the three components W_μ^1, W_μ^2 and W_μ^3 of the gauge boson multiplet of $SU(2)_L$ has the form

$$M_W^2 (W_\mu^1 W_\mu^1 + W_\mu^2 W_\mu^2 + W_\mu^3 W_\mu^3). \qquad (1.35)$$

The mixing of W_μ^3 with B_μ then produces the Z_μ and photon. Any variant of the standard $SU(2) \times U(1)$ model which has a custodial $SU(2)$ symmetry (e.g., a model with an arbitrary number of Higgs doublets) will predict $\rho = 1$. As an example of a model without such a symmetry, suppose that we try to break the $SU(2)_L \times U(1)$ symmetry with a mixture of doublets and triplets. The self interaction of the triplet has an $O(3)$ symmetry which breaks to $O(2)$ when one component acquires a VEV. The residual symmetry of the Higgs sector is then no larger than $O(2)$. There is no custodial $SU(2)$ and ρ is not equal to one. In the case of a model with only triplets $\rho = 3/2$.

How well can we expect to be able to measure the radiative corrections in the near future? At LEP and the SLC the Z mass can be measured directly. The error on M_Z is controlled by the accuracy with which the beam energy can be measured. An error $\delta M_Z = 50\ MeV$ would seem to be reasonable.[4,5] The W mass can be measured at LEP from the reaction $e^+ e^- \to W^+ W^-$. The shape of the cross-section and energy distribution of leptons from the decays $W \to e\nu$ can be used. Studies[27] indicate that an error $\delta M_W \simeq 100\ MeV$ should be achievable.

Since it will be several years before the W mass can be measured at LEP, it is reasonable to ask how well one can measure the W mass at hadron colliders. The W mass must be inferred from the transverse momentum distribution of the leptons from the decay $W \to e\nu$. As can be seen from Table 3, the current errors on the W and Z mass are large. Part of the error on the W mass is a systematic error arising from calibration. This error can largely be eliminated once the Z mass has been measured in $e^+ e^-$ annihilation since leptons from Z decay have known transverse momentum and can be used as a calibration. The remaining error is a statistical one which can be reduced as more W's and Z's are produced. The proposal[28] for the $D0$ detector at the Tevatron claims that an error $\delta M_W \approx 100\ MeV$ can be obtained. A value of $\delta M_W \approx 300\ MeV$ appears to be achievable in the near future. Such an error implies a sensitivity to t quark masses greater than $90\ GeV$ (see Fig. 8).

Other tests of the electroweak theory can arise from measurements of asymmetries at LEP of the SLC.[29] I shall concentrate my discussion on those asymmetries measured at the Z^0 resonance where the event rates are large and a good statistical sample can be obtained. The total cross-section for $e^+ e^- \to Z \to all$ is approximately $40\ nb$.

Measurements of asymmetries can have much smaller errors than measurements of rates themselves. This is because certain systematic errors, for example in the luminosity measurement, will cancel out. The first asymmetry that I will discuss is the forward-backward asymmetry for the process $e^+ e^- \to Z \to f\bar{f}$, where f is a fermion

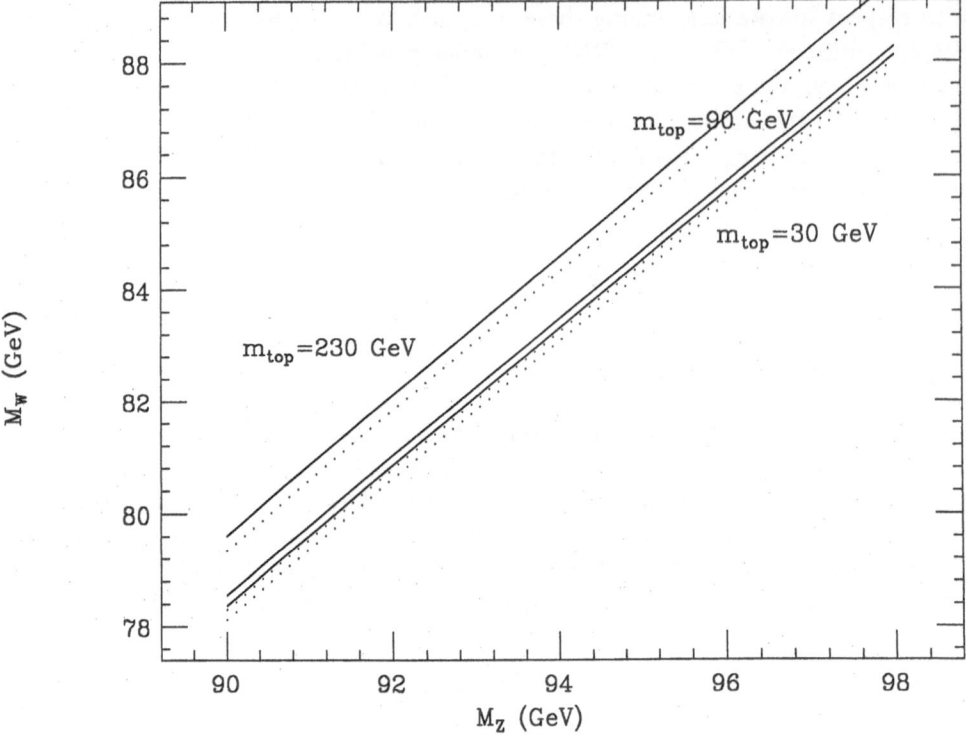

Figure 8: The dependence of M_W upon M_Z for several choices of m_t The solid lines are for $m_H = 10 GeV$ and the dotted for $m_H = 1000 GeV$.

$$A_{FB} = \frac{\int(\sigma(f, \theta) - \sigma(\bar{f}, \theta))d(\cos\theta)}{\int(\sigma(f, \theta) + \sigma(\bar{f}, \theta))d(\cos\theta)}. \tag{1.36}$$

Here $\sigma(f, \theta)$ $(\sigma(\bar{f}, \theta))$ is the cross-section for the production of $f(\bar{f})$ at angle θ to the e^- beam. In lowest order this asymmetry is given by

$$A_{FB} = 3\frac{v_e v_f a_e a_f}{(v_e^2 + a_e^2)(v_f^2 + a_f^2)}. \tag{1.37}$$

The quantities v_i and a_i are given in Table 1. In order to measure this asymmetry it is necessary to distinguish the f from the $\bar{f}$. This is not possible if f is an up, down or strange quark. It may be possible for c and b quarks where the semileptonic decay produces a lepton whose charge is correlated with that of the quark. Clearly, the cleanest final state occurs if f is a muon. I will specialize my discussion to this case. Figure 9[29] shows A_{FB} as a function of the Z mass. Three curves are shown: the lowest order prediction and the value radiatively corrected for $m_t = 30\ GeV$ with $m_H = 10\ GeV$ and $m_H = 1\ TeV$.

How well can A_{FB} be measured? At an e^+e^- luminosity of $10^{31}cm^{-2}sec^{-1}$, there are approximately 1000 $\mu^+\mu^-$ events per day. If we neglect systematic errors, a LEP experiment with an exposure of 200 days can achieve $\delta A_{FB} \approx 0.002$. An experiment at SLC, with its expected lower luminosity, is likely to have an error which is at least three times larger.

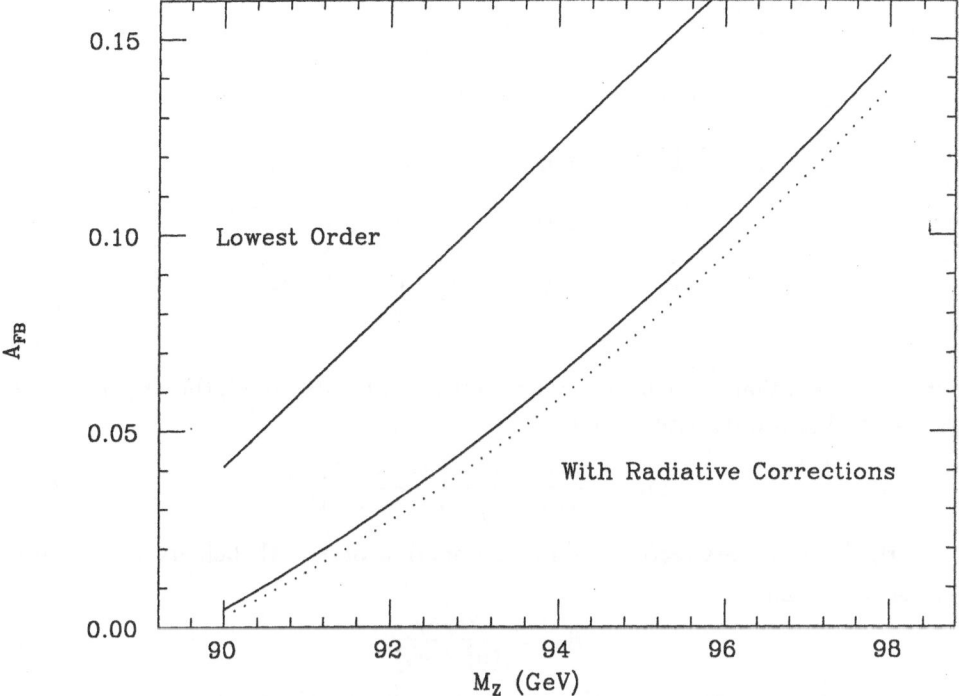

Figure 9:   The dependence of A_{FB} for the process $e^+e^- \to Z \to \mu^+\mu^-$ for unpolarized e^+e^- beams as a function of M_Z. The solid lines are for $m_H = 10 GeV$ and the dotted for $m_H = 1000 GeV$.

It is clear that, with an accuracy of this order, an experiment can detect the difference between a calculation in lowest order and one including radiative corrections. Most of these radiative corrections arise from known physics, such as the coupling of the electron to the W, Z and photon. Figure 10 shows the contribution to A_{FB} from an additional quark doublet. As in the case of corrections to the W mass, the correction becomes large in the region of large quark masses. The curves are shown as a function of the mass of the charge 2/3 quark for a fixed value of the ratio of the quark masses in the doublet. I have indicated regions on the figure which can be excluded by other measurements. If the charge 1/3 quark has mass less than $M_Z/2$ it will be observed directly in Z decay so that the region above the dot-dashed line is probed. If the W mass is within 300 MeV of its predicted value, the region above the dotted line will be excluded; an error of 100 MeV rules out the region above the dashed line. I have indicated a $\pm 2\sigma$ error bar for the LEP scenario discussed above. I am forced to conclude that A_{FB} is not a sufficiently sensitive quantity to be used as a probe of new physics.

Figure 11 shows the contribution to A_{FB} from a doublet of squarks such as will occur in a supersymmetric version of the standard model. Notice that if the up and down squarks are degenerate the contribution to A_{FB} is zero at large squark masses. This is an example of decoupling since the squarks can have equal, non-zero, masses even if the $SU(2) \times U(1)$ symmetry is unbroken and a large degenerate mass does not imply a large Yukawa coupling. If the ratio of the squark masses is large, then there is no decoupling since the splitting violates $SU(2)$ symmetry and must arise from the vacuum expectation value of Higgs fields.

15

Table 4: The error estimated on the left-right asymmetry as a function of the number of produced Z^0's (N) and the accuracy of the measurement of the electron polarization $(\Delta P/P)$[30].

$\Delta P/P$	$N = 10^4$	$N = 10^5$	$N = 10^6$
5%	0.025	0.013	0.010
3%	0.023	0.009	0.006
1%	0.022	0.007	0.003

If the polarization of the outgoing fermion f can be measured, then a polarization asymmetry A_{pol} can be determined

$$A_{pol} = \frac{\sigma(h=1) - \sigma(h=-1)}{\sigma(h=1) + \sigma(h=-1)},$$ (1.38)

where $\sigma(h)$ is the cross-section for the production of f with helicity h. In lowest order A_{pol} is given by

$$A_{pol} = \frac{2v_f a_f}{(v_f^2 + a_f^2)}.$$ (1.39)

The only particle whose polarization can be measured is the tau lepton. In the decay $\tau \to \pi\nu$, the momentum spectrum of the π is sensitive to the tau helicity

$$\frac{dW}{dX_\pi} = 1 - h(2X_\pi - 1)$$ (1.40)

where $X_\pi = 2E_\pi/\sqrt{s}$. The branching ratio $\tau \to \pi\nu$ is only 10% or so. The error on A_{pol} from such a measurement is unlikely to be small enough for one to be sensitive to new physics.

If the electron or positron beam can be polarized then one can measure

$$A_{LR} = \frac{\sigma(L) - \sigma(R)}{\sigma(L) + \sigma(R)}.$$ (1.41)

Here $\sigma(L)$ $(\sigma(R))$ is the cross-section for producing a Z from a left (right) polarized electron and an unpolarized positron. A_{LR} is given by

$$A_{LR} = \frac{2v_e a_e}{(v_e^2 + a_e^2)}.$$ (1.42)

Since there are no plans for polarization at LEP, I will discuss the SLC where a polarized electron source is under construction.[30] Since the total Z production rate is used in the measurement of A_{LR}, the statistical errors are smaller. There is a systematic error due to the measurement error on the polarization of the electron beam $(\Delta p/p)$. Table 4, extracted from the proposal to measure A_{LR}, shows the error on A_{LR} as a function of $\Delta p/p$ and of the number of produced Z^0's. For orientation, at a luminosity of $10^{30} cm^{-2} sec^{-1}$ it takes approximately one year of running to produce 10^6 Z^0's. The value of A_{LR} is shown in Fig. 12 as a function of M_Z in three different scenarios, all of which are consistent with current data. In order to establish a 3σ effect which discriminates between $m_t = 30\ GeV$ and $m_t = 180\ GeV$, it will be

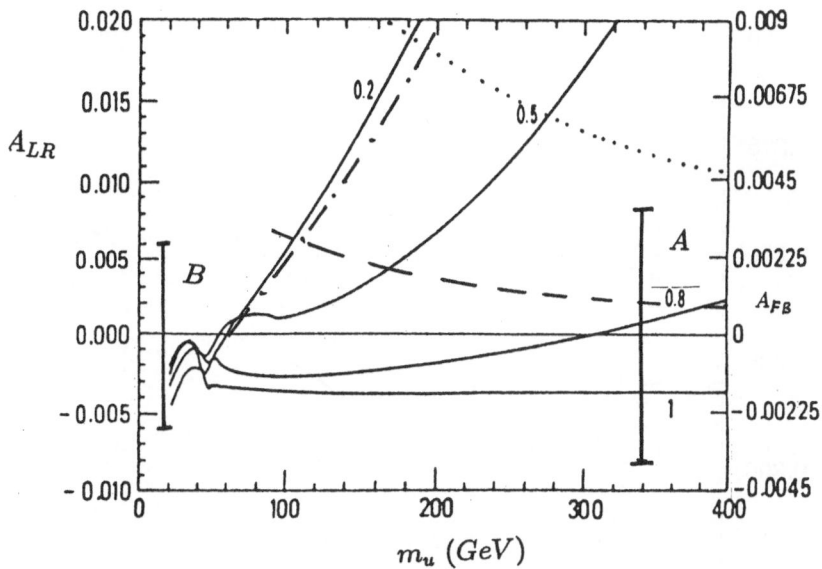

Figure 10: The contribution to A_{FB} and to A_{LR} from the presence of an additional quark doublet as a function of m_u, the mass of the charge 2/3 member of the doublet. The curves are labelled by the ratio m_u/m_d, where m_d is the mass of the charge 1/3 member. The region above the dot-dashed line can be probed directly since m_d is low enough for the Z to decay to $d\bar{d}$. If M_W is within 300(100)MeV of it predicted value, the region above the dotted (dashed) line is excluded. The error bar A applies to A_{FB} and B to A_{LR} (see text).

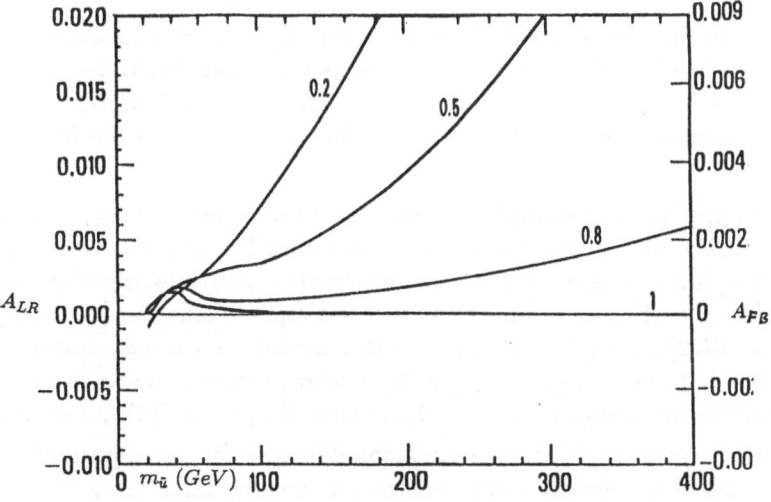

Figure 11: The contribution to A_{FB} from a squark doublet as a function of the up squark mass for fixed ratios of the up to down squark masses.

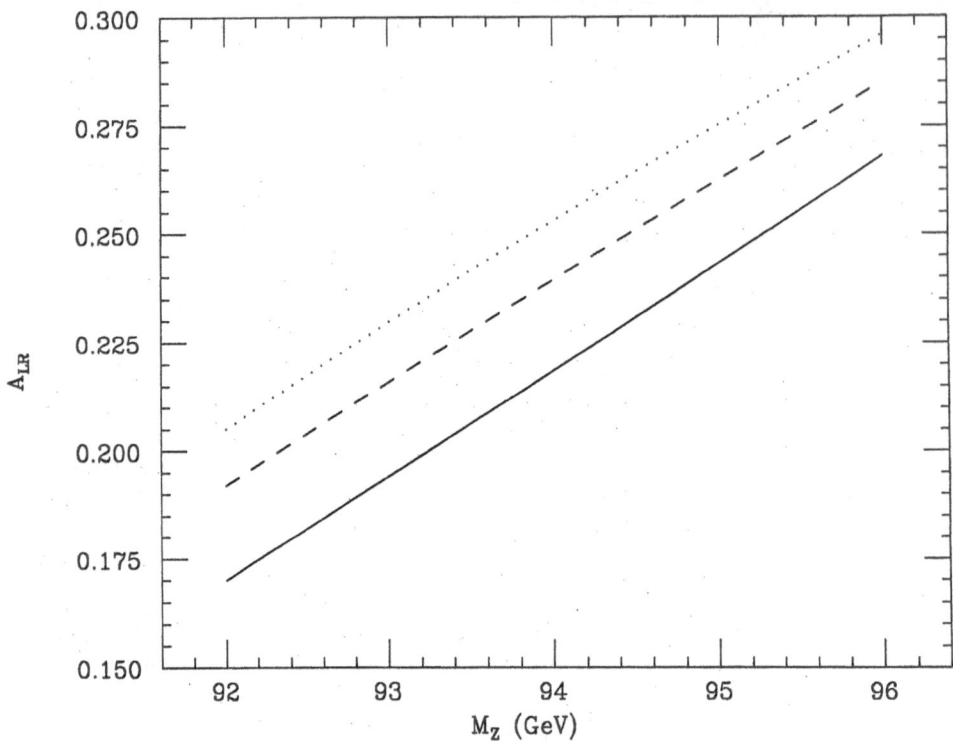

Figure 12: The quantity A_{LR} as a function of M_Z. The solid line has $m_t = 30$ GeV and $m_H = 100$ GeV. The dashed line has $m_t = 180$ GeV and $m_H = 100$ GeV. The dotted line has $m_t = 30$ GeV and a modified Higgs sector such that $\rho = 1.01$.

necessary to measure the polarization to better than 2% and have more than 10^5 produced Z^0's. Figure 10 shows δA_{LR} due to a new quark doublet. I have indicated a $\pm 2\sigma$ error bar corresponding to $\Delta p/p = 1\%$ and to 10^6 produced Z^0's. It may be somewhat easier to establish an effect than in the case of the forward-backward asymmetry.

Before leaving this subject, I would like to comment briefly upon the effect of a more radical modification of the standard model.[31] The recent upsurge in string theory has provided a motivation for considering models where the gauge group is extended. I shall discuss one particular example where the low energy group is $SU(3) \times SU(2)_L \times U(1)_y \times U(1)_{y'}$. In this model the charged current structure is unaffected but there are changes in the neutral current due to the presence of an additional neutral gauge boson associated with the group $U(1)_{y'}$. I shall assume that the coupling constant g' of this group is equal to g_1, a choice supported by these string motivations. If the electric charge operator Q has the same value as in the standard model, viz., $Q = T_3 + y/2$, then the photon will be the same linear combination of B and W_3 as in the standard model. The two massive neutral gauge bosons will be linear combinations of the standard model Z and B', the gauge boson of the $U(1)_{y'}$ group.

The mass matrix of the neutral bosons will depend upon the structure of the Higgs sector and will have the following form

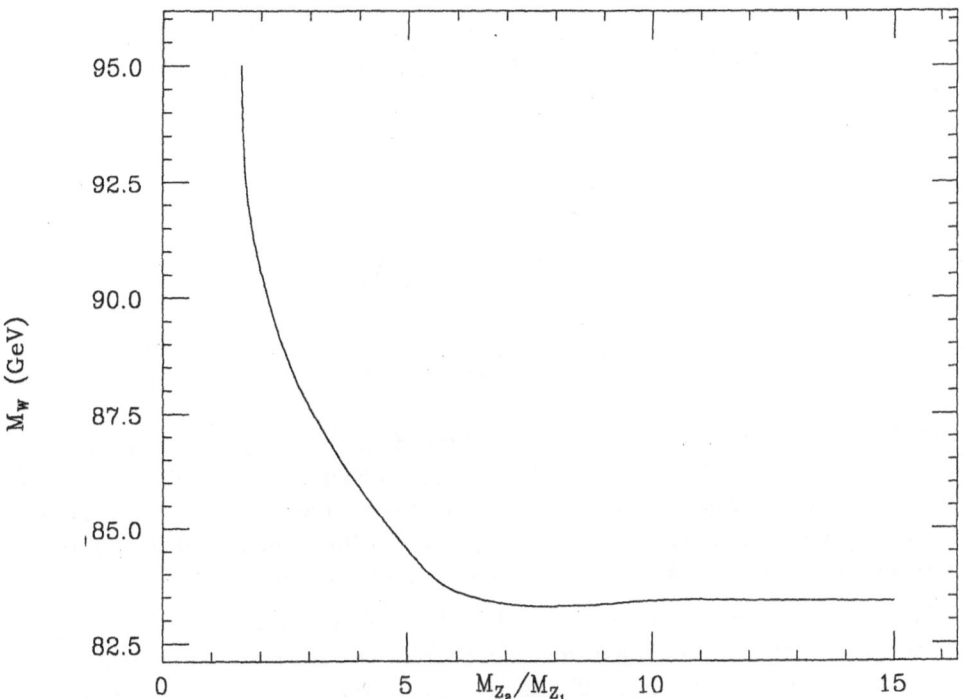

Figure 13: The dependence of M_W upon M_{Z_2}/M_{Z_1} in the model based on $SU(2) \times U(1)_y \times U(1)_{y'}$ for $M_{Z_1} = 94\ GeV$ and $\rho_{y'} = 1/3$.

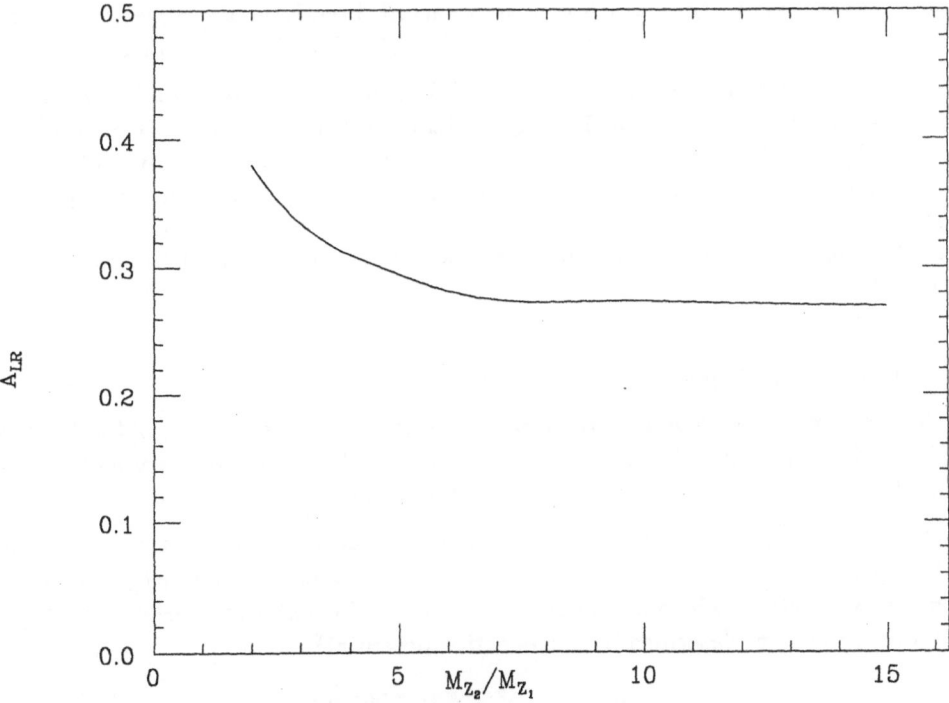

Figure 14: The left-right asymmetry A_{LR} as a function of M_{Z_2} for $\rho_{y'} = 1/3$ in the $SU(2) \times U(1)_y \times U(1)_{y'}$ model. The asymmetry is evaluated at $\sqrt{s} = M_{Z_1}$ which is assumed to be $94\ GeV$.

$$|M|^2 = \begin{vmatrix} (g_2^2 + g_1^2)\,A & (g_2^2 + g_1^2)^{1/2}g_1\,B \\ (g_1^2 + g_2^2)^{1/2}g_1\,B & g\,C \end{vmatrix} \qquad (1.43)$$

where

$$A = \sum_i \left\langle \phi_i T_3^2 \phi_i \right\rangle,$$

$$B = \sum_i \left\langle \phi_i \frac{T_3}{2} y_i' \phi_i \right\rangle,$$

$$C = \sum_i \left\langle \phi_i \frac{y_i'^2}{4} \phi_i \right\rangle.$$

Here the sum i runs over Higgs representations ϕ_i with $U(1)_{y'}$ charge y_i'. T^3 is the neutral generator of $SU(2)_L$. In the case of the standard model $y^i = 0$ and $A = v^2/4$. The eigenvalues are M_{Z_1} and M_{Z_2}. If we assume that there are only doublets under $SU(2)_L$ and singlets under $SU(2)_L \times U(1)_y$, then in the limit $y' = 0$ we will recover the standard model with $\rho = 1$ and a non-minimal Higgs sector.

The model now has five parameters. The three of the standard model, α, G_F and M_{Z_1} together with M_{Z_2} and a parameter describing the Higgs structure analogous to ρ, $\rho' = B/A$. The W mass is predicted in terms of these parameters; it is shown in Fig. 13. A measurement of M_W to an accuracy of $300\ MeV$ is sensitive to the mass of the second massive neutral gauge boson provided that it is lighter than $300\ GeV$.

The left-right asymmetry measured on the Z_1 resonance is shown in Figure 14. Again it would appear that the best experiments are sensitive to $M_{Z_2} \lesssim 300\ GeV$. This value is close to that which can be observed directly at the Tevatron collider from the production of the Z_2 boson followed by its decay into e^+e^- or $\mu^+\mu^-$.

What can we conclude about the potential of measurements of radiative corrections? As we have seen it will be very difficult for an experiment to be sensitive to new physics; the best hopes seem to lie with a precise determination of the W and Z masses and with the left-right asymmetry. It is very important to emphasize that if an effect is seen in measurements of radiative corrections it may be very difficult to discern its origin. Only the direct observation of new particles can resolve ambiguities.

2. Where is the Higgs?

There is very little experimental information about the Higgs sector of the $SU(2) \times U(1)$ model other than that it must have a custodial $SU(2)$ symmetry so that ρ is equal to one at tree level. Do we know anything from theoretical studies?

Since $m_H^2 = \frac{2\lambda}{g_2^2} M_W^2$, it would appear that m_H could be made arbitrarily small by reducing λ. This is not the case since for very small λ one must consider the effect of gauge interactions which induce Higgs self interactions at higher order. The Higgs self interactions are described by the effective potential[¶]

$$V_{eff}(\Phi) = -\mu^2 \Phi^+ \Phi + \lambda (\Phi^+ \Phi)^2. \qquad (2.1)$$

[¶] Although the effective potential is not gauge invariant, its minimum is. Consequently the subsequent discussion which relates only to the minimum is physically meaningful.

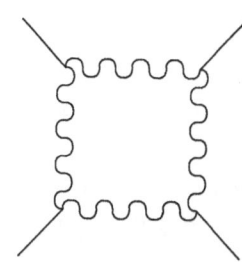

Figure 15: Feynman diagram showing a contribution to the effective potential for the Higgs field due to interactions of the Higgs with W bosons.

Radiative corrections from Feynman diagrams of the type indicated in Fig. 15 modify this potential. At one loop $V_{eff}(\Phi)$ becomes[32]

$$V(\Phi) = -\mu^2(\Phi^+\Phi) + \lambda(\Phi^+\Phi)^2 + c(\Phi^+\Phi)^2 \log((\Phi^+\Phi)/M^2) \qquad (2.2)$$

where

$$c = \frac{1}{16\pi^2 v^4}(3(2M_W^2 + M_Z^2) + m_H^4 - 4\sum_j m_j^4)$$

and M is a renormalization scale. The Higgs mass m_H is given by

$$m_H^2 = \frac{\partial^2 V}{\partial \Phi^2}\Big|_{\Phi=\langle\Phi\rangle} \cdot \qquad (2.3)$$

In general V_{eff} will have more than one minimum. If we require that the minimum with $\langle\Phi\rangle \neq 0$ is lower than that with $\langle\Phi\rangle = 0$ (the phase in which the W boson remains massless), so that this phase will be the true ground state, then a bound on λ, and hence m_H, can be obtained since all the other quantities in Eq. (2.2) are known. We have[33]

$$m_H \gtrsim 7 \ GeV.$$

A more detailed study which requires that the universe not be trapped at $\langle\Phi\rangle = 0$ for too long[34] gives $m_H \gtrsim 10 \ GeV$. This bound is extremely model dependent. A similar bound will exist in models with different Higgs sectors.[35] In models with an arbitrary number of Higgs doublets there must be at least one physical Higgs boson with a mass greater than this bound.

As λ is increased m_H increases. Eventually λ will become too large for the perturbative formula for the Higgs mass to be valid. We can estimate this value naively by requiring that $\lambda^2/4\pi$ be less than one. This implies $m_H \lesssim 600 \ GeV$. In order to be more precise it is necessary to consider the effects of the constraints imposed by partial wave unitarity.[36]

Consider the S matrix for a two-particle scattering process $a+b \rightarrow c+d$. Unitarity requires that

$$S^+S = 1. \qquad (2.4)$$

Writing $S = 1 + iT$, we have

$$-ImT = T^+T. \qquad (2.5)$$

The scattering matrix T is given by

$$T = (2\pi)^4 \delta^4(p_a + p_b - p_c - p_d) \frac{1}{(2\pi)^6} \frac{1}{s} |M_{ab \to cd}|^2. \tag{2.6}$$

Here p_i is the momentum of particle i and M_i the invariant matrix element obtained, for example, by calculating a set of Feynman diagrams. M may be decomposed as follows:

$$M(s, \cos\theta) = 16\pi \sum_{J=0}^{\infty} (2J+1) A_J(s) P_J(\cos\theta). \tag{2.7}$$

θ is the center-of-mass scattering angle between particles a and c, $P_J(\cos\theta)$ is a Legendre polynomial and $A_J(s)$ is some function. Equation 2.5 implies that

$$Im\, A_0 \geq |A_0|^2. \tag{2.8}$$

We can expand A_0 as a perturbation series in some coupling constant g

$$A_0(s) = a_1(s)g^2 + a_2(s)g^4 + \dots \tag{2.9}$$

If the perturbation expression is reliable then

$$g^2 < a_1/a_2. \tag{2.10}$$

The Born term $a_1 g^2$ is real, hence Eq. 2.8 implies that

$$-Im(a_2 g^4) > (a_1 g^2). \tag{2.11}$$

But $|A| > |ImA|$ for any A, so that the requirement that perturbation theory be reliable implies that

$$\left| a_1 g^2 \right| < 1. \tag{2.12}$$

Now this result can be applied to the process $H + H \to H + H$. If we assume that $m_H >> M_W$, the relevant Feynman diagrams are shown in Fig. 16 and give

$$A_0(HH \to HH) = -\frac{G_F m_H^3}{8\pi\sqrt{2}} \left[3 + \frac{9m_H^2}{s - m_H^2} - \frac{2m_H^2}{s - m_H^2} \log(s/m_H^2 - 3) \right]. \tag{2.13}$$

Requiring $|A_0| < 1$ (see Eq. (2.12)) in the limit $s \to \infty$ implies that

$$m_H < 1.7\ TeV.$$

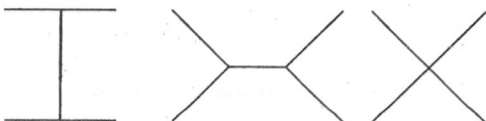

Figure 16: Feynman diagram for the process $HH \to HH$ which dominate in the limit $m_H >> m_W$.

A stronger bound is obtained by considering the coupled channel problem: $HH \to ZZ, HH \to WW, WW \to ZZ, HZ \to HZ, HW \to HW$. In this case one has[36]

$$m_H < \sqrt{\frac{8\pi\sqrt{2}}{3G_F}} = .98 \; TeV. \qquad (2.14)$$

This bound indicates that there must be a scalar particle of mass less than $1 \; TeV$ or the Weinberg-Salam model will contain a strong, non-perturbative coupling. The presence of such a coupling implies that there must be non-perturbative structure in the WW or ZZ channel for WW or ZZ invariant masses of order $1 \; TeV$. (Recall that the longitudinal components of the W and Z come from the Higgs fields.) General arguments which apply to strongly coupled systems can be used to predict these effects.

The basic argument that I have just outlined contains the essential features which justify the choices of energy and luminosity for the SSC. In order to probe the nature of the interactions responsible for the breakdown of the $SU(2) \times U(1)$ symmetry it is necessary to probe the WW and ZZ system with invariant masses of order $1 \; TeV$.

A similar argument can be used to constrain the masses of heavy quarks or leptons which are proportional to Yukawa couplings.[23,37] Consider the scattering of a quark or lepton F_i of mass m_i: $F_i\bar{F_i} \to F_j\bar{F_j}$. There are contributions to the scattering amplitude from exchanges of Z and Higgs bosons in the s and t channels. In the limit $s \to \infty$ with $m_i >> m_H, M_Z$, M has the following form

$$M = \sqrt{2}G_F m_i m_j \delta_{\lambda\bar{\lambda}}\delta_{\lambda'\bar{\lambda}'}[1 - \lambda\lambda' - 2\delta_{ij}]. \qquad (2.15)$$

Here $\lambda(\lambda')$ labels the helicity of the fermion $i(j)$ and $\bar{\lambda}$ labels the helicity of the anti-fermions. The constraint of partial wave unitarity implies that

$$m_i^2 + m_j^2 < \frac{8\sqrt{2}\pi}{G_F}. \qquad (2.16)$$

If $m_1 = m_2$ this implies that $m_1 \lesssim 530 \; GeV$. In the case of a heavy lepton of mass m_L in a doublet with a massless neutrino: $m_L \lesssim 1.2 \; TeV$. Quarks of masses larger than this cannot be discussed within the context of perturbation theory.[‖]

Let us now turn to the possible experimental signatures for Higgs bosons. The Higgs can decay to fermion anti-fermion WW and ZZ final states with the following partial widths:

$$\Gamma(H \to f\bar{f}) = \frac{G_F m_f^2 m_H}{4\pi\sqrt{2}}(3)(1 - 4m_f^2/m_H^2)^{3/2},$$

$$\Gamma(H \to W^+W^-) = \frac{G_F m_H^3}{32\pi\sqrt{2}}(4 - 4\epsilon + 3\epsilon^2)(1 - \epsilon)^{1/2}, \qquad (2.17)$$

$$\Gamma(H \to ZZ) = \frac{G_F m_H^3}{64\pi\sqrt{2}}(4 - 4\epsilon' + 3\epsilon'^2)(1 - \epsilon')^{1/2}.$$

‖An exception to this can occur in modifications of the standard model which contain fermions whose mass is not controlled by $\langle\Phi\rangle$. For example, a right-handed neutrino can have an arbitrarily large Majorana mass.

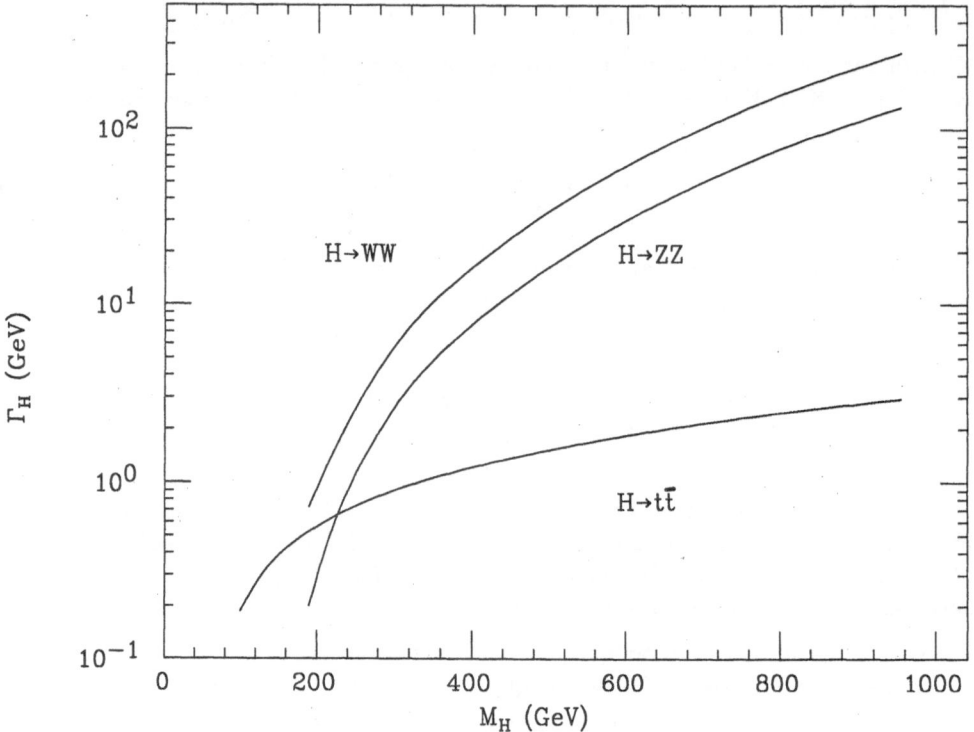

Figure 17: The partial widths $H \to t\bar{t}$ (solid lines), W^+W^- (dashed line) and ZZ (dotted line) as a function of m_H.

with $\epsilon' = 4M_Z^2/m_H^2$ and $\epsilon = 4M_W^2/m_H^2$. The factor of 3 is included in the first expression only if f is a quark. The implications of these formulae are easy to see. If $m_H < 2M_W$, the Higgs will decay dominantly into the heaviest fermion channel which is open. Once m_H is greater than $2M_W$, the decay into two gauge bosons will dominate. This effect is shown in Fig. 17. Notice that the width grows rapidly as m_H is increased. Eventually $\Gamma/m_H \sim 0(1)$: this is another manifestation of the breakdown of perturbation theory at large values of m_H.

The Higgs can be produced in e^+e^- annihilation from the decay of a Z through the graph shown in Fig. 18, with a rate shown in Fig. 19. The rate is given by[38]

$$\frac{1}{\Gamma(Z \to \mu^+\mu^-)} \frac{d\Gamma(Z \to H + \mu^+\mu^-)}{dx} = \frac{\alpha}{4sin^2\theta_W cos^2\theta_W}$$

$$\times \frac{(1 - x + \frac{x^2}{12} + \frac{2m_H^2}{3M_Z^2})(x^2 - \frac{4m_H^2}{M_Z^2})^{1/2}}{(x - \frac{m_H^2}{M_Z^2})^2} \qquad (2.18)$$

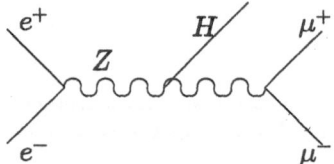

Figure 18: Feynman diagram for the process $e^+e^- \to Z \to H + \mu^+\mu^-$.

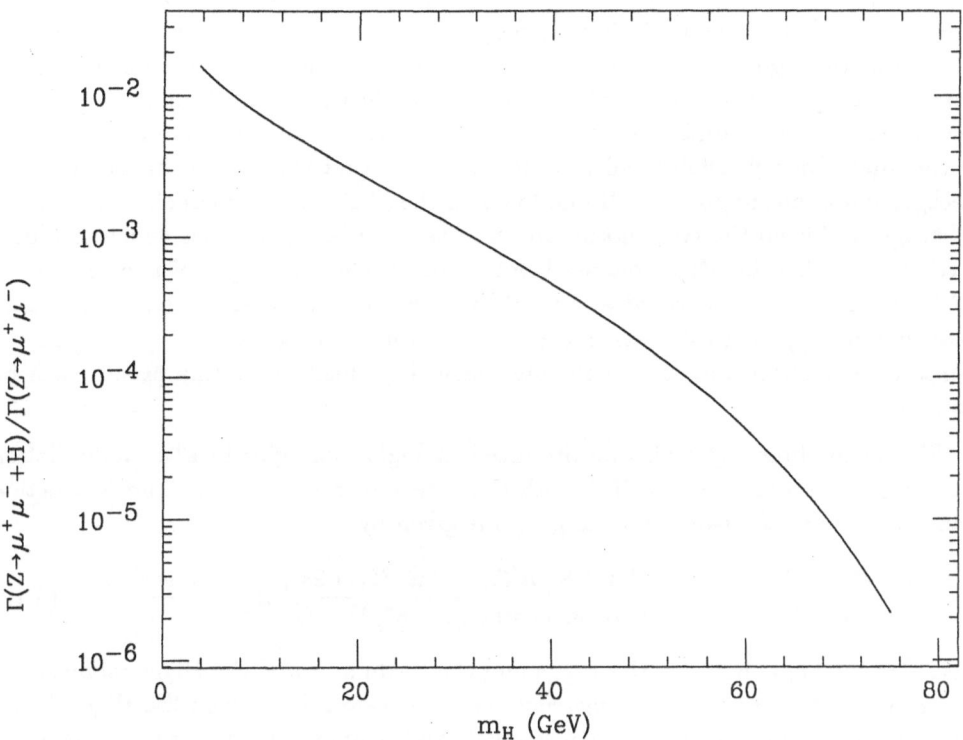

Figure 19: The ratio of widths $\Gamma(Z \to H\mu^+\mu^-)/\Gamma(Z \to \mu^+\mu^-)$ as a function of the Higgs mass.

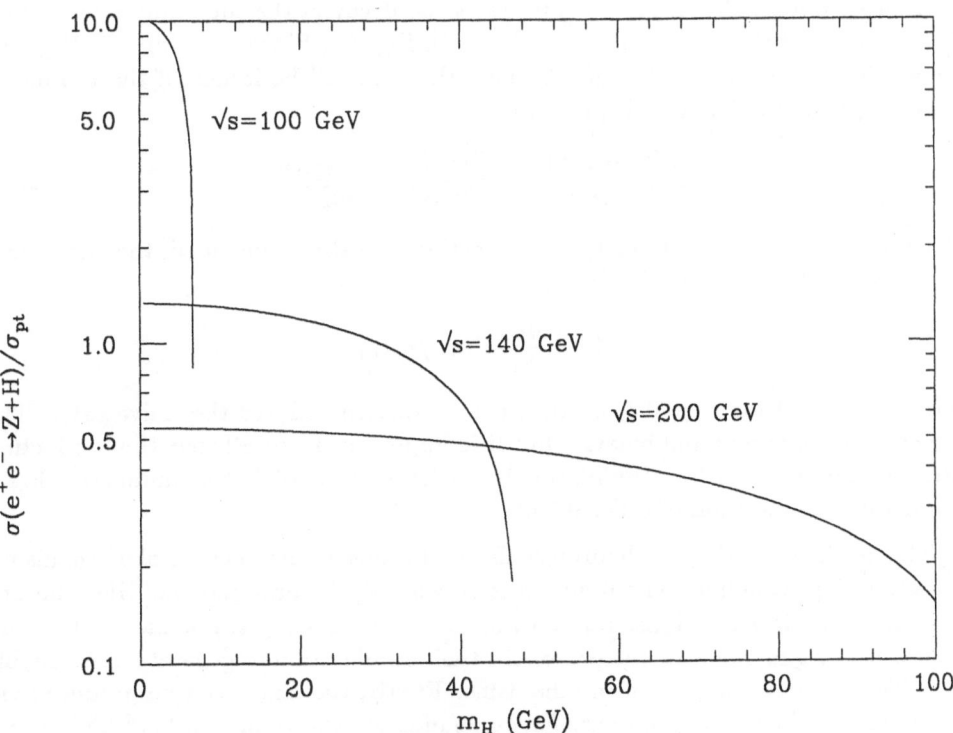

Figure 20: The cross-section for the process $e^+e^- \to Z + H$ as a function of m_H for various values of $\sqrt{s}$.

Here $x = 2E_H/M_Z$ where E_H is the energy of the Higgs boson. The rate is rather small, but the signature is very clean. It is not necessary to reconstruct the Higgs from its decay products; one searches for a peak in the mass recoiling against the lepton pair. Backgrounds arise from the production of a heavy quark pair if both of the quarks decay semileptonically. If we require that the leptons be isolated this background is not important. If one looks in the e^+e^- decay channel, then there is a background from the two photon process $e^+e^- \rightarrow e^+e^- + \text{hadrons}$ which produces a serious problem for Higgs masses below about 8 GeV. A Higgs of mass less than $40\ GeV$ should be discovered at LEP/SLC using this process. If the Higgs mass exceeds $0.6\ M_Z$, then this rate is exceeded by that from $Z \rightarrow H + \gamma$.[39] Again the signal is very clean, but the small rate makes it unlikely that this process will be observed.

The Higgs boson can also be produced at higher energies in e^+e^- annihilation via the process $e^+e^- \rightarrow Z + H$,[40] with the rate shown in Fig. 20. The production cross-section at center-of-mass energy $\sqrt{s}$ is given by

$$\frac{d\sigma(e^+e^- \rightarrow Z + H)}{d(\cos\theta)} = \frac{\pi\alpha^2(1 + 8\sin^4\theta_W - 4\sin^2\theta_W)}{16\sin^4\theta_W\cos^4\theta_W(s - M_Z^2)^2}\frac{2\kappa}{\sqrt{s}}(M_Z^2 + \frac{\kappa^2\sin^2\theta}{2}). \quad (2.19)$$

Here θ is the angle between the Higgs and the beam and κ is the Higgs momentum. Again it is not necessary to reconstruct the final state arising from the Higgs decay. The cross section is not large, particularly if the Z can only be detected via its decay to $\mu^+\mu^-$ or e^+e^-. Nevertheless LEP should be able to probe Higgs masses up to $0.9(\sqrt{s} - M_Z)$ using this mechanism.

Another potentially important process is the decay of the toponium bound state (θ) into $H + \gamma$,[41] the rate for which is shown in Fig. 21. Since coupling of a Higgs to a quark is proportional to the quark mass, the rate will be largest if the top quark mass (m_t) is large. The rate is given by

$$\frac{\Gamma(\theta \rightarrow H\gamma)}{\Gamma(\theta \rightarrow \mu^+\mu^-)} = \frac{G_F m_t^2}{\sqrt{2}\pi\alpha}(1 - \frac{m_H^2}{m_\theta^2})^{1/2}. \quad (2.20)$$

This process has a rather large QCD correction.[42] If this is included, the right-hand side of Eq. (2.20) is multiplied by

$$\left(1 - \frac{4\alpha_s}{3\pi}a(m_H^2/m_\theta^2)\right)$$

Here, $a(x) \sim 10$ for $x \leq 0.8$, so that the correction reduces the naive rate. The branching ratio is reasonably large, but it is important to recall that the production rate for toponium in e^+e^- annihilation is not large. A 80 GeV toponium state has a production cross section of order 0.1 nb.

The product of a Higgs in hadron-hadron collisions occurs via several mechanisms. Since the Higgs coupling to light quarks is very small, the production of Higgs bosons from the annihilation of light quark-antiquark pairs strongly suppressed. There are too few heavy quarks inside the proton for their annihilation to generate a reasonable rate. There are two important mechanisms. Firstly, the Higgs can be produced via gluon-gluon fusion according to the Feynman diagram shown in Fig. 22.[43] This graph contains a vertex coupling the Higgs to a quark-antiquark pair, which is proportional to the quark mass. Consequently, the rate from this process depends sensitively

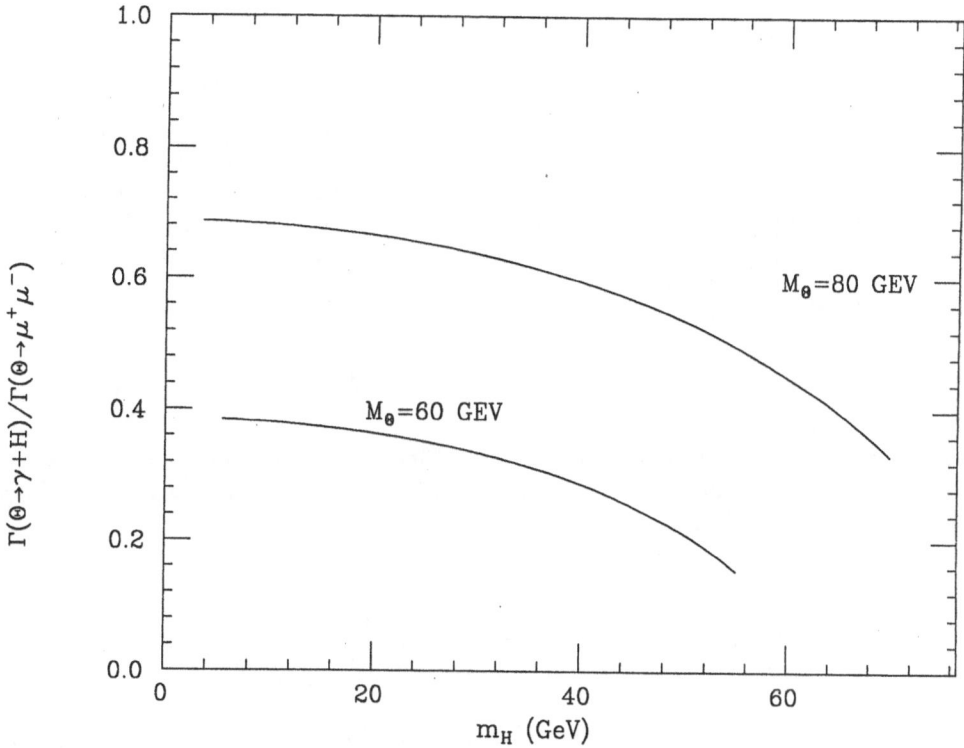

Figure 21: The ratio of decay widths $\Gamma(\theta \rightarrow H\gamma)/\Gamma(\theta \rightarrow \mu^+\mu^-)$ for the decay of the 1^{--} bound state (θ) of $t\bar{t}$. It has been assumed that θ has a mass of 80 GeV.

upon the mass of the top quark. The production rate at center-of-mass energy $\sqrt{s}$ is a proton-proton collision is given by

$$\sigma(pp \rightarrow H + X) = \frac{G_F\pi}{32\sqrt{2}} \left(\frac{\alpha_s}{\pi}\right)^2 \eta^2 \int_{M_H^2/s}^{1} \frac{M_H^2}{s} g(x)g(m_H^2/sx)dx. \qquad (2.21)$$

$g(x)$ is the gluon distribution of a proton (see Sec. 3). Defining $\epsilon_i = 4m_i^2/m_H^2$ for a quark of mass m_i, η is given by

$$\eta = \sum_i \frac{\epsilon_i}{2}(1 + (\epsilon_i - 1)\phi(\epsilon_i))$$

with

$$\phi(\epsilon) = \begin{cases} -[\sin^{-1}(1/\sqrt{\epsilon})]^2 & \epsilon > 1 \\ \frac{1}{4}(\log(\eta_+/\eta_-) + i\pi)^2 & \epsilon < 1 \end{cases}$$

where $\eta_\pm = 1 \pm \sqrt{1 - \epsilon}$.

An alternative mechanism is shown in Fig. 23.[44] At large values of m_H, the rate from this mechanism becomes large due to the large width for $H \rightarrow WW$. The exact formula for this rate is complicated; it simplifies drastically in the so-called effective W approximation. This approximation assumes that the W's are emitted parallel to the incoming quarks and they are treated as if they are on mass-shell. It is similar to the effective photon approximation used to describe two-photon reactions in e^+e^- annihilation where the electron beams are treated as sources of on-shell photons. In this approximation the cross-section for $q + q \rightarrow H + qq$ via intermediate W's is given by[45]

27

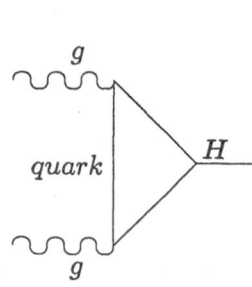

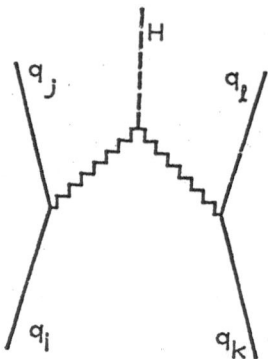

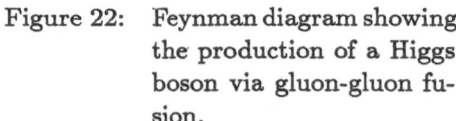

Figure 22: Feynman diagram showing the production of a Higgs boson via gluon-gluon fusion.

Figure 23: Feynman diagram showing the process $qq \to H + qq$.

$$\sigma\left(q_i + q_j \to H + q_j + q_i\right) = \frac{1}{16M_W^2}\left(\frac{\alpha}{\sin^2\theta_W}\right)^3$$

$$\times \left[(1 + m_H^2/\hat{s})\log(\hat{s}/m_H^2) - 2 + 2m_H^2/\hat{s}\right]\theta(-e_i e_j) \qquad (2.22)$$

where $\sqrt{\hat{s}}$ is the center-of-mass energy of the qq system and e_i is the charge of quark of type i. This may be converted into a hadronic cross-section via the parton model (see Sec. 3). In the case of intermediate Z bosons the factor $\theta(-e_i e_j)$ is replaced by $\frac{1}{\cos\theta_W^6}(v_i^2 + a_i^2)(v_j^2 + a_j^2)$ where v_i and a_i were defined in Section 1.

This mechanism will only be important at the SSC; cross-sections evaluated at Tevatron and $S\bar{p}pS$ energies are dominated by the gluon fusion process. The rates for Higgs production are shown in Fig. 24. There are other mechanisms leading to final states with $H + Z, W + Z$ and $H + t\bar{t}$.[46] The rates for Higgs production via these mechanisms are smaller than those discussed above and will be useful only if the additional particles can be used as a tag in order to improve the signal-to-noise ratio.[47]

The signals for Higgs bosons at the SSC are discussed extensively elsewhere.[48,49] At the Tevatron the rates are reasonable only for Higgs masses less than 150 GeV or so. In this mass region the Higgs will decay dominantly to $t\bar{t}$ if the t quark is light enough. There is a large background from the QCD production of $t\bar{t}$ pairs (this will be discussed in the next section) which will make detection difficult even if the t quark can be identified efficiently.

If the $t\bar{t}$ channel is not open, the Higgs will decay to $\tau^+\tau^-$ with a branching ratio of $m_\tau^2/3m_b^2 \sim 4.5\%$. The only background source of τ pairs is Drell-Yan production $p\bar{p} \to \tau^+\tau^- + X$, via a virtual Z or photon. Figure 25 shows the signal and background in the τ pair channel. I have assumed a resolution of 10 GeV in the $\tau^+\tau^-$ invariant mass. It can be seen that the signal to background ratio is rather poor. This figure assumes a top quark mass of 150 GeV. The tau final state can be identified from the one-prong tau decays ($\tau \to e\nu\nu, \mu\nu\nu, \pi\nu$, etc.). Energy is lost into neutrinos so that the resolution in the $\tau^+\tau^-$ invariant mass will be poor. The experiment is clearly very difficult.

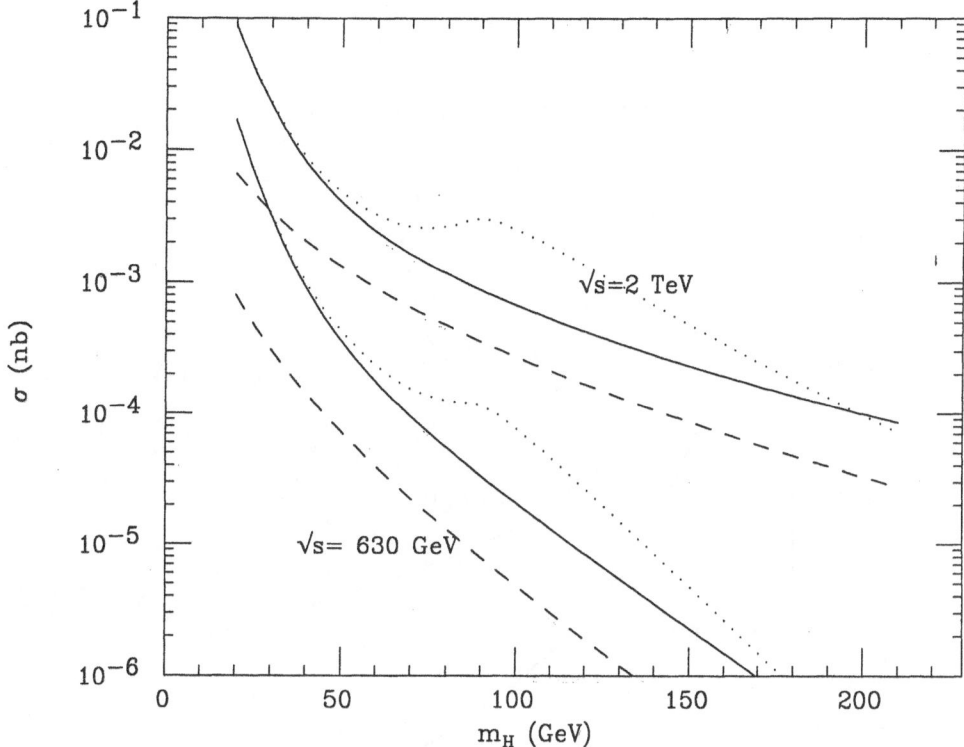

Figure 24: The cross-section $p\bar{p} \to H + X$ as a function of Higgs mass. The solid (dotted) lines correspond to the gluon fusion process of figure 22 with a top quark mass of 150 (40) GeV , and the dashed to the WW fusion process of figure 23.

There is one other possibility. The Higgs can decay to two photons (see Fig. 26) with the branching ratio[50]

$$BR(H \to \gamma\gamma) \sim \frac{m_H^2 \, |A| \, \alpha^2}{6\pi^2 m_b^2}. \tag{2.23}$$

I have assumed that $m_t > m_H/2$. Here A is a number arising from the W, quark and lepton loop diagram. Its value depends upon the masses of the particles involved but it is of order four for Higgs masses around 100 GeV. The background arises from the production of photon pairs via quark-antiquark collisions and is not too large. Unfortunately the branching ratio is so small that there are insufficient events for this decay mode to be useful. It has been suggested [51] as a possible mode at the SSC where the event rates are much larger.

What can we conclude about the prospects for finding the Higgs in the near future? If its mass is less than 40 GeV or so, it should be found in the decay of the Z either at the SLC or at LEP. Masses larger than this can be probed in the decay of toponium, if toponium exists in an accessible mass range. Notice that if $m_t > m_b + M_W$, the top quark will decay too quickly for narrow toponium bound states to exist. Higgs masses up to 100 GeV can be probed in the early 1990's at LEP when the energy is increased to 100 GeV per beam. Higgs bosons of mass greater than this will have to wait for the SSC.

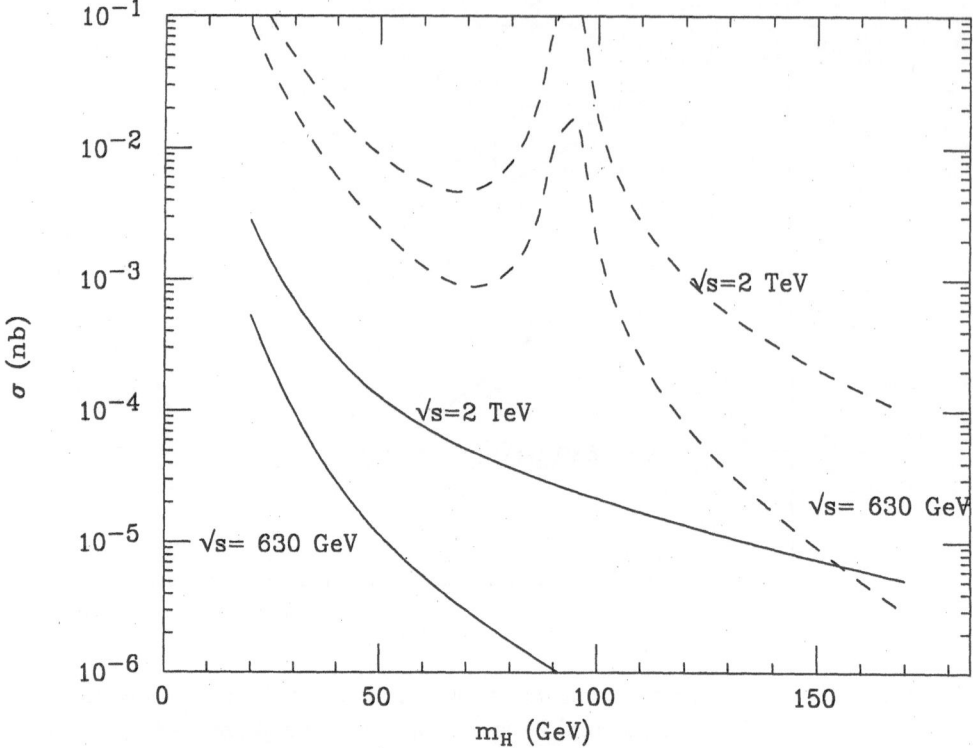

Figure 25: A comparison of the signal and background for the process $p\bar{p} \to H + X \to \tau^+\tau^- + X$. It is assumed that $m_H < 2m_t$. The background is calculated from the Drell-Yan process (see Sect. 3) being $d\sigma/dM \; \Delta M$. The resolution in the invariant mass of the tau pair (ΔM) is taken to be 10 GeV.

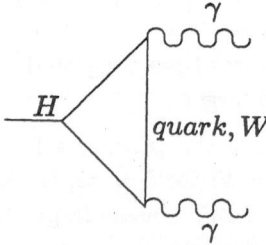

Figure 26: Feynman diagram for the process $H \to \gamma\gamma$.

3. QCD

In this chapter, I shall provide an introduction to perturbative QCD. I shall emphasize the uses of QCD in calculating rates at hadron-hadron colliders. Since QCD processes account for most of the background for new physics at such colliders, it is important to understand the uncertainties in these predicted rates. Given the limited time available I have had to be selective in the topics discussed.* I will begin with a discussion of the one parameter of QCD, namely, its coupling constant. I shall then discuss the parton model in some detail. I will conclude with a discussion of the production of new quarks at hadron-hadron colliders. This discussion will serve as a framework for an analysis of some of the uncertainties in such calculations.

The QCD Lagrangian may be written as follows:

$$-\frac{1}{4}F^i_{\mu\nu}F^i_{\mu\nu} + \sum_j \bar{\psi}_j(i\ \not{D} - m_j)\psi_j \tag{3.1}$$

The sum on j runs over quark flavors and,

$$F^i_{\mu\nu} = \partial_\mu G^i_\mu - \partial_\nu G^i_\mu - igf_{ijk}G^i_\mu G^i_\mu$$

and

$$D_\mu = \partial_\mu - ig\ t^i\ G^i_\mu$$

Here t^i are the 3×3 representation matrices and the structure constant f_{ijk} are given by $[t_i, t_j] = if_{ijk}t_k$.

Apart from the quark masses, which have their origin in the Weinberg-Salam model, the theory has only one fundamental parameter, the coupling constant g. As in the case of the electroweak theory, beyond tree level it is necessary to define a renormalized coupling constant $g(\mu)$. In the case of QED this could be done in terms of the static potential between two electrons. The analogous definition in QCD would be in terms of the inter-quark potential. In the case of light quarks such a definition is impossible in the context of a perturbative theory since QCD is strongly coupled at such low momentum scales. A definition in terms of the potential between two heavy quarks is possible but not particularly convenient. I shall therefore use the modified minimal subtraction scheme discussed in Sect. 1 (see Eq. (1.11)).

Let us calculate a physical process $P(Q^2)$, which depends on some energy scale Q; P could, for example, represent a cross-section. If we neglect quark masses, calculate in n dimensions then

$$P(Q^2) \sim \left[\frac{2A}{4-n} - A\gamma_E + A\log 4\pi - F(\mu, Q^2, g)\right] \tag{3.2}$$

Recall that the scale μ is introduced so that the coupling constant g remains dimensionless in n dimensions, $viz.$,

$$g \to g\mu^{(4-n)/2} \tag{3.3}$$

It is convenient to choose the quantity P to be dimensionless; this can always be done by multiplying it by an appropriate power of Q. Then P must have the form, after subtraction of the $1/(n-4)$, γ_E and $\log 4\pi$ terms

$$P(Q^2) = F(Q^2/\mu^2, \alpha) \tag{3.4}$$

I have replaced g by α: $\alpha \equiv g^2/4\pi$. Now, the scale μ is arbitrary so that a physical quantity cannot depend upon its value

---

*For a more detailed discussion see refs. 52, 53

$$\frac{dP}{d\mu} = 0 \tag{3.5}$$

which implies

$$\left(\mu^2 \frac{\partial F}{\partial \mu^2} + \beta(\alpha) \frac{\partial F}{\partial \alpha} \right) = 0 \tag{3.6}$$

Here $\beta(\alpha)$ is defined by

$$\beta(\alpha) \equiv \mu^2 \frac{\partial \alpha}{\partial \mu^2} \tag{3.7}$$

(Recall that the bare coupling α depends on μ (see Eq. (1.11).) We can introduce a momentum-dependent coupling $\alpha(t)$ via

$$t \equiv \int_\alpha^{\alpha(t)} \frac{dp}{\beta(p)} \tag{3.8}$$

where $t = \log(Q^2/\mu^2)$ Then Eq. (3.6) has the solution

$$F(t, \alpha) = F(1, \alpha(t)) \tag{3.9}$$

Hence the only dependence on the scale Q or t is carried by $\alpha(t)$. We can expand β as a power series in α.

$$\beta = -b\frac{\alpha}{4\pi} - b'(\frac{\alpha}{4\pi})^2 + \ldots \tag{3.10}$$

Hence $\alpha(\mu^2)$ has the following form:

$$\alpha(\mu^2) = \frac{4\pi}{b \log(\mu^2/\Lambda^2)} + \ldots \tag{3.11}$$

Here $b = 11 - 2n_f/3$ where n_f is the number of quark flavors with mass less than μ. We can regard the fundamental parameter of QCD either as $\alpha(Q_0^2)$ or as the scale Λ. Notice that as μ becomes small, α becomes large. Therefore, perturbation theory cannot be used to discuss processes which involve momentum flows as small as a few times Λ.

The value of Λ or $\alpha(Q_0^2)$ which has been obtained is dependent upon the renormalization scheme used. For example, I could have used minimal subtraction, in which case the $\log(4\pi)$ and γ_E would not have been removed. The expression for P written in terms of the new coupling constant $\bar{\alpha}$ can be used to express $\bar{\alpha}$ in terms of α, since the value of a physical quantity cannot depend on the scheme

$$P(\bar{\alpha}) = P(\alpha) \Rightarrow \bar{\alpha} = f(\alpha) \tag{3.12}$$

$$= \alpha + c\alpha^2 + \ldots$$

which corresponds to a new value of Λ

$$\bar{\Lambda} = \Lambda e^{c/2b} \tag{3.13}$$

A physical quantity is, of course, independent of the renormalization scheme. However, if the series is terminated at some finite order in the coupling constant, the values of P (P_N) calculated to this order will differ

$$P_N(\bar{\alpha}) \neq P_N(\alpha) = P_N(\bar{\alpha}) + 0(\alpha^{N+1}) \tag{3.14}$$

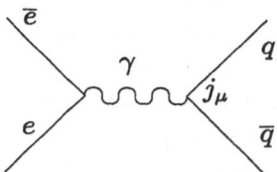

Figure 27: Feynman graph for $e^+e^- \to$ hadrons.

Since the coupling constant of QCD is not very small and most processes are not known to a very high order, these differences can be significant.

As a specific example of QCD process, consider the total cross-section for $e^+e^- \to$ hadrons at center-of-mass energy $\sqrt{s}$. In the one photon approximation (see Fig. 27) this is given by

$$\sigma_{had} = \frac{8\pi\alpha_{em}^2}{3s^2} \sum_n (2\pi)^4 \delta(q - q_n) \langle 0|j_\mu|n\rangle \langle n|j_\mu|0\rangle \tag{3.15}$$

where j_μ is the electromagnetic current of the quarks

$$j_\mu = \sum_i e_i \bar{\psi}_i \gamma_\mu \psi_i \tag{3.16}$$

If we introduce the photon self-energy function $\Pi^{\mu\nu}$

$$\Pi_{\mu\nu}(q) = i \int d^4x e^{iqx} \langle |T(j_\mu(x)j_\nu(0))| 0\rangle \tag{3.17}$$

Defining $\Pi_{\mu,\nu}(q) = (g_{\mu,\nu}q^2 - q_\mu q_\nu) = \Pi(Q^2)$ then

$$\sigma_{had} = \frac{16\pi^2\alpha_{em}^2}{s} Im\ \Pi(s) \tag{3.18}$$

A dimensionless quantity is $R(s)$ defined by

$$R(s) = \frac{\sigma_{had}}{\sigma(e^-e^+ \to \mu^+\mu^-)} \ . \tag{3.19}$$

Apart from the scale $\sqrt{s}$, R could also depend upon m_i^2/s, where m_i is the mass of a quark of type i. The dependence of R upon m_i can be seen by analyzing the Feynman diagram of Fig. 27. The graph is not singular as m_i tends to zero,

$$R_{m_i \to 0} = \text{const} + \frac{m_i^2}{s} \log m_i^2 \tag{3.20}$$

We can therefore neglect all light quark masses, R is then a function only of s/μ^2, and the previous argument implies that $R = R(\alpha(s))$. If we calculate R using perturbation theory we get

$$R = \sum e_i^2 \left(1 + \frac{\alpha}{\pi} + B\left(\frac{\alpha}{\pi}\right)^2 \dots\right) \tag{3.21}$$

where B is a scheme-dependent constant which is small[54] in the $\overline{MS}$ scheme.

Why are the non-perturbative effects irrelevant? After all, the final state consists of hadrons rather than free quarks which are used in the perturbative calculation. This can be understood by considering the time evolution of the final state. At very early times (equivalently large momenta) QCD is a weakly coupled theory so perturbation theory should be reliable. At large q the exponential in Eq. (3.17) is rapidly oscillating so only small values of x_μ contribute to $\Pi(q)$. Hence we are dominated by short distances (times). The hadronization of the final state takes place at later times (or order $1/\Lambda$) and although it can affect detailed properties of the final state, it is incapable of modifying the total cross-section.

In order to discuss processes which involve hadrons in the initial state, we must discuss the parton model.[52] Consider the case of electron-proton scattering, where the cross-section can be written as

$$\frac{d\sigma}{dxdy} = \frac{4\pi\alpha_{em}^2 s}{Q^4}\left[\frac{1+(1-y)^2}{2}2xF_1(x,Q^2) + (1-y)(F_2(x,Q^2) - 2xF_1(x,Q^2))\right]$$

(3.22)

The variables are defined as follows (see Fig. 28): q is the momentum of the exchanged photon and P is not momentum of the target proton and k is that of the incoming electron

$$Q^2 = -q^2$$

$$\nu = \frac{q \cdot P}{M_p}$$

$$x = \frac{Q^2}{2m_p\nu}$$

$$y = \frac{q \cdot p}{k \cdot p}$$

$$s = 2p \cdot k + m_p^2$$

(3.23)

where m_p is the proton mass. I have neglected parity violating effects which arise from the exchange of a Z boson instead of a photon.

In the naive parton model the proton is viewed as being made up of a set of non-interacting partons. The structure functions F_1 and F_2 are related to the probability distribution $q_i(x)$ which represents the probability of finding a parton of type i (quark or gluon) inside the proton with fraction x of the proton's momentum, and the scattering cross-section for such a virtual photon from a parton.

$$F_1 = \frac{F_2}{2x} = \sum_i \int_x^1 \frac{dy}{y}q_i(y)[e_i^2\delta(x/y - 1)]$$

(3.24)

where e_i is the charge of parton of type i. The δ-function appears from the cross-section for $q + \gamma \to q$. Let us consider QCD corrections to this scattering. At next order in α_s, there are contributions from gluon emission which lead to the final state $q + g$ and also from virtual gluons (see Fig. 29). To order α_s (3.24) is replaced by

$$F_1 = \sum_i \frac{dy}{y}q_i(y)\left[e_i^2\delta\left(\frac{x}{y} - 1\right) + \sigma_i\left(\frac{x}{y}, Q^2\right)\right]$$

(3.25)

with

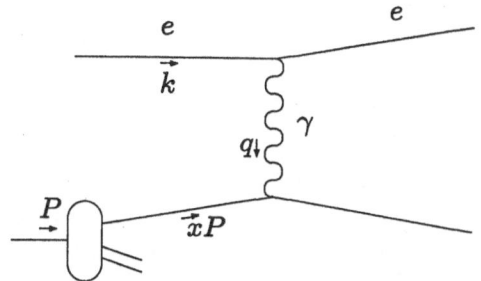

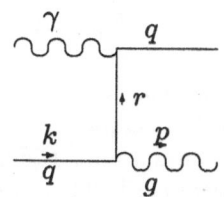

Figure 28: Diagram illustrating the variables in deep inelastic scattering (see Eq. (3.22)): electron + proton → electron + anything.

Figure 29: Diagram contributing to the process $q+\gamma \to X$ at order α_s.

$$\sigma_i(z, Q^2) = \frac{\alpha_s}{2\pi} e_i^2 \left[tP_{qq}(z) + f(z) + 0\left(\frac{1}{Q^2}\right) \right] \tag{3.26}$$

with

$$P_{qq}(z) = \frac{4}{3} \frac{(1+z^2)}{1-z}$$

for $z \neq 1$. Here $t = \log(Q^2/\mu^2)$ and the scale μ has appeared from dimensional regularization (I have dropped terms $1/(n-4)$). The μ dependence arises because σ_i is not finite in four dimensions. In the cases discussed previously, the divergences arise from large momentum flows inside loop diagrams (ultra-violet divergences). In this case these divergences cancel. Individual Feynman diagrams can also have divergences when momentum flows become very small or particles are collinear. The former (soft) divergences cancel between the real and the virtual diagrams but the collinear ones do not. In order to see the origin of the problem consider the graph of Fig. 29 and work in a frame where $k_\mu = (k, k, 0, 0)$.

If the transverse momentum of the gluon(p) relative to k is small then we can take $p = (\eta k + k_\perp^2/2\eta k, \eta k, k_\perp 0)$. (Terms of order $k_\perp^4$ are neglected.) The internal quark line now has invariant mass squared $r^2 = (k-p)^2 = k_\perp^2/\eta$, so that the squared amplitude from the graph will contain $1/k_\perp^4$. Now at very small $k_\perp$ helicity conservation forbids the emission of a real gluon from a quark line, so that one factor of $k_\perp^2$ appears in the numerator. We now have for the total cross-section $q + \gamma \to q +$ anything, a contribution

$$\sigma \sim \frac{\alpha_s}{2\pi} \int \frac{dk_\perp^2}{k_\perp^2} \tag{3.27}$$

which gives rise to a logarithmic singularity. Notice that for a massive quark the singularity becomes $\log(Q^2/m_q^2)$.

We have obtained a result which depends on μ (or contains the large log (Q^2/m_i^2) if quark masses are retained). This is not physically meaningful. But Eq. (3.25) contains the unknown quantity $q_i(y)$. We can define[55]

$$q_i(x, t) = q_i(x) + \frac{\alpha_s t}{2\pi} \int_x^1 \frac{dy}{y} q(y) P_{qq}\left(\frac{x}{y}\right) \tag{3.28}$$

Hence

$$F_1 = \sum_i \int_x^1 \frac{dy}{y} e_i^2 q_i(y) \left[\delta\left(\frac{x}{y} - 1\right) + \frac{\alpha_s}{2\pi} f\left(\frac{x}{y}\right) \right] + 0(\alpha^2) \tag{3.29}$$

The t dependence can be eliminated at the cost of introducing a t-dependent structure function

$$\frac{d}{dt}q(x,t) = \frac{\alpha_s}{2\pi} \int_x^1 \frac{dy}{y} q(y,t) P_{qq}\left(\frac{x}{y}\right) + 0(\alpha_s^2) \tag{3.30}$$

I have so far considered an oversimplification of the true problem. To order α_s there is an additional partonic process, namely $gluon + \gamma \to q + \bar{q}$ (see Fig. 30). This process also contains a log (Q^2/μ) arising from the propagation of the internal quark close to its mass shell. This singularity results in the replacement of Eq. (3.25) and (3.26) by

$$F_1(x,t) = \int_x^1 \frac{dy}{y} \left[\sum_i e_i^2 q_i(y) \left[\delta(\frac{x}{y}) + \frac{\alpha_s}{2\pi} \left[t P_{qq}(\frac{x}{y}) + f_q(\frac{x}{y}) \right] \right] \right.$$
$$\left. + (\sum_i e_i^2) g(y) \frac{\alpha_s}{2\pi} \left[t P_{qg}(\frac{x}{y}) + f_g(\frac{x}{y}) \right] \right] \tag{3.31}$$

with $P_{qg}(x) = 1/2(x^2 + (1-x)^2)$. The t dependence can be absorbed by defining

$$q_i(x,t) = q_i(x) + \frac{\alpha_s}{2\pi} t \int_x^1 (q_i(y) P_{qq}(\frac{x}{y}) + g(y) P_{qg}(\frac{x}{y})) \frac{dy}{y} \tag{3.32}$$

so that the quark and gluon distributions ($q_i(x)$ and $g(x)$) are now coupled.

Given data from which $q_i(x,t_0)$ and $g(x,t_0)$ can be obtained as functions of x for a fixed t_0, the equations for the evolution of $q(x,t)$ and $g(x,t)$ with t (cf. 3.30) can be solved to obtain them for all t.

A vital property of QCD is that the distribution functions defined by (3.32) are universal. In order to illustrate this, consider the Drell-Yan process in proton-proton collisions. In the naive parton model, the cross-section for the production of a $\mu^+\mu^-$ pair of invariant mass M in a proton-proton collision with total center-of-mass energy $\sqrt{s}$ is given by

$$\frac{d\sigma}{dM^2} = \frac{4\pi\alpha_{em}^2}{9M^2 s} \int dx_1 dx_2 [\sum_i q_i(x_1)\bar{q}_i(x_2)e_i^2 \delta(x_1 x_2 - M^2/s) + (1 \leftrightarrow 2)] \tag{3.33}$$

Here $\bar{q}$ is an antiquark distribution. The fundamental process is quark-antiquark annihilation into $\mu^+\mu^-$. Consider the corrections to this at order α_s. As in the case

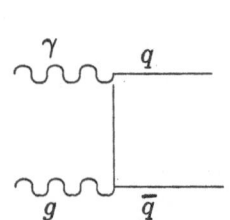

Figure 30: Diagram showing $g + \gamma \to q + \bar{q}$.

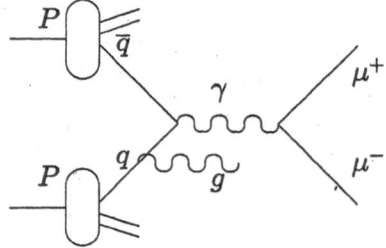

Figure 31: Feynman graph illustrating an order α_s contribution to the Dress-Yan process (see Eq. (3.34)).

of ep scattering these can involve either virtual or real gluons (see Fig. 31). These corrections modify Eq. (3.33), *viz.*,

$$\frac{d\sigma}{dM^2} = \frac{4\pi\alpha_{em}^2}{9M^2s} \int \frac{dx_1\,dx_2}{x_1\,x_2} \left[[e_i^2 q_i(x_1)\bar{q}_i(x_2) + (1 \leftrightarrow 2)]\right.$$

$$\left[\delta(1-z) + \theta(1-z)\frac{\alpha_s}{2\pi}[2P_{qq}(z)t + f'(z)]\right]$$

$$+ \ \left[\sum_i e_i^2(q_i(x_1) + \bar{q}_i(x_1))G(x_2) + (1 \leftrightarrow 2)]\right.$$

$$\left[\theta(1-z)\frac{\alpha_s}{2\pi}[P_{qg}(z) + f''(z)]\right]$$

where $z = M^2/(sx_1x_2)$.[56] The last part of the expression arises from the process $g + q \to \mu^+\mu^- + q$.

If we replace $q(x)$ by $q(x,t)$ defined by Eq. (3.32) then the resulting expression will have no t's appearing explicitly, *viz.*,

$$\frac{d\sigma}{dM^2} = \frac{4\pi\alpha_{em}^2}{9M^2s} \int dx_1\,dx_2[e_i^2 q_i(x_1,t)\bar{q}_i(x_2,t)\delta(x_1x_2 - M^2/s) + (1 \leftrightarrow 2) + \mathcal{O}(\alpha_s(Q^2))] \tag{3.35}$$

where the order $\alpha_s(Q^2)$ terms contain no powers of t. This absorption of the singular terms into $q(x,t)$ is known as factorization; it is a universal property which guarantees that hard processes can be reliably calculated in perturbative QCD.

I would now like to discuss some of the errors and uncertainties present in predictions of the rate for QCD processes in hadron-hadron collisions. I will discuss the production of new heavy quarks. A similar discussion applies to most other processes. The relevant QCD processes for the production of a $Q\bar{Q}$ pair are $gg \to Q\bar{Q}$ or $\bar{q}q \to Q\bar{Q}$. The cross-sections are given by[57]

$$\frac{d\sigma}{dt}(gg \to Q\bar{Q}) = \frac{\pi\alpha_s^2(Q^2)}{8s^2} \left\{ \frac{6}{s^2}(t - M_Q^2)(u - M_Q^2) + \left[\left(\frac{4}{3}\frac{(t - M_Q^2)(u - M_Q^2) - 2M_Q^2(t + M_Q^2)}{(t - M_Q^2)^2}\right.\right.\right.$$

$$+ \frac{3(t - M_Q^2)(u - M_Q^2) + M_Q^2(u - t)}{s(t - M_Q^2)} \left.\right) + [t \leftrightarrow u]\right]$$

$$\left. - \frac{M_Q^2(s - 4M_Q^2)}{3(t - M_Q^2)(u - M_Q^2)}\right\} \tag{3.36}$$

and

$$\frac{d\sigma}{dt}(q\bar{q} \to q\bar{q}) = \frac{\pi\alpha_s^2(Q^2)}{9s^2} \left[\frac{(t - M_Q^2)^2 + (u - M_Q^2)^2 + 2M_Q^2}{s^2}\right] \tag{3.37}$$

Here s,t and u are the usual Mandelstam variables. The rate for $pp \to Q\bar{Q}+$ *anything* when the quark emerges with transverse momentum $p_\perp$ at angle θ to the beam in the center-of-mass frame of the pp system is

$$E\frac{d\sigma}{d^3p} = \frac{1}{\pi}\sum_{ij} \int_{x_{min}}^1 \frac{dx_a}{x_a - x_\perp\left[\frac{x + \cos\theta}{2\sin\theta}\right]} x_a x_b f_i(x_a, Q^2)f_j(x_b, Q^2)\frac{d\sigma}{d\hat{t}}(\hat{s}, \hat{t}, \hat{u}) \tag{3.38}$$

where

$$\hat{s} = x_a x_b s$$

$$\hat{t} = M_Q^2 - x_a x_\perp s \left[\frac{\chi - \cos\theta}{2\sin\theta} \right]$$

$$\hat{u} = M_Q^2 - x_b x_\perp s \left[\frac{\chi + \cos\theta}{2\sin\theta} \right]$$

and

$$x_b = \frac{x_a x_\perp s (\chi - \cos\theta)}{s\sin\theta \left(2x_a - x_\perp \left[\frac{\chi+\cos\theta}{2\sin\theta} \right] \right)}$$

$$x_{min} = \frac{x_\perp (\chi + \cos\theta)}{s\sin\theta \left(2 - x_\perp \left[\frac{\chi-\cos\theta}{2\sin\theta} \right] \right)}$$

$$\chi = \left(1 + \frac{4M_Q^2 \sin^2\theta}{x_\perp^2 s} \right)^{1/2}$$

$$x_\perp = 2p_\perp / \sqrt{s}$$

In order to use this formula we must; fix the structure functions; determine the scale Q^2 appearing in the structure functions; fix the scale appearing in α_s; and define the quark mass. Let us discuss these problems in turn.

The structure functions are extracted at low Q^2 from the scattering of electrons or neutrinos off hadronic targets. These processes can only measure the quark structure functions directily since the gluons have no electroweak interactions. The gluon distributions must be inferred from the Q^2 dependence of the quark distribution (see Eq. (3.32)). This implies that there is a correlation between $g(x)$ and the value of α_s which controls the Q^2 dependence.

For processes at hadron colliders, the required values of Q^2 are larger than those at which the distribution functions are measured; most electron and neutrino scattering experiments have most of their statistics for $Q^2 < 10~GeV^2$. The distribution functions are then evolved up to larger Q^2 using QCD. In this evolution some of the uncertainties tend to wash out. This is illustrated in Fig. 32 which compares two sets of structure functions at different Q^2.[58,59]

In the case of $Q\bar{Q}$ production, the rate from gluon-gluon collisions dominates. There are two reasons for this: the gluon distribution is larger than that for quarks at small x (see Fig. 32) and the process $gg \to Q\bar{Q}$ has a larger rate than $q\bar{q} \to Q\bar{Q}$ due to the higher color charge of the gluon. If, however, in $p\bar{p}$ collisions, we produce quarks of very large mass, the appropriate values of x_a and x_b (see Eq. (3.38)) can become large and we are forced into a region where $g(x) < q(x)$ so that the quark antiquark annihilations can dominate. Notice that at these large values of x the distribution functions and hence the cross-sections are small.

Other data from hadron-hadron collisions can be used to check that the gluon distributions are reasonable. For example, jet production occurs via the processes $qq \to qq, gg \to gg, qg \to qg$, etc. If the measured jet cross-sections are in good agreement with the predicted values, we can have confidence that the distribution functions are reasonable. Such a comparison is shown in Fig. 33. Data from the Sp$\bar{p}$S collider[60,61] are compared with a prediction using distribution functions extracted

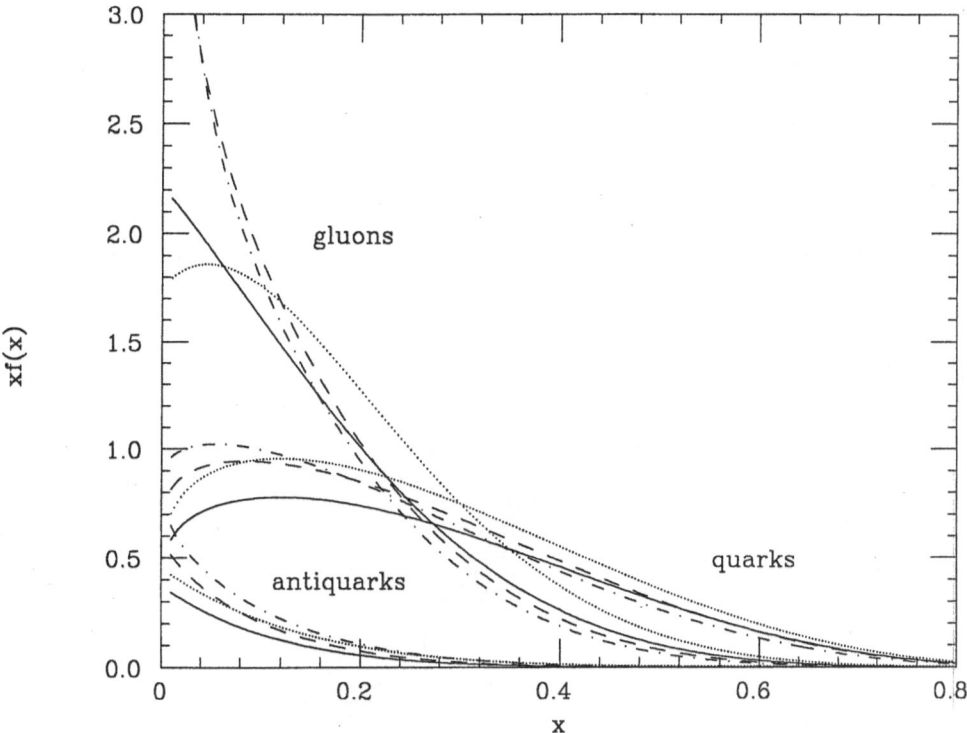

Figure 32: Diagram showing the behavior of the quark and gluon distributions as functions of x for various Q^2. Plotted is $xf(x)$ for gluons, quarks and antiquarks (summed over quark flavors). The solid (dotted) lines correspond to the structure functions of ref. 58 (59) at $Q^2 = 5\ GeV^2$. The dashed (dot-dashed) lines correspond to the structure functions of ref. 58 (59) at $Q^2 = 25\ GeV^2$. The evolution with Q^2 is given by perturbative QCD[48].

from the CDHS neutrino scattering experiment,[59] which were then extrapolated using QCD.[57] Such good agreement leads us to believe that the distribution functions are reliable at the 30% level.

I will now turn to the question of the scale Q^2 appearing in $\alpha_s(Q^2)$ (Eq. 3.36 and 3.37) and in the distribution functions. Suppose we shift the scale in $\alpha_s(M^2)$

$$\alpha_s(xM^2) = \alpha_s(M^2)\left[1 + \frac{(33 - 2n_f)}{12\pi}\alpha_s(M^2)\log x + \mathcal{O}(\alpha^2)\right] \qquad (3.39)$$

It is therefore clear that we cannot decide the question of scale without computing the order α_s^3 in Eqs. (3.36) and (3.37). A bad choice of scale is likely to result in large α_s^3 corrections. In the absence of such corrections we can only guess what the scale should be. Common sense dictates that it should be of order the quark mass M_Q. However, if the quark is being produced at large transverse momenta ($p_\perp$), then something like $\sqrt{M_Q^2 + p_\perp^2}$ is probably appropriate.

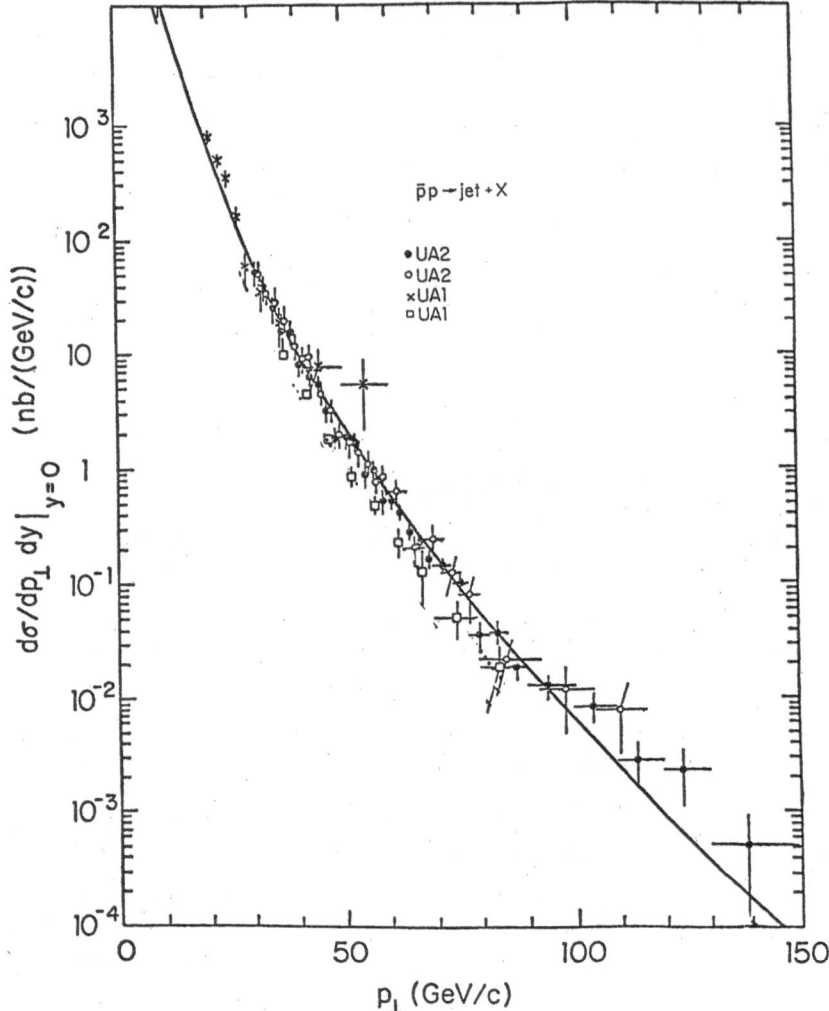

Figure 33: A comparison of the cross-section $p\bar{p} \rightarrow$ jet $+ X$ with a calculation using perturbative QCD for the subprocess $qq \rightarrow qq, gg \rightarrow gg$, etc. The structure functions are those of Ref. 59 evolved in Q^2 up to the relevant scale $(Q = p_\perp/2)$. The data are from the UA1[60] and UA2[61] collaborations.

To claim that the value of M_Q introduces an ambiguity may seem absurd. But suppose we are calculating the production rate for charm or bottom quarks; we must decide what value to use. The total production cross-section is a very strong function of M_Q, it varies roughly as M_Q^{-4}. What value of the charm quark mass should be used? This question is not easy to answer. The threshold for $c\bar{c}$ production opens when there is sufficient energy in the partonic collisions to produce a $D\bar{D}$ meson pair. This could suggest that one should use $M_Q = M_D$. However, the quark mass which appears in other calculations, such as that for the energy levels of the ψ system, is usually less than this. The uncertainty induced by M_Q becomes irrelevant for quarks heavier than the b.

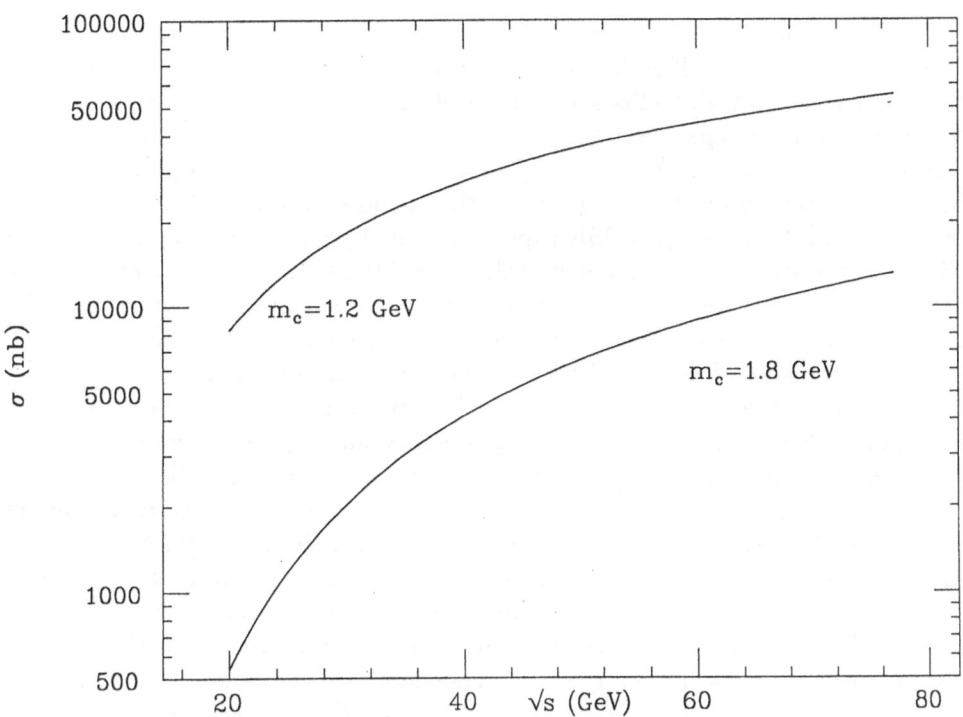

Figure 34: The cross section for the production of a charmed quark pair in proton-proton collisions as a function of $\sqrt{s}$.

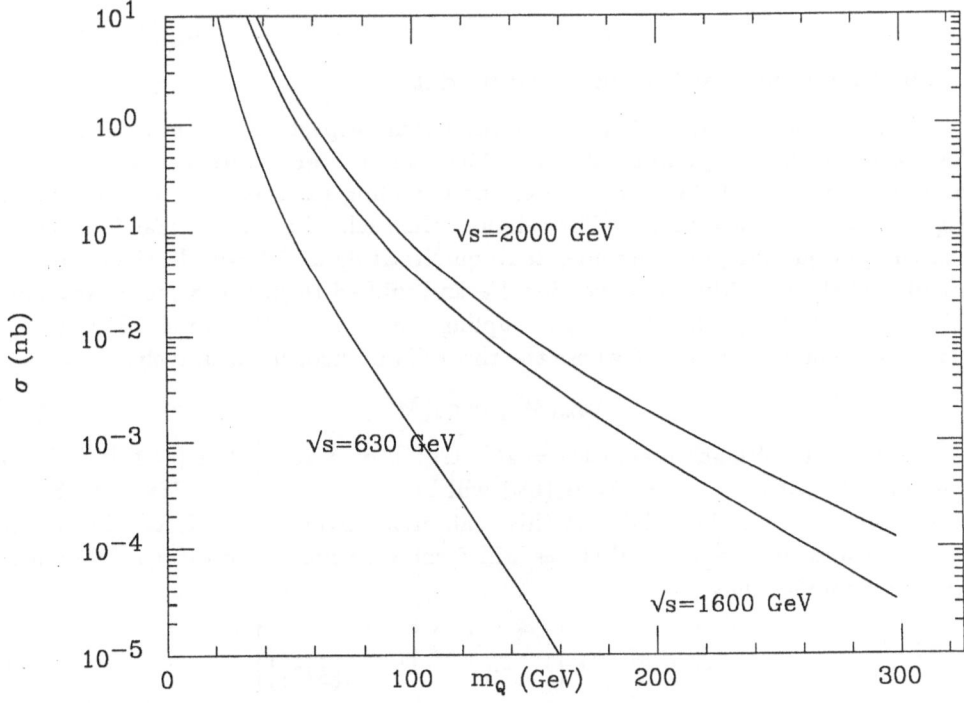

Figure 35: The cross section for the production of a heavy quark pair in proton antiproton collisions as a function of the heavy quark mass.

Figure 34 shows the total cross-section for the production of charmed quark pairs at small values of $\sqrt{s}$. This figure illustrates the longstanding problem of charm production rates. Cross-sections measured at the ISR[62] ($\sqrt{s} \approx 50~GeV$) usually gave rates which are approximately a factor of 10 larger than QCD expectations. Measurements of $\sqrt{s} \approx 25~GeV$[63] using protons on a stationary target gave results which were closer to QCD expectations. New results at $\sqrt{s} = 43~GeV$[64] are also close to the QCD values. The ISR experiments had poor acceptance and it is now difficult to reconcile their large values with those obtained at $\sqrt{s} = 43~GeV$. It seems fair to conclude that the QCD model works reasonably well[65], and while there may be some need to invoke new mechanisms to explain the rapidity distribution[66] of the produced charmed particles, no drastic modifications are required.

The cross section for heavy quark production at the $Sp\bar{p}S$ colliders is shown in figure 35. If we consider the detetion of heavy quarks we can be forced into a kinematical regime where the production cross section is not described by the pair production processes. The mean value of the transverse momenta of the quarks produced by the mechanisms discussed above is of order the quark mass. If detection considerations force us to look only at new quarks which are produced at tranvserse momenta which are much larger than this, then other mechanisms can become more important.[67] If quarks are produced at high transverse momenta via the processes of equations 3.37 and 3.38 then the transverse momenta of the quark and antiquark will balance each other. The process $g + g \rightarrow g + Q + \bar{Q}$ can yield a final state where the quark and antiquark are moving in the same direction, their transverse momenta being balanced by the final state gluon. Although this process is of order α_s^3 It can dominate if the transverse momentum is large enough.

4. The inadequacy of the standard model

In the previous sections I have discussed some aspects of the standard model. Despite the lack of experimental data which fails to agree with this model, most theorists find it unsatisfactory. One of the troubling features of this model is the origin of the electroweak scale. The origin of the scale of strong interactions, either Λ or the proton mass, can be understood qualitatively as follows. In the context of any unified theory, either a conventional grand unified theory or a more exotic one based on superstrings, all the gauge coupling constants of the standard model are related at some large scale M where the theory is unified. Qualitatively,

$$\alpha_{em}(M^2) \approx \alpha_s(M^2) \tag{4.1}$$

M is of order the Planck mass ($M_P = 10^{19}~GeV$), or possibly less ($\sim 10^{15}~GeV$) in some models. At some scale Q_0, $\alpha_s(Q_0)$ will become large and QCD perturbation theory will no longer be valid. At this scale non-perturbative effects will become important and hadronic bound states will form with mass of order Q_0. Requiring $\alpha(Q_0) = 1$ implies that

$$Q_0^2 \simeq M^2 exp\left[\frac{(33 - 2n_f)}{12\pi}\left[1 - \frac{1}{\alpha(M^2)}\right]\right] \tag{4.2}$$

It is easy to see that for $\alpha(M^2) \approx 1/40$, $Q_0/M \approx 10^{-15}$ and the large hierarchy of scales between hadron masses and M can be explained. The presence of the exponential in Eq. (4.2) guarantees that the large ratio M/Q_0 will be generated.

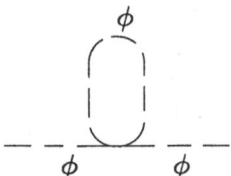

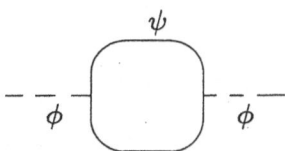

Figure 36: Feynman diagram showing a contribution to the Higgs mass renormalization in the Weinberg-Salam model.

Figure 37: Feynman diagram showing contributions to the mass renormalization of the scalar field ϕ (dashed lines) from its self interactions and those with the fermion ψ (solid lines) in the toy model of equation 4.5.

A similar argument cannot be used to explain the small ratio M_W/M_P. There is a dimensionful parameter μ in the $SU(2) \times U(1)$ Lagrangian, which is chosen to have a value of order M_W. Consider a radiative correction to this mass term due to the $\lambda(\Phi^+\Phi)$ interaction (see Fig. 36)

$$\delta m^2 = i\lambda \int \frac{d^4k}{(2\pi)^4} \frac{1}{(k^2 - \mu^2)} \tag{4.3}$$

This integral is quadratically divergent: cut it off at scale Λ. Then the value of μ^2 computed at one loop is given by

$$\mu^2_{1loop} = \mu^2 + \lambda\Lambda^2 \tag{4.4}$$

In theories such as those described by David Gross[3] in his lectures, new physics enters at the Planck mass and cuts off the divergence. In this case Λ is of order 10^{19} GeV. μ^2_{1loop} must be of order M_W^2, so that Eq. (4.4) implies that the bare mass μ must be adjusted to some 18 significant figures. This fine tuning must take place at all orders in perturbation theory. This unnatural fine tuning is usually referred to as the hierarchy problem and is present in theories with quadratic divergences.

This hierarchy problem can be solved in theories where the quadratic divergences are ameliorated on scales of order M_W. Since these divergences are associated with scalar fields,[68] the scalar sector of the theory must be modified.

One strategy is to introduce some new fields which cancel the divergence. Consider the following toy model consisting of a scalar field ϕ and a (two-component) fermion ψ:

$$\mathcal{L} = \partial_\mu \phi^+ \partial_\mu \phi - m_s^2 \Phi^+ \Phi - \bar{\psi}(i\,\partial\!\!\!/ - m_f)\psi + \lambda_1 \phi^4 + \lambda_2 \bar{\psi}\psi\phi \tag{4.5}$$

In this theory, there are two one-loop diagrams contributing to the one-loop calculation of the scalar mass (see Figs. 36 and 37). The scalar loop gives

$$\delta m^2_{s_1} = i\lambda_1 \int \frac{d^4k}{(2\pi)^4} \frac{1}{(k^2 - m_s)^2} \tag{4.6}$$

and the fermion loop,

$$\delta m^2_{s_2} = \frac{-i\lambda_2^2}{4} \int \frac{d^4k}{(2\pi)^4} \, \text{Tr} \left(\frac{1}{(\not{k} - m_f)(\not{k} + m_f)} \right) \tag{4.7}$$

The quadratic divergence is removed if $\lambda_2 = \lambda_1$ with the result

$$\delta m^2 = \lambda_1(m_f^2 - m_s^2) + \log \text{ divergence}$$

The soft logarithmic divergence is no problem, so the hierarchy problem is solved if $m_f^2 - m_s^2 \approx 0(M_W^2)$. This toy model is a supersymmetric theory in which the supersymmetry is softly broken (via $m_s \neq m_f$). Supersymmetric extensions of the standard model will have this property and so solve the hierarchy problem. Notice that this argument implies that the mass of the new particle predicted by these models must be less than 1 TeV or so. Supersymmetric theories are no help if the the new particles have masses of order the Planck Mass. I shall not discuss the phenomenology of supersymmetric models in these lectures. The reader may consult one of the recent review articles.[69]

Another solution to the hierarchy problem involves the introduction of a new interaction. If the Higgs boson were not elementary but were a bound state of a fermion and an antifermion with binding energy Λ, then the loop integral of Eq. (4.3) will be modified when $k \gtrsim 0(\Lambda)$. If there are no fundamental scalars, only bound states constructed from fermions, the bad divergences associated with elementary scalars will be removed.

The simplest implementation of this idea is in the technicolor theories.[26] I will sketch the basic mechanism here; one of the many review articles70 can be consulted for more details. Suppose we have a non-abelian gauge theory (called technicolor) with two flavors of fermion call U and D which transform as a doublet with respect to $SU(2)_L$ (cf., QCD with two flavors, up and down). If these fermions are massless, the underlying theory has a chiral symmetry $SU(2)_L \times SU(2)_R \times U(1)$. When the theory confines the chiral symmetry may break spontaneously via the formation of condensates $\langle \bar{U}U \rangle, \langle \bar{D}D \rangle \neq 0$. The symmetry will break to $SU(2) \times U(1)$ resulting in the appearance of three Goldstone bosons (these are analogous to the pions of QCD which would be massless in the absence of bare quark masses). Since the U and D form a doublet of $SU(2)_L$ then the W boson can couple to $U\bar{D}$ and hence to the "pion". The graph of Fig. 38 will generate a self energy for the W viz.,

$$\Pi_{\mu\nu} = \frac{g_2^2}{4}(k^2 g_{\mu\nu} - k_\mu k_\nu)\Pi(k^2) \tag{4.8}$$

where $\Pi(k^2)$ represents the $U\bar{D}$ bound state.

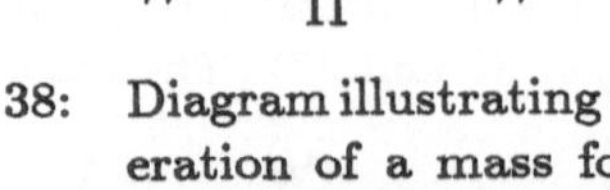

W $\qquad$ Π $\qquad$ W

Figure 38: Diagram illustrating the generation of a mass for the W boson from its coupling to the Goldstone boson in a technicolor theory. See eqn. 4.8.

If there are no Higgs fields then the W is massless and its propagator (calculated to lowest order in g_2) is

$$\frac{g^{\mu\nu} - k^\mu k^\nu / k^2}{k^2} \tag{4.9}$$

The effect of $\Pi^{\mu\nu}$ is to modify the propagator thus:

$$\frac{g^{\mu\nu} - k^\mu k^\nu / k^2}{k^2 \left[1 + \frac{g_2^2}{4} \Pi(k^2) \right]} \tag{4.10}$$

Since the lowest lying $U\bar{D}$ bound state is massless $\Pi(k^2) = f_\pi/k^2$ as $k^2 \to 0$. Hence (4.10) has no pole at $k^2 = 0$ but instead acquires a pole at $M_W^2 = \frac{g_2^2}{4} f_\pi^2$. The "pion" is eaten and becomes the extra polarization state of the massive W boson. If this mechanism is to yield the correct value for M_W, then we need $f_\pi \approx 250 \; GeV$. The neutral pions cause mixing between the "W_3" (i.e., the $I_3 = 0$ member of the $SU(2)_L$ gauge boson multiplet) and B, the gauge boson of $U(1)_y$. The couplings of "π^0" to B and W_3 are determined by the weak charges of the U and D quarks. The resulting mass matrix has the following form:

$$\frac{f_\pi^2}{4} \begin{pmatrix} g_2^2 & g_2 g_1 \\ g_2 g_1 & g_1^2 \end{pmatrix} \tag{4.11}$$

This has one zero eigenvalue corresponding to the photon and one non-zero with mass

$$M_Z^2 = \left(\frac{g_1^2 + g_2^2}{4} \right) f_\pi^2 = \frac{M_W^2(g_1^2 + g_2^2)}{g_2^2} \tag{4.12}$$

Notice that this corresponds to $\rho = 1$ (see Eq.(1.27)). This theory possesses a custodial $SU(2)$ symmetry since there is a chiral symmetry $SU(2)_L \times SU(2)_R \times U(1)$ which breaks to $SU(2) \times U(1)$ in the binding.

Such a theory can be expected to have more states than are present in the Weinberg-Salam model. For example, there will be spin-1 bound states $U\bar{D}, D\bar{U}, U\bar{U} + D\bar{D}, U\bar{U} - D\bar{D}$ analogous to the ρ and ω mesons in QCD. These will have mass of order a few times f_π and will decay to WW, WZ and ZZ (recall that $\rho \to 2\pi$ in QCD and that the "π" in this technicolor theory has been absorbed into the W and Z). One should also get technibaryons made from UUU, etc.

This simple technicolor theory which I have described cannot explain quark and lepton masses. The electron mass arises in the Weinberg-Salam model from the coupling

$$\lambda \Phi \psi_L \psi_R \tag{4.13}$$

which generates $m_e = \lambda \langle \Phi \rangle$. The analogous term in a technicolor theory would be

$$\bar{U}_L U_R \bar{\psi}_L \psi_R \tag{4.14}$$

Such a term is non-renormalizable and is not allowed. The problem is solved by introducing yet another set of interactions which have massive gauge bosons. These bosons mediate interactions between the quarks and techniquarks. The interaction (4.14) then arises as the low energy limit of this theory, in the same way that the Fermi-interaction arises as a low energy limit of the Weinberg-Salam model. Such extended technicolor theories have a rich phenomenology which is discussed in detail in review articles to which the reader is referred.[70,72]

I shall discuss one other variant of the standard model — the idea that quarks and leptons may be composite. This idea is rather unfashionable.[73] It implies that there must be another fundamental scale in physics, that on which the forces responsible for the binding of quarks and leptons are strong. This would be one other scale for theorists to explain. However, experimentalists should be aware that this option exists. Historically it has been the correct route, the chain being molecules, atoms, nuclei, quarks. It is also a natural extension of technicolor models where only the scalars are composite.

In a typical bound state model, the splitting between energy levels is the same order as the binding energy and the levels are fairly uniformly spaced. If quarks and leptons are composite this cannot be the case. It is known that, if the electron has substructure, the scale of the dynamics responsible for the binding must be 1 TeV or more (I will discuss these limits below). Hence the electron mass is much smaller than the binding energy. When confronted with such a problem, theorists search for a symmetry which would guarantee the presence of almost massless states. Two ideas have been suggested: chiral symmetry and/or pseudo-Goldstone fermions.

Under a chiral symmetry, a fermion field ψ transforms like $\psi \rightarrow e^{i\alpha\gamma^5}\psi$. A fermion mass term $m\bar{\psi}\psi$ is not invariant under such a symmetry. Hence fermions which have a chiral symmetry must be massless. If the quarks and leptons are constructed as bound states of some preons and the underlying preonic theory has a chiral symmetry which is preserved in the binding, then there will be massless fermionic bound states. The simplest example of such a preonic theory is one where the preons are fermions bound by some gauge theory. This type of theory is similar to QCD where approximately massless quarks are bound by QCD into hadrons. However, this binding produces no massless fermions; the proton is not massless.

How can we decide whether such a theory can have a massless bound state? The answer is provided by the 't Hooft consistency conditions.[74] In general, the underlying preon theory will have some global symmetries (for example, the chiral symmetry itself). Some of these symmetries may be anomalous. Consider a contribution to the three-point function of some of the currents associated with these chiral symmetries (see Fig. 39). To make the example concrete assume that all these currents are axial,

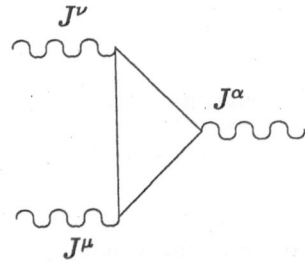

Figure 39:   Triangle diagram showing
the contribution of fermions
to the three point function
of the chiral currents. See
Eqn. 4.18.

e.g.,

$$J_\mu = \bar{\psi}\gamma_\mu\gamma_5\psi \qquad (4.15)$$

These currents are connected with the global symmetries and must be conserved at the classical level:

$$\partial_\mu J^\mu = 0 \qquad (4.16)$$

The explicit calculation of the three-point function of three of these currents ($\Gamma^{\mu\nu\alpha}$) may reveal the presence of an anomaly.

$$q_{3\mu}\Gamma^{\mu\nu\alpha} = A\epsilon^{\nu\lambda\alpha\beta}q_{2\lambda}q_{1\beta} \qquad (4.17)$$

with A non-zero, so that the current is no longer conserved. If the three-point function is evaluated at $q_1^2 = q_2^2 = q_3^2 = q^2$ it must have the form[75]

$$\Gamma^{\mu\nu\lambda} = \frac{1}{q^2}A[\epsilon^{\mu\nu\alpha\beta}q_1^\alpha q_2^\beta q_3^\lambda + \text{permutations}] + \text{terms finite as } q^2 \to 0 \qquad (4.18)$$

If the chiral symmetries are described by some group G and the preons are in a representation R with representation matrices $(\lambda^a)_{ij}$, then the currents (J_μ^a) contained in the $R \times \bar{R}$ representation. In particular, they can be in the adjoint representation. The coefficient A is proportional to $\text{Tr}[\lambda^a\lambda^b\lambda^c]$.

If the momentum scale $\sqrt{q^2}$ of the probing currents is much less than the scale Λ of the preonic binding, then the relevant spectrum of states is not that of the preons but rather that of the bound states. It follows that if the anomaly is calculated using the bound states, the same result must be obtained. In particular, the bound states must reproduce the singularity at $q^2 = 0$ in Eq. (4.18). This can be done in one of two ways.

The chiral symmetry could be spontaneously broken by the binding. In this case there must be a massless Goldstone boson which is coupled to the currents as indicated by Fig. 40. The boson propagator yields a factor of $1/(q^2 - m_{boson}^2)$, which generates the correct dependence on q^2 at small q^2 since m_{boson} is zero.

If the chiral symmetry is maintained by the binding, there will be massless bound state fermions. These fermions, when inserted in the loop of Fig. 39, can generate a contribution which has the correct q^2 dependence. Notice that the massive states bound states (of mass M) are not relevant since they will generate terms of order

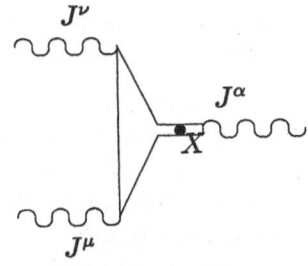

Figure 40: Diagram showing the contribution of a massless fermionic bound state (X) to the three point function of Eqn 4.18.

47

$1/M^2$ at small q^2. The value of A will depend upon the number of such massless bound states and their coupling, which is determined by the symmetry. Hence it may not be possible to get the same value of A from the massless bound states as from the preons. If the anomalies do not match then the chiral symmetry must be broken in the binding and there will be no massless fermions.

This matching condition is very difficult to satisfy. Let us take the familiar case of QCD as an example. Assume that there are only two massless quarks (the up and down). The quark theory now has a global symmetry of the type

$$SU(2)_L \times SU(2)_R \times U(1)$$

The subscripts L and R refer to the fermion helicity states. A linear combination of these $SU(2)$'s is the familiar strong isospin. The quarks are in the $\underline{2}$ representation of each of the $SU(2)$'s. The bound states must be singlets of $SU(3)_{color}$ which contain three quarks. These bound states must transform according to the $\underline{2} \times \underline{2} \times \underline{2}$ representation. It is therefore clear that the anomaly calculated from the quarks (a doublet) can match that calculated from the bound states consisting of a single doublet. It is therefore possible for the ground state of the theory to have two massless fermions.

Let us contrast this with what happens if there are three massless flavors. In this case, the global symmetry is

$$SU(3)_L \times SU(3)_R \times U(1)$$

The fermionic bound states must again contain three quarks and must therefore be contained in the $\underline{3} \times \underline{3} \times \underline{3}$ representation of the global $SU(3)$. The possibilities are $\underline{10}, \underline{8}$ and $\underline{1}$. None of these generates the same anomaly as the $\underline{3}$, so that there cannot be any massless composites and the chiral symmetry must be spontaneously broken in the binding. In the real case of QCD the strange quark is light; it may be this fact which forces chiral symmetry to be broken.

Notice that the consistency conditions cannot predict whether or not the chiral symmetry is unbroken if they are satisfied. Dynamical considerations may still favor the phase where it is broken even if the conditions are satisfied.

The second mechanism for providing massless bound states is the quasi-Goldstone fermion method.[76] The spontaneous breaking of a continuous symmetry will give rise to massless Goldstone bosons. If the theory has an unbroken supersymmetry these bosons must have degenerate fermionic partners. If a supersymmetric theory can be arranged so that during the binding process such a global symmetry is broken, then there will be massless fermionic bound states. Some toy models that have been constructed have both protection mechanisms for the fermion masses.[77]

Even after a model of massless fermions has been constructed there remains a potentially more serious problem. The quarks and leptons have small masses. A way must be found to generate this mass without destroying the protection and so promoting the masses up to the scale of binding. This can be done in the chiral models by giving the preons a small mass (much less than Λ) so that although the chiral symmetry is not exact, it is a very good approximation. The ground state will now consist of almost massless bosons or almost massless composite fermions. The former case arises in QCD where the small quark masses give rise to a non-zero pion mass.

In the pseudo-Goldstone fermion models the problem is more difficult. There are no light elementary scalars, so that as well as generating a small quark and lepton mass, the scalars must be given masses in excess of 20 GeV or so.

There is only one model of composite quarks and leptons which is realistic. This is a variant of the Weinberg-Salam model in which the weak interactions are assumed to be confining.[78] The preons consist of the usual matter representations, fermions which I will call pre-leptons and pre-quarks and scalar Higgs bosons. The ground state is assumed to be that where the Higgs vacuum expectation values is zero and the $SU(2)_L$ interactions are taken to be confining on a scale Λ of order 100 GeV. The candidates for massless fermionic bound states consist of a scalar and a pre-quark or pre-lepton. These states are singlets with respect to $SU(2)_L$ and can be shown to satisfy the consistency conditions. These bound states are the usual left-handed quarks and leptons. The right-handed quarks are elementary, being identical to those in the standard model. (They are singlets of $SU(2)_L$ and are therefore not bound.)

If this binding occurs one should expect the existence of spin-1 and spin-0 bound states made of elementary scalars. Such states should have mass of order Λ. These states play the role of the physical Higgs and the W and Z bosons of the standard model. The form of the coupling of these states to the quarks and leptons is the same as that in the standard model. This model is not inconsistent with any data, nevertheless it has some features which make me uncomfortable.*

The coupling between a W and a quark is known and is small. It seems odd that two bound states should couple so weakly. One would also expect a large number of massive states with masses of the same order as the W boson. In order to reconcile the model with the data the mass of these states must exceed 120 GeV or so. The degeneracy in this model between the W and Z mass is lifted by electromagnetic corrections which mix the Z and the photon. It is rather difficult to imagine this mixing being large enough to generate the observed mass splitting between the W and Z bosons.

Fortunately, it is possible to analyze the phenomenological consequences of compositeness without reference to a specific model. This is done by placing phenomenological constraints on the possible effective operators[80] which involve quarks and leptons and which an arise as a result of the interactions which are responsible for the binding. In the strongly coupled Weinberg-Salam model, it is one of these operators which plays the role of the Fermi coupling.

One of the most accurately known quantities in high energy physics is the anomalous magnetic moment of the muon. A theory in which the muon is composite is likely to have more massive states with the same quantum numbers (called μ^*). An interaction is possible involving the photon, the muon and the excited muon,[81] viz.,

$$\frac{e}{2\Lambda_*}\bar{\mu}\sigma^{\mu\nu}(a - b_{\gamma_5})\mu^* F_{\mu\nu} \tag{4.19}$$

I have included a factor of e since the coupling is electromagnetic. The appropriate power of Λ_* is determined by dimensional analysis.

The parameter Λ_* is of order the scale of the binding of the muon constituents. Its exact value is model dependent. Such an interaction will give a contribution to

---

*A recent reappraisal of this model is given in Ref. 79

$(g-2)$ at one loop shown in Fig. 41.

$$\delta(g-2) = m_\mu \frac{(a^2 - b^2)}{\Lambda} + \frac{(a^2 + b^2)m_\mu^2}{\Lambda^2} \qquad (4.20)$$

In making this estimate, it has been assumed that the mass of μ^* is of order Λ. If the underlying preonic theory is chiral, as it must be if chiral symmetry is used to guarantee the appearance of light bound states, $a = b$ and the contribution to $(g-2)$ is suppressed by m_μ^2/Λ^2. The constraint on Λ is rather poor

$$\Lambda \gtrsim 100 \ GeV \qquad (4.21)$$

Instead of writing an interaction involving μ^*, I could have written an interaction involving only the muon, *viz.*,

$$\frac{m_\mu}{\Lambda_*^2} \bar{\mu} \sigma^{\mu\nu} \mu F_{\alpha\beta} \qquad (4.22)$$

Again, Λ_* is of the order of the composite scale; it has been assumed that the binding forces are chiral, hence the factor of m_μ.

I shall analyze the remaining consequences of compositeness in this manner so that I shall not have to make reference to the more massive states, such as the μ^*.

It is possible that the muon and electron could share constituents, in which case processes like $\mu \to e\gamma$ will be expected to occur. Such transitions will be suppressed by powers of Λ_*.[†] The lowest dimension operator which can contribute to the decay process will be the one which dominates unless it has a very small coefficient. In this case the operator is a four-fermion operator

$$\frac{g^{*2}}{\Lambda^{*2}} \bar{\mu} \gamma^\alpha \frac{(1 - \gamma_5)}{2} e \ \bar{e} \gamma^\alpha \frac{(1 - \gamma_5)}{2} e + (\mu \leftrightarrow e) \qquad (4.23)$$

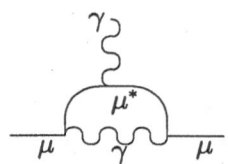

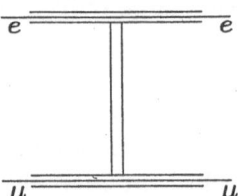

Figure 41: A contribution to $(g-2)$ of the muon from the interaction of muons, excited muons and photons of Eqn. 4.19.

Figure 42: Diagram showing the origin of the four-fermion operator of Eqn. 4.23, due to the exchange of some exotic state with mass of order Λ^*.

[†]Recall that in a grand unified theory proton decay is mediated by interactions which operate on the scale of grand unification and hence the proton decay rate is suppressed by powers of this large scale

In this expression I have inserted a factor of g^{*2}, where g^* is the coupling constant of the composite theory. The effective operator (4.23) arises from a diagram of the type indicated in Fig. 42. The diagram involves the exchange of some particle of mass order Λ_* with coupling of order g^*. Again, the exact value will depend on the particular model. Since the binding is expected to occur at strong coupling, it is usual to take $g^{*2} = 4\pi$ so that $\alpha_* = g^{*2}/4\pi = 1$. Notice that this is not the case in the composite model discussed above where the W and Z are composite. In that model the coupling is $g_W \approx 0.68$.

The operator of Eq. (4.23) can also be bounded from the failure to observe the process $\mu \rightarrow 3e$. There is one other lepton number violating operator which can occur

$$\frac{g^{*2}}{\lambda_*^2} \bar{\mu}\gamma^\alpha \frac{(1 - \gamma_5)}{2} e\bar{q}\gamma_\alpha \frac{(1 - \gamma_5)}{2} q \tag{4.24}$$

Table 5: Limits of contact terms from rare processes. The interaction type assumed for each rare process is shown along with the resulting limit on the compositeness scale Λ^*.

Process	Contact Interaction	Limit on Λ^* (TeV)
$(g - 2)_e$	$\frac{m_e}{\Lambda_*^2} \bar{e}\sigma^{\alpha\beta} e F_{\alpha\beta}$	.03
$(g - 2)_\mu$	$\frac{m_\mu}{\Lambda_*^2} \bar{\mu}\sigma^{\alpha\beta} \mu F_{\alpha\beta}$	.86
$\mu \rightarrow e\gamma$	$\frac{4\pi}{\Lambda_*^2} \bar{\mu}\gamma^\alpha \frac{1}{2}(1 - \gamma_5) e\bar{e}\gamma_\alpha \frac{1}{2}(1 - \gamma_5)e + (\mu \leftrightarrow e)$	60
$\mu \rightarrow 3e$	$\frac{4\pi}{\Lambda_*^2} \bar{\mu}\gamma^\alpha \frac{1}{2}(1 - \gamma_5) e\bar{e}\gamma_\alpha \frac{1}{2}(1 - \gamma_5)e$	400
$\mu N \rightarrow eN$	$\frac{4\pi}{\Lambda_*^2} \bar{\mu}\gamma^\alpha \frac{1}{2}(1 - \gamma_5) e\bar{d}\gamma_\alpha \frac{1}{2}(1 - \gamma_5)d$	460
$K_L \rightarrow e^\pm \mu^\mp$	$\frac{4\pi}{\Lambda_*^2} \bar{s}\gamma^\alpha \frac{1}{2}(1 - \gamma_5) u\bar{e}\gamma_\alpha \frac{1}{2}(1 - \gamma_5)\mu$	140
$K^+ \rightarrow \pi^+ e^- \mu^+$	$\frac{4\pi}{\Lambda_*^2} \bar{s}\gamma^\alpha \frac{1}{2}(1 - \gamma_5) u\bar{e}\gamma_\alpha \frac{1}{2}(1 - \gamma_5)\mu$	210
$\Delta M(K_L - K_S)$	$\frac{4\pi}{\Lambda_*^2} \bar{s}\gamma^\alpha \frac{1}{2}(1 - \gamma_5) d\bar{s}\gamma_\alpha \frac{1}{2}(1 - \gamma_5)d$	6100

where q is a quark. Such an operator could contribute to the deep inelastic scattering process $\mu + \text{nucleon} \rightarrow e + \text{hadrons}$. Such processes have not been observed. The bound on Λ arising from such lepton number violating processes is shown in Table 4.

The interactions involving quarks may give rise to flavor-changing neutral currents. The experimental constraints on the absence of such currents are very strong. A $\Delta S = 2$ transition mediated by the operator

$$\frac{g^{*2}}{\lambda_*^2}\bar{s}\gamma^\alpha\frac{(1-\gamma_5)}{2}d\bar{s}\gamma_\alpha\frac{(1-\gamma_5)}{2}d \tag{4.25}$$

will give a contribution to the $K_L - K_S$ mass difference. The mass difference arises in the standard model from the second order weak process involving the exchange of two W bosons. If such a new contribution is present, Λ_* must be larger than 6 TeV! Table 5 extracted from the lectures of Eichten[82] gives a summary of the compositeness limit arising from the absence of rare processes. These limits are given assuming that $g^{*2} = 4\pi$. They are less stringent in models where the composite states are weakly coupled.

If the composite interactions do not yield flavor changing interactions between quarks and leptons, then the limits from the indirect effects of compositeness are rather weak. In this case one must turn to more direct searches. In the case of composite electrons, one expects that there will be a four electron operator of the form[80]

$$g^{*2}/2\Lambda_*^2 \quad \left[\eta_{LL}\bar{e}\gamma^u\frac{(1-\gamma_5)}{2}e\bar{e}\gamma^u\frac{(1-\gamma_5)}{2}e \right.$$

$$\eta_{RR}\bar{e}\gamma^u\frac{(1+\gamma_5)}{2}e\bar{e}\gamma^u\frac{(1+\gamma_5)}{2}e \tag{4.26}$$

$$\left.2\eta_{LR}\bar{e}\gamma^u\frac{(1-\gamma_5)}{2}e\bar{e}\gamma^u\frac{(1-\gamma_5)}{2}e\right]$$

This term will interfere with the usual one-photon exchange process and cause a modification of the Bhabha scattering cross-section $e^+e^- \rightarrow e^+e^-$, viz.,

$$\frac{d\sigma}{d\Omega} = \frac{\pi\alpha^2}{4s}[4A_0 + A_+(1+\cos\theta)^2 + A_-(1-\cos\theta)^2] \tag{4.27}$$

$$A_0 = \left(\frac{s}{t}\right)^2\left|1+\frac{g_Rg_L}{e^2}\frac{t}{t_z}+\frac{\eta_{RL}t}{\alpha\Lambda_*^2}\right|^2, \quad A_- = \left|1+\frac{g_Rg_L}{e^2}\frac{s}{s_z}+\frac{\eta_{RL}s}{\alpha\Lambda_*^2}\right|^2$$

$$A_+ = \frac{1}{2}\left|1+\frac{s}{t}+\frac{g_R^2}{e^2}\left(\frac{s}{s_z}+\frac{s}{t_z}\right)+\frac{2\eta_{RR}s}{\alpha\Lambda_*^2}\right|^2 + \frac{1}{2}\left|1+\frac{s}{t}+\frac{g_L^2}{e^2}\left(\frac{s}{s_z}+\frac{s}{t_z}\right)+\frac{2\eta_{LL}s}{\alpha\Lambda_*^2}\right|^2$$

Here $t = -s(1-\cos\theta)/2$, $s_z = s - M_z^2 + i\Gamma_z M_z$, $t_z - t - M_z^2 - iM_zM_z$, $g_R = e\tan\theta_w$ and $g_L = -e\cot 2\theta_w$. In this formula I have set $g^{*2} = 4\pi$, a reasonable value for a strongly coupled theory. Current data can be used to set a limit of $\Lambda_* \lesssim 2.5$ TeV.[83]

In the case of quark compositeness an effect will be seen in the production of jets in a hadron collider since one of the relevant partonic processes is quark-quark scattering. The presence of a four-quark operator of the form (4.26) with a quark replacing an electron will yield a modification of the predicted jet cross-section in hadronic collisions. If we take $\eta_{LR} = \eta_{RR} = 0$ then this interaction leads to a modification of the cross-section for quark-quark scattering

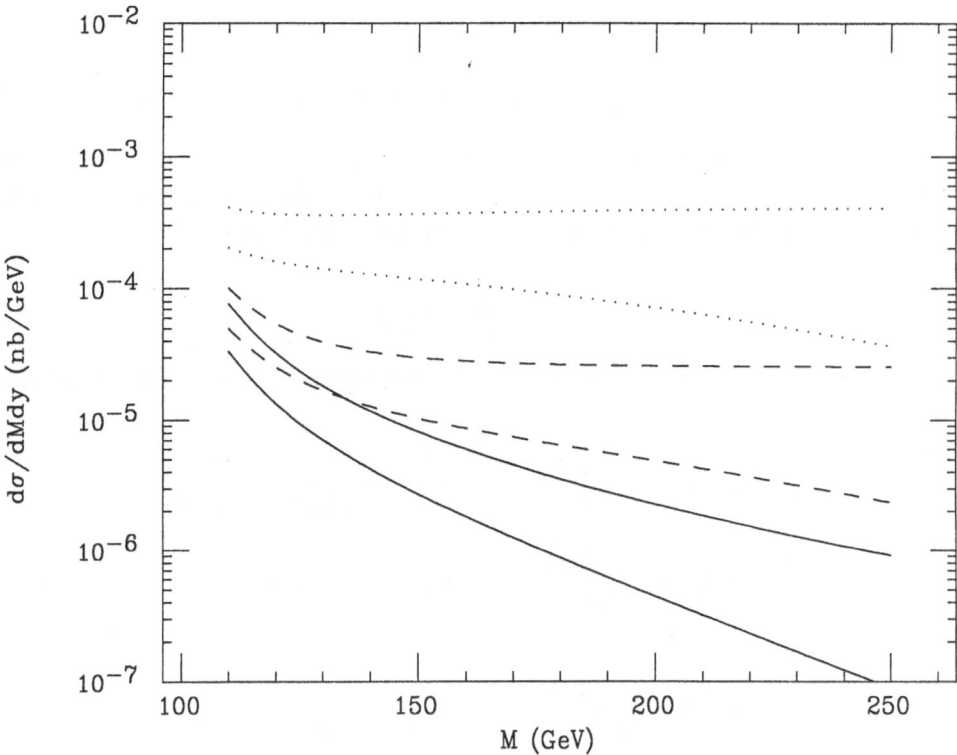

Figure 43: The cross section $d\sigma/dp_t dy$ for the production of jets of hadrons transverse meomentum p_t and rapidity $y = 0$. The curves include the effects of compositeness given by Eqn. 4.28 with $\eta_{LL} = 1$ and $\Lambda^* = 300$ GeV (dotted lines), 1000 GeV (dashed lines) and 5000 GeV (solid lines). The curves are for proton-antiproton interactions as $\sqrt{s} = 630$ GeV (lower lines) and 2000 TeV.

$$\frac{d\sigma}{dt}(q_i q_i \rightarrow q_i q_i) = \frac{4\alpha_s^2}{9s^2}\left[\frac{u^2+s^2}{t^2} + \frac{s^2+t^2}{u^2} - \frac{2}{3}\frac{u^2}{st}\right]$$

$$+ \frac{8\eta_{LL}\alpha_s}{9\Lambda_*^2}\left(\frac{s^2}{t} + \frac{s^2}{u}\right)$$

$$+ \frac{\eta_{LL}^2}{\Lambda_*^4}\left(u^2 + t^2 + \frac{2s^2}{3}\right) \qquad (4.28)$$

$$\frac{d\sigma}{dt}(q_i q_j \rightarrow q_i q_j) = \frac{4}{9}\frac{\pi\alpha_s^2}{s^2}\left[\frac{u^2+s^2}{t^2} + \frac{9}{4\alpha_s^2}\frac{\eta_{LL}^2 u^2}{\Lambda^{*4}}\right]$$

Again I have taken $g^{*2} = 4\pi$. It is to be remarked that these effects are not visible in quark-quark scattering until a higher value of s than was the case for Bhabha scattering. This is simply because the dominant contribution to the jet cross-sections arises from gluon-gluon and gluon-quark scattering which are not modified since the gluon is not composite. The effects of compositeness are not visible until s/Λ^2 is sufficiently large for the composite effects to raise the quark-quark scattering rates above those for gluon-gluon and gluon-quark. The effect is shown in Fig. 43. Current data from the Sp$\bar{p}$S collider place a limit of order 450 GeV.[84]

If quarks and leptons share constituents then operators of the type

$$\eta_0 \frac{g_*^2}{4\pi} \bar{q} \gamma^u \frac{(1-\gamma_5)}{2} q \bar{e} \gamma^u \frac{(1-\gamma_5)}{2} e \qquad (4.29)$$

The effect of such a term can be seen in two different ways. First, it will affect the production rate for lepton pairs in a hadron-hadron collision. This is controlled by the process $q\bar{q} \to e^+ e^-$ for which the cross-section has the form

$$\frac{d\sigma}{dt}(q_i \bar{q}_i \to l^+ l^-) = \frac{\pi \alpha^2}{s^2} \left[A_i(s) \left[\frac{u}{s} \right]^2 + B_i(s) \left[\frac{t}{s} \right]^2 \right] \qquad (4.30)$$

where the quark flavors are $i =$ up, down. The coefficients A_i and B_i may be written as

$$A_i(s) = \left| Q_i - \frac{L_i L_i}{4x_W(1-x_W)} \frac{s}{s - M_Z^2 + iM_Z\Gamma_Z} - \frac{\eta_0 s}{\alpha \Lambda^{*2}} \right|^2$$

$$+ \left| Q_i - \frac{R_i R_i}{4x_W(1-x_W)} \frac{s}{s - M_Z^2 + iM_Z\Gamma_Z} \right|^2 \qquad (4.31)$$

$$B_i(s) = \left| Q_i - \frac{R_i L_i}{4x_W(1-x_W)} \frac{s}{s - M_Z^2 + iM_Z\Gamma_Z} \right|^2$$

$$+ \left| Q_i - \frac{L_i R_i}{4x_W(1-x_W)} \frac{s}{s - M_Z^2 + iM_Z\Gamma_Z} \right|^2$$

where the chiral couplings of the neutral weak current are, as usual, $L_i = \tau_3 - 2Q_i x_W$ and $R_i = -2Q_i xW$. Here the weak mixing parameter is $x_W = \sin^2 \theta_W$ and τ_3 is twice the weak-isospin projection of fermion i.

The cross-sections $d\sigma/dMdy|_{y=0}$ for the reaction

$$p\bar{p} \to l^+ l^- + \text{anything}$$

are shown in Fig 44. A contribution of the type (4.29) will also shown up in deep inelastic scattering where it will cause a modification of the proton structure function $F_2(x, Q^2)$. This is discussed in the lectures by Gunter Wolf.[85] The values of Λ which can be probed at HERA are similar to those reached using the Drell-Yan mechanism at the Tevatron collider.

Acknowledgments

The work was supported by the Director, Office of Energy Research, Office of High Energy Physics, Division of High Energy Physics of the U.S. Department of Energy under Contract DE–AC03–76SF00098.

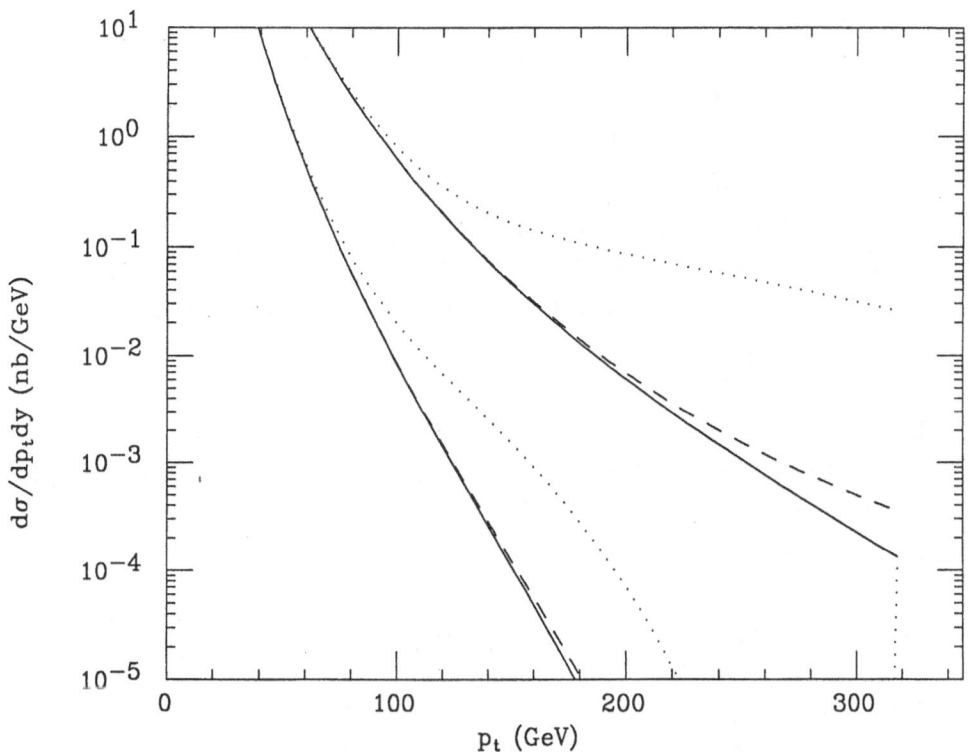

Figure 44: The cross section $d\sigma/dM\,dy$ for the production of a Drell-Yan
pair of mass M and rapidity $y = 0$. The curves include the
effects of compositeness given by Eqn. 4.31 with $\eta = 1$ and
$\Lambda^* = 500$ GeV (dotted lines), 1000 GeV (dashed lines) and
5000 GeV (solid lines). The curves are for proton-antiproton
interactions as $\sqrt{s} = 630$ GeV (lower lines) and 2000 TeV.

REFERENCES

1   S. Glashow, *Nucl. Phys.* 22:579 (1961);
     S. Weinberg, *Phys. Rev. Lett.* 19:1264 (1967);
     A. Salam, *in:* "Elementary Particle Theory," W. Svartholm, ed., Almquist and
     Wiksell, Stockholm (1968).

2   L. Di Lella, lectures at this school.

3   D. Gross, lectures at this school;
     A. Savoy-Navarro, lectures at this school.

4   Physics at LEP, J. Ellis and R. Peccei, eds., CERN 86-02 (1986).

5   Proceedings of the SSC Workshop, SLAC-PUB-267 (1982).

6   G. 't Hooft and M. Veltman, *Nucl. Phys. B* 44:189 (1972);
     C.G. Bollini, J.J. Giambiagi and A. Gonzáles Domínguez, *Nuovo Cim.* 31:551
     (1964);
     J. Ashmore, *Nuovo Cim. Lett.* 4:289 (1972);
     G.N. Cicuta and E. Montaldi, *Nuovo Cim. Lett.* 4:329 (1972).

7   W.A. Bardeen et al., *Phys. Rev. D* 18:3998 (1978).

8   See, for example, J.D. Bjorken and S.D. Drell, "Relativistic Quantum Mechanics," McGraw-Hill (1964).

9   E.R. Williams and P.T. Olsen, *Phys. Rev. Lett.* 62:1575 (1979).

10   W. Marciano, *Phys. Rev. D* 20:276 (1979).

11   A. Sirlin, *Phys. Rev. D* 22:971 (1980).

12   W. Marciano and A. Sirlin, *Phys. Rev. D* 29:945 (1984).

13   Ch. Llewellyn Smith and J. Wheaten, *Phys. Lett. B* 105:486 (1981).

14   G. Arnison et al., *Phys. Lett. B* 166:484 (1986);
     J. Appel et al., *Z. Physik C* 30:1 (1986).

15   L. S. Durkin and P. Langecker, *Phys. Lett. B* 166:436 (1986).

16   L. DiLella, *in* "Proceedings of the 1985 International Symposium on Lepton and Photon Interactions at High Energies," Kyoto, Japan (1985).

17   L. A. Ahrens et al., Univ. of Penn. preprint E-734-85-1 (1985).

18   C. Prescott et al., *Phys. Lett. B* 84:524 (1979);
     J.E. et al., *Rev. Mod. Phys.* 53:211 (1981).

19   F. Bergsma et al., *Phys. Lett. B* 147:481 (1984);
     L.A. Ahrens et al., *Phys. Rev. Lett.* 54:18 (1985).

20   M.A. Bouchiat et al., *Phys. Lett. B* 134:465 (1984);
     E.W. Fortson and L.L. Lewis, *Phys. Rep.* 113:289 (1984).

21   H. Abramowicz et. al, *Phys. Rev. Lett.* 57:298 (1986);
     J.V. Allaby et al., CERN preprint CERN-EP-86/94 (1986).

22   W. Marciano and A. Sirlin, *Nucl. Phys. B* 189:442 (1981).

23   M.S. Chanowitz, M. Furman and I. Hinchliffe, *Nucl. Phys. B* 153:402 (1979).

24   M. Veltman, *Phys. Lett. B* 70:253 (1977).

25   T. Appelquist and J. Carrazone, *Phys. Rev. D* 11:2856 (1975).

26   S. Weinberg, *Phys. Rev. D* 19:1277 (1979);
     L. Susskind, *Phys. Rev. D* 20:2619 (1979).

27   G. Barbiellini et al., in Ref. 4.

28   B. Pifer et al., D0 Design Report.

29   B.W. Lynn, M. Peskin and R.G. Stuart, in Ref. 4.

30   D. Blockus et al., "Proposal for Polarization at the SSC."

31   B.W. Lynn and M. Cretic, SLAC-PUB-3900 (1986);
     F. Gilman and P. Franzini, SLAC-PUB-3932 (1986).

32   S. Coleman and E. Weinberg, *Phys. Rev. D* 7:1888 (1973).

33   S. Weinberg *Phys. Rev. Lett.* 36:294 (1976);
     A. Linde, *JETP Lett.* 23:64 (1976).

34   A. Guth and E. Weinberg, *Phys. Rev. Lett.* 65:1131 (1980).

35   R. Flores and M. Sher, *Ann. Phys.* 168:95 (1983).

36   C. Quigg, D.W. Lee and H. Thacker, *Phys. Rev. D* 16:1519 (1977).

37   M. Veltman, *Acta Phys. Polon. B* 8:475 (1977).

38 J.D. Bjorken, *in*: "Proceedings of the 1976 SLAC Summer Institute," M. Zipf, ed.

39 R.N. Cahn, M.S. Chanowitz and N. Fleishon, *Phys. Lett. B* 82:113 (1979).

40 B.L. Ioffe and V.A. Khoze, Leningrad report LINP-274 (1976).

41 F. Wilczek, *Phys. Rev. Lett.* 39:1304 (1977).

42 M. Vysotsky, *Phys. Lett. B* 97:159 (1980).

43 H.M. Georgi et al., *Phys. Rev. Lett.* 40:692 (1978).

44 R.N. Cahn and S. Dawson, *Phys. Lett. B* 136:196 (1986);
    S. Petcov and D.R.T. Jones, *Phys. Lett. B* 84:660 (1979).

45 M. Chanowitz and M.K. Gaillard, *Phys. Lett. B* 142:85 (1986).

46 Z. Kunszt, *Nucl. Phys. B* 247:339 (1984).

47 J.F. Gunion et al., *Phys. Rev. Lett.* 54:1226 (1985).

48 E. Eichten et al., *Rev. Mod. Phys.* 56:579 (1986).

49 J.F. Gunion and M. Soldate, *Phys. Rev. D* 34:826 (1986).

50 J. Ellis, M.K. Gaillard and D.V. Nanopoulos, *Nucl. Phys. B* 106:292 (1976).

51 G. Kane University of Michigan preprint UM TH 86-16 (1986), F. Gilman, J. Gunion and P. Wiesman, In preparation.

52 G. Altarelli *Phys. Rep.* 81:1 (1982).

53 A. H. Mueller Lectures at the 1985 TASI (yale University), published by Wordl Scientific Publishing (Singapore).

54 M. Dine and J. Saperstein *Phys. Rev. Lett.* 43:688 (1979), W Celmaster and R J. Gonsalves *Phys. Rev. Lett.* 44:560 (1979).

55 G. Altarelli and G. Parisi, *Nucl. Phys.* B126:298 (1977).

56 G. Altarelli, R. K. Ellis and G. Martinelli, *Nucl. Phys.* B143 521 (1978).

57 B. L. Combridge, *Nucl. Phys.* B151:429 (1977).

58 F. Bergsma et al., *Phys. Lett.* 123B:269 (1983).

59 H. Abramowicz et al., *Z. Physik* C13:199 (1982), C17:283 (1983).

60 P. Bagnaia et al., *Z. Physik* C20:117 (1983),

61 G. Arnison et al., *Phys. Lett.* 123B:115 (1983),

62 D. Drijard et al., *Phys. Lett.* 85B:425 (1979).

63 R. Ammer et al., *Phys. Lett.* 135B:237 (1984).

64 R. Ammer et al., CERN EP/86-122 (1986).

65 J.R. Cuddell and F. Halzen, *Phys.Lett.* 175B:227 (1986).

66 G. VanDalen and A. Kernan, *Phys. Rep.* 106:297 (1984).

67 F. Herzog and Z. Kunszt, *Phys. Lett.* 157B:430 (1985).

68 K.G Wilson, *Phys. Rev.* D3:1818 (1973).

69 See for example J. Ellis, CERN–TH–4255/85 Lectures presented at the 28th Scottish Universities Summer School in Physics, Edinburgh. Scotland. or H. Haber and G. L. Kane *Phys. Rep.* 117:75 1985.

70 E. Farhi and L. Susskind, *Phys. Rep.* 74:277 (1981).

71   E. Eichten and K. Lane, *Phys. Lett.* 90:125 (1980).

72   E. Farhi and L. Susskind, *Phys. Rev.* D20:3404 (1979).

73   R.D. Peccei, DESY 86-010 (1986).

74   G 'tHooft *in* Recent Developments in Field Theory, Ed G 'tHooft, Plenum (1980).

75   S. Coleman and B. Grossman, *Nucl. Phys.* B203:205 (1982).

76   W. Buchmuller, R. D. Peccei and T. Yanigida, *Phys. Lett.* 124B:67 (1983).

77   W. Buchmuller, R. D. Peccei and T. Yanigida, *Nucl. Phys.* B244:186 (1984).

78   L. Abbott and E. Farhi, *Nucl. Phys.* B189:547 (1981).

79   M. Claudson, E. Farhi and R. L. Jaffe, *Phys. Rev.* D34:877 (1986).

80   E. Eichten, K. D. Lane and M. Peskin, *Phys. Rev. Lett.* 50:811 (1982).

81   F. M. Renard, *Phys. Lett.* 116B:264 (1982).

82   E. Eichten, Lectures at the Yale University TASI summer school,Ed by F. Gursey, World Scientific Publishing Company (1986).

83   W. Bartel et al., *Z. Physik* C19:197 (1983)

84   W. Scott *in* Proc. of the 1986 International Conferance on High Energy Physics. Ed. S. Loken, World Scientific Publishing Company (1986).

85   G. Wolf, Lectures at this school.

AN INTRODUCTION TO STRING THEORY[*]

DAVID J. GROSS

Joseph Henry Laboratories
Princeton University
Princeton, New Jersey 08544

1.   INTRODUCTION

In these lectures I shall review the structure of superstring theory, which has attracted so much attention recently as a framework for the final "theory of everything". String theory is by now an enormous field, involving new concepts and advanced mathematical techniques. In a few lectures I can only skim the surface of this rapidly developing theory, pointing out the main features and explaining, at a pedestrian level, some of the important ideas.

High energy physics is, at present, in an unusual state. It has been clear for some time that we have succeeded in achieving many of the original goals of particle physics. We have constructed theories of the strong, weak and electromagnetic interactions. The "standard model" is remarkably successful and seems to be an accurate and complete description of all of physics, at least at energies below a Tev. Indeed, as we have heard in the experimental talks at this meeting, there are at the moment no significant experiments that cannot be explained by the color gauge theory of the strong interactions (QCD) and the electro-weak gauge theory. New experiments continue to confirm the predictions of these theories and no new phenomenon has appeared.

This success has not left us sanguine. Our present theories contain too many arbitrary parameters and unexplained patterns to be complete. They

---

[*] Supported in part by the National Science Foundation Grant PHY 80/19754

do not satisfactorily explain the dynamics of chiral symmetry breaking or of CP symmetry breaking. The strong and electroweak interactions cry out for unification. Finally we must ultimately face up to the task of including quantum gravity within the theory. However, we theorists are in the unfortunate situation of having to address these questions without the aid of experimental clues. Furthermore, extrapolation of present theory and early attempts at unification suggest that the natural scale of unification is at least 10^{16} Gev, tantalizingly close to the Planck mass scale of 10^{19} Gev. It seems very likely that the next major advance in unification will include gravity. I don't mean to suggest that new physics will not appear in the range of Tev energies. Almost all attempts at unification do, in fact, predict a multitude of new particles and effects that could show up in the Tev domain (Higgs particles, Supersymmetric partners, etc.), whose discovery and exploration is of the utmost importance. But the truly new threshold might lie in the totally inaccessible Planckian domain.

What strategy are theorists to adopt under these circumstances. One (defeatist) strategy is to do nothing until experiment provides more clues or paradoxes. The other (dangerous) is to push forward and attempt to make a bold leap into the unknown, searching for new symmetries and unifying principles. We are, however, not leaping totally into the dark. Any attempt at further unification is highly constrained by the requirement that it be consistent with the very large amount of understanding that we have of 'low energy' (≤ 1 TEV) physics. This provides powerful phenomenological constraints. Furthermore, we do have one window into Planck mass physics, namely we know much about gravity, a force whose typical energy scale is the Planck mass. The neccesity of including gravity in our theoretical framework provides powerful theoretical constraints.

What are the guidelines in the search for higher unification? The pioneering attempts were modest in scope. Based on the striking similarities of the strong and electroweak gauge interactions and the way that the quark and lepton multiplets seemed to be related, one tried to enlarge the gauge group of nature and construct at a high energy scale a single grand unified gauge interaction. These attempts have largely failed. First, the simple and striking predictions that they made, in particular a proton lifetime of $10^{32} - 10^{33}$ years, have not been verified. Second, although much good physics came out of these attempts (a scenario for proton and the cosmological production of baryons, the study of massive neutrinos, the idea of inflation,etc..), it also became clear that their scope was too limited. Realistic GUT models contained even more arbitrary parameters and required more fine tuning than the standard theory and the mechanism for spontaneous symmetry breaking had to be put in by hand.

GUT theories also brought to the forefront a new and serious issue –the *hierarchy* problem, namely why is the ratio of the W mass to the GUT mass scale (M_G) so small? This is not a new problem, it had worried Dirac many years ago (why is the proton mass so much smaller than the Planck mass?), but in the context of GUT theories it is more worrisome. The point is that in such theories the only natural scale is the unification scale, so that all massive paritcles should have a mass of M_G. The only known way to avoid this is to imagine a symmetry principle that forbids a mass from developing. If this symmetry is broken at a scale much smaller than M_G then much smaller masses can naturally arise. In the case of fermionic matter (quarks and leptons) this scenario is easy to arrange. Chiral symmetries protect spin one-half fermions from developing masses (a mass term, $\bar{\psi}\psi$, is not invariant under chiral transformations). Such symmetries are indeed a property of the observed world (the V-A electroweak interactions are not chirally symmetric). If we can protect this symmetry from breaking until the scale of M_W then one can understand the smallness of both the W-mass and the quark and lepton masses (in this game one assumes that the mass ratios of quark and lepton masses to M_W can be explained, by powers of small couplings). The problem is that the breaking of chiral gauge symmetries is generated by the vacuum expectation values of Higgs particles. But these are scalar particles whose vanishing mass is not protected by chiral symmetries. So what ensures that these don't develop expectation values of order M_G, leading to $M_W \approx M_G$?

The most elegant approach to solving this problem is by means of supersymmetry. Supersymmetry relates fermions to bosons, quarks to Higgs particles. It puts the scalar particles in the same multiplet as the fermions and protects masses from developing as a consequence of supersymmetry. One then has to imagine that supersymmetry remains unbroken until low energy (of order 1 TEV). This isn't too hard to imagine since, in fact it is quite difficult to break supersymmetry at all. In many theories, once introduced, it can only be broken nonperturbatively. Nonperturbative effects are energy dependent (as in QCD where confinement "turns on" at large distances), and are governed by dimensional couplings that vary logarithmically with energy. Thus one can plausibly imagine that $\frac{M_W}{M_G}$ arises as $exp[-\frac{1}{g^2}]$, where g is some small gauge coupling at the unification scale.

This possible solution of the hierarchy problem is the strongest physical reason for trying to build supersymmetry into a unified theory. There are, of course, other reasons. If there is one lesson to be learned from the success of the last decades, it is that the secret of nature is symmetry. If one is to make new progress in unification one must discover new symmetries of nature. This is not trivial. In order to discover truly new symmetries one must discover new degrees of freedom of nature, on which the new symmetries can act, as well as dynamical mechanisms that hide or break these symmetries. After all, if the symmetries were manifest they would be known already. Supersymmetry is the most attractive of the new symmetries that one can imagine. One can think of this symmetry as a consequence of the enlarging of configuration space from ordinary space to a "superspace" which contains fermionic coordinates. Supersymmetry is then a consequence of a natural generalization of ordinary space time symmetries to superspace. It also has a natural local version-supergravity. This marvelous new symmetry is very powerful. I find it's most attractive feature to be the possibility of it yielding an "explanation" of the existence and nature of matter. In the standard model gauge particles (gluons, W's and Z's) are automatic consequences of local gauge symmetry. Quarks and leptons must be put in by hand. This is ugly, asymmetrical and leads to much arbitrariness. In supergauge theories, one might imagine that the matter is in supersymmetric multiplets together with gauge mesons, and thus exists as a consequence of gauge symmetry. One could therefore hope that supersymmetry would predict the nature of fermionic matter, and also the spectrum of scalar (Higgs) particles, in a way that could solve the hierarchy problem.

This would still leave us with the question as to why nature chooses the particular ($SU_3 \times SU_2 \times U_1$) gauge group that we observe. A possible answer to this question might be provided by the revival of an old idea that was put forward by Kaluza in 1921! Kaluza, shortly after Einstein developed his theory of gravity, considered the possibility that there might exist more than three spatial dimensions. He noted, that under certain circumstances, the extra dimensions could lead naturally, in the framework of a purely gravitational theory, to Maxwell's theory of electromagnetism and the emergence of a conserved and quantized electric charge. This possibility has been generalized to non-abelian gauge groups and explored in great detail in recent years. It is now clear that the question of how many spatial dimensions there are is an experimental one, and that there is no conflict with present observation if the extra dimensions are compactified into a small (less than $10^{-15} cm$.) manifold. Kaluza's original idea was a five-dimensional world with pure gravitational interactions. One can find solutions of of Einsteins equations with the toplogy of four dimensional Minkowski space times a circle of small radius R. At low energies, compared to $\frac{1}{R}$ one would not notice the extra dimension, which requires energies of $\frac{1}{R}$ to excite. However there would exist a new conserved quantity, P_5, the component of the energy-momentum in the fifth direction. Now, full five dimensional Poincare invariance is broken by our asymmetric vacuum, however one still has a symmetry of translations about the hidden circle. In fact this is a local gauge symmetry. Consequently P_5 is conserved (and, since the circle is compact, quantized) and furthermore it is coupled to a massless gauge boson (which corresponds to one of the polarizations of the graviton along the hidden dimension). We can, following Kaluza, identify P_5 with a multiple of the electric charge and the gauge boson with the photon.

We thus have "derived" electromagnetism and the quantization of electric charge from five dimensional gravity.

This idea is easily generalized to theories with more than five dimensions, in which case non-abelian gauge groups can emerge from compactified vacua. If we combine this idea with supersymmetry one might dream of a higher dimensional supergravity theory which produces, upon dynamical compactification the low energy gauge groups of nature while at the same time determining the matter content-in other words a predictive theory of everything. These ideas have been pursued at length over the last decade. However, it appears that in the context of ordinary quantum field theory, which heretofore has served us so well, they cannot be made to work. This is for many reasons, but three stand out. It is useful to recall these, if only to compare with the early successes of string theory .

The first problem is that the quantum field theory of gravity arrived at by straightforward quantization of Einstein's theory is sick (non-renormalizable) at high energies. This sickness, which is not cured by supersymmetry or other modifications of the theory, means that the physics must be modified at energies of order the Planck mass. It is clearly dangerous to try to construct a unified theory on such shaky foundations. One might however argue that the high energy modifications will not affect low energy physics and proceed to construct realistic models. One found however, at least in the context of pure Kaluza-Klein supergravity, that one could not construct realistic models, for two reasons. First, it appeared to be impossible to understand the emergence of chiral fermions. Second, the background energy of the vacuum, otherwise known as the cosmological constant Λ , always came out to be of order M_P^4. This is in violent contradiction to observation, which indicates that $\Lambda \leq 10^{-120} M_P^4$. For these and other reasons no one ever came up with realistic field theoretic models based on these ideas.

This brings us to string theories, in particular the heterotic string, which appears to suffer from none of these problems. It yields a finite and consistent theory of gravity, in which chiral fermions and realistic gauge interactions can naturally emerge and in which, for most of the existing solutions, the cosmological constant vanishes.

2.   STRING THEORIES

String theories provide a way of realizing the potential of supersymmetry, Kaluza-Klein unification and much more. They represent a radical departure from ordinary quantum field theory, but in the direction of increased symmetry and structure. They are based on an enormous increase in the number of degrees of freedom, since in addition to fermionic coordiantes and extra dimensions, the basic entities are extended one dimensional objects instead of points. Instead of fields, or wave functions, that are functions of coordinates x^μ we have fields that are functionals of loops $\Psi[x^\mu(\sigma)]$. Whereas a quantum field describes naturally one particle at a time, a string field will describe an infinite number of particles, with a degeneracy that increases exponentially with mass. You might think that such an increase in the size of configuration space is wasteful and unnecesary. After all we only observe a finite number of particles in nature, why do we need so many degrees of freedom? The answer is that corresponding to this increase there is an increase in the the symmetry group of nature , in a way that we are only beginning to comprehend. At the very least, this extended symmetry group contains the largest group of symmetries that can be contemplated within the framework of point field theories–those of 10 dimensional supergravity and super Yang-Mills theory.

The origin of these symmetries can be traced back to the geometrical invariance of the dynamics of propagating strings. Traditionally string theories are constructed by the first quantization of a classical relativistic one dimensional object–whose motion is determined by requiring that the invariant area of the world sheet it sweeps out in space-time is extremized. In this picture the dynamical degrees of freedom of the string are the coordinates, $x_\mu(\sigma, \tau)$ (plus fermionic coordinates in the superstring), which describe its position in space time. The symmetries of the resulting theory are all consequences of the reparametrization invariance of the σ, τ parameters which label the world sheet. As a consequence of these symmetries one finds that the free closed string contains a massless spin two meson, which can be identified as the graviton, whereas the open string, which has ends to which charges can be attached, yields massless vector mesons, which can be identified as Yang-Mills gauge bosons.

Let us consider the first quantization of the original bosonic string. It is constructed by a natural genererealization of the quantization of a relativistic point particle, but with the rather surprising result that the theory will only be consistent in 26 space-time dimensions and will neccesarily contain massless gauge bosons.

$$S = Invariant\ Length = \frac{m}{2} \int_{\tau_1}^{\tau_2} d\tau \sqrt{\left(\frac{dx^\mu(\tau)^2}{d\tau} \right)}.$$ (1)

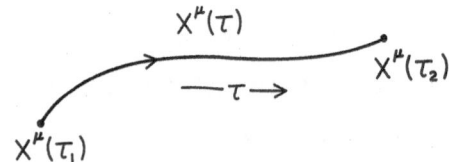

Figure 1. The world line of a point particle

The classical dynamics of a relativistic particle is determined by demanding that it move along a world line $x^\mu(\tau)$ in, say, Minkowski space in such a fashion that the relativistically invariant length of the path be extremal (of course the solution will be rectilinear motion). Thus the dynamics is determined by looking for trajectories that go from x_1, at proper time τ_1, to x_2, at time τ_2, (see Figure 1) and which extremize the action

The cannonical momentum, p^μ, conjugate to x^μ, is $p^\mu = m\frac{dx^\mu}{d\tau}/\sqrt{[\frac{dx^\mu(\tau)}{d\tau}]^2}$. The Euler-Lagrange equations of motion are

$$\frac{dp^\mu(\tau)}{d\tau} = 0, \tag{2}$$

where p^μ is subject to the constraint $(p^\mu)^2 = m^2$.

The origin of this constraint, which tells us that not all the momenta are independent dynamical variables, is the reparametrization invariance of the action (under $\tau \to f(\tau)$) which tells us that not all the coordinates are physical degrees of freedom. In fact we can use the freedom of relabeling the *proper time* variable τ so as to choose it equal to x^0 (note that this breaks the manifest relativistic invariance.) Having made this *gauge choice*, $x^0 = \tau$, we have an unconstrained system of variables x^i and p^i. The conjugate momentum to x^0, which has been eliminated, is the Hamiltonian, which can be expressed, using the constraint, in terms of the momenta p^i, as $H = p^0 = \sqrt{(p^i)^2 + m^2}$. This system is then easily quantized by demanding that x^i, p^i, H be operators in an appropriate Hilbert space, and that $[x^i, p^j] = \delta_{i,j}$.

Now, let us construct the theory of a free relativistic open string. This is an extended object in space. To describe it in an invariant manner we note that as the string moves it sweeps out a two dimensional strip in Minkowski space, which can be parametrized by $0 \leq \sigma \leq \pi$ and τ, so that as τ runs from τ_1 to τ_2, $x^\mu(\sigma, \tau)$, which describes the string position in Minkowski space, runs from $x_1^\mu(\sigma)$ to $x_2^\mu(\sigma)$.(See Figure 2). The natural generalization of the length of the worldline is the area of the worldsheet swept out by the moving string. Therefore we take the action to be the "Nambu action"

$$S = Invariant\ Area = -T \int\limits_{\tau_1}^{\tau_2} d\tau \int\limits_{0}^{\pi} d\sigma \sqrt{\left(\frac{dx}{d\tau} \cdot \frac{dx}{d\sigma}\right)^2 - \left(\frac{dx}{d\tau}\right)^2 \cdot \left(\frac{dx}{d\sigma}\right)^2},$$

$$\tag{3}$$

and determine the (classical) motion of the string so as to extremize this area. T is the "tension "of the string and will turn out to be of order M^2_{Planck}.

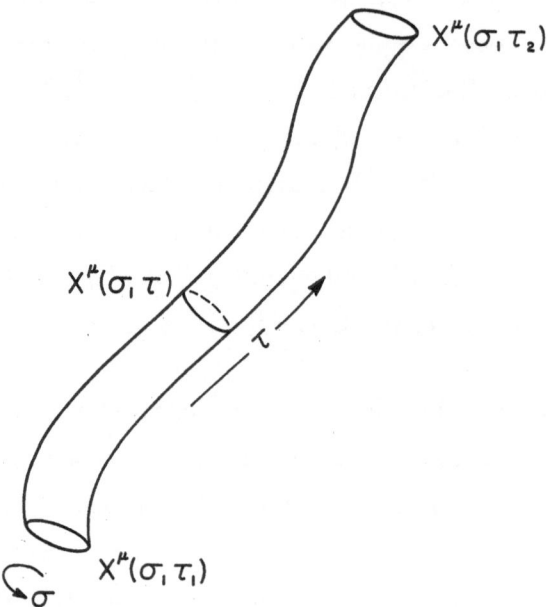

Figure 2. The world sheet swept out by a moving closed string.

(One must also specify boundary conditions at the ends of the string, the appropriate ones being $\frac{dx^\mu(\sigma)}{d\sigma} = 0$, for $\sigma = 0, \pi$.)

This action is invariant under a large group, that of reparametrizations of both world sheet labels, σ and τ. Correspondingly only D-2 of the string coordinates are physical (D is the dimension of the space-time in which the string moves). One coordinate can be identified with the time coordinate and the longitudinal degree of freedom is just a relabeling of the string. This leaves us with D-2 physical degrees of freedom, which describe the oscillations of the string tranverse to itself. At the price of breaking manifest covariance it is sometimes useful to eliminate the unphysical degrees of freedom. To do this we identify one of the string coordinates with the timelike label of the string world sheet, τ. We shall choose not the obvious time coordinate $x^0(\sigma, \tau)$, but rather the "light cone time", $x^+ \equiv \frac{1}{\sqrt{2}}(x^0 + x^{D-1})$, as measured by an observer moving at the velocity of light along the D-1 axis. The advantage of this choice can be seen from the particle case where, if we choose $x^+ = \tau$, then the conjugate momentum $p^- \equiv \frac{1}{\sqrt{2}}(p^0 - p^{D-1})$ which generates light cone time translations (the light cone gauge Hamiltonian) is given by $p^- = \frac{1}{2p^+}[(p^i)^2 + m^2]$, thereby eliminating the nasty square roots which occured before.

Therefore we use the reparametrization freedom to choose

$$x^+(\sigma, \tau) = x^+ + P^+\tau, \quad p^+(\sigma, \tau) = P^+. \tag{4}$$

One then finds that $x^-(\sigma, \tau)$ and the Hamiltonian, P^- ,can be expressed in terms of the transverse degrees of freedom $x^i; i = 1, 2, ...D - 2$

$$P^- = \frac{1}{2P^+} \int\limits_0^\pi d\sigma \left((\frac{dx^i}{d\tau})^2 + (\frac{dx^i}{d\sigma})^2 \right). \tag{5}$$

This is a rather simple Hamiltonian, which can easily be quantized. After all it is just a bunch of harmonic oscillators. The equation of motion for x^i is simply the two dimensional wave equation, $\frac{d^2x^i}{d\tau^2} - \frac{d^2x^i}{d\sigma^2} = 0$, and the general solution can be expanded as

$$x^i(\sigma, \tau) = x^i + p^i\tau + \sum_{n=1}^\infty \left(a_n^i e^{-in\tau} + a_n^{i\dagger} e^{in\tau} \right) \frac{cosn\sigma}{\sqrt{n}}. \tag{6}$$

Upon quantization $a_n^\dagger$ and a_n are creation and annihilation operators for the normal modes of vibration of the string, which obey the cannonical commutation relations, $[a_n^i, a_m^{j\,\dagger}] = \delta_{i,j}\delta_{n,m}$, of harmonic oscillators. Thus the Hilbert space of the free relativistic string is given by states $|p^i; n_1, n_2...\rangle$, in the Fock space of the a_n^i's, where n_i is the occupation number of the ith oscillator. The Hamiltonian can be written as

$$2p^+ p^- = (p^i)^2 + \sum_{n=1}^{\infty} \sum_{i=1}^{d-2} [n a_n^{i\,\dagger} a_n^i + \frac{n}{2}], \tag{7}$$

where the factor of $\frac{n}{2}$ represents the zero point energy of the nth oscillator.

To make sense of this formula we must interpret the divergent sum of zero point energies. A short cut to the answer (which can be more rigorously derived by lengthier methods) is to note that

$$\sum_{n=1}^{\infty} n \equiv Lim_{s \to -1} \sum_{n=1}^{\infty} \frac{1}{n^s} = \zeta(-1) = -\frac{1}{12}, \tag{8}$$

where $\zeta(s)$ is the Riemann zeta function, which is defined by the series for $s > 1$ and for $s = -1$ by analytic continuation! Therefore the spectrum of the free bosonic string has states with mass M given by

$$M^2 = \sum_{n=1}^{\infty} n N_n - \frac{D - 2}{24}, \tag{9}$$

where N_n is the occupation number of the nth oscillator. This formula describes states that have increasing spin (as $\sum_n N_n$) and mass (as $\sum_n n N_n$), and whose degeneracy increases rapidly (as $exp(M)$).

Let us look at the ground state and the first excited state. The ground state is the Fock vacuum, with no oscillator excitations. It has mass equal to $-\frac{D-2}{24}$, so that the lightest string state is a tachyon. What this means is that the bosonic string theory is sick, or at least the vacuumn about which we are perturbing is unstable, since tachyons correspond to runaway modes. Let us ignore this problem, which in any case is avoided for superstrings, and look at the first excited string states, $a^{i\dagger}_n|\Omega\rangle$, where $|\Omega\rangle$ is the Fock space vacuum, annihilated by all the a^i_n. They have the first oscillator excited and therefore a mass of $M^2 = 1 - \frac{D-2}{24} = \frac{26-D}{24}$. There are 24 such transverse modes, $a^{i\dagger}_n|\Omega\rangle$, which transform as a vector under spatial rotations. But now we have a problem with relativistic invariance. A vector particle has D-1 degrees of freedom, whereas we have only D-2. The only exception is when the vector particle is massless, in which case the longitudinal polarization is absent (as is well known in D=4 in the case of the photon). Thus, if the string theory is to be Lorentz invariant it must be that $D = 26$ so that the first excited state be massless! Although Lorentz invariance is not manifest, it is possible to show that when $D = 26$ all the states of the free string yield representations of the Lorenz group.

This is quite remarkable. We have learned, simply by quantizing the free string, that the theory can only be formulated (without breaking Lorentz invariance or otherwise modifying the theory) in 26 space-time dimensions, and furthermore that it neccesarily contains a massless vector meson. This situation is very different than that of a point particle, which can have a nonzero mass and which can be embedded in any spacetime manifold of any dimensionality. It suggests that in string theory the choice of the space-time background is dynamical, so that the theory must be one of gravity, and that Yang-Mills gauge interactions are automatically induced. It is a fact of relativistic quantum mechanics that once you have a massless vector meson that has nontrivial interactions (has nonvanishing couplings at zero momentum) then all of the content of local gauge theory (Maxwell's theory for a single massless vector meson, Yang-Mills theory for massless charged vector mesons). Charges can easily be introduced into the theory of open strings by putting them on the ends of the strings. This resembles the crude picture of a meson in QCD as a string, with the quarks at the ends of a thin flux tube of chromodynamic flux. It is of no surprise that, once one includes interactions of open strings, that such a theory contains within it Yang-Mills gauge interactions.

So far we have considered open strings only. We can, and indeed must, consider closed strings as well. They are described by $x(\sigma, \tau)$, but now x is a periodic funtion of σ. Unlike open strings they have no geometrically invariant points on them, like the ends of open strings, to which charges can be attached. They are inherently neutral objects, that will, as we shall see, describe not gauge interactions but gravity. The closed string coordinates, in light cone gauge, obey the 2 dimensional wave equation, and can be decomposed into right and left moving waves, $x(\sigma, \tau) = x(\sigma - \tau) + x(\sigma + \tau)$. Thus a closed string has twice as many modes, and upon quantization one finds that the mass is given by

$$Mass^2 = \sum_{n=1}^{\infty} \sum_{i=1}^{2} 4 \left(n {a_n^i}^\dagger a_n^i + n {\tilde{a}_n^i}^\dagger \tilde{a}_n^i - 2\frac{n}{2} \right), \tag{10}$$

where $a^\dagger$ ($\tilde{a}^\dagger$) is the creation operator for right moving (left moving) waves, and we must impose the constraint $\sum_{n=1}^{\infty} \sum_{i=1}^{24} n {a_n^i}^\dagger a_n^i = \sum_{n=1}^{\infty} \sum_{i=1}^{24} n {\tilde{a}_n^i}^\dagger \tilde{a}_n^i$. Now we find that the first excited states are ${a_1^i}^\dagger {\tilde{a}_1^j}^\dagger |\Omega\rangle$, with $Mass^2 = \frac{26-D}{12}$. Among these states there is a spin two particle, which again to be Lorentz invariant, must be massless. So D must equal 26, as before, and the closed string neccesarily contains a massless spin two meson- the graviton. Not surprisingly, once interactions are turned on, we will find that the theory contains within it Einstein's theory. In other words, at low energies (compared to $T \approx M_{Planck}^2$) the self interactions of the graviton, as well as its interactions with other particles, will be those of Einstein's theory.

So far we have been discussing free strings. How do we introduce interactions? In the case of pointlike theories one usually reverts at this stage to a second quantized treatment, introducing field operators that create and annihilate particles and couples them so as to produce interacting relativistic theories. One could however continue with the methods of first quantization, most easily by using path integral techniques. Consider again the free particle. The amplitude for this particle to go from $x^\mu(1)$ at proper time τ_1 to $x^\mu(2)$ at

τ_2 is given by summing over all path histories, weighted by the exponential of the classical action

$$\langle x^\mu(2), \tau_2 | x^\mu(1), \tau_1 \rangle = \int\limits_{x^\mu(1)}^{x^\mu(2)} Dx^\mu(\tau) exp[\frac{i}{h} \int\limits_{\tau_1}^{\tau_2} d\tau \mathcal{L}(\tau)], \qquad (11)$$

where $\mathcal{L}(\tau)$ is the classical action. The integration is over all trajectories up to reparametrizations. A similar expression can be written for the free string theory. To introduce interactions we must consider the motion of two particles and modify the rules whenever their trajectories cross, allowing for the particles to combine to form a third particle, or for a single particle to split into two. This is certainly not a unique, or geometrically inspired, procedure and introduces much arbitrariness into the theory of point particles. In the case of strings however there is a natural, geometric way of introducing interactions which is unique. The motion of a single closed string is gotten by summing over all path histories, each one of which looks like a cylinder in σ, τ space. If we add to the cylinder a handle, then the space time process described by such a world sheet looks like a closed string which propagates, then splits into two closed strings, which propagate and then rejoin-namely a self energy (one-loop) correction to the string propagator.

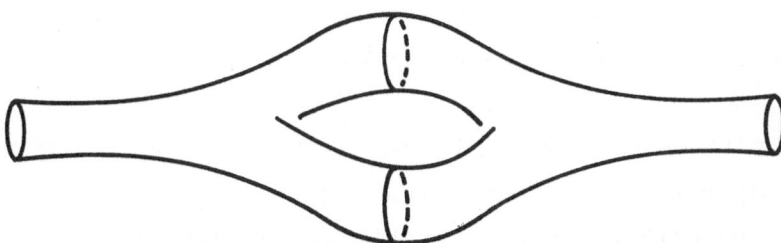

Figure 3. The world sheet describing a string that breaks up and recombines – a one loop self-energy diagram.

To see this we have to follow the motion as a function of τ, and then at a specific value of τ the strings split or rejoin. On the other hand, if we look at the surface as a whole there is no place where anything singular happens (at least for the *Euclidean* surface we get by analytically continuing to complex τ), no place where an interaction is introduced. Thus the complete amplitude in string theory can be written schematically as

$$\sum_{topologies} \int Dx^\mu(\sigma,\tau) exp[\frac{i}{h} \int_{\tau_1}^{\tau_2} d\tau \int d\sigma \mathcal{L}[x^\mu(\sigma,\tau)], \tag{12}$$

where we sum over all toplogies of the two dimensional world sheets connecting our initial and final string configurations.

Such a method for introducing interactions clearly preserves the geometrical invariance of the theory, and thus the ensuing gauge symmetries. Doing the above integrals is somewhat nontrivial, after all these are functional integrals over two-dimensional "fields" $x^\mu(\sigma,\tau)$. Each integral in the above sum is equivalent to constructing a two-dimensional field theory on a surface of a given topology. In general this would be a formidable task. In the case of the bosonic string in flat (26 dimensional) space-time the situation is not so bad—in fact we saw, as in (5), that with the appropriate gauge choice the theory is a free field theory. The path integrals can then be done quite explicitly, the only complications arising from the need to factor out correctly the reparametrization group, and from the fact that one is working on a surface with complicated topology. In the last two years the methods for doing these integrals have been developed to an extraordinary degree, using high powered methods of analytic geometry and complex analysis.

String theories are inherently theories of gravity. Unlike ordinary quantum field theory, however, we do not have the option of turning off gravity. The gravitational, or closed string, sector of the theory must always be present for consistency, even if one starts by considering only open strings, since these can join at their ends to form closed strings. One could even imagine discovering the graviton in the attempt to construct string theories of matter. In fact this was the course of events for the dual resonance models where the graviton (then called the Pomeron) was discovered as a bound state of open strings. In these theory the ordinary gauge couplings, like the fine structure constant, are proportional to the gravitational coupling, so if one turns off gravity one turns off everything. The heterotic string, which is a purely closed string theory, can be regarded as a gravitational theory that produces gauge interactions by a stringy version of the Kaluza-Klein mechanism.

Most exciting is that string theories provide for the first time a consistent, even finite, theory of gravity. The problem of ultraviolet infinities is bypassed in string theories which simply contain no short distance singularities. This is not too surprising considering the extended nature of strings, which softens their interactions. Alternatively one notes that interactions are introduced into string theory by allowing the string coordinates, which are two dimensional fields, to propagate on world sheets with nontrivial topology that describe strings splitting and joining. From this first quantized point of view one does not introduce an interaction at all, one just adds handles or holes to the world sheet of the free string. As long as reparametrization invariance is maintained there are no possible counterterms. In fact all the infinities that have ever appeared in string theories can be traced to infrared divergences that are a consequence of vacuum instability. These arise since all string theories contain a massless partner of the graviton called the dilaton. If one constructs a string theory about a trial vacuum state in which the dilaton has a nonvanishing vacuum expectation value, then infrared infinities will occur due to massless dilaton tadpoles. These divergences however are just a sign of the instability of the original trial vacuum. This is the source of the divergences that occur in one loop diagrams in the old bosonic string theories (the Veneziano model). Superstring theories have vanishing dilaton tadpoles, at least to one-loop order. Therefore both the superstring and the heterotic string are explicitly finite to one loop order and there are strong arguments that this persists to all orders !

String theories, as befits unified theories of physics, are incredibly unique.

In principle they contain no freely adjustable parameters and all physical quantities should be calculable in terms of h, c, and m_{planck}. In practice we are not yet in the position to exploit this enormous predictive power. The fine structure constant α, for example, appears in the theory in the form $\alpha \exp(-D)$, where D is the aforementioned dilaton field. Now, the value of this field is undetermined to all orders in perturbation theory (it has a 'flat potential', see Figure 4). Thus we are free to choose its value, thereby choosing one of an infinite number of degenerate vacuum states, and thus to adjust α as desired. Ultimately we might believe that string dynamics will determine the value of D uniquely, presumably by a nonperturbative mechanism, and thereby eliminate the nonuniqueness of the choice of vacuum state. In that case all dimensionless parameters will be calculable. Even more, string theories determine the gauge group of the world (to be $E_8 \times E_8$ or possibly SO_{32}) and fix the number of space-time dimensions to be ten. This might appear to be disasterous, since the world we observe about us has only four dimensions, and a recognizable gauge group of $SU_3 \times SU_2 \times U_1$. However, as we shall see below the heterotic string theory can have phenomenologically attractive solutions, which could well describe the real world.

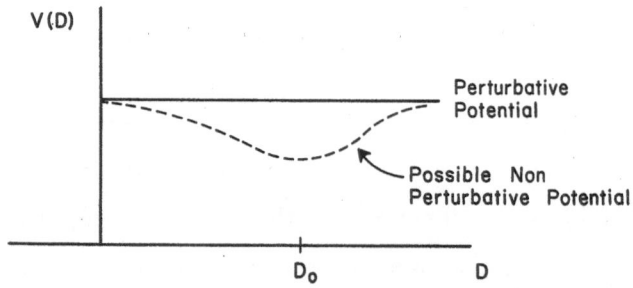

Figure 4. The potential energy of the dilaton field. In perturbation theory it is flat, however nonperturbative effects could determine the ground state value of D.

3.   CONSISTENT STRING THEORIES

The number of consistent string theories is extremely small, the number of phenomenologically attractive theories even smaller. First, there are the closed superstrings, of which there are two consistent versions. These are theories which contain only closed strings which have no ends to which to attach charges and are thus inherently neutral objects. At low energies, compared to the mass scale of the theory, which we can identify as the Planck mass, we only see the massless states of the theory, which are those of ten dimensional supergravity. One version of this theory is non-chiral and of no interest since it could never reproduce the observed chiral nature of low energy physics. The other version is chiral. One might then worry that it might suffer from anomalies, which is indeed the fate of almost all chiral supergravity theories in ten dimensions. Such anomalies are ubiquitous features of theories containing chiral fermions. They are disasterous since they mean that quantum corrections spoil the gauge invariance of the theory. In the standard model, in fact, the would be anomalies cancel only as a consequence of a conspiracy between quarks and leptons. This conspiracy is one of the strongest arguments that nature wants to be chirally asymmetric and goes to some length to achieve this (after all, one could eliminate all possible anomalies by simply imposing manifest chiral symmetry). It is also a strong indication that quarks and leptons are highly correlated, and all by itself suggests unification of the strong and electro-weak interactions.

Remarkably, the particular supergravity theory contained within the chiral superstring is the unique anomaly free pure supergravity theory in ten dimensions. Although consistent, it contains no gauge interactions in ten dimensions and could only produce such as a consequence of compactification. This approach raises the same problems of reproducing chiral fermions that plagued field theoretic Kaluza-Klein models and has not attracted much attention.

Open string theories, on the other hand, allow the introduction of gauge groups by the time honored method of attaching charges to the ends of the strings. String theories of this type can be constructed which yield, at low energies, $N = 1$ supergravity with any Yang-Mills group. These, in addition, to being somewhat arbitrary, were suspected to be anomalous. The discovery by Green and Schwarz that, for a particular gauge group SO_{32}, the would be anomalies cancel, greatly increased the phenomenological prospects of unified string theories.

The anomaly cancellation mechanism of Green and Schwarz also provided the motivation that led to the discovery of a new string theory, the *heterotic* string. This is a theory of closed strings, which generates nonetheless gauge interactions. Quite different from the ad hoc procedure of attaching charges to string endpoints this mechanism determines the gauge group uniquely. There are two manifestations, which we now understand as different states of the same theory, realized as $E_8 \times E_8$ or Spin $32/Z_2$ gauge symmetry groups. The $E_8 \times E_8$ version of this theory offers the best phenomenological prospects for reproducing the real world. In fact the group E_8 was explored seriously as a GUT group by theorists who extrapolated upward from the standard model, so one might hope to be able to proceed in the opposite direction.

4.   THE HETEROTIC STRING

Previously known string theories are the bosonic theory in 26 dimensions (the Veneziano model) and the fermionic, superstring theory in 10 dimensions (an outgrowth of the Ramond-Neveu-Schwarz string). The new heterotic string theory is constructed as a chiral hybrid of these.

The basic idea underlying the construction of the heterotic string is the seperation, for orientable closed strings, between right and left moving modes. The physical degrees of freedom of a closed string, (which can be regarded as two-dimensional massless fields), can be decomposed into right and left movers, i.e. functions of $\tau - \sigma$ and $\tau + \sigma$ respectively. A right moving wave will travel uninterupted and will never turn into a left-moving wave. (We need to consider orientable strings so that one can distinguish right from left). For open strings the boundary conditions mix left- and right- movers, since a right-moving wave is reflected back from the end as a left-moving wave. However, if we consider only closed strings then the right and left movers never mix. All observables can be decomposed into operators acting separately on the left and right handed Hilbert space of the string states. There is only one constraint that ties the two halfs of the closed string together to form a geometrical object, namely that the right- and left- handed number operators are equal, $N_R = N_L$. Apart from this constraint everything else factorizes and we can write Fock space states as a direct product $|\rangle_R \times |\rangle_L$.

This separation is maintained even in the presence of string interactions, as long as we allow only orientable world sheets on which a handedness can be defined. This is because the interactions between closed strings are constructed, order by order in perturbation theory, by simply modifying the topology of the world sheet on which the strings propagate. In terms of the first quantized two-dimensional theory no interaction is thereby introduced; the right and left movers still propagate freely and independently as massless fields.

Thus, there is in principle no obstacle to constructing the right and left moving sectors of a closed string in a different fashion, as long as each sector is separately consistent, and together can be regarded as a string embedded in ordinary space-time. This is the idea behind the construction of the heterotic string, which combines the right movers of the fermionic superstring with the left movers of the bosonic string. The bosonic theory contains, as we have seen, a tachyon. This however is not a sign of *inconsistency*, but rather a sign of *instability*. In any case the tachyon does not show up in the heterotic theory.

We shall now discuss the quantization of the heterotic string, with emphasis on the new degrees of freedom. We shall work in light-cone gauge as before. The physical degrees of freedom of the right-moving sector of the fermionic superstring consist of eight transverse coordinates $x^i(\tau - \sigma)$ $(i = 1, \cdots 8)$ and eight Majorana-Weyl fermionic coordinates $S^a(\tau - \sigma)$. It is the symmetry between these eight fermionic and eight bosonic modes that leads to supersymmetry. The physical degrees of freedom of the left-moving sector of the bosonic string consist of 24 transverse coordinates, which we divide into two classes, $x^i(\tau + \sigma)$ and $x^I(\tau + \sigma)$ $(i = 1, \cdots 8, I = 1, \cdots 16)$. Together they comprise the physical degrees of freedom of the heterotic string. The eight transverse right and left movers combine with the longitudinal coordinates x^+ and x^-, to describe the position of the string embedded in 10 dimensional space. The extra fermionic and bosonic degrees of freedom parametrize an internal space. The fermions are responsible for $N = 1$ supersymmetry, whereas the bosonic coordinates will provide the arena for the gauge symmetries.

The light cone action that yields the dynamics of these degrees of freedom can be derived from a manifestly covariant action, and one can easily quantize it. The only new feature that enters is the compactification and quantization of the extra 16 left-moving bosonic coordinates. It is this compactification, on a uniquely determined 16 dimensional compact space, that leads to the emergence of Yang-Mills interactions.

The extra 16 left moving coordinates of the heterotic string can be viewed as parametrizing an "internal" compact space T. (This interpretation should not be taken too literally, in fact one can equally well represent these degrees of freedom by 32 real fermions). The question then arises as to the nature of the internal manifold T. This is a dynamical question since a string theory is a theory of gravity. The choice of a background spacetime is a dynamical issue— the background must be a solution of the string equation of motion. So far we have always imagined that the string lives in a flat, Minkowski, spacetime. One might instead try to construct the theory about some other spacetime manifold. In fact, in order to describe the real, four dimensional world, we will surely have to do so. If one tries to embed the string in a general spacetime background one finds inconsistencies. Thus the requirement of consistency of the first quantized theory is equivalent to demanding a solution of the equations of motion of the string theory. It was this procedure that fixed the dimension of spacetime to be ten (for the superstring, 26 for the bosonic string) and, in the same fashion, will pick out the allowed backgrounds. It would be nice if the demand of consistency were to pick out a unique solution.

This is certainly not the case, to date, for the ordinary spacetime coordinates. The remarkable feature of the Heterotic string is that the internal coordinates are required to lie in a space T which is completely determined to be a very special 16 dimensional torus (the maximal torus of $E_8 \times E_8$ or Spin $32/Z_2$).

That a torus is a solution is reasonable since a torus is simply a flat space with periodic boundary conditions; however, there are many 16 dimensinal tori. A general 16 dimensional torus is characterized by 136 paramenters (the radii of the 16 circles, whose product is the torus, and the angles between them). What picks out the special torus is the requirement that the coordinates of T are left moving (i.e. functions of $\tau + \sigma$). Consider the expansion of the internal coordinates $x^I(\tau + \sigma)$

$$X^I(\tau + \sigma) = X^I + P^I \tau + L^I \sigma + oscillators. \quad (13)$$

where X^I labels the position of the ceneter of mass, and P^I is the total momentum. The momentum P^I is quantized (since X^I lives on a compact domain) in units of $1/R$ (where R is a radius of T).

The above expansion also contains a term linear in σ, which must be there if X^I is to be a function of $\sigma + \tau$. However X^I is a periodic function of σ so how can it have a term linear in σ? It must be that the coordinate comes back to the same point as we circle the string and σ goes from 0 to π. This is possible since the string lies on a torus and if the term $L^I\sigma$ goes around the torus (in some direction) an integer number of times, as σ goes from 0 to π, then X^I will be a periodic function of σ. Therefore L^I must equal an integer multiple of a radius of T in some direction. A string configuration with nonvanishing L^I represents a soliton, i.e. a string that winds around the torus some number of times, and its winding number is a topological, conserved charge. Such solitons are a new feature of string theories that do not occur in point theories. Now, in the Heterotic string, X^I is a function of $\tau + \sigma$, so that L^I must equal P^I. This clearly restricts the form of the torus. Since $L^I \approx R$ and $P^I \approx \frac{1}{R}$ it means that $R \approx 1$ (in our units this means $R \approx 1/M_{Planck}$). But the form of T is further constrained, since for a general torus it is not possible to identify the winding numbers, which span a lattice Γ, $(T = R_{16}/\Gamma)$, with the momenta, P^I, which lie on the lattice Γ^* dual to Γ. The lattice Γ is a set of vectors $[e_i^I; i = 1, \ldots, 16]$ which define the torus by the statement that points in R_{16} are identified according to (see Figure 5)

$$x^I \equiv x^I + \pi \sum_{i=1}^{16} e_i^I n_i. \tag{14}$$

The momenta lie on the dual lattice, generated by $e_i^{*I}; i = 1, \ldots, 16$; where $\sum_I e_i^{*I} e_j^I = \delta_{i,j}$. This is in order to ensure that the translation operator, $e^{iP \cdot e_i}$, equals unity. It might seem obvious that the lattice and its dual must be identified in the case of the heterotic string. This is indeed so, however one must examine radiative corrections to come to that conclusion. In fact the full consistency of the Heterotic string theory requires that T be such that $\Gamma = \Gamma^*$, i.e. that the lattice defining the torus be *self dual*. If this is not satisfied then "modular invariance" breaks down. This means that the amplitudes, at the one loop level, develop anomalies which destroy reparametrization invariance.

Furthermore, the lattice must be *even*. This means that the length squared of any vector of the lattice be an even integer. This can be seen from the mass formula and the constraint equation for the heterotic string that read

$$Mass^2 = N_R + N_L - 1 + \frac{(P^I)^2}{2}, \quad N_R = N_L - 1 + \frac{(P^I)^2}{2}. \tag{15}$$

The factor of -1 arise from the normal ordering subtraction of the bosonic string, the factor of $\frac{(P^I)^2}{2}$ from the kinetic energy in the internal dimensions.

Now, there exist very few self dual lattices of the appropriate type, in fact they only exist in 8N dimensions! (A familiar hypercubic lattice is selfdual, but not even). In 8 dimensions there is one self dual lattice – Γ_8, the root lattice of E_8. E_8 is the most exceptional of all Lie groups. Cartan classified all Lie groups, and found that most lie in simple families $(SU(N), SO(N)$ and $Sp(2N))$ but there were a finite number of exceptional groups $(F_2, G_2, E_6, E_7,$

and E_8). Although hard to describe as a transformation group, E_8 is a beautiful mathematical object with most remarkable properties. The root lattice of this group is the lattice generated by the 8 dimensional vectors whose components are the eigenvalues of the eight set of commuting generators of E_8. In the case of SU_3 the root lattice is the familiar two dimensional hexagonal lattice, from which one can read the allowed values of isospin and hypercharge. (See Figure 5).

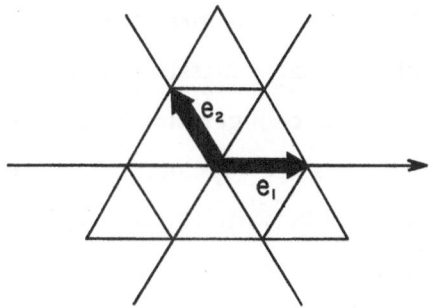

Figure 5. A two dimensional slice of the lattice. The lattice generated by e_1 and e_2 is the root lattice of $SU(3)$.

In 16 dimensions there exist two even self dual lattices, $\Gamma_8 \times \Gamma_8$ and Γ_{16}, where Γ_{16} is the weight lattice of $Spin\, 32/Z_2$. Once we see that the tori on which the heterotic string lies is the maximal torus of these groups it is not surprising that they emerge as symmetries of the theory. To see this let us now examine the massless particles of the heterotic string. Since the theory contains gravity and since it possesses one supersymmetry (the one that transforms the supersymmetric right movers) it will contain the massless multiplet of ten dimensional, $N = 1$, supergravity. We would also expect, in analogy to the standard Kaluza-Klein mechanism, that a compactified string theory will contain massless vector mesons associated with the isometries of the compact space. In Kaluza's original model one circle gave electromagnetism, here we have sixteen circles and so we would expect sixteen copies of electromagnetism. Indeed, since T has an isometry which consists of sixteen $U(1)$'s, this compactification should, and does, yield the 16 gauge bosons of $U(1)^{16}$. The sixteen conserved charges are just the sixteen components of the total momentum, P^I, in the internal dimensions. These will be identified with electric charge, weak hypercharge, etc.

If this was all there was a rather boring world would emerge. However a remarkable feature of closed string theories is that for special choices of the compact space there will exist extra massless gauge bosons. These arise as massless solitons , namely string configurations that wind around the internal torus, whose conserved charges are topological in nature. In the heterotic string there is an identification between the topological charges, the winding numbers L^I and the nonvanishing momenta = winding number = $U(1)^{16}$ charge. If we look at the mass formula, (15), we see that there will exist such a particle for each vector on the lattice with length squared two. There is precisely such a vector for each of the 480 charged generators of the algebra of $E_8 \times E_8$ or SO_{32}. These combine with the Kaluza-Klein gauge bosons to fill out the adjoint representation of a simple group whose rank equals the dimension of T. For the two allowed choices of T these produce the gauge bosons of $G = Spin(32)/Z_2$ or $E_8 \times E_8$.

Once we have the gauge bosons it is not difficult to show that the full gauge symmetry is present. Thus we have seen that the heterotic string automatically produces, in ten dimensions unique gauge interactions. The heterotic string theory has, by now, been developed to the same stage as other superstring theories. Interactions have been introduced, and shown to preserve the symmetries and consistency of the theory, radiative corrections calculated and shown to be finite. In many ways it now appears as the simplest of all superstring theories. It surely provides a most satisfactory explanation for the emergence of specific gauge interactions and, as we shall see, offers much phenomenological promise.

5.   STRING PHENOMENOLOGY

In order to make contact between string theories and the real world one is faced with a formidable task. These theories are formulated in ten flat space-time dimensions, have no candidates for fermionic matter multiplets, are supersymmetric and contain an unbroken large gauge group– say $E_8 \times E_8$. These are not characteristic features of the physics that we observe at energies below a Tev. If the theory is to describe the real world one must understand how six of the spatial dimensions compactify to a small manifold leaving four flat dimensions, how the gauge group is broken to $SU_3 \times SU_2 \times U_1$, how supersymmetry is broken, how families of light quarks and leptons emerge, etc. Much of the recent excitement concerning string theories has been generated by the discovery of a host of mechanisms, due to the work of Witten and of Candelas, Horowitz and Strominger, and of Dine, Kaplonovsky, Nappi, Seiberg, Rohm, Breit, Ovrut, Segre and others, which indicate how all of this could occur. The resulting phenomenology, in the case of the $E_8 \times E_8$ Heterotic string theory is quite promising.

a. Compactification

The first issue to be addressed is that of the compactification of six of the dimensions of space. The heterotic string, as described above, was formulated in ten dimensional flat spacetime. This, however, is not necessary. Since the theory contains gravity the issue of which spacetime the string can be embedded in is one of string dynamics. That the theory can consistently be constructed in perturbation theory about flat space is equivalent to the statement that ten dimensional Minkowski spacetime is a solution of the classical string equations of motion. In Quantum mechanics such a solution yields the background expectation values of the quantum degrees of freedom. We can then ask are there other solutions of the string equations of motion that describe the string embedded in, say, four dimensional Minkowski spacetime times a small compact six dimensional manifold?

At the moment we do not possess the full string functional equations of motion, however one can attack this problem in an indirect fashion. One method is to deduce from the scattering amplitudes that describe the fluctuations of the string in ten dimensional Minkowski space an effective Lagrangian for local fields that describe the string modes. Restricting one's attention to the massless modes, the resulting Lagrangian yields equations which reduce to Einstein's equations at low energies, and can be explored for compactified solutions. Another method is to proceed directly to construct the first quantized string about a trial vacuum in which the metric (as well as other string modes) have assumed background values. In this approach one generalizes the action of (3), or its supersymmetric generalization, so that the string propagates in a curved backround with metric $\eta_{ab}(x)$,

$$S = -T \int d^2 \xi \sqrt{g} g^{\alpha\beta} \eta_{ab} \frac{\partial x^a}{\partial \xi^\alpha} \frac{\partial x^b}{\partial \xi^\beta}, \tag{16}$$

where the string is propagating on a world sheet, labeled by ξ^1 and ξ^2 (σ and τ), whose metric is $g^{\alpha\beta}$. This action doesn't look like the Nambu action of (3), but it is equivalent. One can see this by noting that the equation of motion for $g^{\alpha\beta}$ (which is the matrix inverse of $g_{\alpha\beta}$) is

$$g_{\alpha\beta} \propto \eta_{ab} \frac{\partial x^a}{\partial \xi^\alpha} \frac{\partial x^b}{\partial \xi^\beta}, \tag{17}$$

which, when inserted into (16), yields (3). This form of the string action (sometimes called the Polyakov action, since he was the one to first make serious use of this linearized action in a path integral treatment of strings) has advantages over the Nambu form. It is quadratic in the x's, so that path integrals over x time histories are relatively easy. $\eta_{ab}(x)$ is the metric of the space-time in which the string is embedded. If this is ten dimensional flat space then $\eta_{ab} =$ the Minkowski metric $(+, -, \ldots, -)$ and the action is particularly simple. In general however, for a curved background, η_{ab} is a non trivial function of x^μ and the action is that of an interacting two dimensional field theory. Such field theories are called *non linear σ models*, and are complicated renormalizable quantum field theories.

A consistent string theory can be developed as long as the two dimensional field theory of the coordinates $x^a(\sigma, \tau)$ is invariant under reparametrizations of the two dimensional world surface. After all, it is this symmetry that produces the desired gauge symmetries of the theory and prevents nasty things like violations of Lorentz invariance or unitarity from occuring. Reparametrization invariance is naively assured by the form of the above Lagrangian, which is manifestly invariant under two dimensional general coordinate transformations. However, quantum mechanics can spoil this classical symmetry. It turns out that the only dangerous transformations are *conformal* transformations, i.e. position dependent scale transformations. The requirement of true conformal invariance is a nontrivial requirement, since the theory described by (16) is an interacting nonlinear σ model, wherein quantum effects typically destroy conformal invariance. In fact, even for a flat background the condition was not trivial. It required that the background be 26 dimensional (or ten dimensional for the superstring).

We can think of the condition that the two dimensional theory be conformal invariant as being equivalent to demanding that the string equations of motion are satisfied. Thus one can search for alternative vacuum states by looking for σ models (actually supersymmetric σ models) which are strictly conformal invariant. The standard criterion for a quantum field theory to be conformal invariant is that the β-functions, which determine the rate of variation of the dimensionless couplings with length scale, vanish. In (16) the function η_{ab} as a whole plays the role of the couplings and we demand that the relevant β functions (which are local functions of the metric $\eta_{ab}(x)$ and its derivatives) vanish. (In addition one must check that the anomaly in the commutators of the stress energy tensor is not modified. In the case of a flat background this is the only requirement, since the theory is then free with zero β-function.) Given such a conformal invariant theory one can construct a consistent string theory. If $\eta_{ab}(x)$ describes a curved manifold the string will effectively be embedded in this manifold. We would then calculate string amplitudes by using in the path integral formulation the action of (16) with the particular η_{ab} of the manifold.

Remarkably, there do exist a very large class of conformally invariant supersymmetric σ models, that yield solutions of the string classical equations of motion to all orders in perturbation theory, and describe the compactification of ten dimensions to a product space of four dimensional Minkowski space times a compact internal six dimensional manifold. These compact manifolds are rather exotic mathematical constructions. They are Kahler manifolds (which means that one can define, as in the case of the complex plane, a global complex set of coordinates), admit a Ricci flat metric (which means that they are solutions of Einstein's equations, $R_{ab} = 0$), and they have SU_3 holonomy (which means that when we parallel transport a vector around a closed curve on the six dimensional manifold it undergoes a rotation by an SU_3 transformation.) Such spaces are called "Calabi-Yau" manifolds. There are many (of order tens of thousands) of such manifolds and each has, in general, many free parameters (moduli). These moduli determine the shape and size of the compact space. This is an indication of the enormous vacuum degeneracy of the string theory, at least when treated perturbatively, and leads (at the present stage of our understanding) to many free parameters. This abundance of riches should not displease us. We first would like to know whether there are any solutions of the theory which resemble the real world, later we can try to understand why the dynamics picks out a particular solution. It is very pleasing that there are many solutions for which four of the dimensions are flat, namely the cosmological constant vanishes, whereas the other six are curled up. So even if we don't know why six dimensionsof space neccesarily curl up we learn that they could do so.

b. The Low Energy Gauge Group

We now have to see whether we can find solutions in which the observed gauge group is not $E_8 \times E_8$, but rather that of the standard model. How does this immense group get broken? The first breaking that occurs is a consequence of the spacetime compactification. In the case of the heterotic string it is not sufficient to simply embed the string in a Calabi-Yau manifold. One must also turn on an SU_3 subgroup of the $E_8 \times E_8$ gauge group of the string. This is because the internal degrees of freedom of the heterotic string consist of right-moving fermions, which feel the curvature of space-time, and left moving coordinates which know nothing of the space-time curvature but are sensitive to background gauge fields. Unless there is a relation between the curvature of space and the curvature (field strength) of the gauge group there is a right left mismatch which gives rise to anomalies. Alternatively, one can explore the effective Lagrangian of the massless modes of the string theory. One finds that the equation of motion for the dilaton field, (Φ), is of the form

$$\Delta\Phi \propto R_{\alpha\beta\gamma\delta}R^{\alpha\beta\gamma\delta} - F^{cd}_{\alpha\beta}F^{\alpha\beta}_{cd}, \tag{18}$$

where $R_{\alpha\beta\gamma\delta}$ is the Riemann tensor, and $F^{cd}_{\alpha\beta}$ is the gauge field strength, and Δ is the Laplacian of the internal manifold. For a solution with $\Phi =$ constant we require a cancelation between the curvature of the manifold and that of gauge space. The easiest way to satisfy this is to identify the space-time curvature with the gauge curvature (embed the spin connection in the gauge group). One does this by turning on background gauge fields in an SU_3 subgroup of one of the E_8's (i.e. we set $F^{cd}_{\alpha\beta} = R_{\alpha\beta cd}$ using the standard imbedding of SU_3 in O_6). This has the nice feature of breaking E_8 down to E_6, which is a much better GUT group. This feature of the Heterotic string, namely the tight relation between the curvature of the internal (gauge) space and that of space time, is much more general than these particular compactifications. More generally, we can imagine the background gauge fields being in other subgroups of O_6. This would then allow the possibility of other unbroken gauge groups, O_{10} or SU_5.

This leaves us with an unbroken $E_6 \times E_8'$, which is still much too big. In older GUT approaches one would at this stage introduce by hand a bunch of scalar (Higgs) particles, and arrange the scalar potential so that enough of the scalars get expectation values to break the large symmetry. Here we are not allowed to do this, the theory contains all that there is and we can't add anything. However, there is a very nice and natural mechanism for the breaking of E_6 down to the observed low energy gauge group. This mechanism is more general than string theory, it is a possible symmetry breaking mechanism for any theory which contains compactified dimensions. The point is that the internal six dimensional manifold (which we shall call K) is, in general, multiply connected. This means that K is full of holes, and we can draw closed paths on K that go around these holes and cannot be continuously shrunk to nothing. Now, if K/Z is multiply connected one can allow flux of the unbroken E_6 (or of the E_8' for that matter) to run through it (see Figure 6), with no change in the vacuum energy. The net effect is that when we go around a hole in the manifold, through which some flux runs, we must perform a nontrivial gauge transformation on the charged degrees of freedom. So, even though the flux is running through holes where there is no manifold it has a physical effect. In the case of electromagnetism this effect is known as the Bohm–Aharonov effect, whereby a flux of magnetic field can affect charged particles that are restricted to a multiply connected region of space which has vanishing field strength.

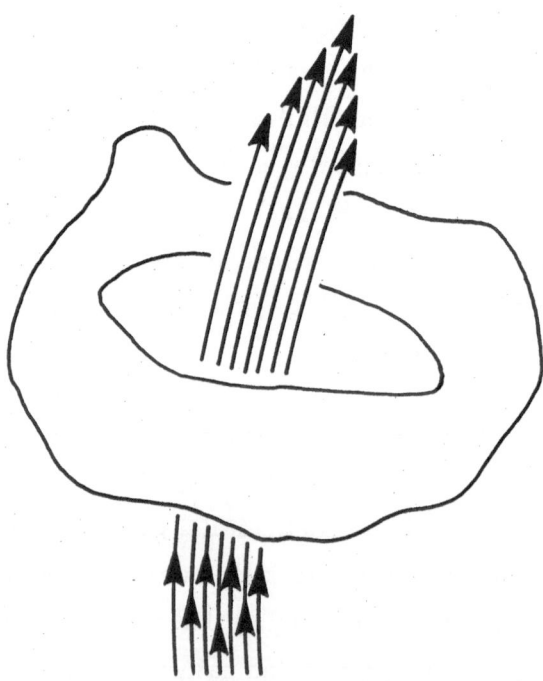

Figure 6. Flux passing through a hole in the manifold.

Another way of describing this is to say that we have nonvanishing Wilson loops around the noncontractible paths on K. These noncontractible Wilson loops act very much like Higgs bosons, breaking E_6 down to the largest subgroup that commutes with all of them. By this mechanism one can, without generating a cosmological constant, find vacua whose unbroken low energy gauge group is, say, $SU_3 \times SU_2 \times U_1 \times$ (typically, an extra U_1 or two). This method of symmetry breaking is quite different in some respects from that induced by explicit Higgs bosons. In particular E_6 is in no sense an approximate symmetry of the theory. It is inherently broken at the compactification scale, and above this scale we have the full E_8 symmetry. This, as I shall argue below is actually quite nice, since it means that various parameters of the theory are not related by E_6 symmetry at the compactification scale.

On the other hand the gauge couplings of the unbroken gauge groups (SU_3, SU_2, and U_1) are all equal at the compactification scale. This is sufficient for us to attempt to repeat the calculation of Georgi, Quinn and Weinberg of the Weinberg angle. Recall that this angle is determined by the ratios of the couplings at low energies, these in turn are equal at the unification scale (in our theory as well), but renormalize differently as one lowers the energies. The precise way in which the different couplings run depends on the matter content of the theory. In fact it can be used to place strong constraints on the matter content as we shall discuss below.

Is this sufficient? What about the other E_8'? It too could be broken by the above mechanism, but it is really not neccessary. An unbroken E_8' is quite unobservable at low energy unless there exist light matter fields with E_8' quantum numbers. As we shall see below there are no such particles for these solutions. In that case E_8' physics is that of a QCD like theory with no quarks. The mass scale of such a theory would be very large since the effective coupling grows faster than the QCD coupling as we go to lower energies. This is because the β function that controls the running coupling is proportional to the Casimir of the gauge group, and this is much larger for E_8 than for SU_3. Thus we would expect to have very heavy E_8' glueballs, and no significant effects on low energy physics. Actually, such a strong E_8' gauge group might be useful, as has been conjectured, as a way of breaking supersymmetry.

In conclusion it is very pleasing that there are many solutions which have the standard model as the observable low energy gauge group (plus perhaps some extra U_1's, which would be welcome, if there, as a way of testing the theory).

c. The Matter Content

These Calabi-Yau compactifications, produce for each manifold K, a consistent string vacuum, for which the gauge group is no larger than $E_6 \times E_8$. Now we must address the issue of the *matter* content of the theory. This is determined by the massless spectrum of the theory. We must determine the quantum numbers of the massless particles and see whether they can correspond to the observed families of quarks and leptons. After Kaluza-Klein compactification the spectrum of massless chiral fermions is determined by the zero modes of the Dirac operator on the internal space. This is because the Dirac equation for a fermion reads $\gamma^\mu D_\mu \Psi(x^\mu) = 0$, where μ runs over the ten spacetime indices. If we have compactified six of the spatial dimensions $(X^n; n = 4, 5, 6, 7, 8, 9)$ it is convenient to expand Ψ in terms of a basis of functions, $\Phi_k(X^n)$, on the compact manifold K, which are eigenfunctions of the internal Dirac operator $D_{int} = \sum_{n=1}^{6} \gamma^n D_n$, i.e. $D_{int}\Phi_k = \lambda_k \Phi_k$. In this basis we can write $\Psi(x^\mu) = \sum_k \Phi_k(X^n)\Psi_k(x^\alpha)$, where $x^\alpha; \alpha = 0, 1, 2, 3$ runs over ordinary Minkowski space. In that case the four dimensional fields, $\Psi(x^\alpha)$, satisfy the four dimensional Dirac equation

$$[\gamma^\alpha D_\alpha + \lambda_k]\Psi_k(x^\alpha) = 0, \tag{19}$$

where the eigenvalue, λ_k, of the internal Dirac operator emerges as the four dimensional mass. For every nontrivial solution of the Dirac equation with $\lambda_k = 0$ in the internal space there will appear a massless fermion in four dimensions.

Now it might appear to be a difficult task to decide how many solutions there are to this equation. The number might depend on grubby details of the internal manifold. This is not the case, one can determine the number of massless fermions by topological arguments! To see this let us recall the chiral properties of the Dirac equation. Remember that in any even number of dimensions, d, there is a matrix γ^{d+1}, analogous to γ^5 in four dimensions, that anticommutes with all the γ matrices, $\{\gamma^{d+1}, \gamma^\mu\} = 0; \mu = 1, \ldots, d$. Now consider the Dirac equation

$$[\gamma^\mu D_\mu + \lambda]\Psi(x^\mu) = 0. \tag{20}$$

If we have a solution Ψ of this equation, with eigenvalue $\lambda \neq 0$, then $\gamma^{d+1}\Psi$ is also a solution, but with eigenvalue $-\lambda$. Thus the solutions with non vanishing eigenvalues neccesarily come in pairs, with equal and oppposite values. This need not be the case when $\lambda = 0$, in which case the solutions can be chosen to be of definite *chirality*, $\gamma^{d+1}\Psi = \pm\Psi$, and need not be paired. Now consider changing the form of the internal manifold K slightly. This will change the Dirac operator and its spectrum. In particular we might imagine that the number of zero eigenvalues could change abruptly. Since we can't lose·a zero eigenvalue by a continuous change in K, it must turn into a nonzero eigenvalue. However the nonzero eigenvalues come in pairs $(\pm\lambda)$. This means that two zero eigenvalues must dissappear at the same time. Furthermore they must have opposite chiralities so that we can form the appropriate linear combinations. We therefore learn from this simple argument that *under continuous changes in the manifold the number of postive chirality, N_+, minus the number of negative chirality zero modes, N_-, is unchanged*. This means that $N_z = N_+ - N_-$ (the *index* of the Dirac operator) is a topological property of K, in other words a property that is invariant under continuous deformations of the manifold.

This is a very nice feature of the zero modes of the Dirac operator. It means that N_z can be calculated in terms of toplogical properties of K alone. Note that this argument only determines N_z and not N_+ and N_- seperately, but that is fine since we are actually only interested in N_z. This is because one positive and one negative chirality fermion can always combine and become a massive fermion (recall that a Dirac mass term does not commute with γ^{d+1} and mixes chiralities). Therefore we would expect, that if, say, $N_z \geq 0$ then N_- fermions will acquire a mass (in string theory typically of order M_{Planck}) and decouple, leaving us with N_z massless fermions. (Actually, in string theory one can also determine the numbers N_+ and N_- seperately by topological arguments.)

Now what are the quantum numbers of possible massless fermions? Since the fermions are originally all in the adjoint representation of $E_8 \times E_8$, the $\mathbf{248} \oplus \mathbf{248}$ representation, the massless fermions that emerge after compactification must appear in decompostions of this representation. Now the adjoint representation of E_8 decomposes as $(\mathbf{1}, \mathbf{78}) \oplus (\mathbf{3}, \mathbf{27}) \oplus (\bar{\mathbf{3}}, \bar{\mathbf{27}}) \oplus (\mathbf{8}, \mathbf{1})$ under $SU_3 \times E_6$. We are interested in the $(\mathbf{27})$, which contains one standard family of quarks and leptons. To see the familiar quarks and leptons let us decompose E_6 under its maximal subgroup $SU(3)_{color} \times SU(3)_{left} \times SU(3)_{right}$. Here we have identified these groups as color and as left and right-handed electroweak interactions respectively. Then the $\mathbf{27}$ can be decomposed as follows

$$\mathbf{27} \to (\bar{\mathbf{3}}, \mathbf{3}, \mathbf{1}) + (\mathbf{3}, \mathbf{1}, \mathbf{3}) + (\mathbf{1}, \bar{\mathbf{3}}, \bar{\mathbf{3}}). \tag{21}$$

A standard family of quarks and leptons can be easily accommodated in this representation. The up and down quarks, together with a new, charge $-\frac{1}{3}$ quark (called g) form an $SU(3)_{left}$ multiplet and their antiparticles an $SU(3)_{right}$ multiplet. The $(\mathbf{1}, \bar{\mathbf{3}}, \mathbf{3})$ contains the electron and the electron neutrino, as well as some new particles, all in all

$$\begin{pmatrix} H_1^u & H_1^d & e^- \\ H_2^u & H_2^d & \nu \\ e^+ & \bar{\nu} & S_2 \end{pmatrix}.$$

Of course, one must suppose that the new particles in this multiplet get large masses, which explains why they have not yet been observed. Note that the particles $H_{1,2}^{u,d}$ have the quantum numbers of Higgs particles, so that their scalar partners could be responsible for the breaking of $SU(2)_L \times U(1)_{EM}$ to $U(1)_{EM}$ at low energy.

Since, for the heterotic string, the gauge and spin connections are forced to be equal one can count the number of chiral fermions by geometrical arguments. This works in the following way. Let N_{27}^L (N_{27}^R) be the number of left (right)-handed massless fermion multiplets transforming as a **27** under E_6 and let $N = N_{27}^L - N_{27}^R$. The number of generations is just equal to $|N|$. Now fermions transforming in the **27** subgroup of E_8 also transform in the **3** of $SU(3)$. Thus N is just the index of the Dirac operator on the internal manifold K acting on spinors in the **3** representation of $SU(3)$. But the $SU(3)$ gauge connection (vector potential) is equal to the spin connection, so that the index can only depend on the toplogy of K. In fact it is equal to one half of the Euler number, χ, of K

$$|N| = \frac{1}{2}|\chi|. \tag{22}$$

So the number of generations is equal to half the Euler character of the manifold (which, if the manifold were two dimensional would simply count the number of handles). For most Calabi-Yau manifolds the Euler character turns out to be quite large. There is , however, a way of reducing it. If there exists a discrete symmetry group, Z, which acts freely on K (which means that the group has no fixed points), one can consider the smaller manifold K/Z (in other words, points on K which are related by a symmetry transformation of Z are identified. If Z acts freely on K this produces a new manifold whose volume is reduced by the dimension of Z. If Z has fixed points one gets a manifold with singular points at the fixed points–an orbifold) On K/Z the Euler character is reduced by the dimension of Z. By this trick, and after some searching, manifolds have been constructed with 1, 2, 3, 4, ... 200 generations.

How many generations can there be? So far we have observed three generations of quarks and leptons, but clearly there might be mnay more if they are sufficiently heavy. There are, however, indirect ways of putting bounds on the number of generations. For example, it is possible to argue on the basis of standard big bang cosmology that there can't be more than ≈ 4 massless neutrinos. More neutrinos would seriously affect the calculation of Helium production, one of the most successful of the predictions of standard cosmology. A direct bound on the number of light neutrinos will shortly be provided by precision measurements of the W and Z masses. From a theoretical point of view the number of generations in a unified theory is bounded, if we are to come up with a successful prediction of the Weinberg angle. The point is that the ratio of the various couplings of the subgroups of a unified gauge group are renormalized seperately once the full symmetry is broken. In string theory, for example, they are all equal at the compactification scale, but differ at low energies where their ratio determines the weak mixing angle. The number of quark and lepton generations affect this differential renormalization, and so affects the calculation of the Weinberg angle. It seems that to be realistic we must restrict to manifolds with three, or perhaps four, generations.

Once again we find an enormous number of possible solutions. It is extrememly pleasing that the theory automatically produces chiral fermions. This was not the case for other Kaluza-Klein theories.It is even more pleasing they can produce families of quarks and leptons with the right quantum numbers.

d. Further Symmetry Breaking

We have shown, so far, that there exist solutions of the heterotic string theory in four dimensional Minkowski space (times a small compact manifold), that the large gauge symmetry can be broken down to the standard model symmetry group and that there are solutions with roughly the right number and kinds of massless fermions to be identified with the observed quarks and leptons. However there is still much to be done before we can directly compare with the real world.

First of all, all of these solutions have exact N=1 supersymmetry. This is good and bad. It is good because we need to have supersymmetry survive all the way down to low ($\approx 1\,\mathrm{Tev}$) energiy if we want supersymmetry to solve the *hierarchy* problem by protecting the masslessness of the Higgs particles. It is bad because supersymmetry is clearly not an exact symmetry of nature, in fact it is so badly broken that no signal of its existence has yet appeared. It is therefore necessary to break the remaining N=1 supersymmetry. However one must be careful. There are easy ways of breaking the supersymmetry, even in perturbation theory. For example, one can with toroidial compactifications introduce twisted boundary conditions that violate supersymmetry. However this has bad consequences. First, it breaks supersymmetry at the compactification scale, which is probably too high. Second, it produces at one loop order a cosmological constant and destabilizes the vacuum. The generation of a cosmological constant is always a potential problem once supersymmetry is broken. It does not seem likely that perturbative supersymmetry breaking could exist without generating an intolerably big cosmological constant. So we must contemplate nonperturbative mechanisms.

For this purpose the extra E_8 gauge group might be useful. Below the compactification scale it yields a strong, confining gauge theory like QCD, but without light matter fields. In general this sector would be totally unobservable to us, consisting of very heavy glueballs, which would only interact with our sector with gravitational strength at low enegies. However there could very well exist in this sector a gluino condensate which can serve as source for supersymmetry breaking. This possibility has been considered, but so far does not seem to work. What goes wrong? The problem appears to be that once supersymmetry is broken the dilaton expectation value can be dynamically fixed, in other words there is now a non-flat dilaton potential. Now there is always one stable point for the dilaton, namely where its expectation value, ϕ, blows up. Since the couplings scale as $e^{-\phi}$ this means that all couplings vanish and the theory is free. When supersymmetry is broken, at least by the mechanism discussed above, the theory tends to relax to a free theory, or, equally bad, to ten dimensional flat space.

This problem, how to break supersymmetry without producing a cosmological constant, is the major obstacle to relating the heterotic string to the real world. It is likely that the solution must await the development of the theory to the point where non perturbative issues can be adressed and answered. If we were to suceed in breaking supersymmetry at low energies, then the remaining ingredients for a sucessful description of the real world are already in place. There exists a natural reason for the existence of massless Higgs bosons which are weak isospin doublets (and could be responsible for the electro-weak breaking at a Tev), without accompanying color triplets.

Thus the Heterotic string theory appears to contain, in a rather natural context, many of the ingredients necessary to produce the observed low energy physics. I do not mean to suggest that there are not many problems and unexplained mysteries. In addition to those discussed above there exists the danger (common to many grand unified models, especially supersymmetric ones) of too rapid proton decay; there is no deep understanding of why the cosmological constant, so far zero, remains zero. Nonetheless the early successes are very reassuring and they give one the feeling that there are no insuperable obstacles to deriving all of low energy physics from the $E_8 \times E_8$ Heterotic string theory.

6. RECENT PROGRESS

I do not want to leave the impression that string theory has brought us close to the end of particle physics. Quite the opposite is the case. There are many unsolved problems and deep mysteries that need to be understood before one can claim success; we have only begun to probe the structure of these new theories. I shall present a list of problems that are the focus of current research and comment on recent progress.

6.1 What is String Theory?

We do not fully understand the deep principles and symmetries that underly string theories. To date these theories have been constructed in a somewhat adhoc fashion and often the formulism has produced, for reasons that are not totally understood, structures that appear miraculous. There is a strong feeling among string theorists that the analogue of the *principle of equivalence* for this extension of general relativity has not yet been discovered.

To date all treatments of string theory have been carried out using the methods of first quantization. This procedure leads to the description of string scattering amplitudes as sums of path integrals of two-dimensional σ models on world sheets with any number of handles. Unlike particle theories, this formulation is very natural and beautiful (for example, interactions are introduced in a geometrical and unique fashion) but one might ask whether first quantization is enough. In my opinion the answer is—probably not! First, although the path integeral formulation is very pretty, it does not manifestly exhibit all the symmetries of the theory; second, it only yields a perturbative expansion which surely is, at best, an asymptotic expansion of the theory; and finally, it is background dependent. In fact, were the first quantized perturbation theory to define the full string theory by itself we would be in deep trouble. This is because there are many properties of all string theories that are true to all orders in perturbation theory but are not true in the real world (exact supersymmetry, massless scalars, etc.).

Most likely a second quantized treatment of strings is required, as in the case of ordinary field theories, where we introduce operator valued string fields, $\Psi(x^\mu(\sigma))$, which are functionals of loops and a Hamiltonian that generates the dynamics. In fact, everything we know to date of the structure of string theories suggests that it is nothing more than an 'ordinary' field theory, albeit one with a very large number of degrees of freedom. This is the *conservative* approach. It is also possible that a novel approach (such as that advocated by Friedan and Shenker, based on 'infinite genus' Riemann surfaces) will prove more fruitful.

A second quantized string field theory has already been developed in light cone gauge, years ago for the bosonic string (by Kaku and Kikkawa, Cremmer and Gervais, and Mandelstam), more recently for superstrings (by Brink,

Green and Schwarz) and now for heterotic strings (by V. Periwal and myself). The advantage of this approach is that it is manifestly unitary, contains only physical degrees of freedom and that it exists. Its disadvantage is that, having fixed a gauge, manifest symmetry (even global Lorentz invariance) is lost and that the formulation is background dependent.

Much effort has been expended recently in attempts to formulate covariant string field theory. One of the most important developments (due mainly to W. Siegel) is the realization that (first quantized) ghost coordintes are not unphysical, but rather natural devices for discussing differential forms on loop space. With this insight free string field theories have been formulated for both the bosonic and superstring theories. As for interactions two separate approaches have been pursued. First, there is the attempt (by Siegel and Zwieback, by Hato, Itoh, Kugo, Kunimoto and Ogawa, and by Neveu and West) to graft the vertex of the light cone field theory onto a covariant approach. This direction suffers from two problems. First, it requires the introduction of an unphysical parameter (which in light cone gauge is identified with P_+) α, whose role is to measure the length of the string.Second, it is not based on any apparent geometrical principle. A different approach has been pursued by Witten, motivated by a generalization of ordinary geometry to "non-commutative geometry" .

Witten constructs an abstract geometrical framework, constructed so that the first-quantized BRST operator, Q, plays the role of an exterior derivative (this is motivated by the fact that $Q^2 = 0$). To this end he defines a (non-commutative) multiplication of string functionals,* , and an integration, $\int$. These obey the usual axions of multiplication (with the exception of graded commutativity), differentiation and integration. Therefore, as the mathematician Connes has showed, they allow for a natural generalization of gauge invariance. Witten argues that the string Lagrangian is just the analogue, for this non-commutative geometry, of the Chern-Simons form

$$S = \int \left[\Psi * Q\Psi + \frac{2}{3}\Psi * \Psi * \Psi \right].$$

$$(23)$$

The resulting equations of motion are simply $F = 0$, where F is the field strength, $F = Q\Psi + \Psi * \Psi$. What could be simpler!

This formulism has been developed quite far. It has been shown explicitly (by Giddings) to reproduce the four point scattering amplitude of the Veneziano model, to yield (using path integral techniques by Martinec and Giddings) the correct measure for the N-point function, and to yield the correct measure (by Giddings, Martinec and Witten) for the loop amplitudes. Jevicki and I have worked out the explicit operator formulation of all of the ingredients of the theory, including the vertex operator, and have explicitly verified the symmetries of the action. Witten has generalized the approach to the open superstring. Much, however, remains to be done. So far this formulation has not been extended to the closed string, including the heterotic string; and to date no nontrivial use has been made of this formulism.

6.2   How Many String Theories Are There?

At the moment one can count four consistent string theories. There are the two forms (chiral and nonchiral) of the closed superstring, there is the SO(32) open string theory of Green and Schwarz and there is the Heterotic string theory. It is now realized that the two manifestations of this theory (with gauge groups $E_8 \times E_8$ or $Spin(32)/Z_2$) are simply two vacuum states of the same theory.

Do there exist more consistent theories than the known five? There have been some recent constructions of new models, but these too are probably just different (likely unstable) vacua of the heterotic string. Do there exist fewer, in the sense that some of the ones we know already are perhaps different manifestations (different vacua) of the same theory? Although there have been speculation along these lines no convincing argument to this effect is known.

What is special about the Heterotic string?

get fixed and thereby the dilaton acquire a mass? Does the vanishing of the cosmological constant survive the physical mechanism that lifts the vacuum degeneracy?

These, in my opinion, are the most important problems facing us. Until we can choose the right vacuum among all the potential candidates we will not be in a position to make direct contact with experiment.

6.7   What is the Nature of High Energy Physics?

By this I mean what does physics look like at energies well above the Planck mass scale? This is a question that is addressable, in principle for the first time, and might be of more than academic interest for cosmology, where is interested in the ' initial conditions' of the universe. Traditionally this means pushing back to arbitrarily early times where the temperature and density were arbitrarily high. At Planckian times, where densities and temperatures are of Planckian values, string physics will be relevant. Does the string undergo a transition to a new phase at high temperatures and densities? Does string theory determine the initial state of the universe? Here no progress has been made. We still await the development of a string cosmology.

Another question of great academic interest is that of singularities. Einsteinian relativity has two severe problems. The first is the nonre normalizability of the quantum theory. This problem appears to be solved by string theory. However there is also the problem of the ubiquitous singularities and the resulting incompleteness of ordinary relativity that already shows up at the classical level. It is perfectly possible that string theory will cure this problem. The above mentioned corrections to Einsteinian relativity, (say, the quartic terms in R_{abcd} in the effective action) reopen the question of singularities. For example, we do not know whether classical string theory contains black holes. Witten and I have, in fact, speculated that the end product of collapsing matter might be nonsingular, dilaton radiation playing the role in the classical theory of Hawking radiation in the quantum theory. One could very well imagine that at very short distances the usual spacetime description

6.3 String Technology

This is not a question but a program of development of techniques for performing calculations within string theory, including control of multiloop perturbation theory and the construction of manifestly covariant and supersymmetric methods of calculation. Much progress has been made in this area during the last year. Of special importance is the development of the superconformal field theory approach to the construction of superstring loop amplitudes, the construction of the fermionic loop amplitudes, the construction of the fermionic vertex (by Friedan, Martinec and Shenker); and the application of sophisticated algebraic geometry to the evaluation of the integrands of multiloop amplitudes (by Belavin and Kniznik and by Manin). Enormous mathematical power is now being brought to bear in this area with remarkably beautiful and suggestive results.

6.4 What is the Nature of String Perturbation Theory?

Does the perturbative expansion of the string theory converge? As I remarked above, probably (and hopefully) not. If not, when does it give a reliable asymptotic expansion? How can one go beyond perturbation theory? Can one develop semi-classical methods (instantons) to treat the nonperturbative physics? Do the non-perturbative effects totally destroy the perturbative picture of the string vacuum? Here there are many questions and no answers.

6.5 String Phenomenology

Here there are many issues that remain to be resolved. They can all be included in the one question—can one construct a totally realistic model which agrees with observation and why is it picked out?

6.6 What Picks the Correct Vacuum

There exist already an enormous number of acceptable vacuum states. What is the dynamics that chooses among all these possibilities? Why don't we live in ten dimensional flat space? How does the value of the dilaton field of physics breaks down, and that string theory avoids the ubiquitous singularities that plague ordinary general relativity already at the classical level. Thus string gravity might be a complete theory and we wouldn't have to resort to hopes of a quantuam cure.

6.8 Is There a Measurable, Qualitatively Distinctive, Prediction of String Theory?

The Heterotic string theory can make, in principle, many postdictions(such as the calculation of mass ratios of quarks and leptons, Higgs masses, gauge couplings, etc.) and it can make many new predictions (such as the masses of the various supersymmetric partners). These would be sufficient to establish the validity of the theory; however one could imagine conventional field theories coming up with similar pre or post dictions. It would be nice to predict a phenomenon which might be accessible at observable energies and is uniquely characteristic of string theory. Perhaps our experimental friends will discover such a phenomenon for us.

EXPERIMENTAL PUZZLES BEYOND THE STANDARD MODEL

A. Savoy-Navarro

DPhPE-CEA,

Saclay, France

1.   INTRODUCTION

The Standard Model is, so far, a successful attempt to combine, within
the gauge invariance

$$SU(3) \otimes SU(2) \otimes U(1) \, ,$$

the theories which try to explain the three fundamental forces between the
elementary constituents [strong, weak, and electromagnetic (e.m.) forces].
In this scheme, the strong interactions are described by quantum chromodynamics (QCD)--the SU(3) gauge theory--and the weak and e.m. interactions
are unified by the Weinberg-Salam (WS) model within $SU(2) \otimes U(1)$ invariance.

How well does this model reproduce the present experimental observations? A good answer to this question is given, for instance, by the study
of the results obtained by the UA1 and UA2 experiments at the CERN $p\bar{p}$ Collider. Among the various kinds of event produced in $p\bar{p}$ collisions at 630 GeV
in the c.m. energy, are the events with 'jets'. They are produced with a
measured cross-section of a few millibarns; the fraction of this cross-
section due to heavy flavours is dominated by $b\bar{b}$ production and the corresponding cross-section is $\sigma^{b\bar{b}} \simeq 6$ μb. Jet events are mainly explained by QCD.
Another type of event is that where intermediate vector bosons W and Z^0 are
produced by the Drell-Yan mechanism in $p\bar{p}$ collisions. At the energy of the
$p\bar{p}$ Collider at CERN, these events give a contribution of 600 pb to the total
cross-section σ^{pp}, and they successfully reproduce the WS model expectations.
DiLella in his lectures[1] gives a very detailed explanation as to how well

these two types of event fit with the Standard Model. In addition to these two main standard processes, which have been measured to be 50 to 60 mb, the most important part of the total cross-section σ^{pp} is provided by the so-called 'minimum bias events'. This means the events which have a rather large average charged multiplicity (say of the order of 30) and a rather small average transverse momentum (say 0.42 GeV/c) at the CERN $p\bar{p}$ Collider energy.

As shown by this example and, in fact, by a general overview of the present experimental results obtained in our field[2], the conclusion by now is: 'Everything fits well with the Standard Model'. So why pursue further; in other words:

Are there any puzzles beyond the Standard Model?

Despite this 'tout va très bien, Mme la Marquise' attitude, we think that we should not feel discouraged, mainly because nothing is perfect in this world; even the Standard Model! There are in fact many signs showing that 'the Standard Model does not qualify as the ultimate theory', such as:
- too many parameters (number of couplings: g_s, g_w, $\sin \theta_w$, fermion mass ratios, mixing matrix, Higgs, etc.),
- many unexplained features (charge quantization, baryon and lepton number conservation, P and C violations, iteration of families),
- many loose ends (Higgs sector, no gravity, hierarchy problem, no unification of forces).

These very important points which remain unexplained or are inexplicable by the Standard Model lead the theorist to look beyond this model. They do this in three different ways.

The first way is to 'extend minimally' the Standard Model by introducing new families (such as new heavy quarks, new heavy leptons, new heavy neutrinos), and new gauge levels:

$$SU(3) \otimes SU(2) \otimes U(1) \otimes G' ,$$

for instance,

$$SU(2)_L \otimes U(1) \quad \text{becomes} \quad SU(2)_L \otimes SU(2)_R \otimes U(1) ,$$

This means, in reality, new W's and new Z^0's.

The second way consists in 'extending maximally' the Standard Model, by introducing a new concept: 'the world is composite'; so quarks and leptons are composite, W and Z^0's are composite and even Higgs are composite (as, for instance, in technicolor)

The last way is even more radical. It considers that to find the really complete explanation we have to consider the whole energy scale (see Fig. 1); so the world is *super* ...; by ... we mean:

... *symmetric* (i.e. supersymmetry)

and also

... *gravitational* (i.e. supergravity with dimension $N \geq 1$)

and also

... *string* (i.e. superstring) .

I refer to the lectures given by Hinchliffe[3] and Gross[4] for the theoretical description of these various approaches. My point of view here will be that of the experimentalist. Let me say first that to talk about 'experimental puzzles beyond the Standard Model', is very much time- and fashion-dependent. Two years ago, on the basis of a few events found at the CERN $p\bar{p}$ Collider during the run of 1983[5], compositeness, new families, unconventional Higgs, supersymmetry and why not technicolor were claimed to be discovered! (Superstring was not yet a 'commercialized' theory at that time.) Nowadays (July 1986), the general consensus of opinion is that the Standard Model agrees with all the experimental observations. So certainly the task is harder and the subject much less fashionable. However, the Standard Model has only been tested at the 5% level in the W-sector (see DiLella's lectures[1]); the top quark has still to be established ($m_t \geq$ 40 GeV/c^2?); the Higgs sector is untested; in the field of perturbative or non-perturbative QCD there are still a lot of remaining puzzles, see 'jet story' of DiLella[1]; there remain a lot of fundamental questions concerning the ν-sector. For instance: Are the ν massive or not? How many are there? Are they Majorana or Dirac particles?, etc.

In the case of the experimentalists, the questions asked as well as the answers provided are not only dependent on the fact that a theoretical idea is correct or not, but also on the capability of the experimental apparatus to check it. So we will, in these lectures, discuss not only results and their possible non-standard interpretation, but also the *experimental techniques* to do this as well as the *improvements* from the detector point of view to achieve a better (and so more accurate) test.

These lectures could become a long and systematic review of all possible theoretical hypotheses and the corresponding experimental upper limits

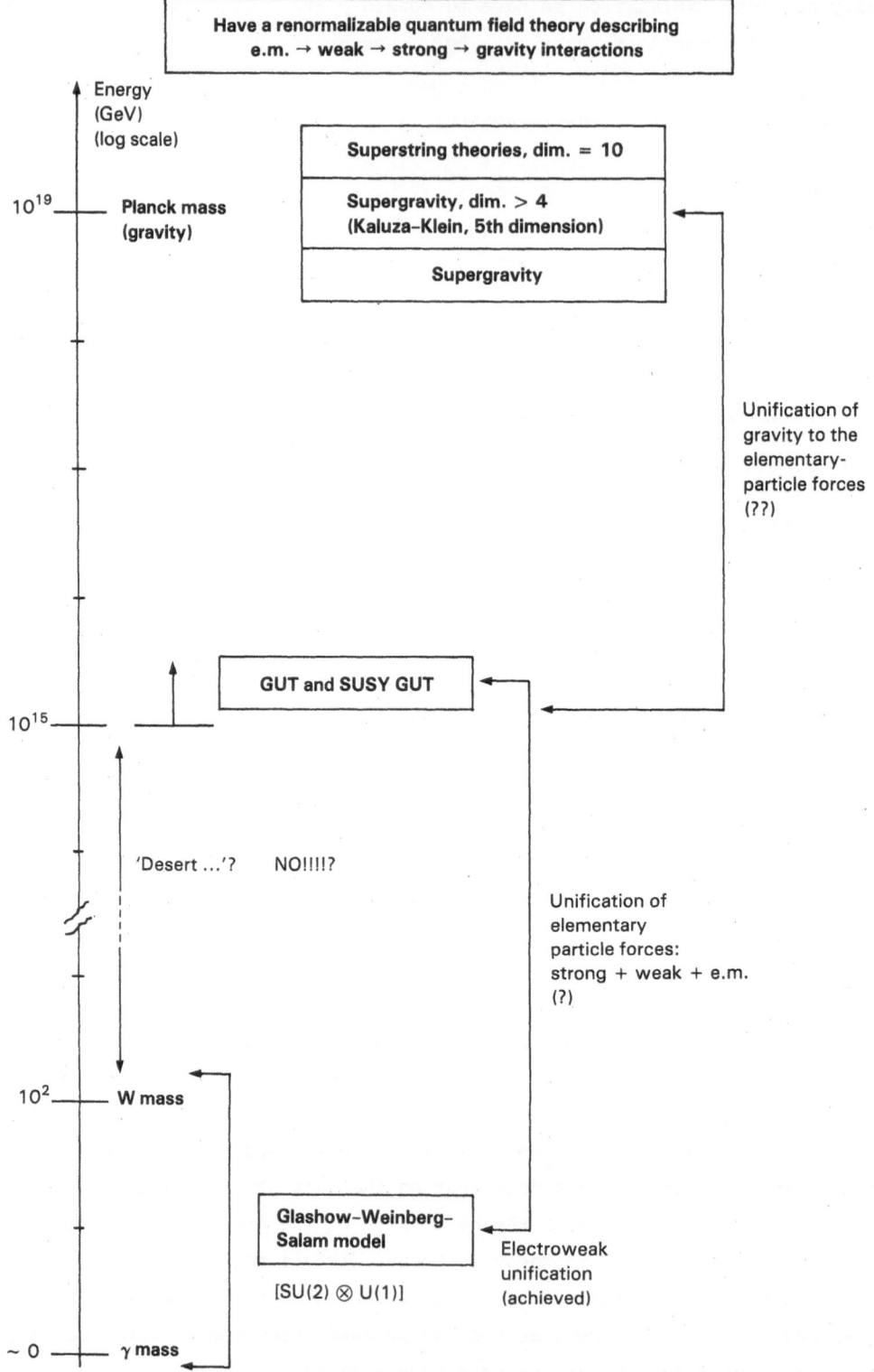

Fig. 1. Description of the full energy scale in high-energy physics: from a few GeV to the Planck mass.

obtained so far. By personal taste, I prefer to focus on some selected topics and study, in more detail, their specific features both from a theoretical and an experimental point of view. So these lectures have been divided into three sections. Section 2 is called 'Superworld?'. The first part of this section is devoted to the search for supersymmetry at e^+e^- and $p\bar{p}$ colliders. For this purpose two 'typical' experiments, namely single photon search at e^+e^- (ASP) and missing energy measurement at $p\bar{p}$ (UA1) are described; the second part is devoted to the discussion of the fifth force to show how antigravity is related to supergravity.

Section 3, called: 'Extended world?' is mainly devoted to a review of the ν physics including a non-accelerator section. The last section, 'Beyond the 1 TeV world' reports on the Higgs sector and includes also a section about new accelerators at very high energy, the related new experimental techniques (i.e. an introduction to the high-energy physics domain at the dawn of the 21st century.)

As you will see, there are still a lot of experimental puzzles beyond the Standard Model: Good for us!

2. SUPERWORLD?

2.1 Supersymmetry at e^+e^- and $p\bar{p}$ Colliders

The N = 1 supergravity model (SUGRA) makes definite predictions at the low-energy level [i.e. the W-mass range, namely from a few GeV/c^2 to $O(100)$ GeV/c^2]. There are a lot of possibilities to test its experimental consequences. This is true both at e^+e^- and pp colliders[6].

We want to describe here two main techniques, used to search for supersymmetric particles at these machines. Each of these techniques is, in some sense, specific to each accelerator. They are the single-photon search at the e^+e^- collider and the missing-energy measurement at the $p\bar{p}$ collider.

The results obtained so far by both experiments are negative. However, we may hope that all the new e^+e^- and $p\bar{p}$ colliders (SLC, LEP, ACOL, Tevatron) are going to completely scan the W-scale and give a final answer about the possible existence of supersymmetric particles in this mass range.

2.1.1 Single photon searches at PEP. I will define the principle of these searches and give a description of the related experiments.

i) Principle

We discuss here the study of the process[*]:

_____

[*] We will come back to this experiment in the next section when discussing it in terms of a neutrino counting experiment [process (3)].

$$e^+ e^- \to \ddot{\gamma}\ddot{\gamma}\gamma$$

by an exchange of a virtual scalar electron $\tilde{e}$. So it can be used to search for $\tilde{e}$ with a mass larger than $\sqrt{s}$. The single-photon differential cross-section is given by:

$$\frac{d\sigma}{dx\,d\cos\theta} = \frac{\alpha^2 s(1-x)}{3x\sin^2\theta}\left[\left(1 - \frac{x}{2}\right)^2 + \frac{x^2\cos^2\theta}{4}\right]\left[\frac{1}{m_{\tilde{e}_R}^4} + \frac{1}{m_{\tilde{e}_L}^4}\right]$$

where $x = 2E_\gamma/\sqrt{s}$. So it peaks at small θ and small E_γ. This implies that it is mandatory to have an e.m. calorimeter down to small θ to perform such searches. Moreover, the main *backgrounds* are the following QED processes:

$$e^+ e^- \to e^+ e^- \gamma \tag{1}$$
$$e^+ e^- \to \gamma\gamma\gamma \ , \tag{2}$$

where $e^+ e^-$ or $\gamma\gamma$ escapes into the small polar angle region not covered by *veto counters* and, also,

$$e^+ e^- \to \gamma\nu\bar{\nu} \ . \tag{3}$$

The predominant quantum electrodynamics (QED) background (1), requires a rejection factor of the order of 10^{-4} to reach a sensitivity to the ν pair production [process (3)]. By using the three-body final-state kinematics, as shown schematically in Fig. 2, we get a way to eliminate the other QED backgrounds [processes (1) and (2)]. For simple kinematical reasons, the transverse energy of the photon, E_T^γ, relative to the beam line, must be balanced by the e^- pair; thus for a detected photon with transverse momentum E_T^γ, at least one of the other particles will be at an angle larger than

$$\sin\theta_{recoil}^{min} = E_T^\gamma/(2E_{beam} - E_\gamma)$$

So θ_{recoil} must be larger than $E_T^\gamma/2E_{beam}$ to achieve the desired degree of rejection.

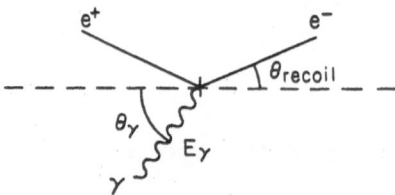

Fig. 2.   Three-body final-state kinematics $ee\gamma$, used to eliminate QED backgrounds.

There only remains then, the two 'conflicting processes':

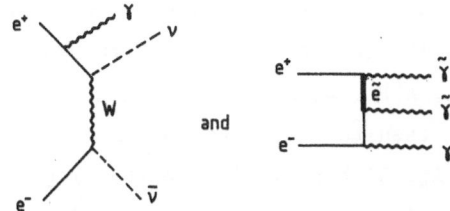

Their cross-section can be parametrized in the following way:

$$\sigma(e^+ e^- \rightarrow \gamma \nu \bar{\nu}) \simeq \alpha \ G_F^2 s \left(1 + \frac{N_\nu}{4}\right)$$

$$\sigma(e\tilde{\gamma} \rightarrow e\tilde{\gamma}) = \frac{8\pi}{3} \alpha^2 \ \frac{s}{m_{\tilde{e}}^4} \quad .$$

So the ratio of these two cross-sections is given by:

$$\frac{\sigma^{\tilde{\gamma}}}{\sigma^\nu} \simeq 50 \left(\frac{40 \ \text{GeV}}{m_{\tilde{e}}^4}\right)$$

Therefore if $m_{\tilde{e}}$ is of the order of a few tens of GeV, the $\tilde{\gamma}$ will escape detection. Moreover, as the final states and the corresponding Feynman diagrams are the same, both cross-sections can be parametrized in the following way:

$$\frac{d^2\sigma}{dxdy} = K \ \frac{1}{x} \ \frac{1}{1-y^2} \ s(1 - x) \left[\left(1 - \frac{x}{2}\right)^2 + \frac{x}{4} y^2\right]$$

where

$$K_{\gamma\nu\bar{\nu}} = \frac{G_F^2 \alpha}{6\pi^2} \ [N_\nu(g_V^2 + g_A^2) + 2(g_V + g_A + 1)]$$

$$K_{\gamma\tilde{\gamma}\tilde{\gamma}} = \frac{4\alpha^3}{2m_{\tilde{e}}^4}$$

$$y = \cos \theta_\gamma \quad .$$

Therefore, the limit on the single-photon cross-section *probes* the sum of the ν-pair and supersymmetric sources; for $m_{\tilde{e}} > \sqrt{2} \ m_W$, the process $e^+ e^- \rightarrow \gamma \nu \bar{\nu}$ dominates. The only way to separate both processes is to do a careful study of the $\sqrt{s}$ dependence of σ, since the weak cross-section has a resonance, while the supersymmetric cross-section does not. Because of this fact, it will not be possible to interpret the results of SLC and LEP studies of the Z^0 width without using the lower limit provided by PEP data.

111

ii) Description of the experiments and their results

Taking into account these different points, two experiments have tried
to perform such a search at PEP: the upgrade of the MAC experiment[7] and a
special-purpose detector, ASP[7]. We now summarize the main features of these
detectors as well as the results they have obtained.

The MAC detector (Fig. 3) has, in its central region, a drift chamber
(CD) and a shower counter (SC) consisting of 32 lead plates interspersed
with proportional wire chambers giving a total of 14 radiation lengths. It
goes down to 40° for photon detection, with E_T^γ > 4.5 GeV. An end-cap hadron
calorimeter (EC) covers the region down to 10°; it consists of 28 1-in.
steel plates, each followed by planar proportional chambers constructed in
30° wedges. The upgrade of the MAC detector has been done essentially by
adding an SAV detector. It is a set of proportional chambers and lead plates
altogether 8.5 radiation lengths thick, and allowing the detection of
photons down to 5° with E_T^γ > 3 GeV.

The ASP detector has been specially built to do a single-photon search
experiment. Figure 4 reproduces a view along the beam axis of this detector.
Drift chambers are placed in the region in front of the low-angle calorimeters, in order to measure the exit angle of charged particles in the three-
body QED final states. By measuring the angles and corresponding energies,

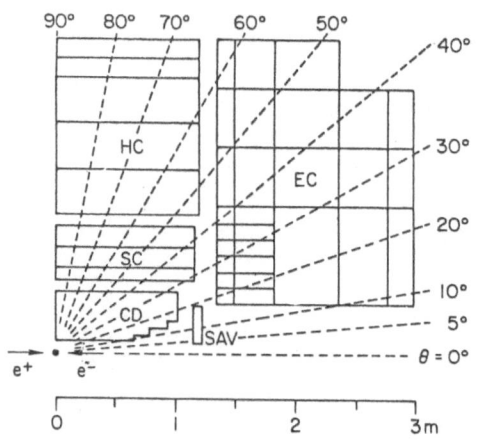

Fig. 3.   Schema of the MAC detector.

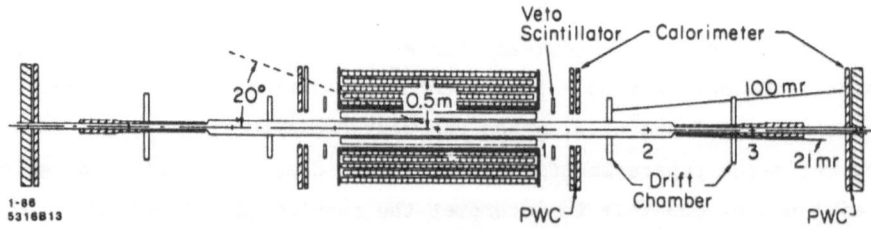

Fig. 4.   View along the beam axis of the ASP detector.

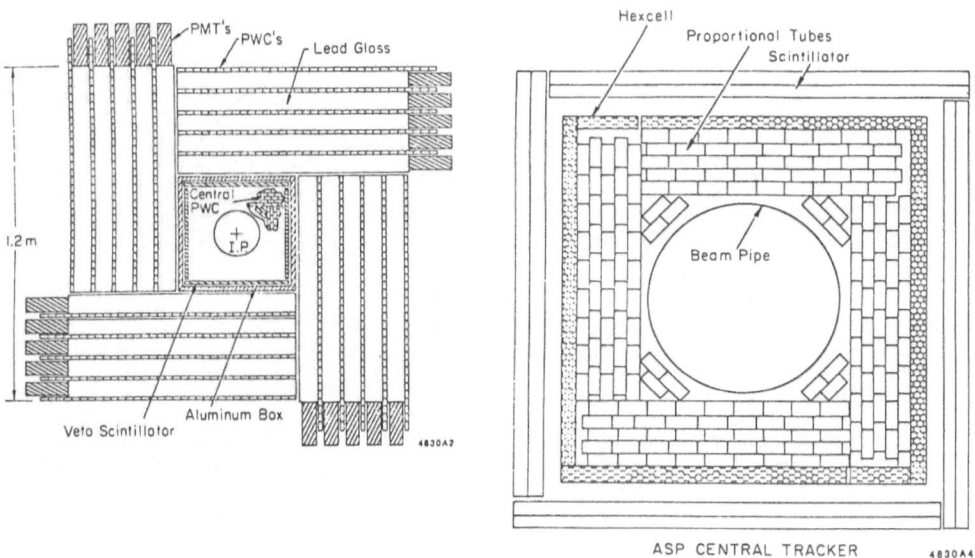

Fig. 5. Transverse view of the ASP detector.

it solves the rejection problem of the eeγ background. Figure 5 shows the
details of the lead-glass array which can detect photons down to 20° and
veto down to 10°. The lead glass has been chosen for its good intrinsic resolution and favourable signal-to-noise ratio. Inside this photon
calorimeter, two systems are designed to reject events with accompanying
charged particles and to distinguish e⁻ and γ. The first of these systems is
a central tracker made of 0.4 in. × 0.4 in. × 88 in. Al tubes
read at each end (so the coordinate is determined by the charge conjuga-
tion); the second of these systems is made of 2 cm thick scintillator
and surrounds the central tracker. Figure 6 shows a typical eeγ event
used to provide a source of constrained single photons.

The main result is provided by the ASP experiment, which allows a much
more accurate measurement than the upgraded MAC experiment. It has been
measured by ASP that:

$$\sigma(ee \rightarrow \gamma + \text{new sources}) < 0.0062 \text{ pb}$$

$$\sigma(ee \rightarrow \gamma + \text{weakly interacting particles}) < 0.094 \text{ pb} ;$$

Fig. 6. Typical eeγ event in
the ASP detector.

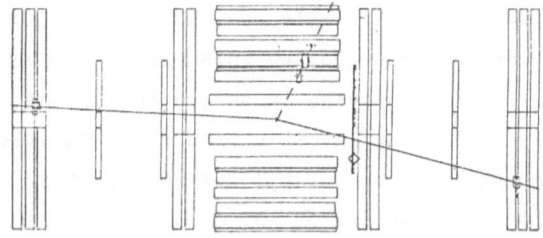

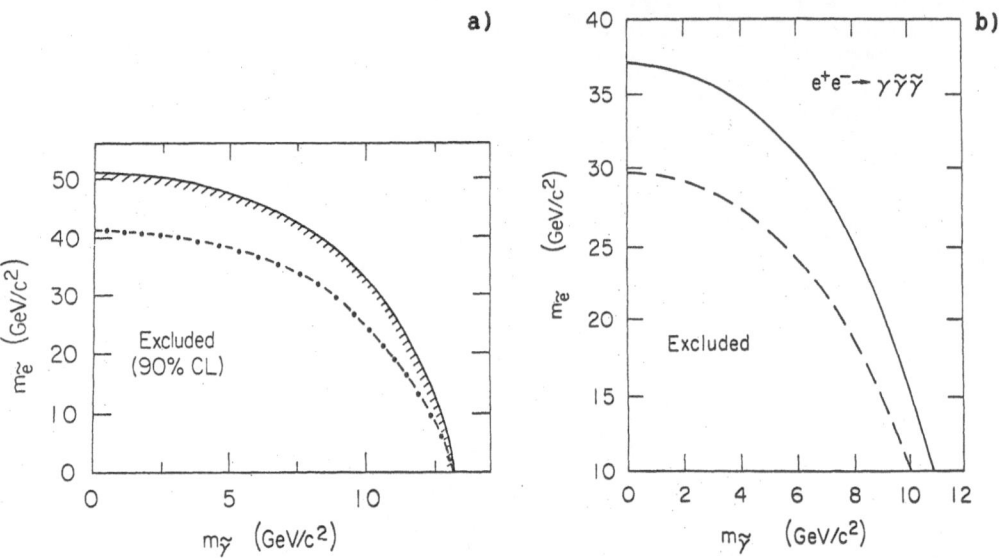

Fig. 7.   Scatter plot of $m_{\tilde{e}}$ versus $m_{\tilde{\gamma}}$:
a) for the ASP detector,
b) for the MAC detector.

and assuming $m_{\tilde{\gamma}} = 0$, this states an upper limit on $m_{\tilde{e}}$ of 50 GeV/c^2 (90%
CL). The corresponding scatter plot of $m_{\tilde{\gamma}}$ versus $m_{\tilde{e}}$ is reproduced in Fig. 7a
for ASP and Fig. 7b for MAC.

2.1.2 <u>Missing-energy measurement at the CERN p$\bar{\text{p}}$ Collider</u>

<u>(UA1 experiment)</u>. The first monojet plus missing-energy events
found in the run of 1983 have triggered a lot of enthusiasm[8]. They were
interpreted as a sign of new physics. They corresponded to a total
integrated luminosity of 130 nb^{-1}. Once the statistics have increased by
about a factor of 5 (UA1 has recorded a total of 715 nb^{-1}), the evidence for
'new phenomena' has dramatically decreased[9]. However, what has been
developed by UA1 in its pioneering work is a special technique. It is based
on the fact that many new phenomena are predicted to produce ν's or 'ν-like'
particles which escape recognition by the detector thus creating missing
energy. Provided the detector has a full angular coverage (4π) and is
hermetic enough (i.e. has reduced cracks and dead regions as much as
possible), it is able to measure quite accurately the energy missed in the
event. Let me describe this method in more detail.

i) Missing-energy technique in the UA1 detector.

The first important remark, when starting this discussion, concerns the
fact that in p$\bar{\text{p}}$ (or pp) colliders, it is not possible to measure properly
the total missing energy of an event, but only the transverse component of

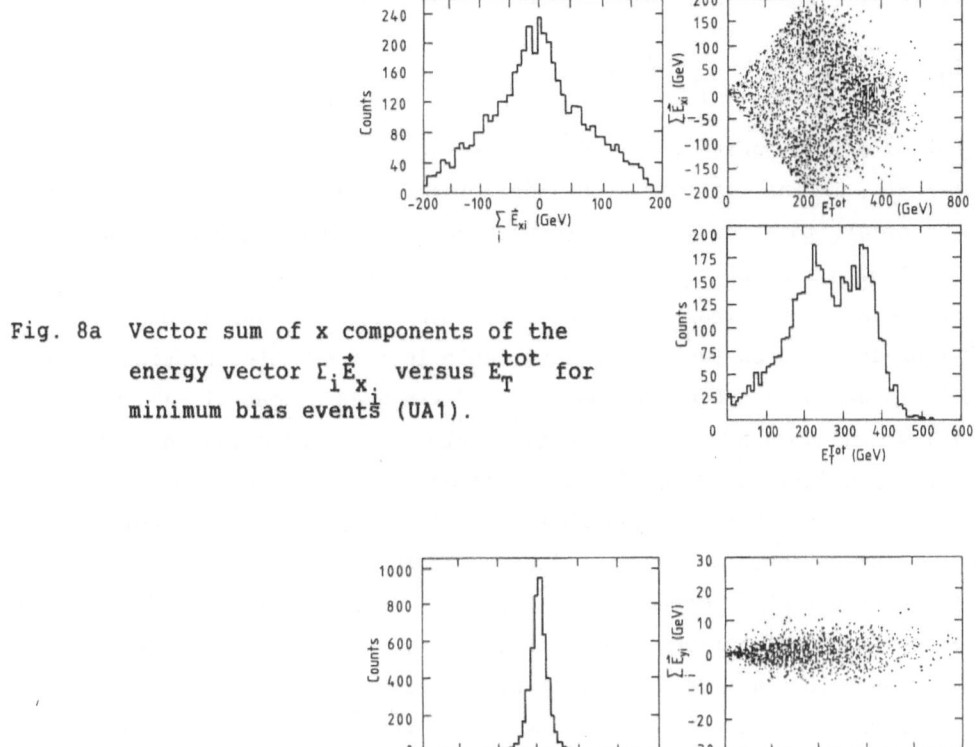

Fig. 8a Vector sum of x components of the energy vector $\Sigma_i \vec{E}_{x_i}$ versus E_T^{tot} for minimum bias events (UA1).

Fig. 8b Vector sum of the y components of the energy vector $\Sigma_i \vec{E}_{y_i}$ versus E_T^{tot} for minimum bias events (UA1).

the missing energy: E_T^{miss}. This is clear when looking at the plots of Fig. 8, where we note that owing to the beam jets produced in such collisions, a large fraction of the energy is lost in the beam pipe. So the component of the energy along the beam axis is very badly measured. We are restricted then to defining a transverse missing energy, by considering the component along the two axes in the plane perpendicular to the beam axis (i.e. components E_y and E_z). By projecting the energy of each calorimeter i-cell onto the transverse plane, we obtain the transverse energy vector $\vec{E}_{Ti}$ given by:

$$\vec{E}_{Ti} = \vec{E}_{yi} + \vec{E}_{zi} .$$

The corresponding total energy E_T^{tot} of the event is defined as the scalar sum of the transverse energy of each i-cell:

$$E_T^{tot} = \sum_i ||\vec{E}_{yi} + \vec{E}_{zi}|| ,$$

and the total transverse missing energy is given by

$$E_T^{miss} = -E_T^{tot} = - \sum_i ||\vec{E}_{Ti}|| \quad .$$

This method may be applied, *if* and *only if* the calorimeter is really 4π
and hermetic. In the case of the UA1 detector, 'good 4π coverage' is ensured
by the central, forward, and very forward calorimetry which covers the range
down to 2° with respect to the beam axis, plus a μ-detector (able to detect
μ's down to about 20° from the beam axis), plus a good handling of the
cracks and dead region (which are inevitable in any 4π-detector technology).
There is in UA1 a main dead region. It is defined as a cone of ±15° around the
vertical up and down axis in the central calorimeter (covering the region in
pseudorapidy η defined by $|\eta| \leq 1.5$). This dead zone is due to the
separation into two gondola hemispheres. Some energy may flow in this region
without being detected by the calorimeter. However, this leakage may be
detected by the central detector (CD) (which is able at least to see the
charged component of this possible energy flow). This information from the
CD may be used to tag a good fraction of the so-called vertical events. This
fact is taken into account to validate or not the total missing transverse
energy measured in an event.

Moreover an estimate of the resolution on E_T^{miss} has been made by
measuring the resolution on E_T^{tot} on a sample of minimum-bias data and of
high-E_T jet triggers. The result is plotted in Fig. 9; the resolution on
E_T^{miss} derived from this measurement is given by $\sigma = 0.7\sqrt{E_T^{tot}}$. This parameter is used to validate the transverse missing energy measured in an
event; a cut, which is a function of σ, is applied to E_T^{miss}. We will consider two types of cut, a loose cut which is equivalent to 2.5σ or 3σ and a

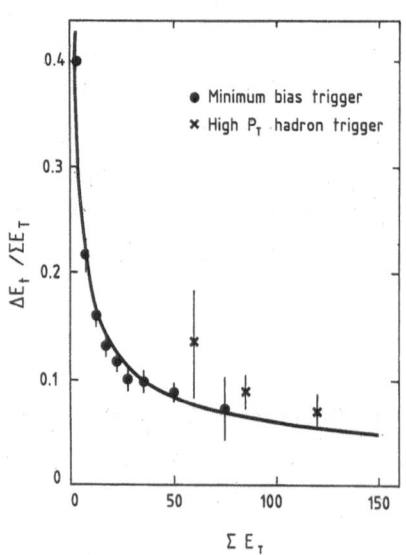

Fig. 9. Resolution on E_T^{tot} in the UA1
experiment for minimum bias
data and jet triggered events.

tight cut which corresponds to 4σ. Most of the results we are discussing later on refer to this tight cut.

Historically, the missing-energy technique has been first used in UA1 to search for W → ev events. The fact of being able to look both for the e⁻ and for the v in such event has been very powerful in identifying the W → ev signal (see L. DiLella's lectures[1]). This technique has then been extended to look for other processes which also produced v('s) or v-like particles. We are considering this case now.

ii) Definition of the missing-energy selection in UA1.

The various filters applied to select missing-energy events in UA1 are presented here. The first level of selection is, of course, given by the trigger.

a) Missing-energy trigger:

Until 1984, no such first-level trigger was set. From 1984 on, a missing-energy has been introduced both at the first level (i.e. in the hardware processor) and at the second level (i.e. using on-line 168E or 3081E emulators). It is done by requesting an imbalance in the total transverse energy between the left and right sides of the central calorimetry of at least 17 GeV, together with at least one jet; in the trigger processor a jet is defined as a minimal amount of 15 GeV of transverse energy deposited in eight adjacent e.m. cells and the two hadronic cells behind. The estimate of E_T^{miss} in this first-level trigger is biased by the left-right symmetry of the central e.m. calorimetry in UA1 (so-called gondolas). At the second-level trigger, the on-line emulators recompute E_T^{miss} by using the information provided by all the calorimeter cells in the apparatus; a cut on E_T^{miss} of 3.5σ (where $\sigma = 0.7\sqrt{E_T^{tot}}$) is applied on these data.

b) Off-line filter:

Apart from this on-line selection of missing-energy events, an off-line filter is applied on all the recorded data (i.e. the 10 millions of triggers which correspond to the total integrated luminosity of 715 nb⁻¹). It is this selection, which finally retains 1 % to 2 % of the data, which provides the complete sample of missing-energy data. It includes all the different types of triggers which may be set by the first-level trigger[10]. It also implies that all these data have been completely preprocessed, which means that the calorimetry information has been reconstructed (including all the constants of calibration, bookkeeping of dead channels, etc.). It retained, independently of the first-level trigger bit, each event which had a computed transverse missing energy of 15 GeV and validated at 2σ (where $\sigma = 0.7\sqrt{E_T^{tot}}$). It also rejects spurious events such as cosmic rays and beam halo, which fake missing-energy events. We show in Fig. 10 a diagram of the complete data selection from the on-line trigger to the off-line filter. The tight

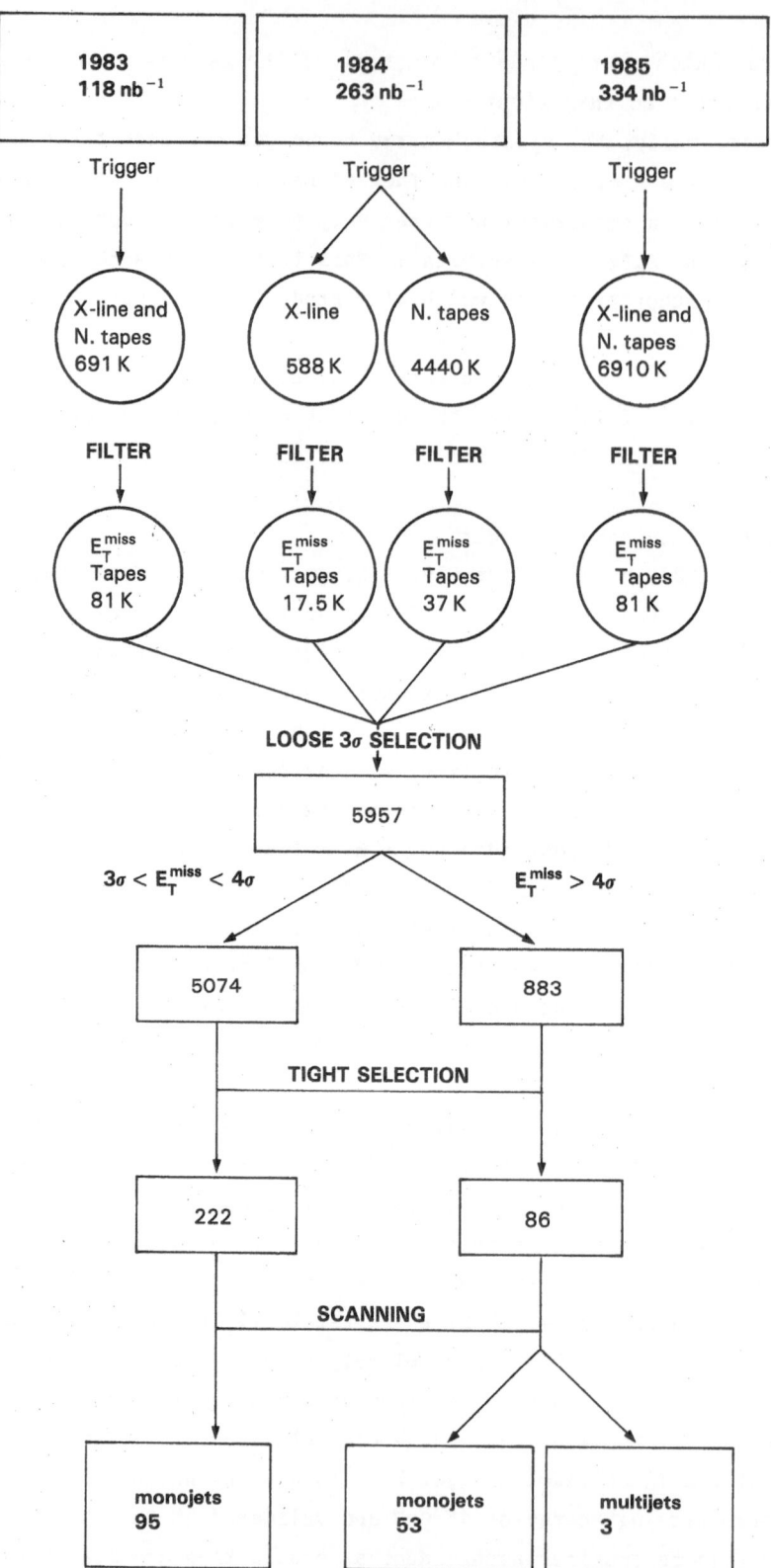

Fig. 10. Flow diagram of the data selection for missing energy events in
the UA1 experiment.

selection obtained by applying a 4σ cut on E_T^{miss} finally gives 56 events out of the initial 10 million triggers); the work on the loose filter selection is still in progress. We will discuss here the results provided by the 4σ filter.

iii) Physical interpretation of the 4σ sample; τ-signal.

When looking at the 56 events obtained in this sample, it is obvious that a clear monojet signal is seen for 53 out of the 56 events (see Fig. 11) and that few multijet events are found: only three bijet events which are not back to back (Fig. 12). These favoured jet topologies (monojets and bi-jets not back to back) are simply due to our peculiar filtering conditions, which forbid the activity back to the jet trigger in the event.

The distribution of the total missing transverse energy in these events (Fig. 13) as well as of the transverse energy E_T of the jet trigger (Fig. 14) show a concentration of events between 15 and 40 GeV. There are few events above 40 GeV. The distribution of the second highest E_T jet of the event is shown in Fig. 15. Most of these jets have E_T below 12 GeV; only three events (the bijet candidates) have the second jet above this value. The rates for such events, given by all possible known sources, have been estimated; they take into account both the UA1 detector limitations and any possible physical process. So there are two types of background: the processes from the Standard Model which may generate such signatures and the so-called QCD jet-jet fluctuations. This last process is the one corresponding to $p\bar{p} \rightarrow$ (q or g) 2-jet QCD events, where one of the two jets is mismeasured because of apparatus limitations and provides a faked missing energy. All these contributions to the background have been calculated using both real data and ISAJET Monte Carlo[11]; they are summarized in Table 1. The predominant background is due to the process $W \rightarrow \tau\nu_\tau$, where τ decays hadronically. So before trying to go further, i.e. to look for new signals, one has to find the rules to get rid of the τ-sample.

As in the case of the search for new signals in e^+e^- (subsection 2.1), we see here that it is mandatory to exclude the background due to standard physics and, in particular, that due to the τ-signal. Perhaps it is even trickier in the case of $p\bar{p}$ collisions. To the question of how to separate τ's from the rest, we will give an answer in two steps; the first step consists in defining a variable called 'τ-likelihood' able to separate τ's from non-τ events. The second step tries to make use of all the information provided by the detector and to refine the τ-signal. As the identification of τ's is an important issue at pp colliders, let us spend some time to describe it.

The 'τ-likelihood method', considering that no single cut provides adequate separation between τ's and standard QCD jets, tries to find a set

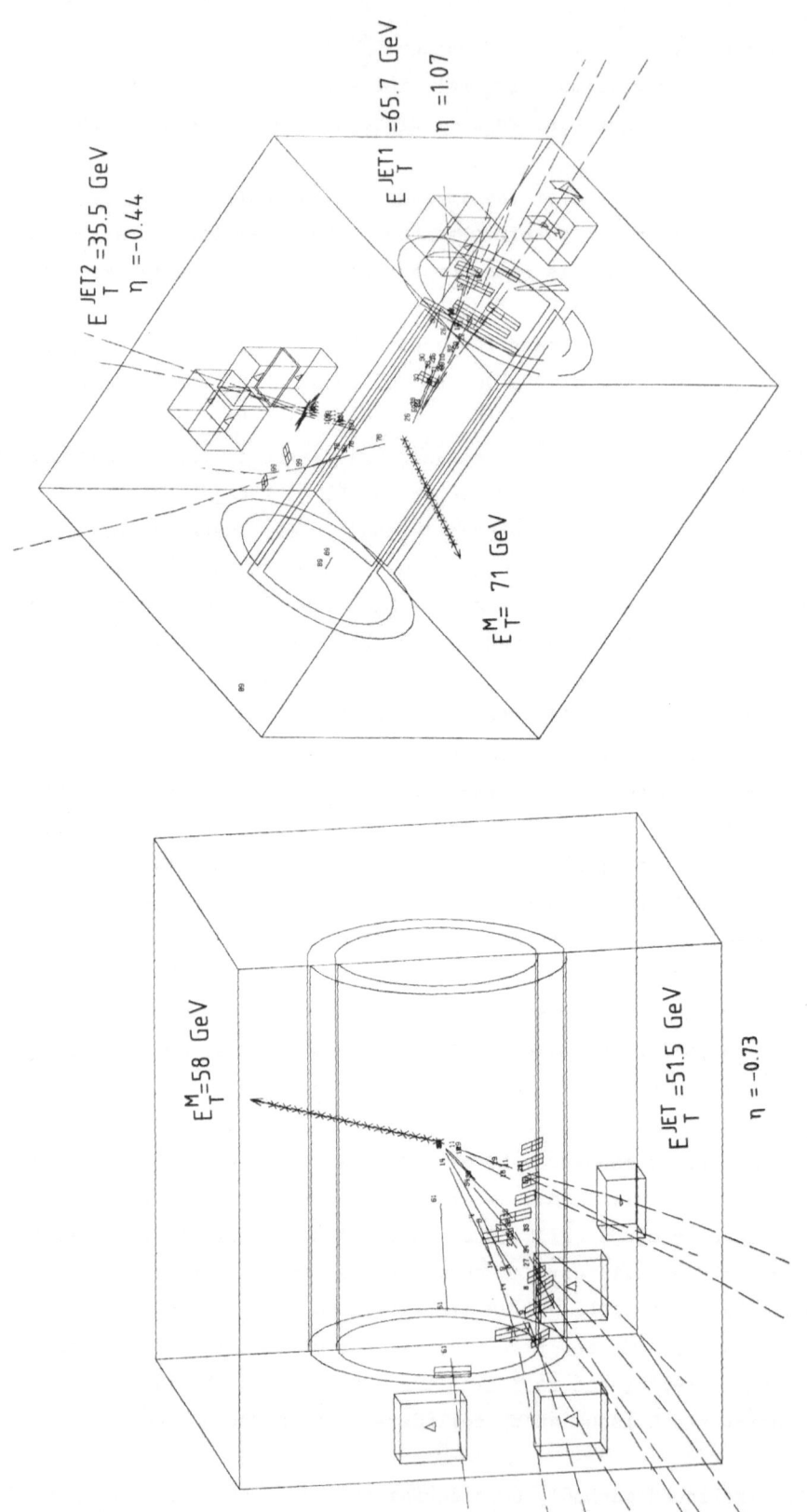

$E_T^{JET2} = 35.5$ GeV

$\eta = -0.44$

$E_T^{JET1} = 65.7$ GeV

$\eta = 1.07$

$E_T^M = 71$ GeV

Fig. 12. View of a typical bijet non back to back and missing energy event in the UA1 experiment (1985 data).

$E_T^M = 58$ GeV

$E_T^{JET} = 51.5$ GeV

$\eta = -0.73$

Fig. 11. View of a typical monojet plus missing energy event in the UA1 experiment (1985 data).

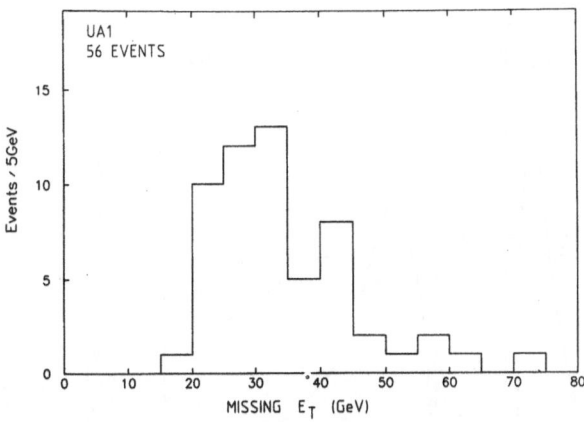

Fig. 13. E_T^{miss} distribution for the 56 events selected in the 4σ sample in the UA1 experiment.

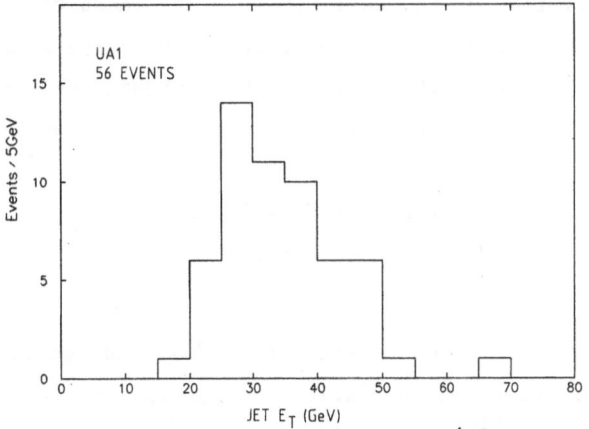

Fig. 14. Distribution of the transverse energy, E_T^{jet}, of the trigger jet for the 56 events selected in the 4σ sample in the UA1 experiment.

Fig. 15. Distribution of the transverse energy of the second E_T jet for the 56 events selected in the 4σ sample in the UA1 experiment.

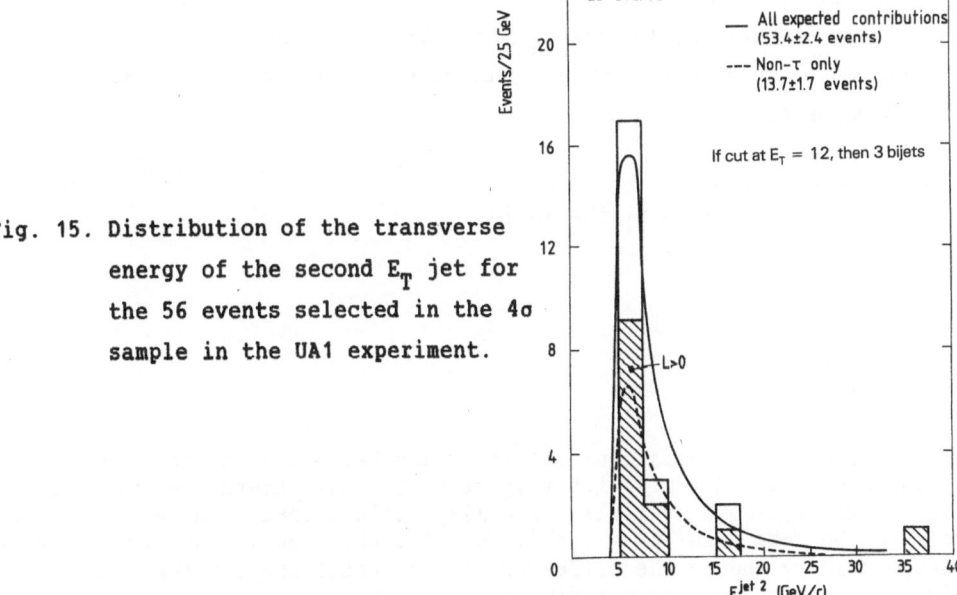

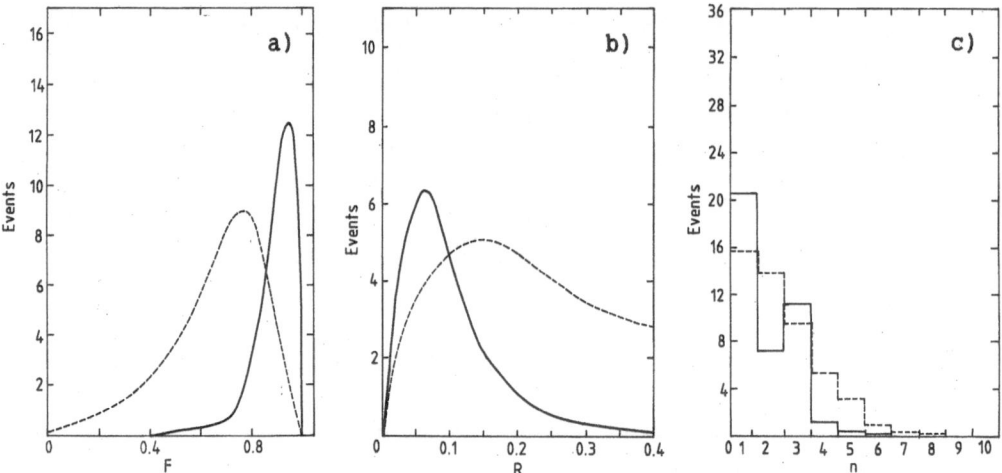

Fig. 16. a) Distribution of the variable F ('narrowness of the jet'), for generated τ's and QCD real jets in the UA1 experiment.
b) Distribution of the variable ΔR ('matching between the central tracking and the calorimeter jets') for generated τ's and QCD real jets in the UA1 experiment.
c) Distribution of the charged multiplicity n_{CD} for generated τ's and QCD real jets in the UA1 experiment.

of independent variables, each of which provides some separation between these two types of event. The variables used in this approach are: F, the 'narrowness' of the jet[*]; ΔR, the 'matching in space' between the highest-p_T track and the calorimeter jet axis; and n_{CD}, the track multiplicity defined here by counting all the tracks which have p_T bigger than 1 GeV/c and are contained in a cone in ΔR less than or equal to 0.4. The plots of these three parameters F, ΔR, and n_{CD}, as obtained in the case of τ's generated by Monte Carlo or QCD jets given by real jet data, are shown and compared in Fig. 16.

---

[*] The variable F is given by the ratio:

$$\frac{\sum_{\Delta R < 0.4} E_T^{cell}}{\sum_{\Delta R < 1} E_T^{cell}} \quad ,$$

where the numerator sums over all the cells belonging to the trigger jet inside a cone in $\Delta R = 0.4$ with respect to the calorimeter jet axis and the denominator sums over all the cells inside a cone in $\Delta R = 1$ with respect to the same axis: $\Delta R = (\Delta \eta^2 + \Delta \varphi^2)^{1/2}$, where $\Delta \eta$ is the difference in pseudorapidity and $\Delta \varphi$ the difference in azimuthal angle between the highest-p_T track and the calorimeter jet axis.

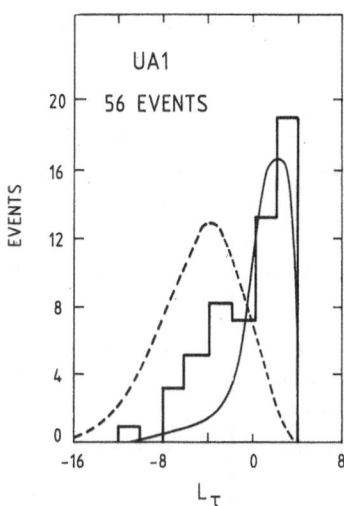

Fig. 17. τ log-likelihood distribution for generated τ's (solid line), QCD
real jets (broken line), and for the 56 events selected in the 4σ
sample in the UA1 experiment.

One then defines a τ log-likelihood variable L in the following way:

$$L = \ln (P_F \, P_{\Delta R} \, P_{n_{CD}}) \, ,$$

where the distributions of F, ΔR and n_{CD} as obtained from the τ-Monte Carlo
have been used to calculate the relative probabilities (P_F, $P_{\Delta R}$, $P_{n_{CD}}$) for
an event to fit the τ-hypothesis. The corresponding L-distributions for τ's as
generated by Monte Carlo and QCD jet data have been used to define a cut at
L = 0 (Fig. 17). If L is negative that means that the events are classified
as $W \rightarrow \tau\nu_\tau$ decay candidates, whereas if L is positive they are classified
as non-τ-candidates. The L-distribution given by the 56 events (Fig. 17),
leads to 32 events classified as τ-candidates and a sample of 24 events considered as non-τ-events.

However, one may make the following remarks concerning this classification.

a)   The 'contamination' of the τ-sample by non-τ-events and vice versa is
      non-negligible (see Fig. 17), as can be seen from the number reported
      in Table 1 which gives an estimate that 9.5 ± 1.0 events out of the 24
      so-called non-τ-events are in fact τ's.

b)   At the c.m. energy of the CERN p̄p Collider the QCD jets show essen-
      tially the same features as the τ's in what concerns the chosen vari-
      ables n_{CD}, ΔR, F (as already pointed out[12] and shown in Fig. 18).

c)   To study the process $W \rightarrow \tau\nu_\tau$, where τ decays into a single charged pion
      plus some possible additional π^0's, it is essential to ensure the
      hadronicity of the charged track, otherwise there is a non-negligible

Background process	Expected contribution (No. of events)	
	for all data (56 events)	for L < 0 data (24 events)
$W \rightarrow \tau \nu_\tau \rightarrow$ hadrons	39.7 ± 1.8	9.5 ± 1.0
All leptonic decays of W or Z^0 (except $W \rightarrow \tau \nu \rightarrow$ hadrons)	9.7 ± 1.3	7.5 ± 1.2
Heavy flavour (b, c) production from $p\bar{p} \rightarrow Q\bar{Q}$ and from W and Z^0 decays	0.2 ± 0.8	0.2 ± 0.8
QCD jet-jet fluctuations	3.8 ± 1.7	3.8 ± 1.7
Total	53.4 ± 2.4	21.0 ± 1.7

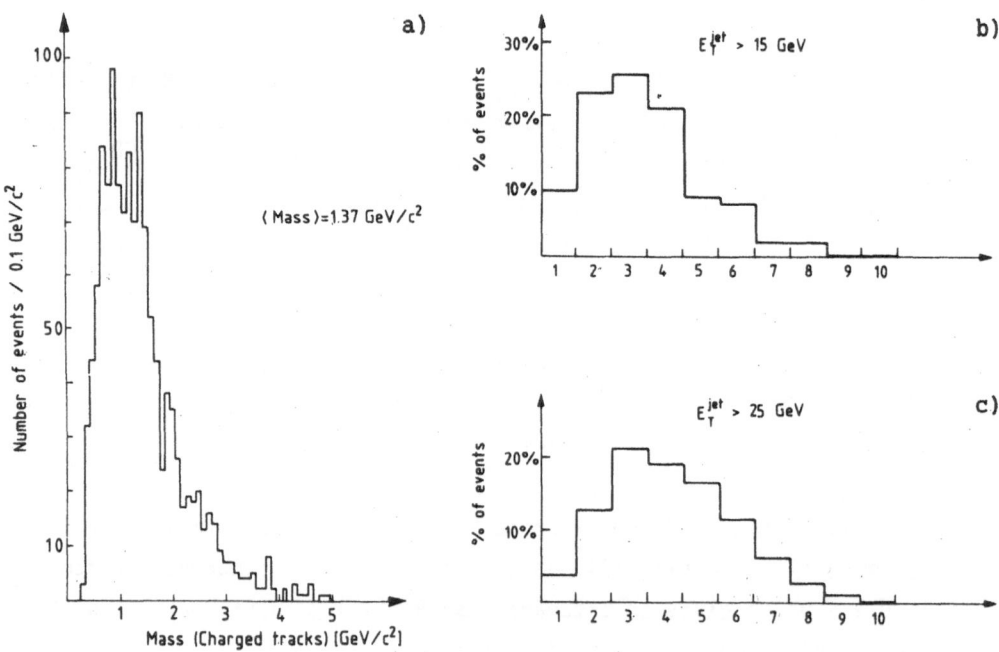

Fig. 18. a) Effective mass of charged tracks of an experimental jet, with E_T > 25 GeV passing the requirements of the τ-selection.
b) Multiplicity spectrum for jets which have E_T^{jet} > 15 GeV passing the requirements of the τ-selection.
c) Multiplicity spectrum for jets which have E_T^{jet} > 25 GeV passing the requirements of the τ-selection.

contamination of W → ev events (estimated to be 7.5 ± 1.2 events out of the 24 events.
This becomes even more important when relaxing the cut on E_T^{miss} (loose filter selection). These facts are taken into account in the second step.

The 'refined τ-selection' imposes stricter requirements on the narrowness of the jet and on the matching between the jets as defined in the calorimeter and in the central tracking device (CD). But the main point concerns the charged multiplicity.

Depending on the charged multiplicity (defined by counting all tracks with p_T larger than 0.5 and not 1 GeV/c) of the event two cases are considered in this approach.

The first case concerns the events where the jet trigger has a charged multiplicity equal to 1. If so, it is required, in addition, that the track should be well isolated both in terms of central tracking and calorimetry, that it should have a momentum larger than 2 GeV/c; this is correlated with a maximum transverse-energy deposit of 7 GeV in the corresponding e.m. cells and a minimum transverse-energy deposit of 15 GeV in the corresponding hadronic cells (by corresponding cells we mean the cells which are in a cone in ΔR < 0.4 with respect to the track). Moreover, strong requirements on the hadronicity of the track are demanded. This is a crucial issue as, contrary to the case of W → ev events, most of the $\tau^{\pm} \rightarrow \pi^{\pm}(n\pi^0\text{'s})v_\tau$ correspond to low p_T tracks (i.e. lower than 10 GeV/c) and E_T in e.m. cells lower than 15 GeV; so the e/π rejection becomes quite tricky. This problem is discussed in more detail in Appendix I.

The second case concerns the events where the track multiplicity is larger than 1. There, to ensure the τ-character of the events (namely the case $\tau \rightarrow \pi^+\pi^-\pi^+ + n\pi^0\text{'s} + v_\tau$: τ_3), a set of requirements are imposed. The multiplicity equal to 3 is not a stringent enough condition; as shown in Fig. 19, at this c.m. energy most of the QCD jets have an average multiplicity of 3 also. So a requirement on the *fragmentation properties* of the jet is defined[13] (following a study reported in Appendix II), which allows τ_3 to be well identified from QCD. Half of the real τ_3 events pass these cuts and only 3% of the QCD jets (including heavy flavour). Owing to the heavy contamination of τ_3 events by standard QCD, such strict cuts are requested. We show in Fig. 20 a typical τ_3 candidate as found in the UA1 detector.

In conclusion, W → τv_τ events may be detected by the UA1 experiment[12,14]. This however requires very accurate handling of the information and the corresponding systematics given by the e.m. and hadronic calorimeter, as well as the central detector which is a major tool also in this game.

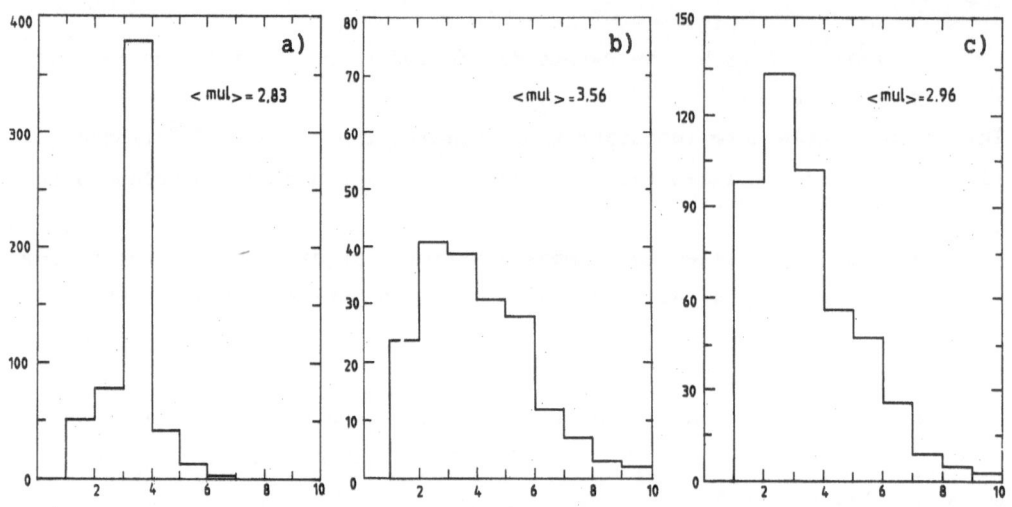

Fig. 19. Multiplicity distributions for

    a) $\tau \rightarrow 3\pi^{\pm} + n\pi^{0} + \nu_{\tau}$, as generated by ISAJET MC.

    b) $p\bar{p} \rightarrow b\bar{b}$ or $c\bar{c}$, as generated by ISAJET MC.

    c) $p\bar{p} \rightarrow g$, u, d, s, real data events.

$W \rightarrow \tau\nu$ (3 PRONG)

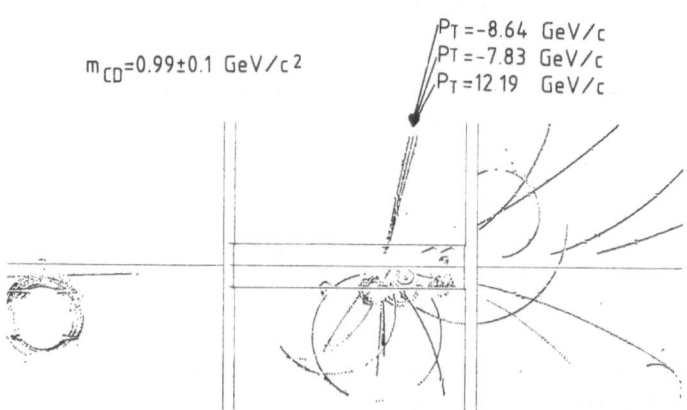

Fig. 20. ($\tau \rightarrow 3\pi^{\pm} + n\pi^{0} + \nu_{\tau}$) candidate in UA1 data.

This process is the dominant standard process which contributes to the missing-energy sample. Moreover the contribution of all the standard backgrounds is estimated to be 53.4 ± 2.4, corresponding to the 56 events selected out of all the data recorded fom 1983 to 1985 in the UA1 experiment. Not much room is left for anything other than a Standard Model explanation for these events.

iv) Supersymmetry at $p\bar{p}$ colliders[15].

One of the main reasons for studying events with missing energy and one or several jets is the search for supersymmetric particles. They are produced mainly through the following three reactions:

$$pp \to \tilde{g}\tilde{g} \qquad\qquad\qquad\qquad\qquad (5)$$

$$pp \to \tilde{g}\tilde{q} \qquad\qquad\qquad\qquad\qquad (6)$$

$$pp \to \tilde{q}\tilde{q} \qquad\qquad\qquad\qquad\qquad (7)$$

The mass range accessible at the CERN $p\bar{p}$ Collider or at the Tevatron is th e W-mass range, namely a mass from a few GeV/c^2 to 100 GeV/c^2. In this case, the minimal supergravity model (SUGRA), provides a complete framework within which the supersymmetric particles are given by simple relations such as:

$$m_{\tilde{q}} = m_{3/2}(1 + 1.67\xi^2)^{1/2}$$

$$m_{\tilde{g}} = 2.88\xi m_{3/2}$$

$$m_{\tilde{\gamma}} = 0.4\xi m_{3/2} \ ,$$

where the quantity $\xi m_{3/2}$ is equal to $m_{\tilde{\gamma}}$ (mass of the photino) and to $m_{\tilde{g}}$ (mass of the gluino), measured at the Planck mass, and the parameter $m_{3/2}$ (mass of the gravitino) is equal to $m_{\tilde{q}}$ (mass of the squark) measured at the Planck mass too. The factors $1.67\xi^2$, 2.88, and 0.4 measure the renormalization effects. Moreover, these models provide two alternatives[16]; the first one assumes that the gluino mass is larger than the scalar quark mass. This is realized if, and only if, ξ is bigger than 0.8 but, in addition, the mass of the gluino must always be smaller than 1.1 times the mass of the scalar quark. Therefore, in this framework, the masses of the gluino and of the scalar quark are almost degenerate. As a consequence, the gluino decays into an antiquark plus a scalar quark, and the scalar quark decays into a quark and a photino; so the processes $p\bar{p}$ (or pp) going to $\tilde{g}\tilde{g}$ or $\tilde{g}\tilde{q}$ or $\tilde{q}\tilde{q}$ have almost the same signature. The corresponding rates have been computed[16]. This first alternative of the SUGRA model is particularly favourable for the low-energy case (i.e. masses less than or equal to 100 GeV/c^2). The second alternative assumes the squark mass to be bigger than the gluino mass; in this case, the discrepancy between both masses can be quite large[16]. The most probable decay mode for the squark is then that the scalar quark decays into the corresponding quark and a gluino. The gluino decays into quark plus antiquark and photino. The photino is assumed to be the lightest supersymmetric particle and therefore is stable. The main consequences of this second picture is that the signatures of the processes (5), (6), and (7) are rather different and that the rates are different too.

Here different supersymmetric scenarios, defined within this framework, have been applied to the UA1 detector. Therefore, events corresponding to the supersymmetric processes (5), (6), and (7) have been generated by the Monte Carlo ISAJET. These events are then simulated through the whole UA1 detector. They are finally analysed in the same way as the missing-energy sample of real data, using, in particular, the same filters.

The various cases considered in this study are summarized in Table 2. The rates corresponding to these different cases, and the effect of the different filters applied, are also listed in Table 2. Table 2a concerns the case where the $\tilde{g}$ and $\tilde{q}$ masses are non-degenerate. In Table 2b we report a case where the $\tilde{g}$ and $\tilde{q}$ masses are very much different (strongly non-degenerate), and in addition the gluino mass is very small: $m_{\tilde{g}}$ = 10 GeV/c^2. In this special case, we have looked at both the dominant decay mode of the $\tilde{q}$ in $q\tilde{g}$ (93% of branching ratio) and the case where $\tilde{q} \rightarrow q\tilde{\gamma}$ (which represents about 7% of the branching ratio). Finally, in Table 2c, we deal with the scenario where $\tilde{g}$ and $\tilde{q}$ are degenerate in mass; in this latter case, the process $\tilde{g}\tilde{q}$ is higher, by a factor of about 5 to 10, than the other two.

Typical supersymmetric events simulated in the UA1 detector are shown in Figs. 21 and 22. The different jet topologies obtained in each scenario, for the various filters, are summarized in Table 3. This table compares, for the case of the process $p\bar{p} \rightarrow \tilde{g}\tilde{q}$, the effect of no filter (i.e. trigger only), and of the 4σ inclusive filter (i.e. where a cut of 4σ is applied on E_T^{miss}), on the jet-structure of the events. Table 3a summarizes the percentage of events which belong to each category, in the case where the $\tilde{g}$ and the $\tilde{q}$ masses are non-degenerate. We note that the multijet structure is then favoured, especially when the masses increase (see for instance the case: $m_{\tilde{g}}$ = 50 GeV/c^2 and $m_{\tilde{q}}$ = 100 GeV/c^2). If the two masses are strongly non-degenerate and if, in addition, the gluino mass is very low (see Table 3b), the bijets back-to-back are the dominant topology, followed closely by $\geq$ 3-jet events. When the cut at 4σ in E_T^{miss} is applied, the back-to-back topologies are dramatically decreased, by about 50%. In the case where $\tilde{q} \rightarrow q\tilde{\gamma}$, the non-back-to-back structures are the most important ones, and remain also when the cut on the missing energy is applied. When the two SUSY particles are degenerate in mass (see Table 3c), the main jet-structures are the non-back-to-back ones (i.e. the monojets and the bijets not back to back); this effect is even enhanced when applying the cut in missing energy. So for the same process, depending on the mass and the decay scenarios, jet topologies may change a lot (and also the effect of the cut on E_T^{miss}).

Table 2a: Non-Degenerate Mass Scenarios, Cross-Sections, Rates, and Effect of the Various Filters

Mass (GeV/c^2)	Process Decay mode	Reaction	σ* (mb)	Expected rates* (No. of events)	Trigger	4σ inclusive	2.5σ non-b/b	4σ non-b/b
$m_{\tilde{g}}$ = 25	$\tilde{g} \to q\bar{q}\tilde{\gamma}$	$p\bar{p} \to \tilde{g}\tilde{g}$	0.40×10^{-4}	28,600	33	3.5	2	0.4
$m_{\tilde{q}}$ = 50	$\tilde{q} \to q\tilde{g}$	$p\bar{p} \to \tilde{g}\tilde{q}$	0.30×10^{-4}	21,450	55	3.9	0.5	0.25
		$p\bar{p} \to \tilde{q}\tilde{q}$	0.70×10^{-6}	500	78	3.9	0.5	0.13
$m_{\tilde{g}}$ = 40	$\tilde{g} \to q\bar{q}\tilde{\gamma}$	$p\bar{p} \to \tilde{g}\tilde{g}$	0.24×10^{-5}	1,716	63	9.3	1.4	0.5
$m_{\tilde{q}}$ = 60	$\tilde{q} \to q\tilde{g}$	$p\bar{p} \to \tilde{g}\tilde{q}$	0.55×10^{-5}	3,932	76	9.8	1.4	0.4
		$p\bar{p} \to \tilde{q}\tilde{q}$	0.24×10^{-6}	171	88	8.6	-	-
$m_{\tilde{g}}$ = 50	$\tilde{g} \to q\bar{q}\tilde{\gamma}$	$p\bar{p} \to \tilde{g}\tilde{g}$	0.40×10^{-6}	286	85	18.3	2.9	2
$m_{\tilde{q}}$ = 100	$\tilde{q} \to q\tilde{g}$	$p\bar{p} \to \tilde{g}\tilde{q}$	0.27×10^{-6}	193	97	17	0.27	0.27
		$p\bar{p} \to \tilde{q}\tilde{q}$	0.52×10^{-8}	3.7	100	19	-	-

* The cross-sections are computed with ISAJET 5.20; the quoted rates are the total expected rates for the total integrated luminosity of 7.15 nb^{-1}; the effect of the various filters is not included in these rates.

Table 2b: Strongly Non-Degenerate Mass Scenario, Cross-Section, Rates and Effect of the Various Filters

Mass (GeV/c²)	Process		σ* (mb)	Expected rates* (No. of events)	Percentage of events passing each filter			
	Decay mode	Reaction			Trigger	4σ inclusive	2.5σ non-b/b	4σ non-b/b
$m_{\tilde{g}} = 10$ $m_{\tilde{q}} = 100$	$\tilde{g} \to q\bar{q}\tilde{\gamma}$ $\tilde{q} \to q\tilde{g}$	$\bar{p}p \to \tilde{g}\tilde{q}$	0.74×10^{-5} (× 0.93)	4,920	89	7.8	0	0
$m_{\tilde{g}} = 10$ $m_{\tilde{q}} = 100$	$\tilde{g} \to q\bar{q}\tilde{\gamma}$ $\tilde{q} \to q\tilde{\gamma}$	$\bar{p}p \to \tilde{g}\tilde{q}$	0.74×10^{-5} (× 0.07)	370	86	64	19.9	17.5

* The cross-sections are computed with ISAJET 5.20; the quoted rates are the total expected rates for the total integrated luminosity of 7.15 nb^{-1}; the effect of the various filters is not included in these rates

Table 2c: Degenerate Mass Scenarios, Cross-Sections, Rates, and Effect of the Various Filters

| Process | | σ^* | Expected rates* | Percentage of events passing each filter | | | |
Mass (GeV/c^2)	Decay mode	(mb)	(No. of events)	Trigger	4σ inclusive	2.5σ non-b/b	4σ non-b/b
$m_{\tilde{g}} = 21$ $m_{\tilde{q}} = 19$	$\tilde{g} \to q\bar{q}$ $\tilde{q} \to \tilde{g}\gamma$	0.04×10^{-3}	286,000	26	3	6.5	2.5
$m_{\tilde{g}} = 42$ $m_{\tilde{q}} = 40$	$\tilde{g} \to q\bar{q}$ $\tilde{q} \to \tilde{g}\gamma$	0.15×10^{-4}	10,725	62	24	10.2	7.1
$m_{\tilde{g}} = 63$ $m_{\tilde{q}} = 57$	$\tilde{g} \to q\bar{q}$ $\tilde{q} \to \tilde{g}\gamma$	0.14×10^{-5}	1,001	83	47	11	7
$m_{\tilde{g}} = 84$ $m_{\tilde{q}} = 80$	$\tilde{g} \to q\bar{q}$ $\tilde{q} \to \tilde{g}\gamma$	0.11×10^{-6}	78.6	95	64	12	10.6

* The cross-sections are computed with ISAJET 5.20. We only consider here the case $\bar{p}p \to \tilde{g}\tilde{q}$ which is the dominant reaction; the other two have a lower cross-section by a factor of 5 to 10, but pass the filters in the same way as the process $\tilde{g}\tilde{q}$. The expected rates do not include the effect of these filters.

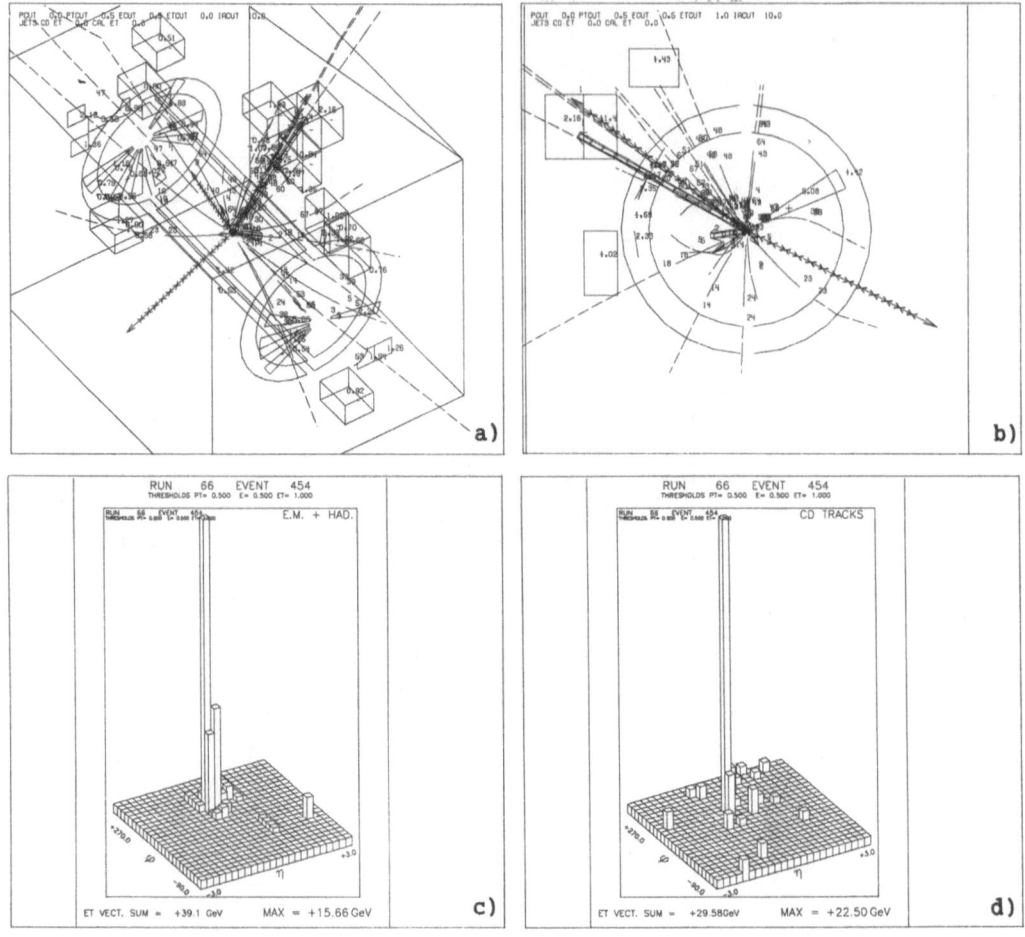

Fig. 21. Monojet plus missing-energy event, generated by pp → g̃g̃ (ISAJET MC) with m_g̃ = 42 GeV/c², m_q̃ = 40 GeV/c², simulated in the UA1 detector: E_T^{miss} = 43 GeV; E_T^{jet} = 40.9 GeV; p_T^{jet} = 28.7 GeV/c. a) Full detector view; b) Transverse view; c) Lego plot from calorimetry; d) Lego plot from the CD.

By comparing the rates obtained for the different jet topologies with the results of UA1 on the 4σ-E_T^{miss} selection, we may exclude the scenario corresponding to $m_{\tilde{g}}$ = 25 GeV/c² and $m_{\tilde{q}}$ = 50 GeV/c² and that corresponding to $m_{\tilde{g}}$ = 42 GeV/c² and $m_{\tilde{q}}$ = 40 GeV/c². The scenario where $m_{\tilde{g}}$ = 84 GeV/c² and $m_{\tilde{q}}$ = 80 GeV/c² and the one with $m_{\tilde{g}}$ = 50 GeV/c² and $m_{\tilde{q}}$ = 100 GeV/c² cannot be excluded.

The case where $m_{\tilde{g}}$ = 10 GeV/c² and $m_{\tilde{q}}$ = 100 GeV/c² is also rejected by present UA1 data. Moreover, if we consider the ($m_{\tilde{g}}$, $m_{\tilde{q}}$) plot we may define a certain number of zones; the first one is defined by $m_{\tilde{g}}$ > 1.1$m_{\tilde{q}}$ and is excluded by the supergravity models. The second area is that defined in the

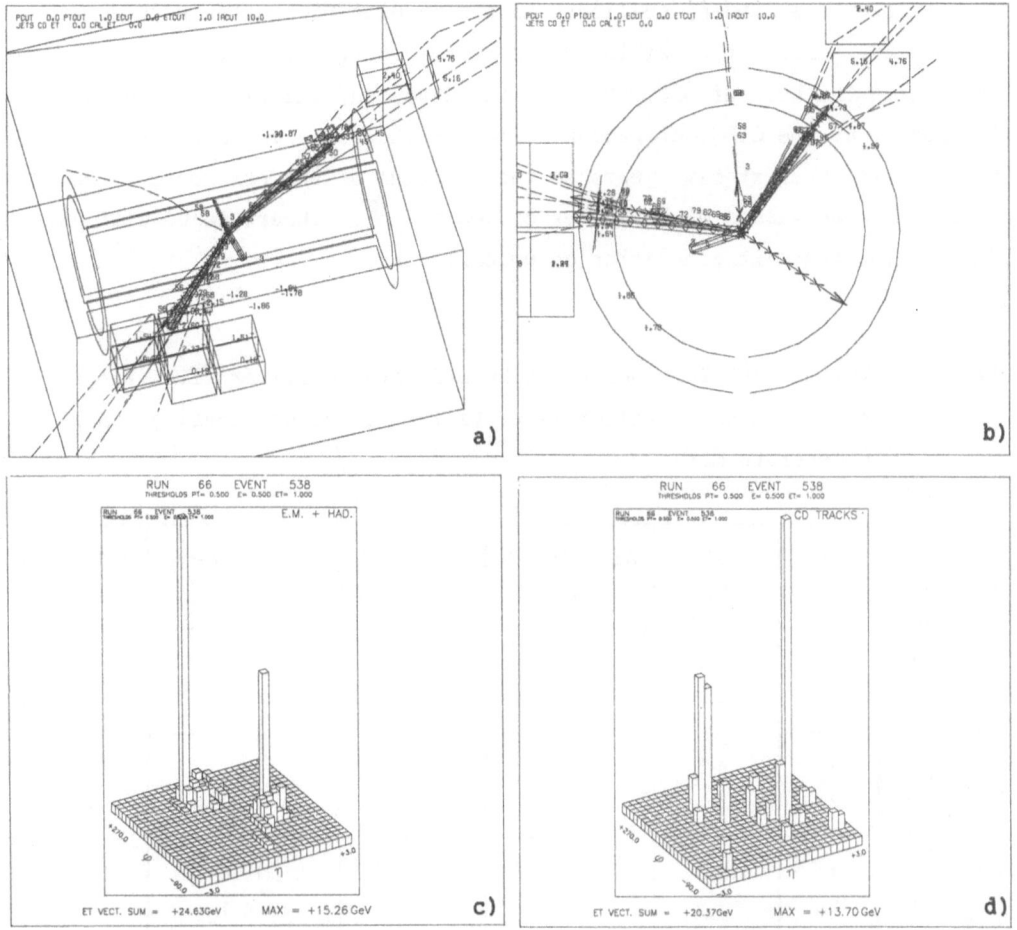

Fig. 22. Non-back-to-back bijet plus missing-energy event generated by $p\bar{p} \rightarrow \tilde{g}\tilde{g}$ (ISAJET MC) with $m_{\tilde{g}} = 42$ GeV/c^2; $m_{\tilde{q}} = 40$ GeV/c^2, simulated in the UA1 detector: $E_T^{miss} = 20.9$ GeV; $E_T^{jet1} = 28.5$ GeV; $p_T^{jet1} = 13.7$ GeV/c; $E_T^{jet2} = 20.8$ GeV; $p_T^{jet2} = 19.2$ GeV/c. a) Full detector view; b) Transverse view; c) Lego plot from calorimetry; d) Lego plot from the CD.

range $m_{\tilde{q}} < m_{\tilde{g}} < 1.1 m_{\tilde{q}}$ (i.e. the degenerate-mass case); in this scenario, with the present UA1 data, we may set an upper limit on both masses around 60 GeV/c^2 (this is using only monojet events; bijet events are compatible with such a limit, but the error made on this estimation is still worse than in the monojet case because of the lack of statistics). In the remaining area, it is less easy to set a definite limit on both masses. Therefore, we may conclude that the present experimental $p\bar{p}$ results allow us to exclude some supersymmetric scenarios, within certain limited regions of the ($m_{\tilde{g}}$, $m_{\tilde{q}}$) plot; these limits are strongly model-dependent, i.e. mass and decay-mode dependent.

Present e^+e^- and $p\bar{p}$ colliders have both done pioneering work concerning the search of possible supersymmetric particles. It is quite sure that, with the help of SLC, LEP, and the Tevatron, and with an increased luminosity at the CERN antiproton collector (ACOL), a definite answer will be given in the next five years to the question of the existence of such particles with masses of the order of 100 GeV/c^2. Otherwise we will have to wait for the next generation of pp colliders.

Table 3a: Various Jet Topologies for Events with no Filter (i.e. Filter 1) or 4σ Inclusive Filter (i.e. Filter 2) and Considering a Non-Degenerate Mass

Mass (GeV)	Filter No.	Monojets (%)	Monojets b/b (%)	Bijets b/b (%)	Bijets non-b/b (%)	$\geq$ 3 jets
$m_{\tilde{g}}$ = 25	1	2.8	7.6	22	20	4.5
$m_{\tilde{q}}$ = 50	2	6.2	25	9.3	9.3	3.1
$m_{\tilde{g}}$ = 40	1	4.5	8.6	22	21	23
$m_{\tilde{q}}$ = 60	2	19	4	13.5	10.8	4
$m_{\tilde{g}}$ = 50	1	1.1	-	1.6	15.8	8.9
$m_{\tilde{q}}$ = 100	2	3.1	3.9	8.7	4.7	20.5

Note that the remaining percentages are rejects and that only the process $p\bar{p} \rightarrow \tilde{g}\tilde{q}$ is considered here

Table 3b: Various Jet Topologies for Events with no Filter (i.e. Filter 1) or 4σ Inclusive Filter (i.e. Filter 2) Applied and Considering Strongly Degenerate Mass and a Small Gluino Mass ($m_{\tilde{g}}$ = 10 GeV; $m_{\tilde{q}}$ = 100 GeV)

Decay Mode	Filter No.	Monojets (%)	Monojets b/b (%)	Bijets b/b (%)	Bijets non-b/b (%)	$\geq$ 3 jets
$\tilde{q} \rightarrow q\tilde{g}$	1	2.6	10.2	39	11	25.5
	2	4.6	15.6	9.3	10.9	12.5
$\tilde{q} \rightarrow q\tilde{g}$	1	27.6	14.5	8.6	25	8.5
	2	33	14	5	26.6	8

Note that the remaining percentages are rejects and that only the process $p\bar{p} \rightarrow \tilde{g}\tilde{q}$ is considered here

134

Table 3c: Various Jet Topologies for Events with no Filter (i.e. Filter 1), or 4σ Inclusive Filter (i.e. Filter 2) Applied, and Considering Degenerate Mass

Mass (GeV)	Filter No.	Monojets (%)	Monojets b/b (%)	Bijets b/b (%)	Bijets non-b/b (%)	$\geq$ 3 jets
$m_{\tilde{g}} = 21$	1	15	10	18.5	8.5	2
$m_{\tilde{q}} = 19$	2	50	–	–	–	–
$m_{\tilde{g}} = 42$	1	14.5	9	12.5	19	6.5
$m_{\tilde{g}} = 40$	2	40.1	10.7	3.2	27.8	3.2
$m_{\tilde{g}} = 63$	1	11.5	9	19	3.7	13.6
$m_{\tilde{q}} = 57$	2	14.8	8.5	9.6	30.8	25.5
$m_{\tilde{g}} = 84$	1	11.8	7.6	14	36	7.8
$m_{\tilde{q}} = 80$	2	13.5	6.3	3.6	45	9.4

Note that the remaining percentages are rejects and that only the process $p\bar{p} \rightarrow \tilde{g}\tilde{q}$ is considered here

2.2 Antigravity and Supergravity

The fifth force, even if it is only a crazy idea, gives an attractive example of how the concept of antigravity may link classical mechanics (Newton and Galileo) to elementary particle physics (K^0 experiments) and to sophisticated theories such as supergravity at $N \geq 2$ dimensions. I have chosen this case mainly for this reason and I would like to warn the reader to take all that is reviewed here with a lot of care (including, of course, Newton's law of universal gravitation!).

2.2.1 Reminder of some fundamental notions of classical mechanics[17]. Newton's law of universal gravitation says that 'The gravitational inter-actions between two bodies can be expressed by an attractive central force, proportional to the masses of the bodies and inversely proportional to the square of the distance between them'; in other words this law can be expressed by the relation:

$$F = G \frac{mm'}{r^2} ,$$

where the proportionality constant G must be determined experimentally. Measuring the force F between two known masses, m and m', at a known distance r, yields the value of $G = 6.67 \times 10^{-11}$ N·m^2·kg^{-2} or

$$G = \frac{hc}{m_p^2},$$

where m_p is the Planck mass and is equal to $1.221045(46) \times 10^{19}$ GeV/c^2.
The experiment of Cavendish measured this according to the schema of Fig. 23.

Moreover, the principle of equivalence, which expresses the fact that
the inertial and gravitational masses are the same for all bodies, gives
rise to an important result: 'All bodies at the same place, in a gravitational field, experience the same acceleration'[*].

One example is that all bodies fall to Earth with the same acceleration.
To verify the above statement, we note that in a place where the gravitational field is G, the force on a body of mass m is:

$$F = mG$$

and the body acceleration is $a = F/m = G$, which is independent of the mass m
of the body subject to the action of the gravitational field. Because of
the principle of equivalence, the laws of Nature must be written in such a
way that it is impossible to distinguish between a uniform gravitational
field and an accelerated frame of reference. This is a statement which constitutes the basis of the 'general principle of relativity' proposed by
Einstein in 1915.

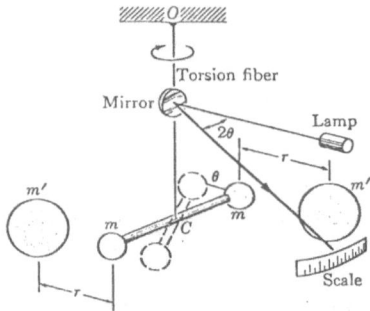

Fig. 23. Schema of the Cavendish measurement. When masses m' are placed
close to masses m, their gravitational attraction produces on the
horizontal rod a torque that results in a torsion of the fibre
(OC). Equilibrium is established when the gravitational and
torsional torques are equal. The torsional torque is proportional
to the angle θ, which is measured by the deflection of a ray
reflected from a mirror attached to the fibre. By repeating the
experiment at several distances r, and using different masses m and
m', we can verify the law: $F = G(mm'/r^2)$.

---

[*] Remember this principle when discussing in particular the K^0-$\bar{K}^0$ system.

2.2.2 <u>How experimental tests of gravity lead to antigravity.</u> A set of experiments have measured the constant G, both by redoing the experiment of Cavendish (G_{Cav}) or by measuring G within a mine (geophysical experiments: G_{mine})[18]. The results obtained by each experiment is reported in Fig. 24; a discrepancy is seen between G_{Cav} and G_{mine}:

$$G_{Cav} < G_{mine} \; .$$

Why? A way to explain this is by adding to the potential of gravity, $V_0 = V_{gravity} = -g(m_1 m_2/r)$, another potential ΔV, which is a Yukawa potential:

$$\Delta V \propto \frac{1}{r} \; e^{-\mu r} \; ,$$

such that in fact $V = V_0 + \Delta V = -g(m_1 m_2/r)(1 + \alpha \, e^{-\mu r})$; so:

$$G_{Cav} = g(1 + \alpha) \quad \text{and} \quad G_{mine} = g \; .$$

As $G_{Cav} < G_{mine}$, this implies that α must be negative so the additional force corresponding to ΔV must be *repulsive* for like objects; this means it is a *vectorial*-type force.

So, in addition to gravitational forces, there exists an antigravitational force also called the fifth force.

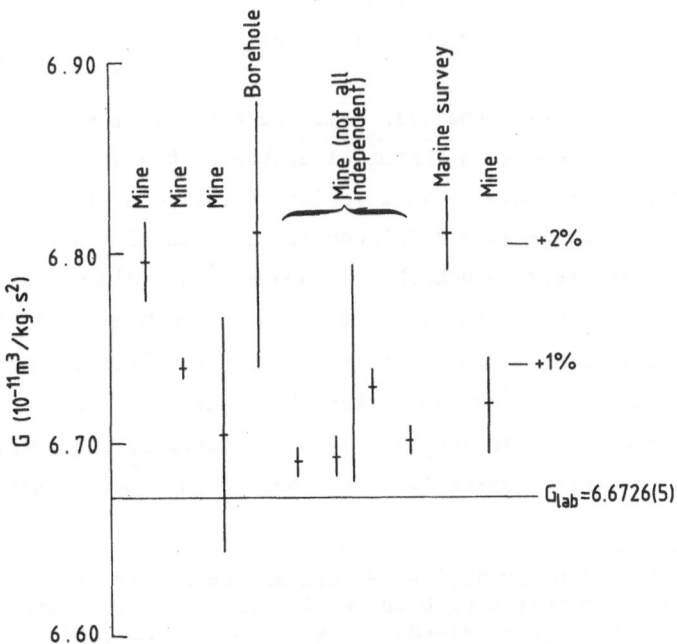

Fig. 24. Modern geophysical measurements of G.

2.2.3 <u>How can the fifth force be related to particle physics?</u> If this extra force exists, it is mediated by a *light vector boson* that we call *graviphoton* and represent by the symbol λ [*].

To take into account this extra force in particle physics, there are two ways. One is conventional, the second one is fancy. I will discuss the second one first, as it offers a way to *relate antigravity* to extended supergravity theories. Certainly the ideas summarized here are still controversial from the theoretical point of view. (Extended supergravity theories are far from being in their definite form and still have no realistic formulation; it is not the purpose of these lectures to discuss the real validity of such an approach.) However from a didactic point of view, they can be used to show how, ultimately, the high-energy physicist should be able to relate all forces of Nature within a global and unique description. Note that what I report below is *directly* derived from the inspired lectures[19] given by Scherk in 1979. [At that time the fifth force or antigravity was not at all fashionable as it has been the last few months[20] and supergravity, of which Scherk was one of the pioneers, was still in its infancy.]

i) <u>Antigravity and extended supergravity</u>

I reproduce here the arguments developed by Scherk[19], since I am interested in the attempt he made at relating a very abstract idea (such as supergravity) to real life[**].

Discussing antigravity, Scherk remarks that 'antigravity is an inescapable consequence of extended supergravity; many attempts to get rid of this "crazy photon" have failed so that it may be wiser to learn to live with it'.

This is due to the fact that, in extended supergravity models, gravity will proceed not only through a tensorial exchange, but also through vectorial and scalar exchanges. This is indeed the case, if one takes for instance the N = 2 supergravity model coupled to a massive scalar multiplet but in which supersymmetry is unbroken. Zachos[19] has discovered that a remarkable phenomenon occurs in such a model. In the weak field limit, the potential between a pair of massive particles can be obtained from the Born diagrams where both a graviton and a graviphoton can be exchanged. If several scalar multiplets are introduced, with masses m_i and coupling constants g_i to the vectorial particle, one finds in the static limit the amplitude:

---

*) This symbol has been invented by Scherk and Nahm. According to them, it is an Egyptian hieroglyph with phonetic value šn (shen) which is an inversion of the Egyptian hieroglyph γ, phonetic value šš (shess).

**) More recent developments of these ideas have been proposed, for instance by Barbieri and Bars.

$$A = 4\kappa^2 \, \frac{m_i m_i'}{q^2} \left(m_i m_i' - \frac{g_i g_i'}{\kappa^2} \right) .$$

The resulting potential in the static limit is given by:

$$V(r) = - \frac{\kappa^2}{4\pi r} \left(m_i m_i' - \frac{g_i g_i'}{\kappa^2} \right) .$$

In $N = 2$ supergravity, $g_i = \varepsilon \kappa m_i$ where $\varepsilon = +1$ for a particle and -1 for an antiparticle. So one finds:

$$V(r) = - \frac{\kappa^2}{4\pi r} \, m_i m_i' \, (1 - \varepsilon\varepsilon') ,$$

which vanishes between a pair of particles and a pair of antiparticles, whilst the gravitational attraction is doubled between a particle-antiparticle pair. Therefore, the extra-vectorial force is repulsive between like particles and attractive between unlike particles. This phenomenon is called 'antigravity'. Once we understand the inevitability of antigravity in extended supergravity, we may ask also, Why do we not see it?

Scherk gives the beginning of an answer to such a question in the following way. Derived from the previous equation, in the static limit, if a graviton is exchanged between two particles, we find the amplitude:

$$A = - \frac{4\kappa^2}{q^2} \, m^2 m'^2 \, (1 - \varepsilon\varepsilon')^2 .$$

If we now consider the static force between two protons which are $u\bar{u}d$ bound states, as the *graviton* couples not to the masses but to $T_{\mu\nu}$, it *sees* the total energy of the quarks as *the mass of the proton*; this mass is mostly the kinetic energy of the quarks and the gluons. So its contribution to the force is given by $4\kappa^2 \, m_p^2/r^2$, where m_p is the proton mass.

The graviphoton λ is coupled to $2\kappa m \, \bar{\chi}\gamma^\mu\chi$, where m is the mechanical mass of the quark in question and $\bar{\chi}\gamma^\mu\chi$ is the conserved e.m. current. So λ sees directly the quark mass and not the proton mass and its contribution is:

$$- \frac{4\kappa^2}{r^2} \, (2m_u + m_d)^2 .$$

The ratio of the graviphoton contribution to that of the graviton is then of the order of $3m_u^2/m_p$, i.e. 10^{-4}; so it is much smaller.

By applying then the principle of equivalence (see Appendix III, or do it as an exercise!) and if we consider the fall of K^0, $\bar{K}^0$ towards the Earth one finds that the relative difference in acceleration between K^0 and $\bar{K}^0$ is:

$$\frac{a(K^0) - a(\bar{K}^0)}{a(K^0)} \simeq 6 \frac{m_u}{m_p} \frac{m_s}{m_{K^0}} \simeq 10^{-2} \ !$$

So to save antigravity, Scherk assumes that the graviphoton λ *acquires a mass*, presumably through radiative corrections. It is rather likely since is universally coupled to scalar fields through the action

$$L = -\frac{1}{4} F_{\mu\nu}^\lambda \ F^{\mu\nu\lambda} - \frac{1}{2} \left| \gamma_\mu - 2i\kappa m A_\mu^\lambda \right| \ \phi^2 - \frac{1}{2} m^2 \phi^* \phi \ ,$$

if there are no radiative corrections:

$$V(\phi) \approx \frac{1}{2} m^2 |\phi|^2 \quad \text{at } m_\lambda = 0 \ .$$

The radiative corrections are proportional to g_{strong}^2, or α or g^2; so $m_\lambda = 2\kappa m_\phi \langle\phi\rangle$. If $\langle\phi\rangle$ is due to $SU(2) \otimes U(1)$ breaking, $\langle\phi\rangle$ is equivalent to 1 to 100 GeV; so with $m_\phi \simeq 1$ GeV we get

$$m_\lambda \simeq 10^{-19} \ \text{GeV/c}^2$$

and the associated Compton wavelength is thus $\lambda_\lambda \sim 1$ km. This means that the graviphoton, if it exists, is a light object associated with a short-range force[21].

ii) Other approaches

Let us come back now to a more conventional approach. The question is, if such a force exists, mediated by a light vector boson, to *what current does this light vector boson couple?* and thus what is the associated charge.

In the case of a vectorial boson with vectorial current[*]: $j_\mu = A_\mu j^\mu$, the corresponding charge is:

$$Q = d^3x \ j_0(x) \ .$$

Apart from choosing the weak hypercharge Y as first candidate for Q, the most general way to define the charge is by taking any combination of B - L and $L_i - L_j$ (where $L_i - L_j$ is successively: $L_e - L_\mu$, $L_e - L_\tau$, $2L_e - L_\mu - L_\tau$). On macroscopic stable matter:

---

[*] Warning: Remember that if we want to build an anomaly-free theory, we have to choose a current which is *vectorial* (i.e. couples equally to *left* and *right* particles) and not *chiral*. Otherwise the required anomaly cancellation is model dependent.

$$B \equiv P + N \equiv N + Z, \quad L_e \equiv -Z \ .$$

So the most general linear combination is given by

$$'Q' = const \times (cos \ \theta \ N + sin \ \theta \ Z)^{22} \ .$$

Another way to formulate these ideas is by considering the alternative proposed by Fayet[23]. In 1979, he introduced a new light neutral spin 1 gauge boson, called U. This extra U(1) factor responsible for spontaneous supersymmetry breaking, has vector couplings. They could lead to a new macroscopic force of intermediate or long range, usually repulsive, which would manifest itself as a derivation from Newton's law of gravitation. The vector in the U current purely originates from Z^0-U mixing, which is a linear combination of the conserved (B-L) and electromagnetic currents:

$$J_U^\mu \ vector = \left[(1 - x^2)/(1 + x^2)\right] \left[- \frac{1}{4} J_{B-L}^\mu + \left(\frac{1}{2} - sin^2 \ \theta\right) J_{em}^\mu\right] \ .$$

The U charge is then defined as follows:

$$Q^U \equiv Q^Z = (B - L) + (4 \ sin^2 \ \theta - 2)Q \ .$$

In particular, for a nucleus, $Q^U = Q^Z = N + (4 \ sin^2 \ \theta - 1)Z$; this coincides with the weak charge which may be measured in atomic physics experiments on parity violation.

An upper bound on the relative strength of the U force compared with gravitation is calculated as a function of its range λ_U; Fayet found it to be $\lesssim 0.16/\lambda_U^2$. Therefore, only short-range gravity experiments at various distances between a few centimetres and ~ 10 m should give really measurable effects. Otherwise the antigravitational force becomes too small compared to gravity.

To summarize, if an antigravitational force exists, from the high-energy physics point of view, this means that there is a *light vector boson* (that we call λ denoting a *graviphoton*) which mediates this force. It couples to a charge, which in its more general form can be associated with any linear combination of the conserved B - L and e.m. currents. Its mass is of the order of 10^{-19} GeV/c^2 and it is a short-range force as the associated Compton wavelength is at most O(1 km).

2.2.4 <u>Experimental tests of antigravity.</u> We have assumed so far that the antigravitational force exists and so we have tried to characterize its properties and define its consequences concerning particle physics.

We should, however, come back to the fundamental problem which is in fact: *Do we have any serious experimental proof of the existence of such a fifth force?*

We should say that, at the present time, the answer to this question is no.

The interest in such a subject has been revived a few months ago by the results of a re-analysis of the Eötvös experiment.

i) Re-analysis of the Eötvös experiment[20]

The Eötvös experiment[24], performed in 1922, used a torsion balance to compare the acceleration of different materials in a gravitational field (Fig. 25). The Earth was used as a source of gravitational field and the conclusion was that there was no sign of antigravity.

Fishbach, Aronson et al., motivated by results obtained in a $(K^0 - \bar{K}^0)$ experiment[25] (see next subsection) and the geophysical results (Subsection 1.1.2), re-analyse Eötvös's results. They argue about the fact that this measurement is, according to them[26], sensitive to 'local horizontal mass anomalies' (buildings, mountains, etc.); so they say that the Earth should not be considered as an 'idealized uniform density sphere, but that one has to introduce also the effects of local matter distribution'.

By applying such a trick they find that the relative acceleration between two objects made of different materials is different from zero (in contradiction with the equivalence principle) and moreover that it varies with the materials.

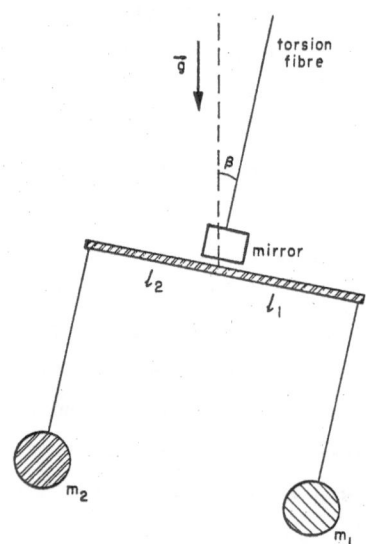

Fig. 25. Schema of the Eötvös experiment: the torsion is used as a balance. It compares the acceleration of different materials in a gravitational field (the Earth).

ii) <u>Antigravity and anomalies in the $(K^0-\bar{K}^0)$ system</u>

Aronson et al.[25] performing a K^0 regeneration experiment have found that the characteristic parameters of the $(K^0-\bar{K}^0)$ system, such as the mass difference Δm between K_L and K_S, the K_S lifetime τ_S, and the CP non-conserving parameter η_{+-}, present an anomalous behaviour with energy. This means that instead of being constant with respect to energy, they vary. Between two different ranges in energy, 2 to 10 GeV and 30 to 130 GeV, the mean corresponding values differ by 2 to 9σ.

They interpret these results by stating that no simple long-range field can account for this difference; so there is room for an intermediate range force (among possible explanations).

But these results have not been confirmed by the E617/731 experiment at FNAL[27], which measured no such discrepancy for η_{+-} at $\langle E_K \rangle = 65$ GeV.

Two experiments should be able to clarify this point quite soon; the E621 experiment at FNAL, which could measure τ_S up to 200-300 GeV and the NA31 experiment at CERN[28] which should be ideal for measuring $(K-\bar{K})$ parameters at 100 GeV.

iii) <u>Experimental tests of the existence of the fifth force</u>

The results presented in (i) and (ii) have triggered a renewed interest in experimentally proving the existence or non-existence of antigravity.

The main ways to test this hypothesis remain the classical measurements such as the balance experiment (i.e. Cavendish measurement), the torsion balance (à la Eötvös or Dicke, where either the Earth or the Sun are used as gravitational field), and Galileo's experiment. In this last case, the experiment consists in comparing the times of flight of different test masses dropped from the same height.

To achieve a sensitivity within desired requirements, say $\Delta a/g \simeq O(10^{-10})$, it is necessary to measure the time of flight to within 0.1 ns over a distance of 10 m. This should be feasible. An impressive list of proposals to perform one of these three types of measurement has been proposed since a few months (of the order of 20 experiments). Let us hope that some time they will give a final answer, as 'Few experiments are simpler in principle, harder in practice and so far reaching in implication' (Dicke[29]).

In addition to these classical measurements, high-energy physics may provide another way to test this hypothesis than the one provided by looking at the $(K^0-\bar{K}^0)$ system. It consists in a direct search for the graviphoton γ[30] by the study of a possible process such as:

$$K^{0/+/-} \to \pi^{0/+/-}\gamma .\tag{4}$$

The graviphoton may be produced also in $K^0 \rightarrow 2\pi^0\lambda$ but it cannot compete with the corresponding background provided by $K^0 \rightarrow 2\pi^0\gamma$; whereas the decays for γ corresponding to reaction (4) are strictly forbidden by angular momentum conservation since spin of massless γ is necessarily perpendicular to γ-π orbital momentum. However, as λ is massive, there is an additional ($S_z = 0$) spin degree of freedom which allows process (4).

As the decay of λ in 3γ is highly suppressed, the signature for the process $K \rightarrow \pi\lambda$ is one single π of fixed energy in the c.m. of the K and nothing else. So it is equivalent to an experiment in a K^+ stopping beam looking for $K^+ \rightarrow \pi^+\nu\bar{\nu}$ and $K^+ \rightarrow \pi^+$ + axion. Recently the branching ratio $Br(K^+ \rightarrow \pi^+a^0)$ has been measured to be less than 4.6×10^{-8} at 90% CL [31]. Soon BNL experiments[32] will be able to reach 10^{-10} or 10^{-11}. No such limits exist for the K^0 case; there is only an old limit[33], $Br(K_L^0 \rightarrow \pi^0\gamma) < 3 \times 10^{-3}$. This indicates that the Compton wavelength λ_ϱ for a possible λ is ≤ 25 m. Note that the Eötvös and geophysical experiment give $\lambda \lesssim 200$ m.

2.2.5 <u>Conclusion on the existence of the fifth force?</u> There are three as yet unexplained classes of phenomena:

i) The slight discrepancies between the values obtained for the Newtonian constant of gravity from laboratory measurements and geophysical data.

ii) The results of the Eötvös experiment, which seem to show that the observed accelerations of various bodies to the Earth are not universal but depend on the ratio of their hypercharges (or baryon numbers) to their masses.

iii) The energy dependence of the parameters of the neutral K system.

While each one of these phenomena may not be statistically very significant, the combination of the three may be taken as a first sign of evidence for a graviphoton. However, it is clear that the real and definite answer will be given by the experiments of classical mechanics.

3. EXTENDED WORLD?

We have seen, in the Introduction, that a theoretical way to go beyond the Standard Model is to introduce new families, i.e. new heavy leptons, heavy quarks, etc. We will show in this section how this can be done experimentally; but our main concern will be to concentrate especially on one object which is a wonderful probe to see whether or not the standard world must be extended. This probe is the *neutrino*.

It was in 1933 that the neutrino ν was predicted for the first time, by Pauli, at the Solvay conference. In 1942, J.S. Allen, following the suggestion of K.C. Wang, gave the experimental evidence by observing ^{7}Li recoil after the K-capture in ^{7}Be. The same year, the idea of ν_μ was theoretically introduced by Sakata and Inoue. In 1958, Feinberg suggested that if the weak interaction is mediated by intermediate vector bosons, then, two neutrinos must be introduced to explain the small mass of the ratio:

$$\frac{\mu \rightarrow e + \gamma}{\mu \rightarrow e\nu\nu} \leq 10^{-8} \ .$$

In 1959, Reines and collaborators observed directly the neutrino in the reaction $\bar{\nu}_e + p \rightarrow \beta^+ + n$ at a reactor experiment. In 1962, both Columbia and CERN groups observed, for the first time, a μ neutrino ν_μ in the reaction $\pi \rightarrow \mu\nu$. In 1980, the existence of three neutrinos, ν_e, ν_μ, ν_τ, with the following mass limits, was known: $m_{\nu e} \leq 60$ eV, $m_{\nu\mu} \leq 570$ keV and $m_{\nu\tau} \leq 250$ MeV. The same year at ITEP, Lyubimov et al. measured, in the reaction $^3T \rightarrow {}^3He + \nu_e + e^-$, the following mass limit for the neutrino ν_e:

14 eV $\leq m_{\nu e} \leq 46$ eV at 99% CL .

This result is still very controversial.

This 'historical' review is to show that the neutrino is an old pal. But despite that, the fundamental questions concerning the nature of the neutrinos still remain unanswered. These questions are:

Are neutrinos massive?

Is ν_e a mixture of several neutrinos?

What is the precise relationship between ν_e, ν_μ, and ν_τ?

Are neutrinos Dirac particles (so that ν is different from $\bar{\nu}$) or

Majorana particles (so that ν and $\bar{\nu}$ are the same particle)?

Are there more than three neutrinos?

The Standard Model claims that ν's are massless and that there are only three neutrinos: ν_e, ν_μ, and ν_τ; that is why neutrinos are a splendid probe to look beyond the Standard Model: it is, so far, a real experimental puzzle.

Many of the questions listed above are going to be answered by *non-accelerator* experiments; the main experimental ways to study the nature of ν's are single and double β-decay and ν-oscillation experiments (using, as the source of neutrinos, reactor, atmospheric or solar ν's). The different corresponding techniques and results will be reviewed here. It will be shown that neutrino physics is always a very active and attractive area of high-energy physics as well as a very original one.

Apart from the section on neutrino physics, another short section will be devoted to the search for fourth-generation heavy charged or neutral leptons in e^+e^- and $p\bar{p}$ colliders; it is another example of extending the Standard Model.

3.1 The Nature of Neutrinos and Non-Accelerator Physics[34]

We are reporting here the main technique used in this experimental field essentially to determine the mass of the neutrinos, and their possible couplings with ν_1 and ν_2 neutrino states. This search is done by studying single or double β-decay reactions or possible ν oscillations.

3.1.1 **Single β-decay experiment.** The single β-decay is a direct way to measure the mass of the ν_e and to test whether or not it is coupled with several neutrinos.

i) **Principle of this method**

The nuclear β-decay is defined by the process

$$(A, Z) \rightarrow (A, Z \pm 1) \ .$$

The electron e^- can couple with several neutrinos with different coupling constants; suppose for instance, that the e^- couples with two neutrinos, ν_1 of mass m_1 and ν_2 of mass m_2; we write ν_e as a linear combination of both neutrinos ν_1 and ν_2:

$$\nu_e = \nu_1 \cos \theta + \nu_2 \sin \theta \ .$$

By convention, we choose $|\sin \theta| \leq |\cos \theta|$, so

$$\nu_e = \nu_1 + \varepsilon \nu_2 \ ,$$

which means that ν_e couples mostly with ν_1; in other words the branching ratio ε of ν_e in ν_2 is small.

Experimentally, one measures the energy spectrum of the emitted e^- in the reaction

$$(A, Z) \rightarrow (A, Z \pm 1) + e^{\stackrel{-}{+}} + \stackrel{(-)}{\nu}_e \ .$$

The Fermi theory gives the corresponding β spectrum: $dN(E_\beta)/E_\beta$, where E_β is the kinetic energy of the β-electron. Instead of plotting directly the distribution $dN(E_\beta)/E_\beta$, it is preferable to use the so-called Kurie plot, which is the best way to illustrate the dependence of the spectrum shape on the neutrino mass.

This plot is derived from the $dN(E_\beta)/E_\beta$ plot by the following transformation:

$$\frac{dN(E_\beta)}{E_\beta} \rightarrow \frac{dN(E_\beta)}{\sqrt{F(Z,E_\beta)p_\beta E_\beta dE_\beta}} \propto \sqrt{W_0 - E_\beta} \left[(W_0 - E_\beta)^2 - m_\nu^2\right]^{1/4} \ ,$$

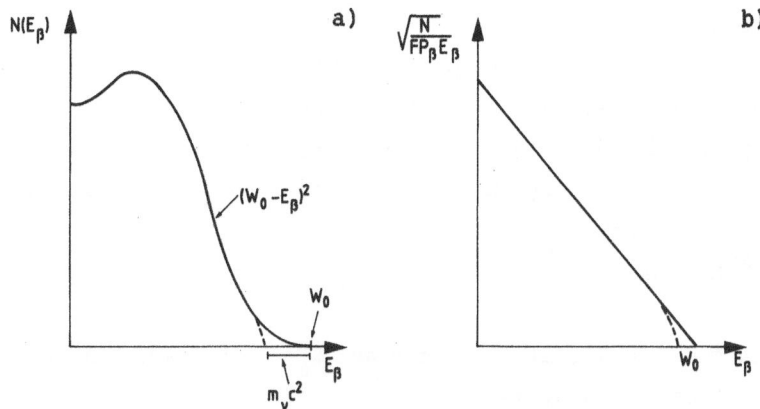

Fig. 26. Equivalence between the Fermi plot (a) and the Kurie plot (b)
obtained for a single-β-decay reaction.

where $F(Z,E_\beta)$ is a Coulomb factor

p_β is the momentum of the β-electron

$W_0 = m_i c^2 + m_f c^2 - m_e c^2 - E_R$

(with m_i and m_f being, respectively, the initial and final
nuclear masses and E_R the recoil energy), so W_0 is the end-point
energy.

Figure 26 shows the main features of the Kurie plot.

ii) Experimental situation

The measurement of single β-decay processes is rather difficult. There
exist mainly two techniques: one using spectrometers, the other using calorimetry.

The ITEP experiment[35], using a radioactive source of valine ($C_5H_{11}NO_2$)
containing 18% of tritium, has shown evidence for an electron antineutrino
mass between 14 and 46 eV, with a best fit at 35 eV. However, the use of a
tritiated valine source gives systematic uncertainties which are comparable
to the size of the neutrino mass observed; the main systematic uncertainty
is due to the energy given up in final-state excitations of the molecule
following the β-decay of one of the tritium atoms. It is difficult to
evaluate this effect in a molecule as complex as valine.

A way to get rid of these problems is by using calorimetry. Simpson[36],
for instance, implanted tritium in a Si(Li) detector. Because the whole
energy in each decay is measured (β-energy and X-rays), the spectrum recorded in the crystal always corresponds to ground-to-ground-state transition.
He studied the reaction

$$^3H \rightarrow {}^3He + e^- + \bar{\nu}_e \ ,$$

with a corresponding end energy W_0 = 18.6 keV. He found that ν_e can be analysed as a linear combination of two neutrinos ν_1 and ν_2: $\nu_e \simeq \nu_1 + \varepsilon\nu_2$, with a measured value of $\varepsilon \simeq$ 2 to 4% and a mass of ν_2 of 17 keV.

This result (despite many controversies) seems in total disagreement with those competitors using the same technique but looking at the reaction:

$$^{35}S \rightarrow {}^{35}Cl + e^- + \bar{\nu}_e .$$

This process has a much larger end energy: $W_0 \simeq$ 166.8 keV. No ν_2 of 17 keV was found by these experiments[37].

At the present time, the most common agreement on the limit of the $\bar{\nu}_e$ mass as given by single β-decay experiments may be summarized by the results obtained by Bowles et al.[38]. They measure the $\bar{\nu}_e$ mass in free molecular tritium β-decay (to get rid of the problems when using more complex molecules), using a sophisticated spectrometer as shown in Fig. 27. In the case of a tritium molecule, final-state effects may be accurately calculated. The Los Alamos group[38] has estimated that these effects do not generate a spurious neutrino mass greater than 1 eV. They find, with present restricted statistics, an upper limit on the mass of the $\bar{\nu}_e$ of 29.3 eV at the 95% CL or 25.4 eV at the 90% CL. The Kurie plot (Fig. 28) does not support the existence of a neutrino ν_2 'à la Simpson'. It does not support the central value reported by Lyubimov and collaborators, but neither does it exclude the lower part of the range 17 to 40 eV. Improvements to the apparatus transmission and resolution now in progress are expected to result in a sensitivity to neutrino mass in the vicinity of 10 eV.

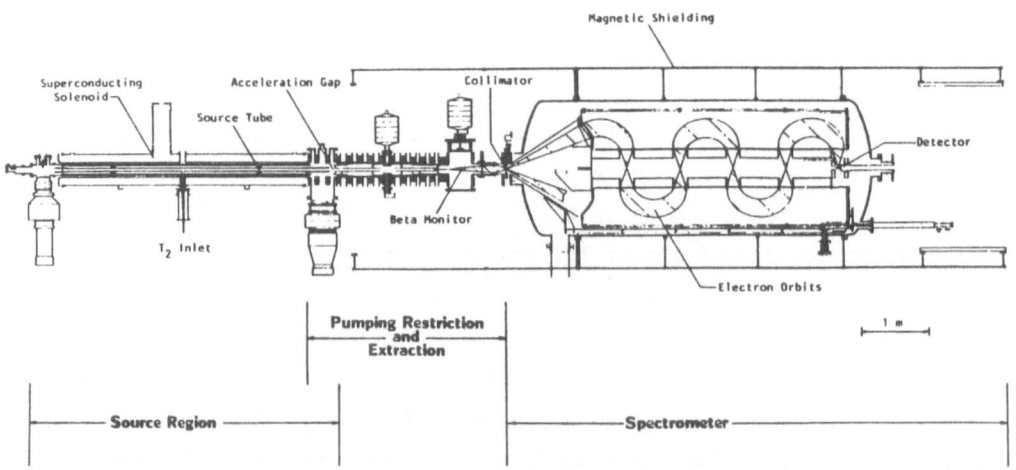

Fig. 27. Spectrometer for a single β-decay experiment. Cross-sectional view of the Los Alamos gaseous tritium beta-decay experiment[38]).

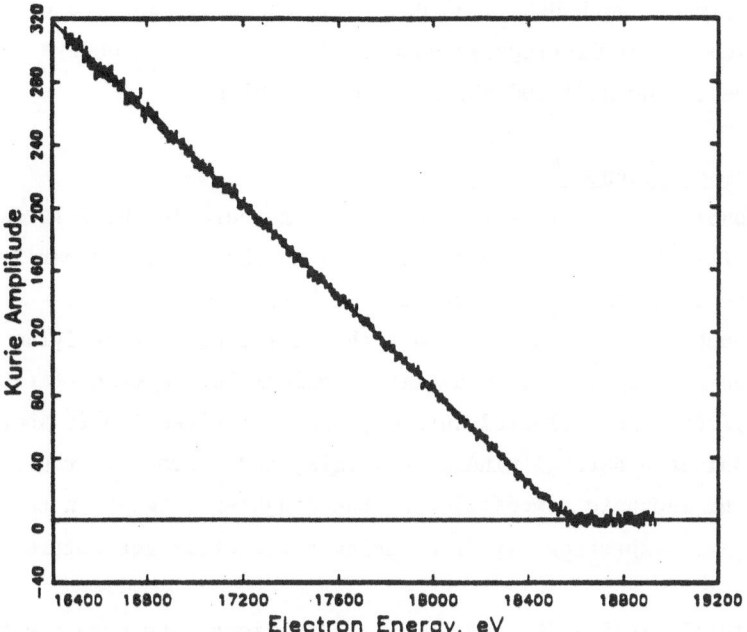

Fig. 28. Kurie plot for the molecular tritium β-decay experiment[38]. The
line through the data in this plot is the calculated best fit for
$m_v = 0$.

At the present time, the β-decay experiments do not give any evidence of
the neutrinos being massive and a mixture of several v_i's ($v_{i=1,2,...}$), but
a lot of work is in progress.

3.1.2 <u>Double β-decay (DBD) and Majorana neutrinos.</u> The double β-decay
corresponds to the processes

$$(Z, A) \rightarrow (Z + 2, A) + 2\beta^- + 2\bar{v}_e$$

$$(Z, A) \rightarrow (Z - 2, A) + 2\beta^+ + 2v_e ,$$

where two neutrinos are produced. We call it DBD_{2v}. However, if the v is
a Majorana and massive particle, the neutrinoless double β-decay reaction
DBD_{0v} is also possible:

$$(Z, A) \rightarrow (Z + 2, A) + 2\beta^-$$

$$(Z, A) \rightarrow (Z - 2, A) + 2\beta^-$$

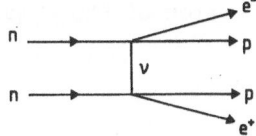

149

There are two ways of measuring these processes: the *direct counting experiments*, where both $DBD_{2\nu}$ and $DBD_{0\nu}$ are possible; and the *geochemical method*, which cannot distinguish between $DBD_{2\nu}$ and $DBD_{0\nu}$ decays. Let me describe the two methods and their present results.

i) <u>Geochemical method</u>[39]

The double β-decay is a rare process; 'favourable' half-lives are of the order of 10^{21} years, which corresponds to about 1 decay per 3 months in a 1 g source of a DBD-active isotope.

The geochemical method overcomes this difficulty by applying the *accumulation principle*. By accumulation principle, we mean, firstly, that the decay product is collected during geological times (up to about 10^9 years) in suitable natural mineral crystals; and, secondly, their integral effect on the isotopic composition of the daughter element in the sample is detected by mass spectrometry (i.e. using radiometric age determination methods).

Geochronologically, Te and Se mineralizations have occurred from the early Precambrian to the Quaternary period. But very ancient minerals are particularly scarce. So it is very difficult to provide the 'ideal' sample of a DBD experiment; 'ideal' means here old, rich in Te, low in U, well-crystallized, coarse-grained, and plentiful.

The measurement is done by getting the mass spectrum of Xe extracted from an old tellurium ore (Fig. 29). The large ^{130}Xe anomaly observed in this spectrum indicates the double β-decay of ^{130}Te in ^{130}Xe. The decay constant $^{130}\lambda_{DBD}$ of such a decay of ^{130}Te is

$$^{130}\lambda_{DBD} = \frac{N(^{130}Xe)}{N(^{130}Te)} \frac{1}{t} ,$$

where t is the accumulation time determined by the radiometric age determination method, $N(^{130}Xe)$ is measured by mass spectrometry, and $N(^{130}Te)$ is obtained by standard chemical analysis. The results obtained by the

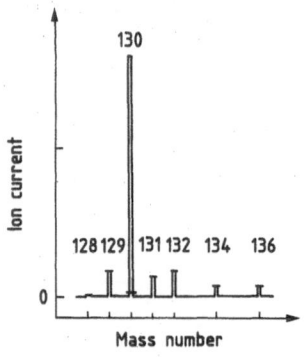

Fig. 29. Isotopic contribution of xenon extracted from a native tellurium ore. The horizontal lines indicate the maximum contribution of the atmospheric xenon.

Table 4.	Half-Lives ($T^{1/2}$) from Geochemical Experiments

Isotope	$T^{1/2}$ (years)
^{130}Te	$(2.55 \pm 0.2) \times 10^{21}$
^{128}Te	$> 8 \times 10^{24}$
^{82}Se	$(1.7 \pm 0.3) \times 10^{20}$

Table 5. Limits on the Mass of the Neutrino $\langle m_v \rangle$ and the Parameter η, from Geochemical Experiments

Isotope	$\langle m_v \rangle$ (eV)	η
^{82}Se	≤ 33	$< 4 \times 10^{-5}$
^{130}Te	≤ 100	$\leq 15 \times 10^{-5}$
^{128}Te	< 8.7	$\leq 3.5 \times 10^{-5}$
^{130}Te and ^{128}Te combined	< 5.6	$\leq 2.4 \times 10^{-5}$

geochemical experiments are summarized in two tables. Table 4 gives the half lives $T_{1/2}$ for ^{130}Te, ^{128}Te, and ^{82}Se. Table 5 gives the limit on the mass of the neutrino $\langle m_v \rangle$; as found by the geochemical method, and on the value of the parameter η. This parameter η is the amplitude corresponding to an admixture of a right-handed (V+A) component, which happens in the case where v is a Majorana particle. In such a case the lepton-number-violating DBD may occur. This method cannot measure the v-mass to better than 1 or 2 eV, and cannot really be much more improved, so let us direct our attention now to the other DBD method.

ii) Direct counting experiments in DBD[40]

The direct counting experiments in DBD are mainly based on the study of the double β-decay of ^{76}Ge which follows the schema given below.

Natural germanium contains 7.76% (in atoms) of ^{76}Ge, which decays on the ground state of ^{76}Se (0^+-0^+) transition with a transition energy ε_0 equal to 2040.71 ± 0.52 keV. An alternative decay corresponds to the transition 0^+-2^+ to the excited state of the ^{76}Se at 559.1 keV with a transition energy of 1481.6 ± 0.5 keV.

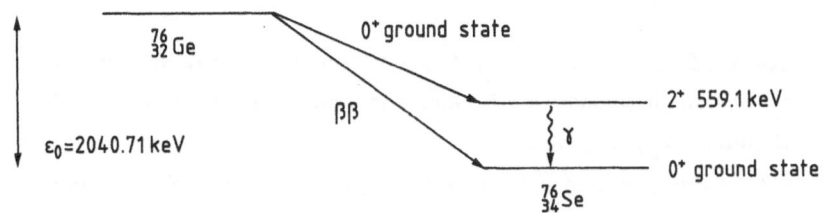

So the DBD_{0v} would be revealed by the presence, in the background spectrum of a germanium detector, of peaks corresponding to the 0^+-0^+ and 0^+-2^+ transitions. The spectra obtained in the region (0^+-0^+) DBD_{0v}

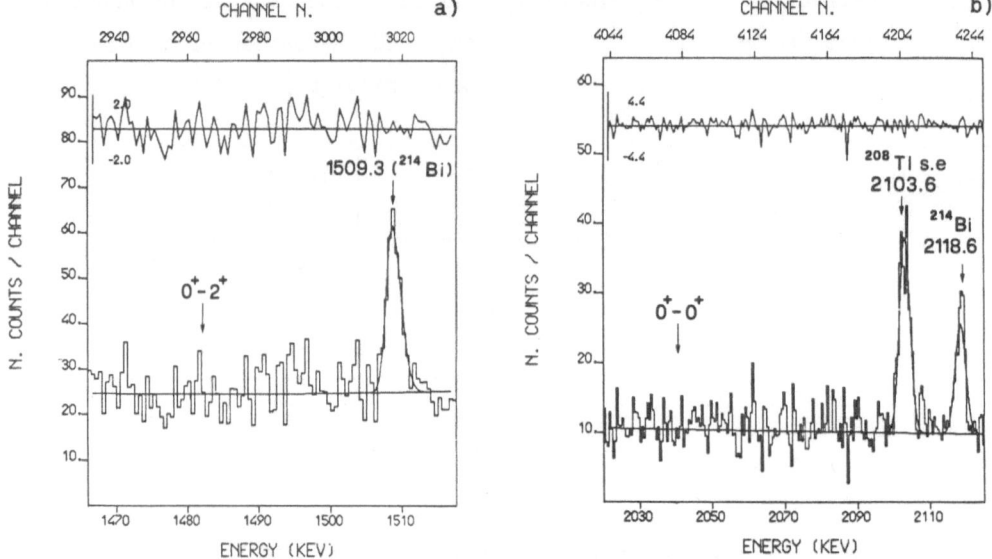

Fig. 30. a) Spectra in the regions of 0^+-2^+ neutrinoless double β-decay in the Milan experiment.

b) Spectra in the regions of 0^+-0^+ neutrinoless double β-decay in the Milan experiment.

(Fig. 30a) and the region (0^+-2^+) DBD_{0v} (Fig. 30b) as measured by the Milan group[41] do not show such evidence.

In conclusion, the present limit provided by DBD measurements on m_v are confined to a few electronvolts: $m_v \lesssim 6$ eV. This is somehow in contradiction with single β-decay results, which at the present stage of accuracy give $m_v \gtrsim 10$ eV. There are prospects for improving this limit using the DBD direct counting method. New experiments are planned using very sophisticated technologies, such as ^{136}Xe inside a TPC[42] and an ^{81}Kr source in an ionization resonance spectrometer using a yag laser[43] or superconducting tunnel junctions.

But if one wants to look for lower mass regions, certainly other methods must be used.

3.1.3 <u>Study of neutrino oscillations.</u> The principle of such a measurement is based on the hypothesis of neutrino oscillations between the different flavours: v_e, v_μ, v_τ.

The neutrinos v_e, v_μ, v_τ, eigenstates of the weak interactions, are not eigenstates of the mass, but linear combinations of v_1, v_2, and v_3, which are the eigenstates of the mass:

$$|v_1\rangle = \sum U_{1i}|v_i\rangle$$

where

$$l = e, \mu, \tau \quad \text{and} \quad i = 1, 2, 3 .$$

The fraction of neutrinos of flavour m at a distance L from the source of neutrinos of flavour l can be written as:

$$P(\nu_l \rightarrow \nu_m) = \sum U^2_{li} U^2_{mi} + \Sigma U_{li} U_{mi} U_{lj} U_{mj} \times \cos (2\pi L/l_{ij})$$

where $l_{ij}(m)$, the oscillation length, is

$$l_{ij}(m) = 4\pi p_\nu / |m^2_i - m^2_j| .$$

If there is only a mixture of two flavours, the U matrix is, in fact, a rotation matrix of angle θ and

$$P(\nu_e \rightarrow \nu_e) = 1 - \frac{1}{2} \sin^2 2\theta (1 - \cos 2\pi l/l\nu) ,$$

where l is the vacuum oscillation length and:

$$l = 2.53 \text{ m} \frac{p}{1 \text{ MeV/c}} \frac{1 \text{ eV}^2}{\Delta m^2} \quad \text{and} \quad \Delta m^2 = m^2_1 - m^2_2 .$$

So there remain only two parameters to characterize the neutrino oscillations, namely $\sin^2 2\theta$ and Δm^2.

The neutrino oscillations are studied with nuclear reactors or atmospheric ν's or solar ν's. We are now going to discuss each case.

i) Nuclear reactor experiments

The nuclear reactor is a pure source of $\bar{\nu}_e$ emitted isotropically from β-decay of fusion products. The experimental area has to be well shielded from the reactor core and associated activities. There are two possible options. The first one is to push the detector as far as possible from the reactor so that one gets cleaner conditions but lower statistics; this is the choice made by the Gösgen experiment[44]. The second option is to keep the apparatus near the reactor, as is required for high statistics, optim- izing the signal/background ratio as well as possible; this is the option taken by the Le Bugey experiment[45]. The last results from the Gösgen experi- ment contradict the previous results from the Le Bugey group who claimed experimental evidence for ν oscillations.

The Gösgen experiment has measured the positron spectrum from the reaction $\bar{\nu}_e + p \rightarrow e^+ + n$, at distances of 64.7 m, 45.9 m, and 37.9 m from

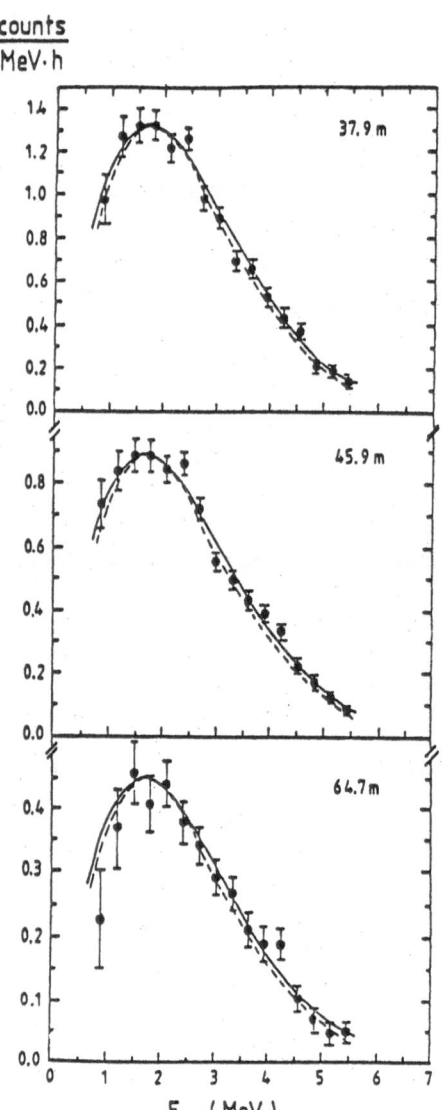

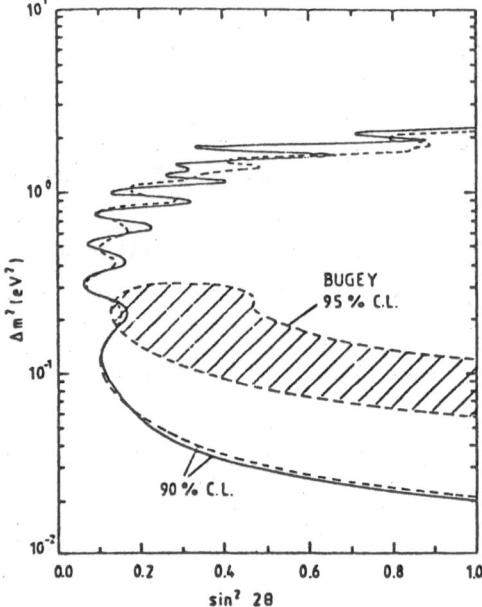

Fig. 31. Positron spectrum measured by the Gösgen experiment at the following distances from the reactor: 64.7 m, 45.9 m and 37.9 m.

Fig. 32. Scatter plot of Δm^2 versus $\sin^2 \theta$ as measured by the Gösgen experiment and comparison with the results from the Le Bugey experiment.

the core of the reactor. The three spectra are equivalent (Fig. 31); so there is no evidence for ν oscillations. Moreover, the oscillation parameter region ($\sin^2 2\theta$, Δm^2) favoured by the Le Bugey experiment is not supported by the Gösgen data (Fig. 32). The Le Bugey group are redoing their measurement.

ii) <u>Atmospheric neutrinos</u>

Neutrinos produced in the atmosphere can also be used to look for ν oscillations. These ν's are produced by π- and μ-decays by interaction of cosmic rays with the atmosphere. Below 10 GeV, these ν's prevail over the extraterrestrial ν-flux. They are a well-known source of background for proton-decay experiments. The IMB experiment has taken advantage of this and has measured possible ν oscillations[46].

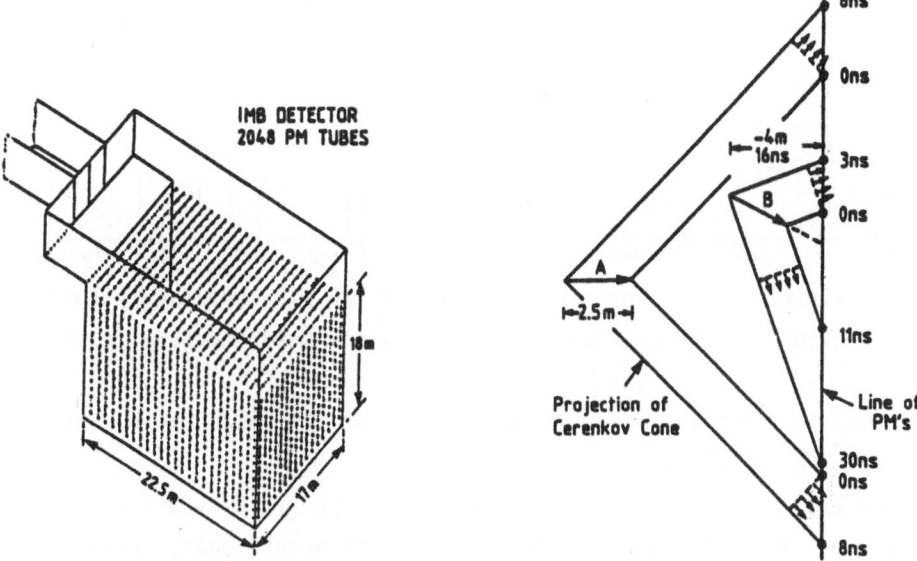

Fig. 33. Detection of Cherenkov light in the IMB experiment. The pattern of
the hit PMs plus the timing allow us to reconstruct the direction;
v_μ interactions produce less light (for CC) than v_e interactions.

The IMB detector[46] is a large imaging water Cherenkov detector, with
the largest fiducial mass (3300 t) available at present, well shielded at a
depth of 610 m (1600 m water equivalent). The detector is based on the
detection of Cherenkov light produced by 2048 photomultipliers (see
Fig. 33). The distance a neutrino has travelled is a function of its zenith
angle:

$$L = \left[r^2 - (r - d)^2 \sin^2 \theta \right]^{1/2} - (r - d) \cos \theta \ ,$$

where r is the Earth's radius ($\sim$ 6380 km), d is the scale height of atmosphere $\simeq$ 10 km. In the lower hemisphere: $L \simeq -2r \cos \theta$. If v oscillations
occur, there will be a difference between v's in the upward-going fifth of
the solid angle (as sketched in Fig. 34), which travel a mean distance of

Fig. 34. Schema of upward versus downward
atmospheric neutrinos.

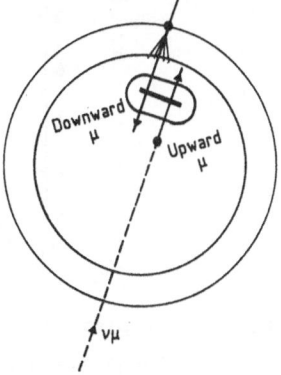

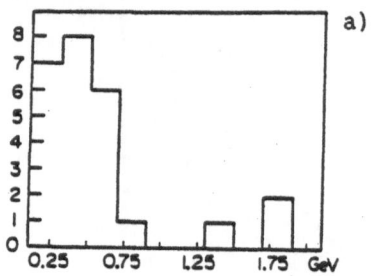 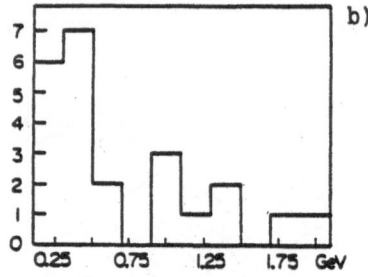

Fig. 35. Visible energy spectrum measured in the IMB experiment, for
a) upward atmospheric neutrinos, b) downward atmospheric neutrinos.

10^7 m, and those in the downward-going fifth of the solid angle, which have
only travelled $\sim 10^4$ m.

The IMB experiment has collected 135 atmospheric ν events in 420 days
of running. By comparing the visible energy of upward-going neutrino interactions (Fig. 35a) and the downward-going neutrino interactions (Fig. 35b),
the analysis cannot reject the fact that the two distributions are just the
same. The corresponding neutrino energy E_ν is defined by

$$E_\nu \simeq 0.758 \ E_{vis} + 410 \ \text{MeV} ,$$

which gives an average value for E_ν of about 920 MeV. From the comparison
of the two spectra above, an exclusion contour is obtained in the (Δm^2,
$\sin^2 2\theta$) plot (Fig. 36). No evidence for ν oscillations is observed in this
experiment.

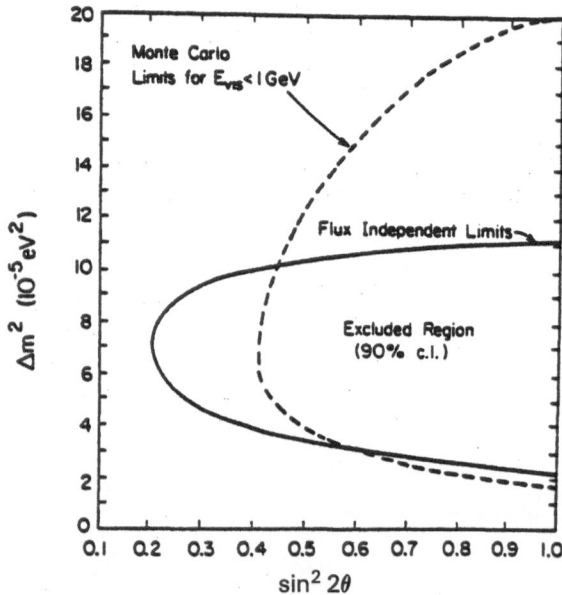

Fig. 36. Scatter plot of Δm^2 as a function of $\sin^2 \theta$ as measured by the IMB
experiment (exclusion contour).

iii) Solar neutrinos

A new field starts to be actively investigated; it is the range of solar neutrinos. As shown in Fig. 37, they allow one to go a step further in the study of the (Δm^2, $\sin^2 2\theta$) plot, by reaching Δm^2 values smaller still (i.e. $\lesssim 10^{-8}$ eV^2).

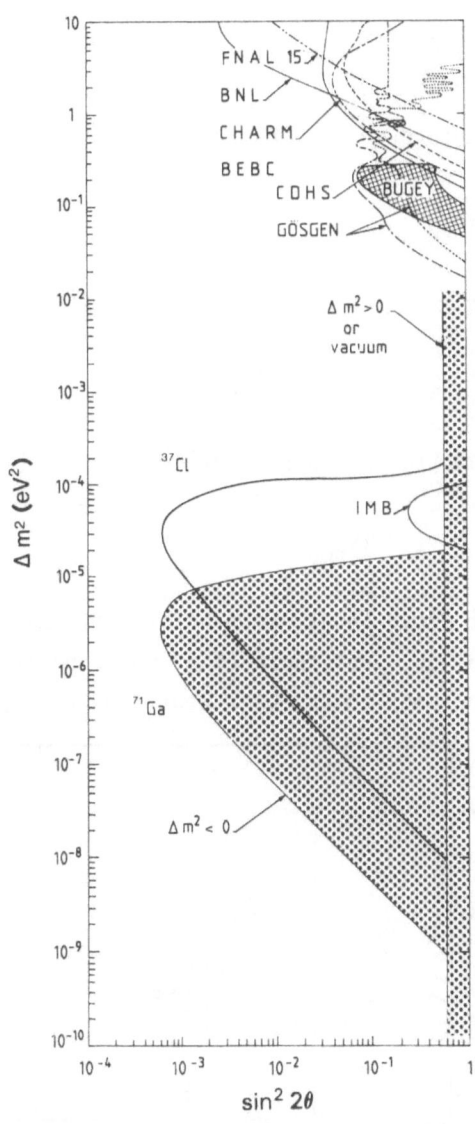

Fig. 37. Δm^2 range available for each technique to measure ν-oscillations.

a)   <u>Origin of solar neutrinos</u>

Most of the energy produced in the Sun comes from the pp fusion reaction chain:

$$p + p \to {}^2H + e^+ + \nu_e \quad \text{(twice)} \quad (\nu_{pp})$$

$${}^2H + p \to {}^3H + \gamma \quad \text{(twice)}$$

$${}^3He + {}^3He \to {}^4He + p + p$$

or

$${}^3He + {}^4He \to {}^7Be + \gamma \; .$$

Electron or proton capture on the ${}^7Be$ may also give:

$$e^- + {}^7Be \to {}^7Li + \nu_e \qquad (\nu_{Be})$$

$$p + {}^7Be \to {}^8B + \gamma$$

$${}^8B \to {}^8Be^* + e^+ + \nu_e \qquad (\nu_B)$$

A small fraction comes from the pep reaction:

$$p + e^- + p \to {}^2H + \nu_e \quad (\nu_{pep}) \; .$$

The neutrino flux, at Earth level, as predicted by the Standard Solar Model (SSM) is shown in Fig. 38. But there is a 'solar neutrino puzzle'. Some experiments have found many fewer solar neutrinos than predicted by the SSM.

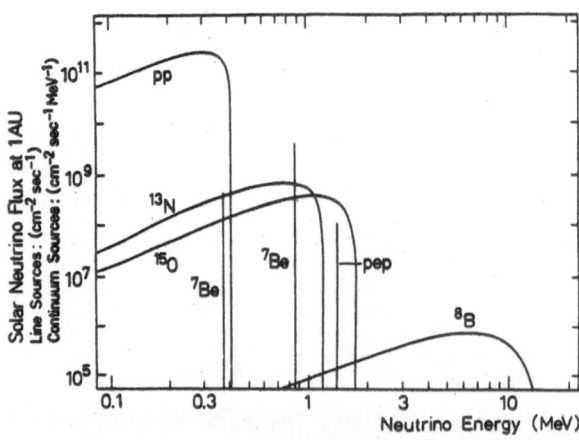

Fig. 38. Neutrino flux diagram as predicted by the Standard Solar Model.

Mikheyev and Smirnov[47] have explained this fact by showing that electron neutrinos, above a certain minimum energy E_m, may all be converted into μ-neutrinos on their way out through the Sun. The results of Mikheyev and Smirnov have been derived in a different but equivalent way by Bethe[48]; he draws additional conclusions from this phenomenon concerning the masses of μ, e, and τ neutrinos, Δm^2, and a very minor restriction on the value of the neutrino mixing angle.

So there also the theoretical situation is very unclear as to whether the Sun is 'standard' or 'non-standard'. Again experiments may clarify the situation. Let us see how. So far, the Brookhaven experiment[49] is the only one to have observed solar neutrinos by detecting ^{37}Ar atoms produced via $\nu_e + ^{37}$Cl $\rightarrow e^- + ^{37}$Ar in a 600 t tank filled with C_2Cl_4. There is now an impressive list of very important experimental projects.

b) <u>'New generation' of solar neutrino experiments: GALLEX</u>

Somewhere in the middle of Italy, there is the Cave of 'Zichi Baba' also known as the Gran Sasso Laboratory, Fig. 39. It is an impressive experimental area. Several huge experiments are planned to run there by the end of the 80's (MACRO, GALLEX, etc.).

We take as an example[*)] the GALLEX experiment[50]. It is based on the counting of ^{71}Ge atoms produced in the reaction

$$\nu_e + ^{71}\text{Ga} \rightarrow e^- + ^{71}\text{Ge} ,$$

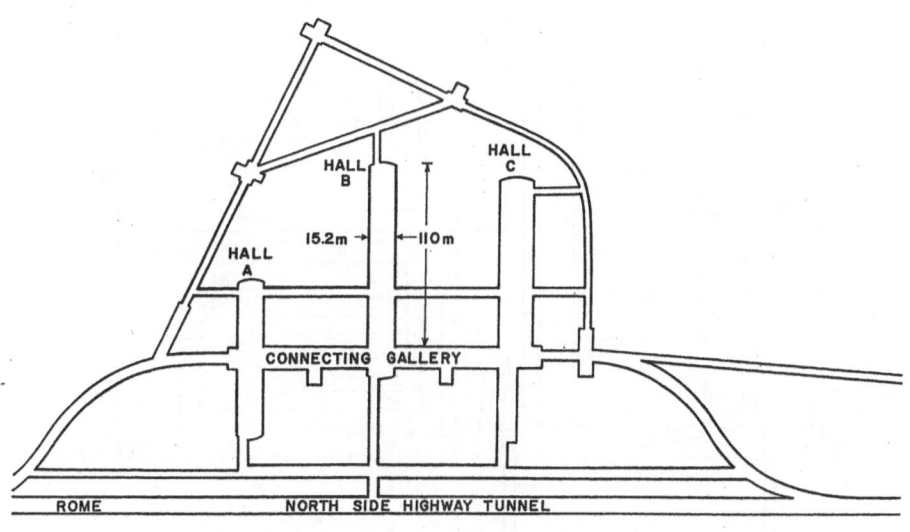

Fig. 39. Schema of Zichi Baba's cave (Gran Sasso).

*) Another gallium experiment is going to be performed at the Baksan Neutrino Observatory (USSR), at an approximate depth of 3800 m of rock, using 50 tons of solid gallium[50].

which has a threshold of 233 keV and is sensitive to almost the whole spectrum of solar neutrinos.

The main goals of the GALLEX experiment are:

- Experimental verification (for the first time) that nuclear fusion is the principal solar energy source.
- Search for ν-flavour oscillations down to ν-mass difference as small as $\Delta m^2 \simeq 10^{-12}$ eV^2.
- Test of the Standard Solar Model versus the non-standard one, which has the potential of being able to affect drastically the fluxes of ^8B and ^7Be neutrinos (cf. Standard Model results).

The schema of the detector is shown in Fig. 40. The main phases of this experiment can be summarized in the following way:

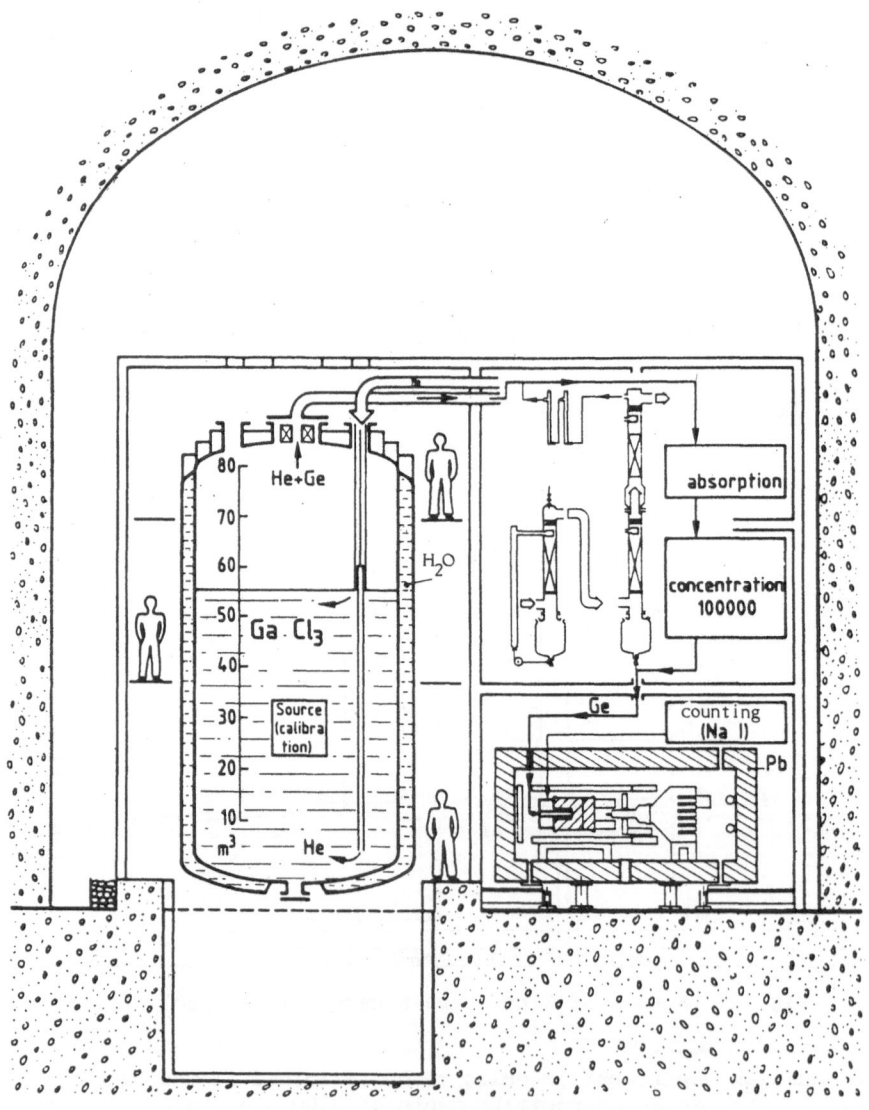

Fig. 40. Schema of the GALLEX experiment.

- <u>The target:</u> This is a single tank of about 55 m³. The gallium target is a highly concentrated $GaCl_3$ solution. As soon as the Ge atom is formed in this solution, it gives $GeCl_4$ (germanium chloride), which is highly volatile; so it is easily swept out from the solution by a circulating stream of air and helium.
- <u>The extraction of the [71]Ge:</u> This is done with gas scrubbers. The [71]Ge is confined in a small volume and reduced to gaseous germane (GeH_4); so it is admixed with the counting gas of proportional counters.
- <u>Detection of the [71]Ge:</u> This is done through the schema shown in Fig. 41; via its decay by Compton effect (EC), it comes back to [71]Ga in extremely background-free counting devices. We show in Fig. 42 some results on the study of the rejection efficiency of Compton background, as simulated by a ^{60}Co source in [71]Ge.

    This experiment also requires:

- <u>Accumulation of sufficient statistics:</u> This means 2 to 4 years of running with about 25 runs per year.

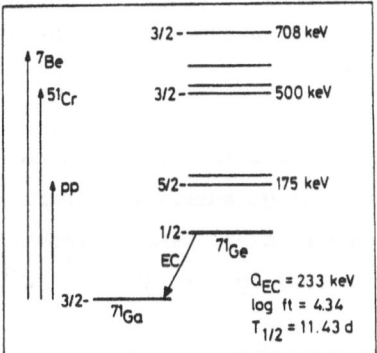

Fig. 41. Schema of the [71]Ge decay process.

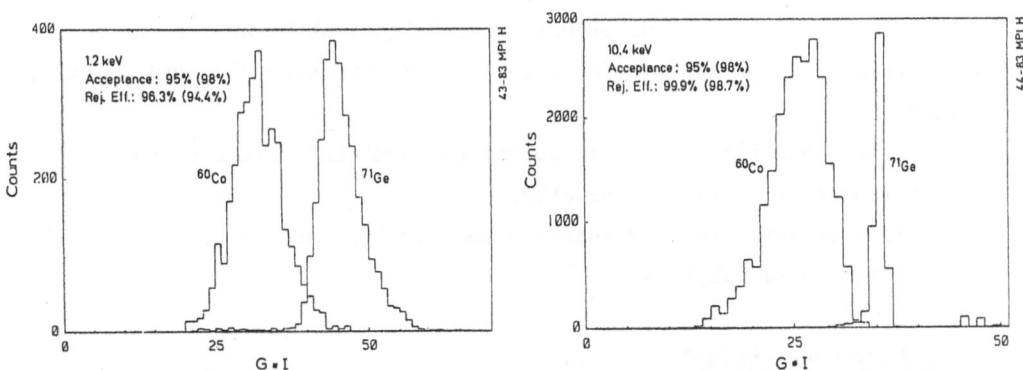

Fig. 42. Results obtained by the GALLEX Collaboration when studying the rejection efficiency of the Compton background, as simulated with a ^{60}Co source in [71]Ge counting (the pulse-form parameter G·I is a function of the pulse height).

- Constant monitoring: This is a crucial issue in order to demonstrate the combined performance of all experimental components and processes; it is done by means of a strong artificially-produced low-energy neutrino source, inserted into the target tank. The best choice is considered to be a ^{51}Cr, 600 kCi source. The ^{51}Cr is produced by activation of ^{50}Cr. It decays by e^- capture ($T_{1/2}$ = 27.7 days) into ^{51}V. It emits mono-energetic neutrinos of 745 keV (90%) and 426 keV (10%). The actual source is made of 120 kg of natural Cr powder in a sealed cylindrical container of 1 m length and 10 cm radius; it is surrounded by a Pb shield of 10 cm thickness. One source could be used, during about 60 days, for four to six Ge extractions.

This experiment is expected to have started running by 1989. This domain of high-energy physics is certainly very challenging. From the point of view of size, required performance, and complexity of the technology applied, these non-accelerator physics experiments are very comparable with the experiments running at present at $p\bar{p}$ colliders (CERN $p\bar{p}$ Collider or the Tevatron), or even with those projected for the next generation of pp colliders (see last section).

3.1.4 Conclusion. We may make the following concluding remarks:
i) We have been reviewing a large number of experiments which allow scanning of a wide range in mass; namely from a few tens of electron-volts (with β-decay and DBD) down to 10^{-4} or 10^{-5} eV with solar neutrino experiments.
ii) An impressive variety of experiments are implied, all working in the *non-accelerator* domain. They are either 'simple detectors' in a laboratory (such as a magnetic spectrometer experiment) or quite huge detectors inside enormous experimental halls (GALLEX at Gran Sasso).
iii) All these experiments are far reaching from the technological point of view.
iv) At the present time, the main questions about the nature of the neutrinos still remain unanswered.
So a lot of activity, and research and development, is going on in this domain of high-energy physics.

3.2 How Many Generations?
If there are more than the three generations stated by the Standard Model, this would imply, in particular, that there are sequential heavy charged (L) or neutral leptons (v_L) of the fourth generation. Both e^+e^- and pp colliders may perform such searches, as will be described below.

3.2.1 Heavy-charged-lepton search.

i) In e^+e^- experiments the main way to look for such new objects is to consider the process

$$e^+e^- \rightarrow L^+L^- \; ,$$

where L then decays, via charged current, into:

$$L^{\pm} \rightarrow \nu_L W^{\pm}$$

where $W^{\pm}$ decays leptonically or hadronically. Such an explanation (among others) has been proposed for the eight events observed by the MARK J experiment at PETRA[51]. These events were produced by the reaction:

$$e^+e^- \rightarrow \mu + \text{hadrons} \; ,$$

during the run with a c.m. energy of $46.3 \leq \sqrt{s}$ (GeV) ≤ 46.78 and with an average luminosity of 50 nb^{-1} per day. The running conditions corresponded, in the case of PETRA, to a relatively high background. The Lego plots characterizing these events are reproduced in Fig. 43. Their main features are a broad e.m. and hadronic energy flow, and muons isolated from the main energy flow with a low average momentum of about 3 to 4 GeV. The missing energy in these events is small, about 10 to 20% of the total c.m. energy. The energy flow is broad and could be divided into three or more jets and clusters. In most of these events, the e.m. shower is about 10 GeV. Such events were not observed by the other PETRA experiments (but no such high-energy beam was available any longer). So we have to wait for SLC, LEP, and $p\bar{p}$ results to solve this puzzle.

ii) At $p\bar{p}$ Colliders, the search for such an object may proceed by looking at the process

$$p\bar{p} \rightarrow W \rightarrow L\bar{\nu}_L$$

where

$$L \rightarrow q\bar{q}'\nu_L \; .$$

This is in fact another 'by-product' of the missing-energy measurement.

Such an analysis has started at the UA1 experiment and very preliminary results have been obtained[52]. Again the main limitation, at the present time, is the lack of statistics.

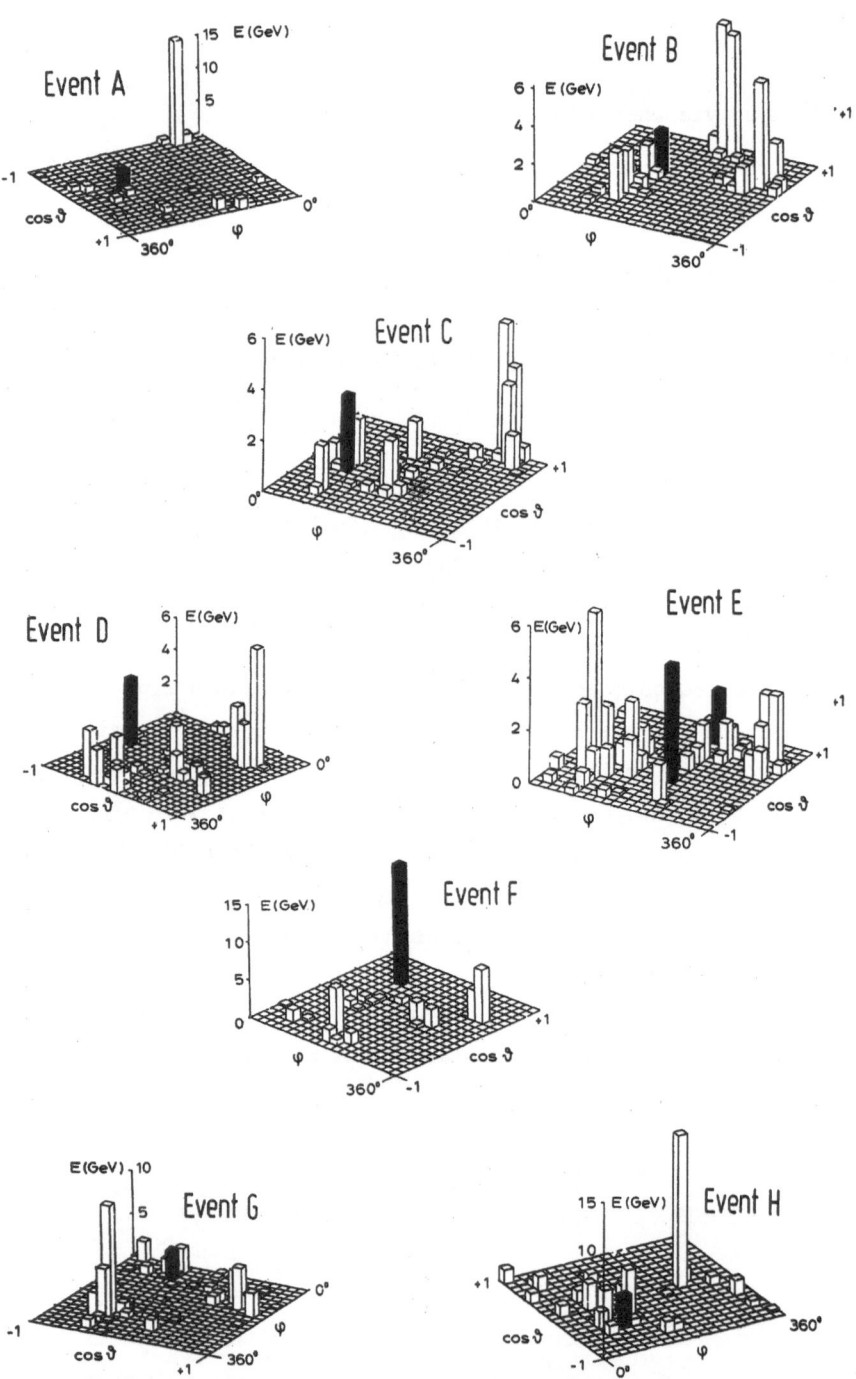

Fig. 43. Lego plots of the 8 MARK J events characterized by one muon
(indicated by a black tower) and hadronic activity. The energy
scale is different for each event. Event E has two muons.

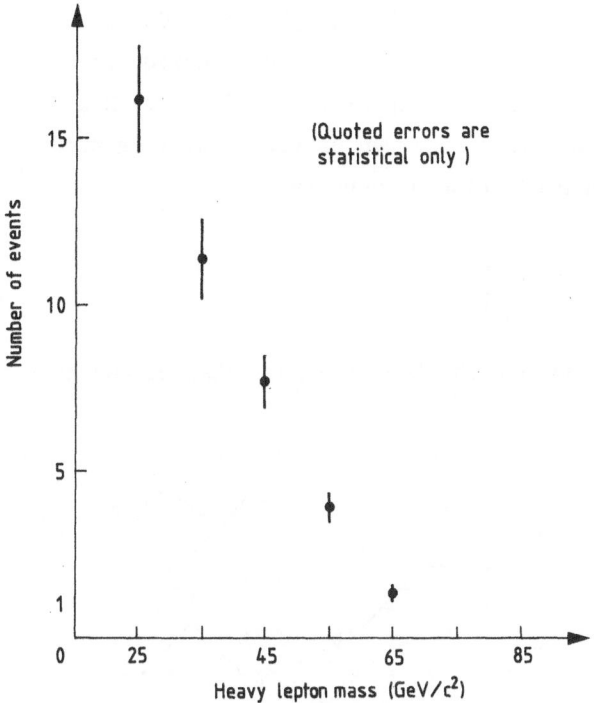

Fig. 44. Expected number of monojets plus missing-energy events
produced by $W \rightarrow Lv_L$ in the UA1 experiment, for an
integrated luminosity of 715 nb^{-1}, as a function of
the decay lepton mass.

The main experimental signatures of such events are monojet plus
missing energy or bijets non back-to-back plus missing energy. An estimate
of the number of expected heavy leptons corresponding to the total integrated
luminosity recorded on tape by the UA1 experiment (715 nb^{-1}) has been made
as a function of the mass of the possible new heavy lepton L. It includes a
detailed Monte Carlo generation of the corresponding decay of the W, a full
detection simulation, and a 4σ filter as the one applied to the data. The
corresponding result is reported in Fig. 44. An upper limit on the mass
of a possible heavy lepton L is calculated by comparing the expected number
of monojet events (coming from heavy leptons) with the number of monojets
obtained in the data (after subtracting the standard background). This
gives an upper limit on the mass of this possible heavy lepton:

$$m_L \gtrsim 41 \text{ GeV/c}^2 \text{ at } 90\% \text{ CL} .$$

More definite answers in this mass range again imply a factor of 10 more in
statistics; so we need to wait for ACOL and Tevatron results as well as SLC
and LEP.

3.2.2 <u>Search for the neutral heavy lepton.</u> One way to search for such
a new particle is by the neutrino-counting technique at e^+e^- colliders. It
is based on the following principle; in the Standard Model, events with a
single photon and no other observed particles will be produced by radiative
corrections to the production of ν-pairs:

$$\sigma(e^+e^- \rightarrow \gamma\nu\bar{\nu}) \propto G_F^2 \, s\left(1 + \frac{N_\nu}{4}\right) .$$

This process goes through the two diagrams (charged and neutral currents)
given below:

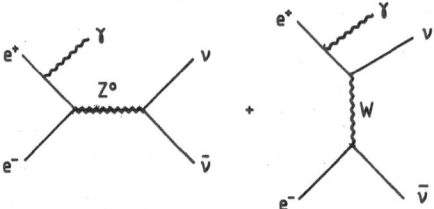

So it is a good way to count the number of neutrinos. It has the advantage
of being insensitive to the masses of the associated charged leptons.

Present results are given by two experiments already described in
Section 1, namely MAC and ASP; ASP has been especially designed for such a
search. The results are an upper limit on the number of neutrinos,
amounting to 41 for MAC and 14 for ASP at the 90% CL (see Fig. 45).

Proton-antiproton colliders also have various ways to furnish this
number, among which is the study of the process $Z^0 \rightarrow \nu\bar{\nu}$. This is also a

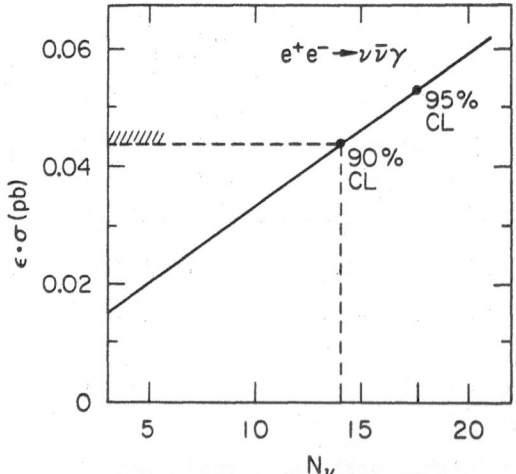

Fig. 45. Corrected $\nu\bar{\nu}\gamma$ cross-section as a function of the number of neutrino
generations, as obtained by the ASP experiment.

by-product of the missing-energy analysis. The statistics are at the present time much too low to really give some plausible information. I refer to the lectures of DiLella for the discussions of the other possibilities[1].

We could review many other examples of experimental puzzles beyond the Standard Model which refer to this section on an 'extended world', such as the search for new W's and new Z^0's which could be a sign of a (L-R)-handed symmetry etc.

However, we have chosen to restrict ourselves to the case of the neutrinos, since this is still a very puzzling domain in many respects and involves mainly non-accelerator physics and a lot of different and special techniques. In addition, the example we have quoted, concerning the search for heavy leptons, shows the other aspect which essentially implies pp and e^+e^- colliders. Again there, most of the work remains to be done.

4. BEYOND THE 1 TeV/c^2 MASS RANGE: NEW WORLD?

The question as to whether or not there is a threshold for new physics, somewhere around 1 TeV/c^2 in the mass scale of high-energy physics represented in Fig. 1, is certainly one of the most if not the most puzzling point that has to be clarified. This new physics scale is directly connected with the *Higgs sector* and the *hierarchy problem*. This problem is discussed in detail in Hinchliffe's lectures at this school[3]. Let me summarize some facts.

The Standard Model expects a scalar Higgs boson H^0, with a mass somewhere between a few GeV/c^2 and about 1 TeV/c^2, decaying into a fermion and antifermion pair ($f\bar{f}$). Whereas the Higgs is the 'inseparable' companion of the intermediate vector boson W and Z^0 discovered at the CERN $p\bar{p}$ Collider, it has not yet been found: the *Higgs sector remains untested*. Moreover, the presence of these fundamental scalars leads to the *hierarchy problem*. In the Standard Model, all masses are fixed by the Higgs potential mass scale, in the following way:

$$V(\varphi) = -\mu^2 \; \varphi^+\varphi + \lambda(\varphi^+\varphi)^2 + \text{loops} ,$$

where $\langle\varphi\rangle$ is equal to $\sqrt{-\mu^2/\Lambda}$, that is to say it is equivalent to $G_F^{-1/2} \simeq$ 300 GeV. The W and Z masses are related to $\langle\varphi\rangle$ by $m_{W,Z} \simeq g_w\langle\varphi\rangle$. The masses of the fermions f are related to $\langle\varphi\rangle$ by $m_f \simeq g_{\varphi f\bar{f}} \langle\varphi\rangle$, and the Higgs mass is $\sqrt{\lambda}\langle\varphi\rangle^2 = \sqrt{2} \mu$.

Now $\langle\varphi\rangle \sim \mu$ is very much smaller than the Planck mass m_p; but even if μ^2 is small at the tree level[53], in general the loops which are quantum-radiation corrections can become divergent too. A new symmetry at $\mu^2 = 0$ could enforce $\delta\mu^2 \simeq O(\mu^2)$; but it is not possible to do so with scalars in the Standard Model. Only supersymmetry is able to solve this problem, the so-called hierarchy problem, by introducing a new symmetry which relates bosons and fermions.

As the Standard Model cannot extend its validity up to the Grand Unification scale (i.e. 10^{15} GeV) or to the Planck mass scale (i.e. 10^{19} GeV) and as the *great desert* is implausible:

    'Something must happen in between'!

(See how in Gross's lectures.) The parameter Λ we have just introduced above can then be interpreted as the scale of new physics beyond the Standard Model; and it is natural to expect: $\Lambda \simeq O(1 \text{ TeV})$: *so new phenomena are expected thereon.*

The Higgs sector, predicted by the Standard Model, and still totally unexplored, is a very fundamental sector, at the border line of the Standard Model and new thresholds: *It is a main concern* for the experimentalists: we have to search for it *everywhere* (i.e. with all possible types of experiments and machines and in a very wide mass range). As we shall see, it is not an easy task.

This introduction to the 1 TeV/c^2 mass sector explains why we will consider two points in this section; the first subsection will concentrate on how to explore the whole Higgs sector. We will start from the very-low-mass Higgs range (say $m_{H}0 \lesssim 20$ GeV/c^2), then the range in mass which goes up to twice the W mass, and finally we will look at the high-mass Higgs scale, say $m_{H}0 > 2m_{W}$ and mainly around 1 TeV/c^2. This last point will connect immediately with the second subsection. There, we will review what are the means we need, in terms of machines and detectors, to scan this high-mass region.

4.1 Experimental Exploration of the Higgs Sector

As we have seen, the Higgs mass can extend from a few GeV/c^2 to about 1 TeV/c^2. To explore experimentally this tremendously wide range we have to go step by step, as a function of the accelerators and the detectors we have or will have at our disposal. So we have divided this subsection into three parts. The first part is concerned with presently running machines and detectors and is mainly restricted to the scanning of the very-low-mass region: $m_{H}0 \lesssim$ few GeV/c^2. The second part will discuss the capabilities of new e^+e^- colliders (TRISTAN, SLC, LEP I and II, and the Tevatron) to scan the mass range from a few tens of GeV/c^2 up to twice the W mass. The last part will be devoted to the high-mass range, where the Higgs decays into intermediate vector boson pairs and has a mass larger than twice the W mass.

4.1.1 <u>'Prehistorical search for Higgs'</u>: $m_{H^0} \lesssim 10$ GeV/c². In summer 1984, the Crystal Ball experiment[54] claimed experimental evidence for a narrow state with a mass of 8.3 GeV/c² called ζ and observed in the decay of Y(1s) → ζ + γ. The corresponding branching ratio of Y(1s) into ζ + γ was measured to be 0.5%. It is 100 times higher than the one expected for a minimal Higgs H⁰. So ζ was thought perhaps to be a non-standard H⁰. (Remember, this was the year when everything was non-standard physics!). Moreover, the decay of Y(2s) into H⁰γ was not seen. In autumn 1984[55], with higher integrated luminosity at DORIS II, the experiments CESR and CUSB did not see ζ at the Y(1s) resonance. This result follows a well-known experimental law: 'as the statistics increase, the evidence decreases'. Later, the CLEO, CUSB, ARGUS and Crystal Ball experiments studied the process of the production of a minimal Higgs at the Y resonance. The branching ratio for this decay of the Y (where there is emission of a monochromatic γ) is related to the branching ratio of the decay of Y into two μ's by

$$B(Y \to \gamma H^0) = B(Y \to \mu\mu) \frac{G_F m_b^2}{\sqrt{2}\pi\alpha} \left(1 - \frac{m_{H^0}^2}{m_Y^2} \right) ,$$

and is of the order of 10^{-4} if the mass of the minimal H⁰ is small (without including QCD radiative corrections). To improve the energy resolution on photons, the calorimeter of CUSB has been upgraded. The NaI modules have been partially replaced by BGO counters. The energy resolution for photons achieved by the BGO counters, $\sigma(E_\gamma)/E_\gamma = 0.018/E_\gamma^{1/4}$, has to be compared with $\sigma(E_\gamma)/E_\gamma = 0.038/E_\gamma^{1/4}$, which is the energy resolution for photons obtained with NaI. The measured photon-energy spectra after event topology cuts both for BGO and NaI data are shown in Figs. 46a and b. No monochromatic photon

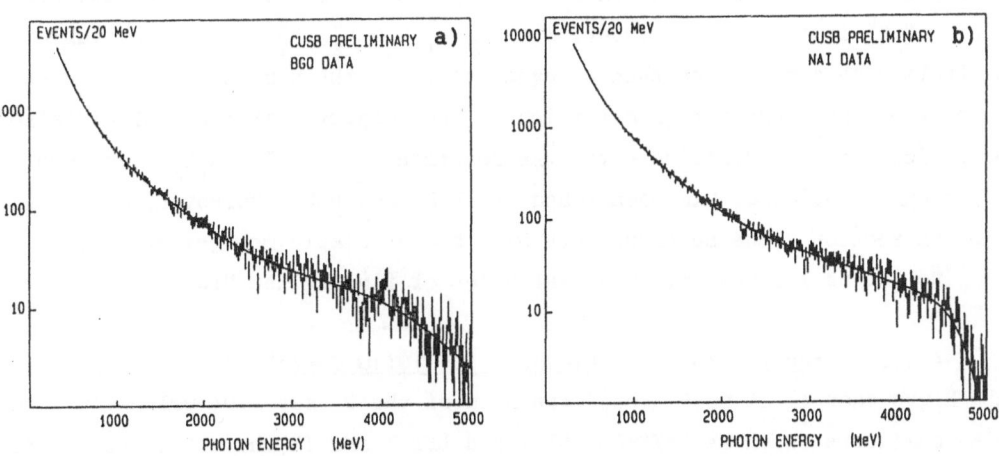

Fig. 46. Inclusive photon spectra obtained by CUSB for photons (a) in the BGO array and (b) in the NaI array.

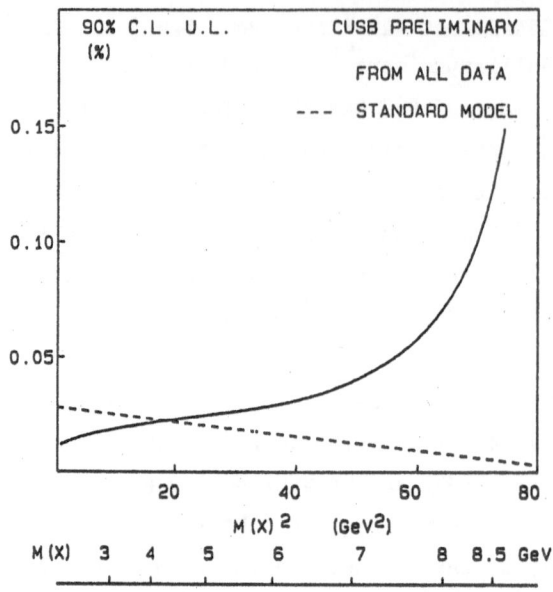

Fig. 47. The CUSB 90% CL upper limit of the branching fraction for radiative
Υ decay into minimal Higgs bosons.

signals are observed for a photon with an energy between 0.7 and 5 GeV. So
combining NaI and BGO results, CUSB has estimated an upper limit for the
branching ratio of Υ in γH^0, as a function of m_{H^0} and at the 90% CL, which
is reproduced in Fig. 47. This excludes the mass range which goes from 1.2
to 4.2 GeV/c^2, for a minimal Higgs. However, as pointed out by Vysotsky,
the QCD radiative corrections, which have not been taken into account in
this analysis of the CUSB experiment, can be as important as 50%. So the
limits obtained by CUSB are no longer valid; more data are needed.

   Another example of the search for Higgses of very small mass is the
one motivated by the 'supermonojet' events found in 1983 by the UA1
experiment[8]. Glashow and Manohar[56] interpreted these events as a possible
decay of Z^0 into two non-standard Higgses h_1^0 and h_2^0; h_1^0 would be a stable
particle with a mass less than or equal to twice the μ mass; h_2^0 would decay
into $f\bar{f}$ or $f\bar{f}h_1^0$ (where f is a fermion). Such objects, if true, should also
be produced in e^+e^- colliders via the reaction $e^+e^- \rightarrow Z^0 \rightarrow h_1^0 h_2^0$. No such
experimental evidence has been found at PETRA and PEP. Moreover, as already
seen in Section 2 the supermonojets have become simple monojet events.

   So let us see what are the real hopes of finding the Higgs.

   4.1.2 <u>Search for low-mass Higgses, i.e. $O(10 \; GeV/c^2) \lesssim m_{H^0} \lesssim 2m_{W}.$</u> If
the Higgs mass is at most around the mass of the W, say $\lesssim O(100) \; GeV/c^2$, the
e^+e^- colliders, such as TRISTAN, SLC, and LEP I and II, have the best chance
of discovering them. Depending on the luminosity, SLC will be able to
discover a Higgs with a mass of the order of 20 GeV/c^2 [57]; LEP I will

certainly be able to look for Higgses with masses of the order of 30 or 40 GeV/c^2; LEP II will be able to scan the 100 GeV/c^2 mass range. They will do it by looking at the process:

$$e^+e^- \to Z^0 + H^0 ,$$

where Z^0 or H^0 will be virtual; Z^0 decays into l^+l^- and H^0 into $f\bar{f}$ where f = c, b, τ, and Q. For very low mass Higgses--say < 20 GeV/c^2--$\tau^+\tau^-$, $c\bar{c}$, and $b\bar{b}$ decay modes are the most important ones; $\tau^+\tau^-$ gives the cleanest signature. If the top mass is around 40 GeV/c^2, $t\bar{t}$ becomes the dominant decay mode once the Higgs mass is larger than twice the top mass, say 100 GeV/c^2.

Despite the fact that e^+e^- colliders are certainly the best ones to look for Higgses in this mass range, pp colliders may also actively participate in this search. A study is under way[58] to see how the Tevatron $p\bar{p}$ collider with a c.m. energy of 2 TeV or a pp collider with higher c.m. energy, say 10 TeV, would be able to search for this object. At these energies the dominant mode to produce H^0 is through the qq or gg fusion mechanism; WW fusion starts to become of the same order when dealing with 10 TeV or 18 TeV pp machines and a Higgs mass of 200 GeV/c^2. Another production mechanism is through the hadroproduction:

$$u\bar{d} \to W \to W + H^0 \quad \text{or} \quad d\bar{d} \to Z^0 \to Z^0 + H^0 .$$

This type of process is lower by a factor of at least 10 than the dominant mode, but if we consider the decay of W into $q\bar{q}$ or Z^0 into $\nu\bar{\nu}$ we may then obtain quite interesting signatures. This is shown in Fig. 48, where some displays (Lego plots) corresponding to such events are presented. These events have been submitted to a simulation program which reproduces the Collider Detector Facility (CDF) at Fermilab. In Table 6 we present the cross-section for different Higgs masses, say 20, 50, and 70 GeV/c^2 and different luminosities of the Tevatron. We consider only the case where the Higgs decays into $\tau^+\tau^-$ and each τ decays into a charged pion; this gives a rather clean signal in the case of pp collisions, much cleaner than if the Higgs decays into $b\bar{b}$ (which is certainly the dominant decay mode at this mass range). The numbers quoted in this table include all the branching ratios and are computed using PYTHIA Monte Carlo calculations[59]. In Table 7 we quote the case of a 10 TeV pp collider working with a luminosity of 10^{32} and look at the possibility of producing Higgses of 50 or 100 GeV/c^2. The rates obtained are not unreasonable. However, much more work is needed because, as we will see in the next subsection, the main problem with pp

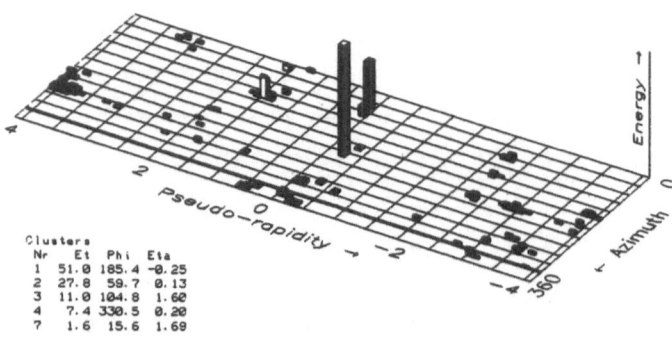

a) $p\bar{p} \rightarrow W^{\pm} \rightarrow W^{\pm} + H^0$
at $\sqrt{s}$ = 2 TeV and
m_{H^0} = 20 GeV/c^2;
$W^{\pm} \rightarrow q\bar{q}$ and $H^0 \rightarrow \tau^+\tau^-$;

Run   1 Event   6   FILE C$ADAT:CDF_WH_1.DAT   5-MAY-1986 03:47

DAIS E total   Eta-Phi LEGO Plot
Etotal = 404.3 GeV   Et(scale)= 130.0 Gev
Max tower E = 49.9 Min tower E= 0.50
N clusters   5   Et(miss)= 18.4 at Phi= 172.5 Deg.

Clusters
Nr   Et   Phi   Eta
1   37.1   1.5 -1.35
2   31.7 237.7 -1.55
3   23.8   93.3 -0.43
4   4.7 189.2 -2.12
6   1.0 232.5   2.01
3   23.8   93.3 -0.43

b) $p\bar{p} \rightarrow Z^0 \rightarrow Z^0 + H^0$
at $\sqrt{s}$ = 2 TeV and
m_{H^0} = 70 GeV/c^2; $Z^0 \rightarrow$
$\nu\bar{\nu}$ and $H^0 \rightarrow \tau^+\tau^-$;

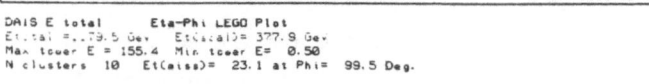

Run   1 Event   8   FILE C$ADAT:CDF_ZH_3.DAT   5-MAY-1986 03:20

DAIS E total   Eta-Phi LEGO Plot
Etotal = 272.5 GeV   Et(scale)= 123.3 GeV
Max tower E = 52.4 Min tower E= 0.50
N clusters   5   Et(miss)= 39.5 at Phi= 325.3 Deg.

Clusters
Nr   Et   Phi   Eta
1   51.0 185.4 -0.25
2   27.8   59.7   0.13
3   11.0 104.8   1.60
4   7.4 330.5   0.20
7   1.6   15.6   1.69

c) $p\bar{p} \rightarrow H^0 \rightarrow W^{\pm}W^{\mp}$ at
$\sqrt{s}$ = 10 TeV and m_{H^0} =
200 GeV/c^2; $W^{\pm} \rightarrow q\bar{q}$
and $W^+ \rightarrow 1^+\nu_1$.

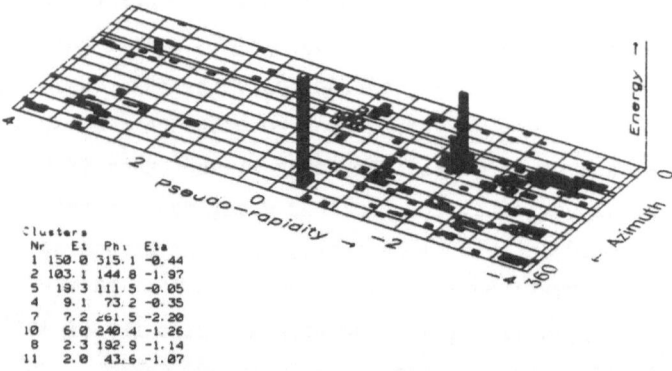

Run   1 Event   9   FILE C$ADAT:PPTEN_WW.DAT   7-MAY-1986 02:32

DAIS E total   Eta-Phi LEGO Plot
Etotal = 779.5 Gev   Et(scale)= 377.9 Gev
Max tower E = 155.4 Min tower E= 0.50
N clusters 10   Et(miss)= 23.1 at Phi= 99.5 Deg.

Clusters
Nr   Et   Phi   Eta
1   150.0 315.1 -0.44
2   103.1 144.8 -1.97
5   18.3 111.5 -0.05
4   9.1   73.2 -0.35
7   7.2 261.5 -2.20
10   6.0 240.4 -1.26
8   2.3 192.9 -1.14
11   2.0   43.6 -1.07

Fig. 48. Pictures of Higgs events simulated in the CDF detector and
generated by the PYTHIA MC according to different processes.

Table 6. Production of Low-Mass Higgses at the $p\bar{p}$ Tevatron Collider
($\sqrt{s}$ = 2 TeV): Estimate of Cross-Section and Rates
(corresponding to an integrated luminosity L for 1 year
of run of 10^{36} or 5×10^{38} cm^{-2})

Process	Mass (H^0) (GeV/c^2)	σ * (pb)	Rates (No. of events per year)	
			L = 10^{36}	L = 5×10^{38}
$p\bar{p} \to H^0 \to \tau^+\tau^-$	20	5.5	1	500
$p\bar{p} \to H^0 \to \tau^+\tau^-$	50	0.3	-	85
$p\bar{p} \to H^0 \to \tau^+\tau^-$	70	0.16	-	40

* The cross-sections quoted in this table are computed with PYTHIA M.C. and
include the branching ratio of H^0 in $\tau^+\tau^-$ (or W^+W^-) and the branching ratio
of each τ in $\pi^{\pm}(n\pi^0)\nu_{\tau}$.

Table 7. Production of Low-Mass Higgses at a pp Collider
($\sqrt{s}$ = 10 TeV): Estimate of Cross-Section and Rates
(for an integrated luminosity per year of 10^{39})

Process	Mass (H^0) (GeV/c^2)	σ * (pb)	Rates (No. of events per year)
$pp \to H^0 \to \tau^+\tau^-$	50	3.2	600
$pp \to H^0 \to \tau^+\tau^-$	100	0.15	40
$pp \to H^0 \to W^+W^-$	200	4	4000

* The cross-sections quoted in this table are computed with PYTHIA M.C. and
include the branching ratio of H^0 in $\tau^+\tau^-$ (or W^+W^-) and the branching ratio
of each τ in $\pi^{\pm}(n\pi^0)\nu_{\tau}$.

colliders is the very high background provided by standard processes which
mimic well the signals.

Concerning ep colliders, very similar remarks to those made for pp
colliders are applicable; moreover their main constraint is the e beam
energy which is difficult to get higher than 100 GeV energy (lectures of
Wolf[60]).

4.1.3 <u>Search for high-mass Higgses: $m_{H^0} > 2m_W$</u> [61]. It may well happen
that, even in the Standard Model, the Higgs is a heavy object; by heavy we
mean $\geq 2m_W$.

In this case, the dominant decay modes for the Higgs is $H^0 \to W^+W^-$ or
Z^0Z^0. Moreover, to produce such massive Higgses a very-high-energy machine
is needed. LEP 200 (if built by 1995) would be at the limit of producing a

Higgs of about 200 GeV/c^2 mass; we know that with our present technological knowledge it is not possible to build a 1 TeV e$^+$e$^-$ collider, whereas we are able to design a very-high-energy pp collider such as an SSC (i.e. 40 TeV in c.m. energy). So the 1 TeV/c^2 range becomes 'the mass scale' of such high-energy pp colliders.

This domain is, at the present time, only reserved for prospects. But it is clearly a very active field as it is an introduction to the physics and the technologies of the end of this century in high-energy physics. So we will outline, in this subsection, the ways these studies are being made to prepare the work that will be done in about 10 years from now.

Computations from Ref. 61, as summarized in Table 8, give the Higgs-mass discovery limit as a function of the c.m. energy and the luminosity in the reaction pp → H^0 → W$^+$W$^-$. The schema reproduced in Fig. 49 gives the different mass ranges accessible to the various machines which should be running by the end of this century. It shows that a machine such as the SSC pp collider, i.e. high energy and high luminosity would be able to scan masses in the range from 200 GeV/c^2 to 1 TeV/c^2 with a favoured range around 1 TeV/c^2.

Table 8.   Higgs-Mass Discovery Limit as a Function of
√s in pp → H(→ W$^+$W$^-$)

∫ L dt (cm^{-2})	√s (TeV)	m_{H^0} discovery limit (GeV/c^2)
10^{38}	20	–
	40	200
10^{39}	20	250
	40	400
10^{40}	20	700
	40	1000

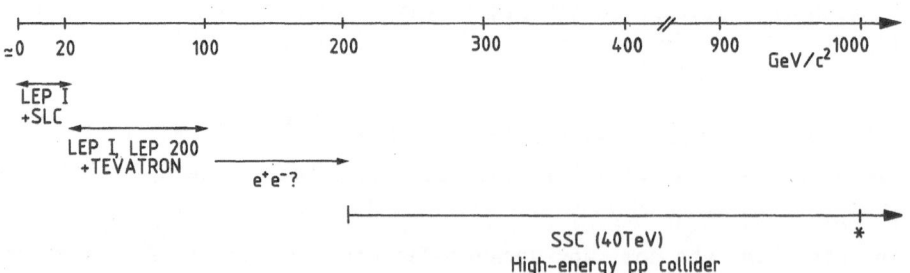

Fig. 49. Higgs mass range accessible to all the machines which are built or are going to be built by the end of this century.

Table 9. Main Scenarios of the Search for the High-Mass Higgs at a pp Collider ($\sqrt{s}$ = 40 TeV)

Process	Rate			Backgrounds
	m_H = 150 (pb)	m_H = 300 (pb)	m_H = 800 (pb)	
$q\bar{q} \rightarrow H \rightarrow W^{\pm}W^{\mp}$ where $W^{\pm} \rightarrow 1^{\pm}\nu$ $W^{\mp} \rightarrow q\bar{q}$	-	1.32	0.156	$pp \rightarrow q\bar{q}$, qg, gg + W $pp \rightarrow W^+W^-$, WZ $pp \rightarrow W$ + jets
$q\bar{q} \rightarrow H \rightarrow Z^0 Z^0$ where $Z^0 \rightarrow \nu\bar{\nu}$ $Z^0 \rightarrow 1^+1^-$	-	0.081	0.0095	$pp \rightarrow ZZ$, WZ $pp \rightarrow Z$ + jets $pp \rightarrow qg$ + jets
or both Z^0's $\rightarrow 1^+1^-$	-	0.013	0.0016	
or $Z^0 \rightarrow 1^+1^-$ $Z^0 \rightarrow q\bar{q}$	-	0.22	0.026	
$q\bar{q} \rightarrow H + W$, Z where W, Z $\rightarrow q\bar{q}$ $H \rightarrow \tau^+\tau^-$	0.0052	-	-	$pp \rightarrow W$ + jets and/or $pp \rightarrow Z$ + jets
or $H \rightarrow t\bar{t}$	6.9	-	-	

The main scenarios to search for Higgs at high-energy pp colliders are listed in Table 9. For the time being, people are mainly concentrating on the study of the following two processes:

$$pp \rightarrow H^0 \rightarrow W^+W^- , \qquad (8)$$

where one W decays leptonically and the second one decays hadronically; and

$$pp \rightarrow H^0 \rightarrow Z^0 Z^0 , \qquad (9)$$

where one Z^0 decays leptonically and the second one into $\nu\bar{\nu}$.

We should like to review briefly what are the main difficulties of detecting such particles in a high-energy pp collider such as the SSC. This will serve as an introduction to the last part of this section. All the results given in this subsection are from Ref. 62.

i) **Main characteristics of these events**

They are very rare processes, with cross-section less than or equal to 1 nb. They are hidden by large backgrounds (typically the ratio signal/background $\lesssim 10^{-3}$), mainly due to Standard Model processes which mimic the signal quite well. They may be produced in pp interactions through different processes, but all these processes give signatures characterized by jet(s), missing energy, and lepton(s); plus the fact that the H^0 decays into W or Z^0 pairs which are *longitudinally polarized* (this imposes an important additional constraint as we will see later). So, apart from this last point, the search for Higgs particles will have very similar points to the search for other new particles such as heavy supersymmetric particles[63]. The main requirements are high luminosity, sophisticated trigger (with different levels) and a very performing 4π detector.

Moreover, we may estimate *quantitatively* the characteristics of events by generating these processes by a Monte Carlo program[59] and simulating them through a detector.

We report in Table 10, as an example, the results of this estimate on a sample of events reproducing W-pairs, produced in a 4π fine-grained

Table 10. Characteristics of W-pair and W + jets Events as given by PYTHIA[59] and ISAJET

MC Process[*]	$\langle E_T^{miss} \rangle$ (GeV)	$\langle E_T^l \rangle$ (GeV)	$\langle n_{jet}^{E_T > 50\ GeV} \rangle$
a	171.3	194.5	3.2
b	204.5	235.9	3.3
c	62.1	61.8	2.4
	$\langle E_{isol}^l \rangle$ (GeV)	$\langle m_T^{l+\nu} \rangle$ (GeV/c^2)	
a	6.5	70.0	
b	6.2	97.7	
c	2.8	64.5	
	$\langle E_T^{j_1} \rangle$ (GeV)	$\langle E_T^{j_2} \rangle$ (GeV)	$\langle E_T^{j_3} \rangle$ (GeV)
a	371.4	40.8	14.6
b	410.4	69.0	18.8
c	107.2	24.0	11.3
	$\langle m_{j_1 j_2} \rangle^{**}$ (GeV/c^2)	$\langle m_{WW} \rangle$ (GeV/c^2)	$\langle \delta m_{WW} \rangle$ (GeV/c^2)
a	90.3, 82.7, 76.1	992.0	4.0
b	83.7, 76.8, 71.2	796.0	8.1
c	30.7, 23.9, 17.1	40.0	346.6

[*] Process (a) refers to pp $\rightarrow$ $W_\pm^+ W_\pm^-$ as generated by ISAJET.
Process (b) refers to pp $\rightarrow$ $W^+ W^-$ as generated by PYTHIA.
Process (c) refers to pp $\rightarrow$ W + jet as generated by ISAJET.
[**] For different threshold values (0, 0.5, and 1 GeV) on the E_T of each cell

calorimeter. We apply also the same simulated detector to Z^0-pair events. One of the main parameters in this last case is the missing transverse energy produced by the decay of one of the Z^0's into a $\nu\bar{\nu}$ pair. We show in Figs. 50a, b, and c the missing-energy spectra we may expect for Higgs with a mass of 0.3 or 0.6 and 1 TeV/c^2.

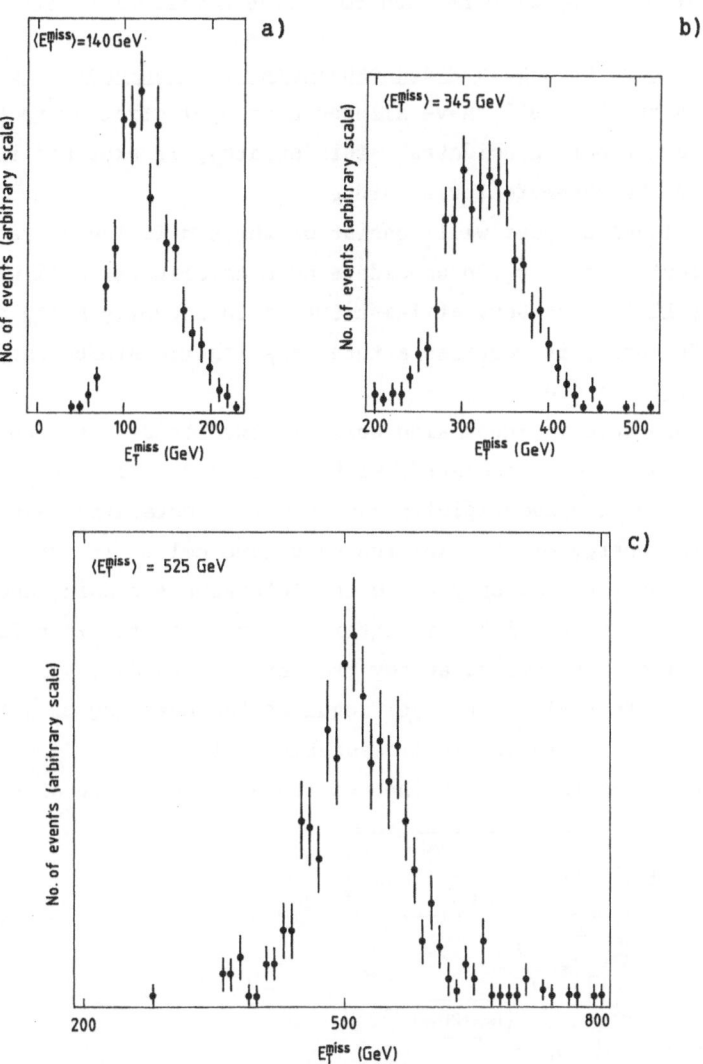

Fig. 50. Transverse missing energy spectrum provided by the process pp →
$Z^0 Z^0$, where one Z^0 → $\nu\bar{\nu}$ and the other Z^0 → l^+l^- at $\sqrt{s}$ = 40 TeV
a) for $100 \leq P_T^{Z^0}$ (GeV/c) ≤ 200, which corresponds to the case of an
H^0 with a mass of ~ 300 GeV/c^2;
b) for $300 \leq P_T^{Z^0}$ (GeV/c) ≤ 400, which corresponds to the case of an
H^0 with a mass of ~ 600 GeV/c^2;
c) for $400 \leq p_T^{Z^0}$ (GeV/c) ≤ 600, which corresponds to the case of an
H^0 with a mass of ~ 1 TeV/c^2.

Such studies give an idea of the spectrum as well as the average value of the main parameters describing each type of event [namely, missing energy, electron momentum, jet momentum, and number of jets in the case of the process (8); missing energy and lepton momentum in the case of the process (9)].

These studies can also be used to define a trigger strategy.

ii) <u>Triggering on high-mass Higgs candidates at high-energy pp colliders</u>

Preliminary studies[64], have allowed a trigger strategy to be defined in order to select W-pair candidates. This strategy is sketched in Fig. 51 and is defined in the three following steps.

A first-level trigger will consist of the two following requirements. First it selects the electron candidate as a calorimeter cell with a transverse energy $E_T^{cell} > 25$ GeV, at least 80% of this energy being electromagnetic. Secondly, it requires a total missing transverse energy of more than 40 GeV for the event.

The second-level trigger also works in two steps. It requires the electron candidate to be isolated, with a surrounding region of ± 5 calorimeter cells in both pseudorapidity and azimuth, containing less than 20% of the transverse energy of the electron candidate cell. It asks, in addition, that in the 'opposite hemisphere' to the (electron + missing energy) system, there is either one jet with a transverse energy greater than 80 GeV or two jets each having a transverse energy greater than 40 GeV.

Finally, a third-level trigger requires the matching of a track with $p_T > 10$ GeV/c with the candidate electromagnetic cell.

Such requirements lead, at the end, to a trigger rate of ~ 1 Hz.

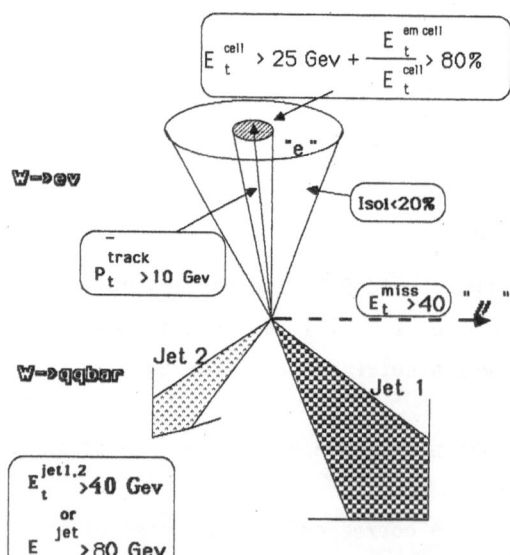

Fig. 51. Schema of the strategy to trigger on high-mass W pairs at the SSC.

iii) <u>Main detection problems</u>

The appropriate detector to perform such a search is certainly a 4π detector, which should be able to make a precise measurement of the total missing transverse energy, to tag electrons and make an accurate measurement of their characteristics (i.e. momentum, charge and very good e/π rejection), and to measure precisely the jet content of the events. Let us now consider this last point.

Most of the physics which will be studied at the SSC implies new particles which decay into $q\bar{q}'$ or $qq\bar{q}$ (plus sometimes additional neutrinos). This is the case, for instance, of new heavy quarks, new heavy leptons, supersymmetric particles, etc. This is why the problem of recognizing tricky jet patterns is a main concern[65].

The present problem, that we take here as an example, is to differentiate events as given by

$$pp \to H^0 \to W^{\pm} W^{\bar{+}} \to (q\bar{q}) + (1^{\bar{+}}v_1)$$

from those given by the following standard backgrounds:

$$\begin{cases} pp \to W^{\pm} + (g \text{ or } q) \text{ jet} \\ \quad \hookrightarrow 1^{\pm}v_1 \\ pp \to W^{\pm} + qg \text{ (where W is radiated by one of the partons)} \\ pp \to W^{\pm} W^{\bar{+}} \to q\bar{q} + 1^{\bar{+}}v_1 \ . \end{cases}$$

It is essential there to measure properly both W's and, in particular, to be able to reconstruct a W from its hadronic decay products. This is not an easy problem. As we see from the plot in Fig. 52, which shows the distribution of the invariant mass of the two jets $m_{1\ 2}$ it is not sufficient to

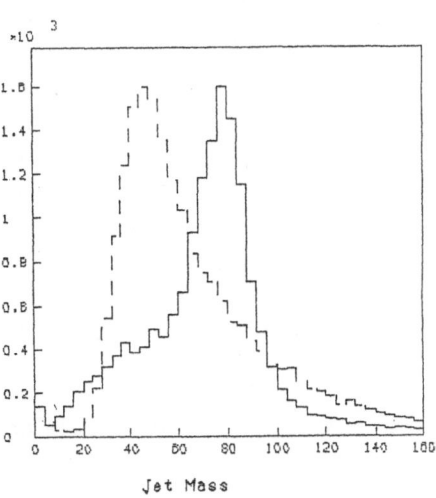

Fig. 52. Distributions of the invariant mass of two jets produced in the case of W's decaying hadronically (solid line) and of a QCD jet (dashed line).

Jet Mass

cut on $m_{1,2}$ around the W mass as the standard background is about 200 times higher than the signal[66]. So a much more tricky game has to be played.

We show here some aspects of this game, namely we will discuss how the energy spectrum of an event may be reproduced and also how to reproduce the polarization properties.

By projecting the transverse energy of each cell of the calorimeter onto the main axis s of the event in the (η = pseudorapidity, φ = azimuth) plane E_T^s we obtain an energy spectrum related to the jet structure of this event. This is shown in Fig. 53a where we have plotted the E_T^s spectrum of

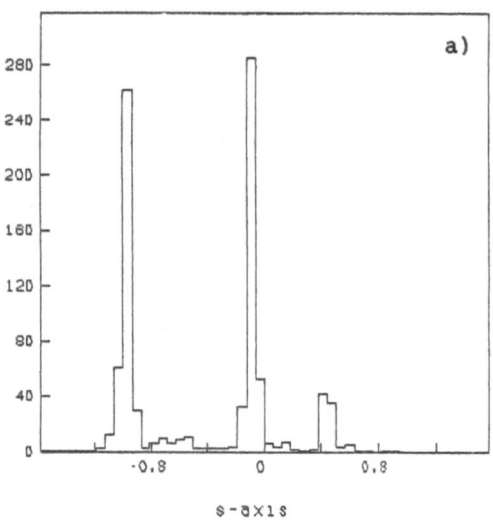

$s-axis$

E_T^s spectrum of event pp → W + jet

THETA= 50   PHI= 30

Corresponding Lego plot

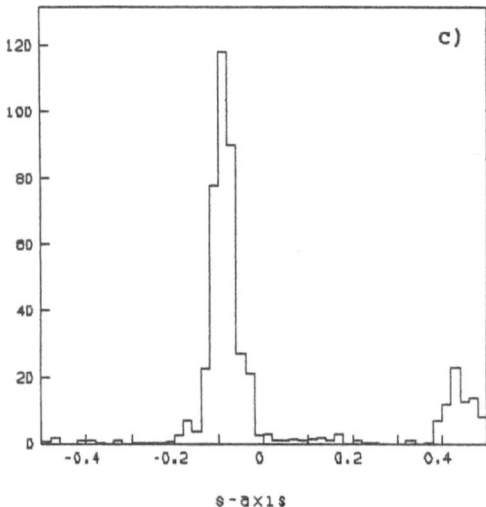

$s-axis$

E_T^s spectrum of a (W → q$\bar{\text{q}}$) system only

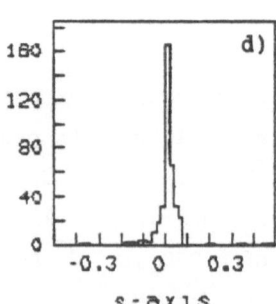

$s-axis$

E_T^s spectrum of a typical QCD jet only

Fig. 53

an event generated according to the process pp → W + jet; the transverse
energy of the recoil jet is required to be > 300 GeV (in order to correspond
to a Higgs mass of ~ 700 GeV/c^2). We clearly see a three-peak structure in
this energy spectrum (Fig. 53a); one peak corresponds to the QCD jet (recoil
jet), the other two correspond to the (W → q$\bar{q}$) system. The Lego plot in
Fig. 53b is another way to represent the three-jet structure of the event we
are discussing. If we now expand on the W system of the same event, we get
the E_T^S spectrum corresponding to the two quarks that are the decay products
of the W (Fig. 53c). It contrasts with the E_T^S spectrum of a typical QCD jet
as shown in Fig. 53d.

However, gluon bremsstrahlung may distort such nice two-jet structure
topologies by adding extra very neigbouring jets to the q$\bar{q}$ system we are
looking at (Fig. 54). So Fig. 55 shows a sample of a typical q$\bar{q}$ E_T^S spectrum
we may get in reality.

We can then conclude that in order to search for hadronic decays, we
may use the 'E_T^S-spectrum technique' to select the events which have the
expected 2-jet structure.

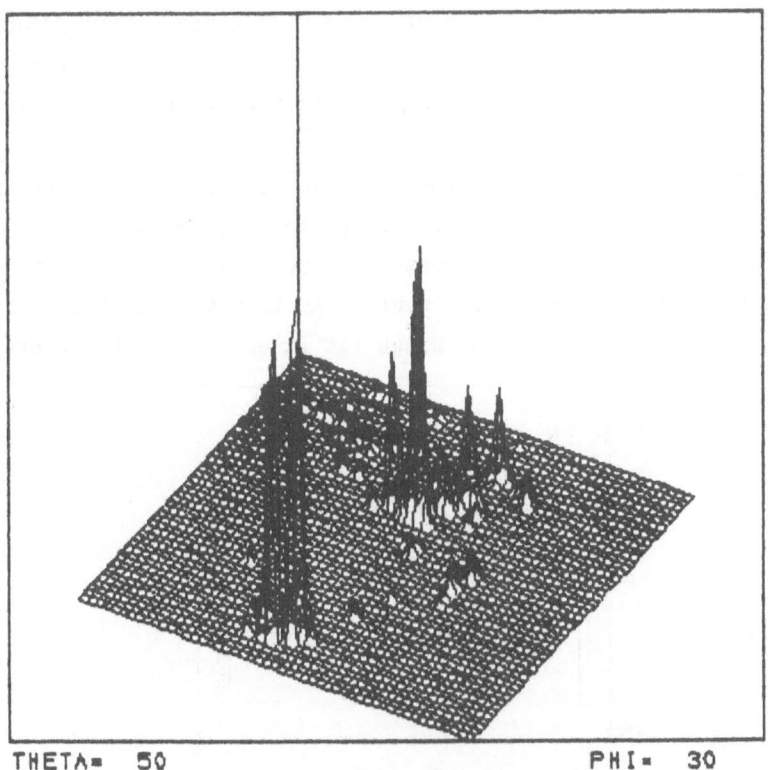

THETA= 50   PHI= 30

Fig. 54. Effect of gluon bremsstrahlung in the display of an event, pp →
W$^{\pm}$W^{+}, where one of the W's decays into c$\bar{s}$ (see corresponding Lego
plot shown in this picture) and each quark, c and $\bar{s}$, emits gluons.

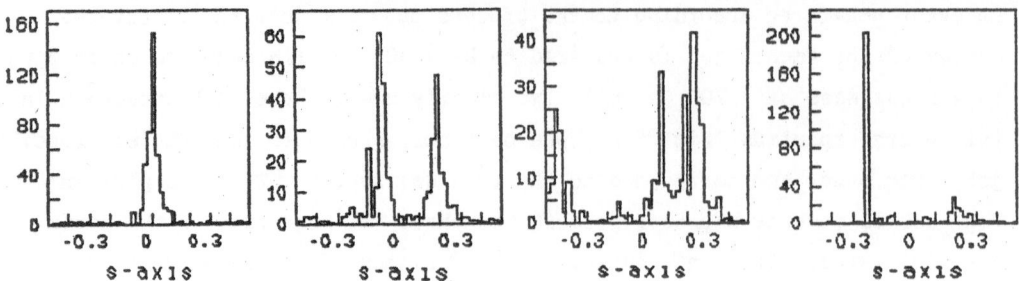

Fig. 55. E_T^s spectrum of different ($W \rightarrow q\bar{q}$) systems.

There are also some ways to recognize an important property of the event we are looking at, namely the polarization effects.

Whereas usually W's are transversely polarized, the ones produced by H^0 decays are *longitudinally* polarized. This provides a very useful piece of information that we can define by a parameter

$$\cos \theta^* = \frac{p_L^1 - p_L^2}{p_L^1 + p_L^2} ,$$

where $p_L^{1,2}$ are the longitudinal momentum of the jets 1 and 2 along the W-axis in the laboratory system. A W produced with a transverse polarization will have an angular distribution in $1 + \cos^2 \theta^*$, whereas a W produced with a longitudinal polarization will have an angular distribution in $\sin^2 \theta^*$, Fig. 56.

These two examples show that the detection of massive Higgses in a high-energy pp collider will require very performing detectors and sophisticated analysis; perhaps this will be possible.

To conclude this subsection, I would like to say that it is very unlikely that we will discover the Higgs (if it exists) with the present

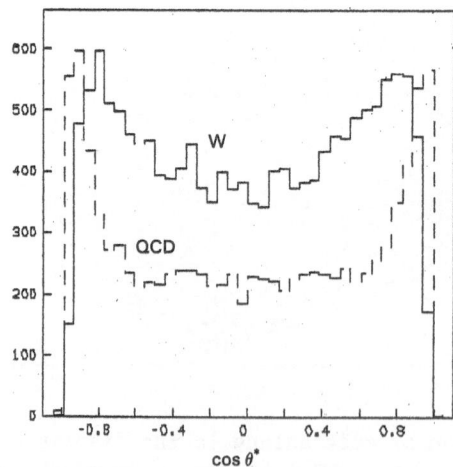

Fig. 56. Distribution of $\cos \theta^* = (p_L^1 - p_L^2)/(p_L^1 + p_L^2)$ as given by the ($W \rightarrow q\bar{q}$) system and standard QCD jets.

machines. We have to wait first for a new generation of e^+e^- colliders and, in particular, for LEP 200, but maybe the Higgs mass is such that it will not yet be discovered by 1995 and will need much-higher-energy machines such as the SSC pp collider. Whereas the Higgs is predicted by the Standard Model, it is in fact intimately related to new thresholds for unconventional physics, staying somewhere beyond the 1 TeV range. For the next generation of high-energy physics, 'puzzles beyond the Standard Model' are related to the construction of new machines and new detectors able to scan the TeV range.

4.2 New Accelerators, New Detectors, and New Technologies for the New Physics beyond the 1 TeV Range

This subsection is essentially to give a brief introduction on how, at the present time, we imagine the high-energy physics will be, from the experimental point of view, in 10 years from now. We could be totally wrong in our predictions; however, we have already to think in terms of this time scale. Ten years is not too much, as we will see, to prepare and design all the future projects.

Figure 57 shows the machines which have been built since the very beginning of high-energy physics and also the ones which are planned.

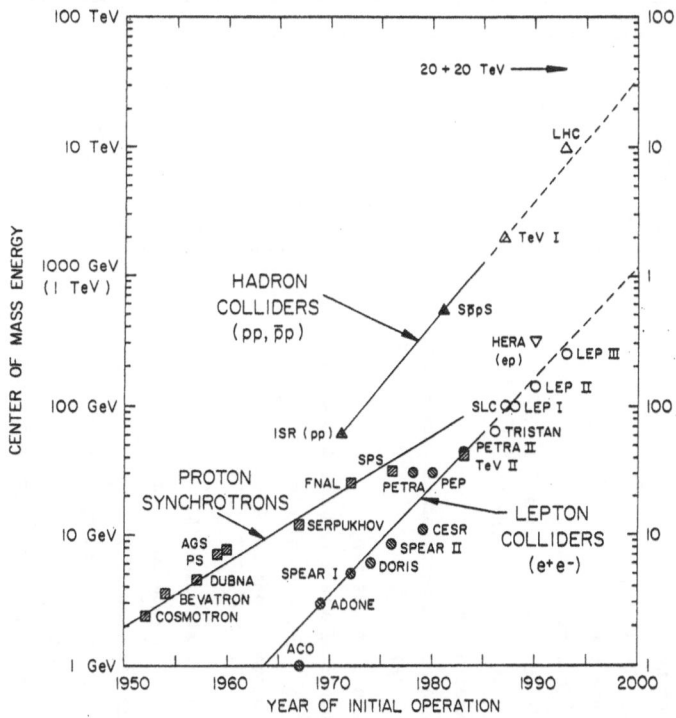

Fig. 57. Centre-of-mass energy accessible to all the machines which have been built up to now or are going to be built by the end of this century.

Table 11. Summary of the SSC Parameters

Parameter	Value
Type of machine	pp collider
Beam energy (TeV)	20
Circumference (km)	82.944
Revolution frequency (Hz)	3614
Luminosity: at $\beta^* = 0.5$ m (cm^{-2}·s^{-1})	10^{33}
at $\beta^* = 10$ m (cm^{-2}·s^{-1})	5.6×10^{31}
Bunch separation (m)	4.8
Nominal interaction point space between magnetic quad ends (m)	20
Number of bunches per ring	1.71×10^4
Average number of reactions per bunch crossing at 10^{33} cm^2·s^{-1} for a total cross-section of 90 mb	1.4
Number of protons: per bunch	7.3×10^9
per ring	1.27×10^{14}
Beam current: peak value (A)	2.0
average value (mA)	73
Beam energy per ring (MJ)	405
Normalized transverse emittance (rad·m)	1.0×10^{-6}
Luminosity lifetime (d)	~1
Synchrotron radiation power (kW per ring)	9.1
Synchrotron radiation energy damping time (h)	12.5
Beam-beam tune shift (max. value)	0.8×10^{-3}
r.m.s. energy spread: at injection	1.75
at 20 TeV	0.5×10^{-4}
β max. (m)	332
β min. in arc (m)	111
Horizontal dispersion: maximum (m)	3.92
minimum in arc (m)	2.36
Crossing angle: typical value (μrad)	75
maximum value (μrad)	150
Distance between adjacent interaction points (km)	2.40
Angle between adjacent interaction points (mrad)	10.6
Magnetic field dipole (T)	6.6
Magnetic radius of curvature (km)	10.1
Dipole length: magnetic (m)	16.54
slot (m)	17.34
Arc quadrupole length: magnetic (m)	3.32
slot (m)	4.32
Number of regular SC dipoles (both rings)	7680 (horizontal dipoles)
Number of quadrupoles (both rings)	1776
Acceleration period (s)	1000
Energy gain per turn per proton (MeV)	5.26
Peak RF voltage (MV)	20
Total RF power ring (MW)	2
RF system slot length (per ring) (m)	25
r.m.s. bunch length (cm)	6.0-7.3

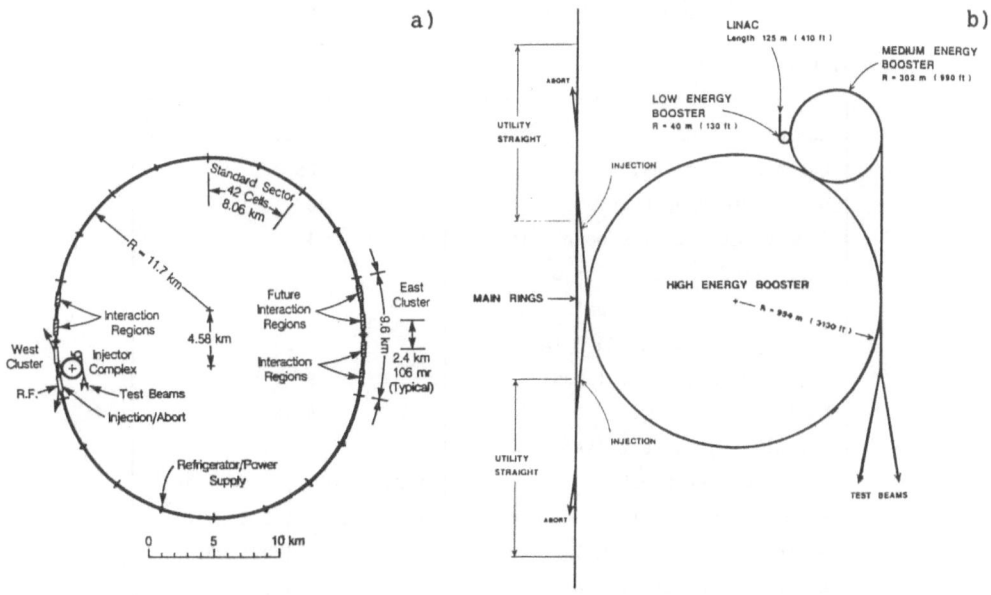

Fig. 58. SSC collider: a) SSC collider ring layout, and b) injector.

Concerning the possibility of scanning the 1 TeV/c^2 mass range, there are two main projects: the LHC pp collider[67] in Europe and the SSC machine[68] in the States.

We reproduce in Fig. 58a a view of the SSC collider ring layout as well as the injector (Fig. 58b) and, in Table 11, the main characteristics of such a machine as defined by the SSC central design group[68]. In Fig. 59 the LHC project is shown as it could be designed inside the LEP tunnel and we report in Table 12 its main characteristics. Both projects are now in their final form and, in particular, the SSC project is waiting for final approval.

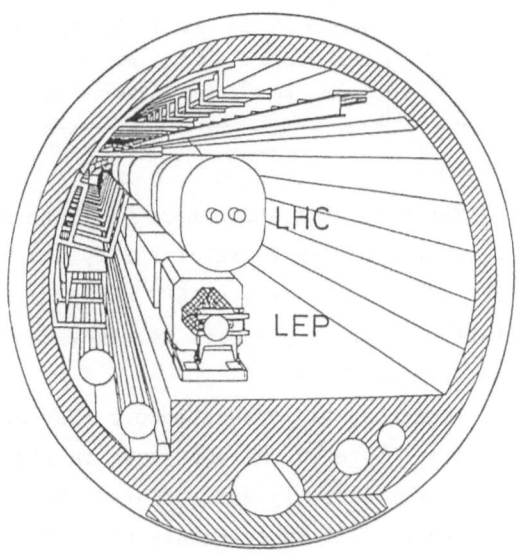

Fig. 59. Transverse view of the
LHC collider in the
LEP tunnel.

Table 12. Main Characteristics of the LHC Project

a) Performance

No. of bunches	3564	
Bunch spacing (ns)	25	
No. of crossing points	8	
β value at crossing points (m)	1	
Normalized emittance $4\pi\gamma\sigma^2/\beta$ (μm)	5π	
r.m.s. beam radius (μm)	11.76	
Full bunch length ($4\sigma_s$) (m)	0.31	
Full crossing angle (μrad)	96	
Maximum energy (TeV)	8.488	
Circulating current (mA)	86	167
Particles per bunch	1.34×10^{10}	2.56×10^{10}
Beam-beam tune shift	1.3×10^{-3}	2.5×10^{-3}
Stored beam energy (MJ)	64.91	126.01
Luminosity half-life (h)	16.6	8.7
Luminosity ($cm^{-2} \cdot s^{-1}$)	4.14×10^{32}	1.56×10^{33}
Luminosity per collision (cm^{-2})	1.03×10^{25}	3.89×10^{25}
$\langle n \rangle$ at Σ = 100 mb	1.03	3.89

b) Magnets

Maximum dipole field (T)	10
Maximum quadrupole gradient ($T \cdot m^{-1}$)	250
Effective gap between dipoles (m)	1.15
Effective gap between dipoles and quads (m)	2
Horiz. dipole corrector strength ($T \cdot m$)	2.5
Vert. dipole corrector strength ($T \cdot m$)	2.5
Maximum sextupole strength ($T \cdot m^{-1}$)	2000
Ramping time (min)	10
Vacuum chamber inner diameter (mm)	~ 40
Coil inner diameter (mm)	50
Horiz. separation between orbits (mm)	180

The main characteristics of these projects are very-high-energy beams: 8.5 TeV for LHC or 20 TeV for SSC, and high luminosity from 4.1×10^{32} to 1.5×10^{33} at LHC or 1×10^{33} at SSC. A lot of activity is going on in relation to these projects to study the physics issues and the detectors to achieve them.

Among the various detectors that are projected to work in the SSC collider, there is the one shown in Fig. 60 which represents a 'classical' 4π detector directly extrapolated from the present detectors such as UA1, CDF, and D0. It is certainly not the final design and we hope that new

Table 12. Main Characteristics of the LHC Project (cont.)

c) RF system

Harmonic No.	35640	
Frequency (MHz)	400.8	
Momentum compaction	4.595×10^{-4}	
Energy	Injection	Operation
	= 450 GeV	= 8.488 TeV
Bunch area (eV·s)	1.0	2.5
Bucket area (eV·s)	1.73	7.5
Circumferential voltage (MV)	16.89	16.89
Synchrotron tune	9.89×10^{-3}	2.28×10^{-3}
Synchrotron frequency (Hz)	111.2	25.6
Bucket half-height	1.21×10^{-3}	2.78×10^{-4}

d) Typical periods

Cell length (m)	119.463	
Bending angle (mrad)	34.195	
Phase advance	$\pi/2$	
No. of quadrupoles	2	
Effective quad. length (m)	2.72	
Number of dipoles	10	
Effective dipole length (m)	9.68	
Maximum β value (m)	202.8	
Minimum β value (m)	35.2	
Maximum dispersion (m)	2.762	
Minimum dispersion (m)	1.329	
Bending radius (m)	2831.4	
Normalized emittance $4\pi\gamma\sigma^2/\beta$ (µm)	5π	
Energy	Injection	Maximum
	= 450 GeV	= 8.488 TeV
Vert. beam radius (4σ) (mm)	2.91	0.67
Horiz. beam radius ($4\sigma + D\delta_b$) (mm)	6.25	1.44

Fig. 60. Design of a conventional
SSC 4π detector (project).

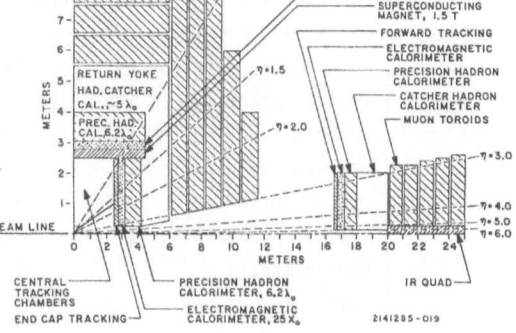

techniques will in fact be used for such a detector. However, we may use it to give a first idea of the sizes and main parameters such as SSC detector will have (Table 13) and also the amount of electronics that will be required.

Table 13.   Summary of the Main Characteristics which Describe a Conventional 4π Detector at the SSC

a) Calorimetry

Calorimeter	e.m	hadronic
Depth (X_0) [λ]	25 [0.8]	6.2
Rapidity range η	-6 to 6	-6 to 6
$\Delta R^{cell} = (\Delta\eta^2 + \Delta\varphi^2)^{1/2}$	0.03	0.06
No. of towers	60,000	14,500
Weight (U + LAr) (t)	~ 375	~ 4000

b) Superconducting magnet

Field (T)	1.5
Length (m)	9
Radius (m)	2.5

c) Tracking parameters

Vertex detector	Silicon	Wire chamber
Rapidity range η	-1.5 to 1.5	-1.5 to 1.5
Length (cm)	10 or 20	100
No. of layers	2	25
No. of strips or sense wires	18,000	8,000
Resolution (μm)	10	100

Tracking	Central	End cap	Forward
Rapidity range $|\eta|$	$\leq$ 1.2	1 to 3	3 to 5
No. of wires	100,000	10,000	12,000
Resolution (μm)	$\leq$ 175	$\leq$ 200	$\leq$ 200

Muon detector (drift tubes)	
Rapidity range η	-5 to 5
No. of wires	87,000
Total weight (t)	30.700
Resolution (μm)	$\leq$ 200

d) Electronics (No. of electronic channels, in thousands)

Wire vertex detector	16	
Central tracking	min 132	max 232
End-cap tracking	20	
Forward tracking	24	
e.m. calorimeter	220	
Hadronic calorimeter	58	
Muon system	184	
Total	min 654	max 754

Another major experimental issue when working at very-high-energy, very-high-luminosity pp colliders is the trigger; this is not the case in e^+e^- colliders. So let us also look at some general features about triggering events at the SSC. In the earlier part of this section we have defined, as an example, a trigger strategy to trigger on W-pair events.

A collider of 40 TeV c.m. energy with $L = 10^{-33}$ cm^{-2} s^{-1} and a corresponding inelastic cross-section of about 100 mb corresponds to a rate of 10^8 collision, per second. This is a rate much too high to be acceptable for a *first-level trigger* so it has to be decreased to 10^4 or 10^5 Hz. The second-level trigger is able then to decrease this rate to 10 or 10^2 Hz. Finally, a third-level trigger refines the rejection of the second-level trigger but does not cut so much on the final rate which is typically between 1 and a few hertz for most of the physics trigger of interest[64]. To achieve these different rejection levels the following strategy is foreseen at present; the first-level trigger relies on very fast signals, coming directly from the calorimetry. The main example is given by the CDF first-level trigger derived from the structure of the calorimetry (4π fine-grained calorimeter)[69]. The energy or transverse energy of each calorimeter cell is considered, provided it is higher than a predefined threshold. Then by adding, subtracting, clustering, and comparing the signals, the following elementary basic trigger levels are built; the missing-energy trigger is defined by calculating the total missing transverse energy in the event as derived from the sum of the energy of each cell and requiring a minimal amount of ~ 50 GeV. The e^- trigger is defined by requiring that a cell has more than a certain amount of transverse energy E_T (say > 25 GeV) and that 80% of this energy is electromagnetic. Clusters are identified in the event by summing each cluster of adjacent hit cells around the highest E_T cell energy. Muon signals are defined by matching the information given by the hadronic cell equivalent to a minimum ionizing particle with a signal in the µ-detector. Total transverse energy triggers are defined by summing all the hit cells in the events and requiring the total transverse energy to be higher than a predefined threshold, say for instance 300 GeV. The total trigger rate corresponding to these types of trigger and associated thresholds has been estimated to be $\simeq 10^5$ Hz [70]. The required thresholds are low enough so that it would be no problem (from the physics requirements point of view) to higher them a little bit such that the trigger rate at the first level can go down to 10^4 Hz.

The ideas for constructing a second-level trigger are mainly based on the use of digitized information, microprocessors for local responsibility, and tracking information to some extent. The tracking information, for instance, quick stiff track reconstruction (as designed for the CDF detector[71]) can be used to ensure a muon track or an electron. There are also proposals to use information from the transition radiation detectors (TRDs) to reconstruct e^- at a second-level trigger[72]. The use of microprocessors allows the refinement of the first-level trigger action and the introduction of sophisticated algorithms using the various sources of digitized information and possibly matching some of them together. This gives refined levels of decision, by access in particular to refined parameters. The third-level trigger could be considered like a park of computers and almost an on-line reconstruction level at least for certain types of events. These are very preliminary ideas, mainly based on our present knowledge and technological capabilities[64]. Certainly a lot of progress is going to be made in this domain, particularly in the next 10 years.

5. CONCLUSION

In these lectures, I have tried to cover a set of specific subjects that are still, nowadays, very puzzling. They include the search for supersymmetric particles in the W-mass range, done at both e^+e^- and $p\bar{p}$ colliders. So far, there is no sign that such particles exist. Some limits may be set, within specific models. However, there is real hope to get a definite answer, in this mass range, by e^+e^- and $p\bar{p}$ machines with higher luminosity, higher c.m. energy and improved detectors. The proof that there is an antigravitational force has still to be given. It is up to the classical experiments (remake of Cavendish's, Eötvös', and Galileo's measurements) to solve the problem. It will certainly be amusing and trivial, then, to verify that antigravity is an unescapable consequence of extended supergravity theories. The β-decay measurements (in the tens of eV range) and the large solar neutrino experiments (in the $\Delta m^2 \sim 10^{-8}$ eV range) should be very active and interesting fields of high-energy physics. They should succeed in filling in the identity card of an old friend, the neutrino (mass, nature, how many relatives, ...?). This last question about the number of neutrinos will also be answered, hopefully soon, by the new generation of e^+e^- and $p\bar{p}$ colliders. Is there a scalar neutral Higgs as predicted by the Standard Model? If yes, what is its mass? Are there other Higgses? The clarification of all these questions related to the crucial Higgs sector may take a certain time and a lot of effort.

These various puzzling aspects of high-energy physics have been expressed here in their experimental form--they address very fundamental questions; they require most of the detection techniques used in our field of physics; they imply the use not only of most of the accelerators we have built or are going to build (in particular the e^+e^- and pp colliders) but also of the sky (non-accelerator neutrino physics). They are very far reaching from the technological point of view.

We are a long way from having answered the main questions of particle physics--and we will need to be really smart to do so. Yes, particle physics is a puzzling field!

Acknowledgements

I wish to thank the organizers of the School, and in particular T. Ferbel. He succeeded in meeting the incredible challenge of creating a stimulating working atmosphere in an environment of coconut palms and coral reefs. I am grateful to S. Aronson, E. Cremmer, A. de Rùjula, P. Fayet, D. Gross, J. Rich, C.A. Savoy, M. Spiro, D. Vignaud, H. Ziock and N. Zaganidis for useful discussions and information.

The Scientific Reports Editing and Text Processing Sections at CERN have realized the typing and presentation of these lectures; I am very much indebted to them for their real professionalism!

REFERENCES

1. L. DiLella, Physics at the CERN p̄p Collider and status of the electroweak theory, these proceedings.
2. Talks on experimental physics given at the Int. Conf. on High-Energy Physics, Berkeley, Calif., 1986.
3. I. Hinchliffe, these proceedings.
4. D. Gross, these proceedings.
5. G. Arnison et al. (UA1 Collab.), Phys. Lett. 139B:115 (1984).
   C. Rubbia, Experimental observation of events with large missing transverse energy accompanied by a jet or a photon(s) in p̄p collisions at $\sqrt{s}$ = 540 GeV, in CERN 84-09, 'Proc. 4th Topical Workshop on Proton-Antiproton Collider Physics', Berne, March 1984, CERN, Geneva (1984), p. 218.
   A. Roussarie, Observation of electrons produced in association with hard jets and large missing transverse momentum in p̄p collisions at $\sqrt{s}$ = 540 GeV, ibid., p. 219.
   P. Bagnaia et al. (UA2 Collab.), Phys. Lett. 139B:105 (1984).
6. For general reviews on supersymmetry, see:
   H. P. Nilles, Supersymmetry, supergravity and particle physics, Phys. Rep. 110:3 (1984).
   D. V. Nanopoulos and A. Savoy-Navarro (eds.), Supersymmetry confronting experiments, Phys. Rep. 105:1 (1984).
   A. Savoy-Navarro, Experimental tests of supersymmetry, Phys. Rep. 105:91 (1984).
   H. E. Haber and G. L. Kane, The search for supersymmetry probing physics beyond the standard model, Phys. Rep. 117:77 (1984).
   J. Ellis, Supersymmetry, supergravity and superstring phenomenology, preprint CERN TH-4255/85, Lectures given at the 28th Scottish Universities Summer School in Physics, Edinburgh, 1985.

7. G. Bartha et al., Search for anomalous single photon production at PEP, Stanford preprint SLAC-PUB-3817 (1985).

   E. Fernandez et al. (MAC Collab.), Phys. Rev. Lett. 52:22 (1984).

   R. Hollebeek, Single photon searches at PEP, Stanford preprint SLAC-PUB-3846 (1985), presented at the SLAC Summer School Institute on Particle Physics, Stanford, Calif., 1985.

8. See all theoretical reports at the Berne (1984) and Leipzig (1985) conferences. A large number of theoretical papers written in 1984 and 1985 tried to find an overall explanation for the 'strange signal' found at the CERN $p\bar{p}$ Collider during the 1983 run: W events à la UA2, monojets and missing-energy events in UA1, unusual Z^0's (both UA1 and UA2 found Z^0's emitting a photon at a large angle). These explanations were essentially expressed in terms of supersymmetry or composite models. These signals were not confirmed by the run of 1984.

9. R. Batley (UA1 Collab.), Missing-energy results from the UA1 experiment, talk given at the 6th Topical Workshop on Proton-Antiproton Collider Physics, Aachen, 1986.

10. A. Astbury et al., The UA1 calorimeter trigger, Rutherford preprint RAL-84-025 (1984).

11. F. E. Paige and S. P. Protopopescu, ISAJET Monte Carlo, Brookhaven report BNL-29777, 1983.

12. A. Savoy-Navarro, Experimental evidence for the decay $W \to \tau\nu_\tau$ of the charged intermediate vector boson at the CERN $p\bar{p}$ Collider, in 'Proc. 5th Topical Workshop on Proton-Antiproton Collider Physics', Saint-Vincent, 1985, World Scientific, Singapore (1985), p. 196.

13. Y. Giraud-Héraud, A. Savoy-Navarro, C. Tao and N. Zaganidis, Using fragmentation to interpret the origin of monojets, internal note UA1-TN/85-86 (1985).

14. UA1 Collaboration, paper in preparation on the study of the $W \to \tau\nu_\tau$ decay.

15. For the search for supersymmetric particles at $p\bar{p}$ colliders, see among others:

    M. J. Herrero, L. E. Ibañez, C. Lopez and F. J. Yndurain, Phys. Lett. 132B:199 (1983); Erratum in 142B:455 (1984).

    G. Altarelli, B. Mele and S. Petrarca, preprint CERN-TH-3822 (1984).

    I. Antoniadis, L. Baulieu and F. Delduc, Z. Phys. C23:119 (1984).

    S. Dawson, E. Eichten and C. Quigg, Fermilab and Berkeley preprint Fermilab-PUB-83/82-THY and LBL-16540 (1984).

    S. Dawson and A. Savoy-Navarro, in Ref. 63.

    J. Ellis and H. Kowalski, Phys. Lett. 142B:441 (1984).

    E. Reya and D. P. Roy, Phys. Lett. 141B:442 (1984) and Phys. Rev. Lett. 53:881 (1984).

    A. Savoy-Navarro, Supersymmetry and (Super)hadron-hadron colliders, 'Proc. 19th Rencontre de Moriond', La Plagne, 1984, Ed. Frontières, Gif-sur-Yvette (1984), vol. 2, p. 95.

    J. Ellis and H. Kowalski, Nucl. Phys. B259:109 (1985).

    A. Savoy-Navarro, Hadron-hadron colliders: SUSY now or later, Saclay preprint DPhPE 86-01 (1986), 'Proc. 3rd CSIC Workshop on SUSY and Grand Unification from Strings to Collider Phenomenology', Madrid, 1985, World Scientific, Singapore (1986), p. 1.

    R. M. Barnett, H. E. Haber and G. Kane, Nucl. Phys. B267:625 (1986).

16. A. Bouquet, J. Kaplan and C. A. Savoy, Phys. Lett. 148B:69 (1984) and Nucl. Phys. B262:299 (1985).

    F. Delduc, H. Navelet, P. Peschanski and C. A. Savoy, Squark pair production mechanism in $p\bar{p}$ collisions, Phys. Lett. 155B:173 (1985).

17. Refer to your usual manual of classical mechanics.

18. The results on geophysical experiments, plotted in Fig. 24, have been
    provided by S. H. Aronson and A. De Rújula (private communications).
    For more information see also:
    F. D. Stacey and G. J. Tuck, Nature 292:230 (1981).
    S. C. Holding and G.J. Tuck, Nature 307:714 (1984).
    F. D. Stacey, Sci. Prog. (Oxford) 69:1 (1984).

19. C. Zachos, Phys. Lett. 76B:329 (1978); PhD Thesis, CalTech (1979).
    J. Scherk, Phys. Lett. 88B:265 (1979);
    J. Scherk, Ecole Normale Sup. preprint LPTENS 79/19, Lecture given at
    the Int. Conf. on Mathematical Physics, Lausanne, 1979;
    J. Scherk, 'Proc. Supergravity Workshop', Stony Brook, NY, 1979, North-
    Holland, Amsterdam (1979), p. 43.
    J. Scherk, 'Proc. Ecole d'Eté de Physique des Particules', Gif-sur-
    Yvette, 1979, IN2P3, Paris (1980), p. 175.
    J. Scherk, Gravitation at short range and supergravity, Ecole Normale
    Sup. preprint LPTENS 80/15, presented at the Europhysics Study Conf.
    on Unification of the Fundamental Interactions, Erice, 1980.

20. E. Fischbach et al., Reanalysis of the Eötvös experiment, Phys. Rev.
    Lett. 56:3 (1986).

21. It is interesting to note that similar results are obtained by the
    'conventional approach' [see S. Aronson's papers in Refs. (25),
    (26) and (30), and references therein].

22. A. De Rújula, private communication.

23. P. Fayet, Phys. Lett. 171B:261 and 172B:363 (1986).

24. R. Eötvös, D. Pekár and E. Fakete, Ann. Phys. (Leipzig) 68:11 (1922).

25. S. H. Aronson et al., Phys. Rev. Lett. 48:1306 (1982) and Phys. Rev.
    D28:476, 495 (1983).
    E. Fischbach et al., Phys. Lett. 116B:73 (1982).

26. C. Talmadge, S. H. Aronson and E. Fischbach, Effects of local mass
    anomalies in Eötvös-type experiments, Prep. 40048-12-N6, presented
    at the 21st Rencontre de Moriond, Les Arcs, 1986.

27. D. P. Coupal et al., Phys. Rev. Lett. 55:566 (1985).

28. D. C. Cundy et al., Measurement of $|\eta_{00}|^2/|\eta_{+-}|^2$, CERN Proposal
    SPSC/81/110, P174 (1981).
    P. Clarke et al., A measurement of the phase difference of η_{00} and η_{+-}
    in CP violation $K \to 2\pi$ decays, CERN proposal SPSC/86/6, P174 Add. 2
    (1986).

29. P. G. Roll, R. Krotkov and R. H. Dicke, Ann. Phys. (NY) 26:442 (1964).
    R. H. Dicke, Sci. Amer. 205:84 (1961).

30. C. Bouchiat and J. Iliopoulos, On the possible existence of a light
    vector meson coupled to the hypercharge current, Phys. Lett.
    169B:447 (1986).
    S. Aronson et al., Experimental signals for hyperphotons, Phys. Rev.
    Lett. 56:1342 (1986).

31. Y. Asano et al., Phys. Lett. 107B:159 (1981) and 113B:195 (1982).

32. An example of a new experiment to study very rare K_L decays at
    Brookhaven can be found in the proposal for the BNL expt. 791, by
    R. D. Cousins et al. (UCLA-Los Alamos-Pennsylvania-Stanford-Temple-
    William and Mary Collab.).

33. M. Banner et al., Phys. Rev. 188:2033 (1969).

34. For general reviews on ν-physics, see
    Ching Cheng-rui and Ho Tso-hsiu, On the determination of the
    neutrino mass - A critical status report, Phys. Rep. 112:1 (1984).
    J. D. Vergados, The neutrino mass and family, lepton and baryon number
    non-conservation in gauge theories, Phys. Rep. 133:1 (1986).
    V. Flaminio and B. Saitta, Neutrino oscillation experiments, Pisa
    preprint INFN PI/AE 85/6 (1985), submitted to Rivista del Nuovo
    Cimento.

      D. R. O. Morrison, Review of neutrino masses and oscillations, preprint CERN-EP/86-44 (1986), lecture given at the 10th Hawai Conference on High Energy Physics, 1985.

      M. Spiro, Particle physics without accelerators (selected topics) CEA-Saclay preprint DPhPE 85-08 (1985), lectures given at the Institut d'Etudes Scientifiques de Cargèse, 1985.

      K. Winter, Neutrino properties, preprint CERN-EP/86-61 (1986), invited talk given at the 2nd ESO/CERN Symposium on Cosmology, Astronomy and Fundamental Physics, Garching, 1986.

35.   V. A. Lyubimov et al., Zh. Exp. Teor. Fiz. 81:1158 (1981), and contribution to the VIth Moriond Workshop on Massive Neutrinos in Particle Physics and Astrophysics, Tignes, 1986.

36.   J. J. Simpson, Phys. Rev. D23:649 (1981) and Phys. Rev. Lett. 54:1891 (1985).

37.   See for instance, the results of the Princeton group in T. Altzitzoglou, Phys. Rev. Lett. 55:799 (1985).

38.   T. J. Bowles et al., A limit on the $\bar{\nu}_e$ mass in free molecular tritium β decay, talk given at the Int. Conf. on Weak and Electromagnetic Interactions in Nuclei, Heidelberg, 1986.

      J. F. Wilkerson et al., The Los Alamos free molecular and atomic tritium β-decay experiment, Los Alamos preprint LA-UR-86-1054, contribution to the VIth Moriond Workshop on Massive Neutrinos in Particle Physics and Astrophysics, Tignes, 1986.

39.   For a review on the geochemical method in double β decay detection, see:

      T. Kirsten, Double β decay detection: geochemical methods, 'Proc. Fifth Workshop on Grand Unification, Providence, RI, 1984, World Scientific, Singapore (1984), p. 268, and references therein.

40.   For a review on direct experiments on double β decay, see

      E. Fiorini, Direct experiments on double β decay, 'Proc. Fifth Workshop on Grand Unification, Providence, RI, 1984, World Scientific, Singapore (1984), p. 283, and references therein.

41.   E. Fiorini, A. Pullia, G. Bertolini, F. Cappellani and G. Rastelli, Nuovo Cimento 13A:747 (1973).

42.   E. Bellotti et al., Xenon time projection chamber for ββ decay, 'Proc. Workshop on the Time Projection Chamber', Vancouver, B.C., Canada, 1983, Amer. Inst. Phys., New York (1984), p. 42.

43.   C. H. Chen, S. D. Kramer, S. L. Allmann and G. S. Hurst, Selective counting of krypton atoms using resonance ionization spectroscopy, unpublished.

44.   V. Zacek et al., Improved limits on oscillation parameters from $\bar{\nu}_e$-disappearance measurements at the Gösgen power reactor, Phys. Lett. 164B:193 (1985).

45.   J. F. Cavaignac et al., Phys. Lett. 148B:387 (1984).

46.   J. M. Losecco et al. (IMB Collab.), Test of neutrino oscillations using atmospheric neutrinos, Phys. Rev. Lett. 54:2299 (1985).

    For a description of the IMB detector, see

    R. Bionta et al., Phys. Rev. Lett. 51:27 (1983).

    T. W. Jones et al., Phys. Rev. Lett. 52:720 (1984).

    H. S. Park et al., Phys. Rev. Lett. 54:22 (1985).

47.   S. P. Mikheyev and A. Yu. Smirnov, contribution to the Tenth Int. Workshop on Weak Interactions, Savolinna, Finland, 1985.

48.   H. A. Bethe, Possible explanation of the solar-neutrino puzzle, Phys. Rev. Lett. 56:1305 (1986).

      M. Cribier, W. Hampel, J. Rich and D. Vignaud, MSW regeneration of solar ν_e in the Earth, CEA-Saclay preprint DPhPE 86-17 (1986).

      W. C. Haxton, The solar neutrino puzzle--comments, Nucl. Part. Phys. 16:95 (1986).

49. J. K. Rowley, B. T. Cleveland and R. Davis, Jr., 'Proc. Int. Conf. on Solar Neutrinos and Neutrino Astronomy', Homestake, 1984, AIP Conf. Proc. No. 126, New York (1985), p. 1.

50. T. Kirsten, Status report on the GALLEX solar neutrino project, contribution to the VIth Moriond Workshop on Massive Neutrinos in Particle Physics and Astrophysics, Tignes, 1986.
   D. Vignaud, The gallium solar neutrino experiment GALLEX, published in the 'Comptes rendus du Congrès de la Société française de physique', Nice, 1985.
   (GALLEX = Heidelberg-Karlsruhe-Milan-Munich-Nice-Rehovot-Rome-Saclay Collab.)
   I. R. Barabanov et al., Pilot installation of the gallium-germanium solar neutrino telescope, 'Proc. Int. Conf. on Solar Neutrinos and Neutrino Astronomy', Homestake, 1984, AIP Conf. Proc. No. 126, New York (1985), p. 175.
   A. E. Chudakov, talk given at the First Symposium on Underground Physics, Saint Vincent, 1985.

51. A. Böhm, Observation of e^+e^- events in a broad energy flow and isolated muons by MARK J, 'Proc. 20th Rencontres de Moriond', Les Arcs, 1985, Ed. Frontières, Gif-sur-Yvette (1985), p. 559.

52. M. Mohammadi, Calculations of the limits on heavy lepton mass, number of neutrino families and cross-section for new physics, internal note UA1-TN/86-71 (1986).

53. R. Barbieri, S. Ferrara, L. Maiani, F. Palumbo and C. A. Savoy, Phys. Lett. 115B:212 (1982).

54. C. Peck et al. (Crystal Ball Collab.), DESY report 84-064 (1984) and SLAC-PUB 3380 (1984).

55. See review of the subject by:
   S. Komamiya, Search for new particles in e^+e^- annihilation, in 'Proc. Int. Symposium on Lepton and Photon Interactions at High Energies', Kyoto, 1985, Kyoto Univ., Kyoto (1986), p. 612.

56. S. L. Glashow and A. Manohar, Phys. Rev. Lett. 54:526 (1985).

57. R. N. Cahn, M. S. Chanowitz and N. Fleishon, Phys. Lett. 82B:113 (1979).

58. This study is done by H. U. Bengtsson, A. Savoy-Navarro and Y. Takaïwa; see A. Savoy-Navarro, report at the Madison Workshop on Physics Simulations at the High Energy, 1986, and report of the Working Group on W, Z and Higgs at SSC at the Summer Study on the Physics of the Superconducting Super Collider, Snowmass, Colo., 1986.

59. H. U. Bengtsson and G. Ingelman, Lund preprint LUTP 84-3 and CERN TH 3820 (1984).
   T. Sjöstrand, Description of the capabilities of PYTHIA, contribution to UCLA Workshop on SSC Physics, Los Angeles, 1986.
   H. U. Bengtsson and T. Sjöstrand, paper in preparation.

60. G. Wolf, HERA: The machine and the physics, these proceedings.

61. J. Gunion and A. Savoy-Navarro, Univ. California (Davis) preprint UCD-86-11, Report of the W/Z/Higgs Working Group at the UCLA Workshop on SSC Physics, Los Angeles, 1986.

62. E. Eichten, I. Hinchliffe, K. Lane and C. Quigg, Supercollider physics, Rev. Mod. Phys. 56:579 (1984).

63. S. Dawson and A. Savoy-Navarro, Report of the Working Group Searching for Supersymmetry at the SSC, 'Proc. Summer Study on the Design and Utilization of the Superconducting Super Collider', Snowmass, Colo., 1984.

64. 'Proc. Workshop on Triggering, Data Acquisition and Offline Computing for High Energy/High Luminosity Hadron-Hadron Colliders', Batavia, 1985, FNAL, Batavia (1985).

65. The development of this technique to recognize the (W → q$\bar{\text{q}}$) system is due to J. Hauptman. The study presented in these lectures has been done by J. Hauptman and A. Savoy-Navarro. For references, see A. Savoy-Navarro, same reports as those in Ref. 58.

66. M. Abud, R. Gatto and C. A. Savoy, Prospects for high energy p$\bar{\text{p}}$ beams: study of hadronic jets and possible intermediate bosons, Ann. Phys. (USA) 122:219 (1979).
   And see difficulties to measure such a W-decay in present UA1 and UA2 experiments: A. Roussarie (UA2 Collab.), talk given at the Int. Conf on High-Energy Physics, Berkeley, Calif., 1986.

67. 'Proc. ECFA-CERN Workshop on Large Hadron Collider in the LEP Tunnel', Geneva-Lausanne, 1984 (report ECFA 84/85, CERN 84-10), CERN, Geneva (1984).

68. SSC Central Design Group, Conceptual design of the Superconducting Super Collider, report SSC-SR-2020 (1986).

69. T. M. Liss, The CDF level 1 and level 2 trigger system, talk given at the VIth Topical Workshop on p$\bar{\text{p}}$ Physics, Aachen, 1986.

70. P. Frahzini et al., Lowest level trigger for SSC general purpose detectors, in Ref. 64, p. 93.

71. L. D. Gladney, N. S. Lockyer and R. Van Berg, The CDF track processor, Prospects for the SSC, in Ref. 64, p. 152.

72. M. Abolins et al., Report of the High Level Trigger Group, in Ref. 64, p. 131.

APPENDIX I: HADRONICITY IN THE UA1 DETECTOR

The main concern in this appendix is to show that to identify a charged
pion, it is necessary to have an accurate measurement of the hadronicity,
i.e. a precise e/π rejection.

To measure the hadronicity in the UA1 detector, one may make use of
various quantities. On the one hand, there are the quantities which are
directly measured by the apparatus. These include, first of all, the energy
deposited in the hadronic cell (E_{had}), behind the e.m. cell candidates. The
energy deposited by an e^- at various energies and different incident angles
has been measured in a beam test (see Fig. I.1). It allows us to calibrate
the corresponding E_{had}. One also uses the energies deposited in each of the
four samples in depth of the central e.m. calorimeter (gondolas); namely:
E_{S1}, E_{S2}, E_{S3}, E_{S4}; moreover the momentum (p^{track}) of the charged track associated to the e.m. cluster and measured by the central tracking is a very
useful piece of information; it is compared to the energy measured in the
corresponding e.m. cell(s).

On the other hand, there are the quantities which are derived from beam
test measurements and/or obtained by a fit. These include the evaluation of

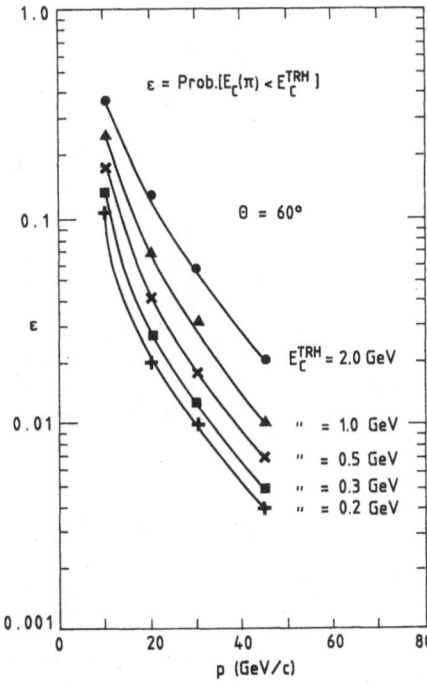

Fig. I.1. Probability ε that a pion, with an incident momentum p (GeV/c) and
an incident angle θ, gives an energy deposit $E_C(\pi)$, in the corresponding C-cell, smaller than a threshold value (E_C^{THR}). (The
quoted values are derived from a beam test measurement.)

the difference (S1) between the energy deposited in the first sample of the e.m. cell and that deposited in this sample by an e$^-$ of same incident energy and angle (this is given by beam test measurements). The corresponding difference obtained in the case of the fourth sample (S4) is also considered. The difference between the position along the beam axis ($\Delta x / \sigma x$) measured in the CD and that measured in the calorimeter, is computed; this is done by taking the barycentre of the energy deposit in each sample, performing a fit of the points obtained. The difference between the azimuthal angle defined by the CD and that defined in the calorimeter ($\Delta \varphi / \sigma \varphi$) is also used; finally, the chi-square value (χ^2_{relic}) which measures the shower development in the calorimetry is computed by taking into account the energy deposit in each sample of the e.m. calorimeter.

These various parameters are sketched in Fig. I.2. They are presented in a projection of the central e.m. calorimeter cells, in the (φ = azimuth,

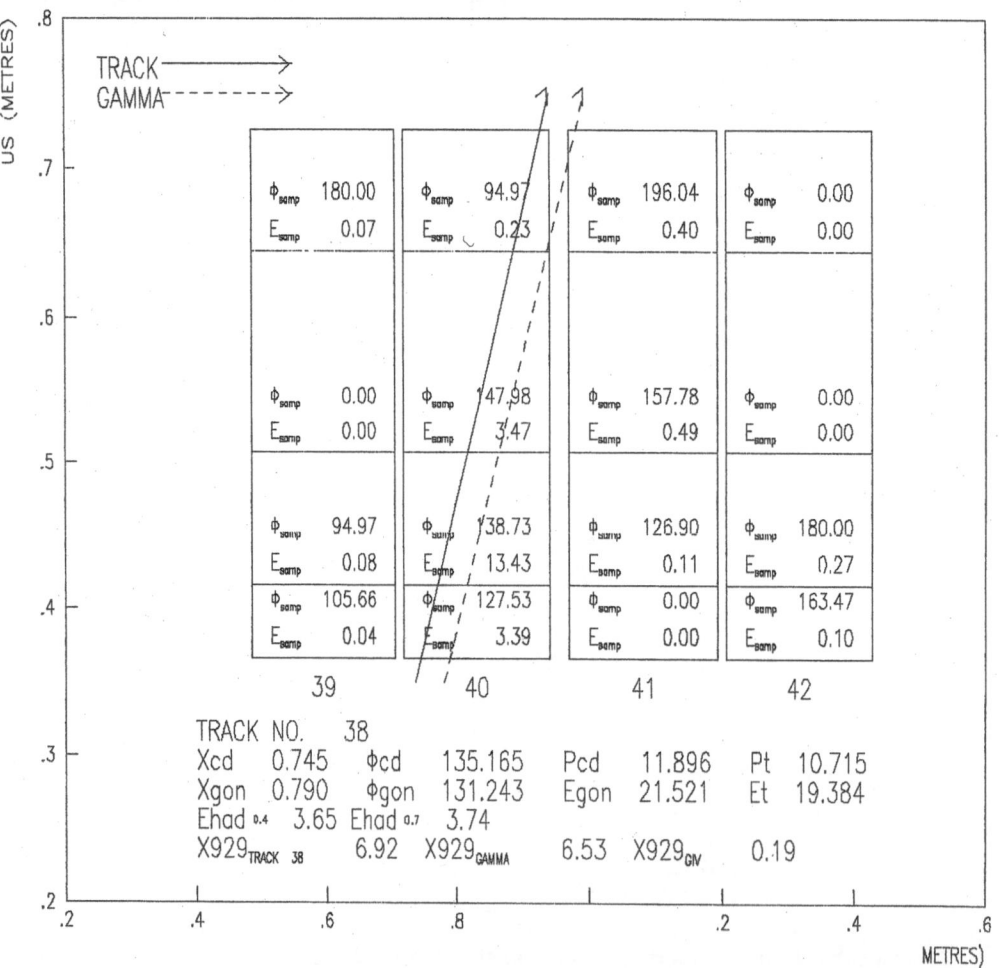

Fig. I.2. Schema of the central e.m. calorimetry, in the plane (x = beam direction, φ = azimuthal angle), with the various parameters defining the 'hadronicity' of an event.

x = beam axis) plane. They are all used to identify the e⁻ candidate in the search for W → ev events, by imposing on them the following set of cuts:

$$E_{had} \leq 500 \text{ MeV (main constraint)}$$

$$\Delta(1/p - 1/E)/\sigma_{p,E} \leq 3$$

$$\Delta x/\sigma x \leq 3$$

$$\Delta\varphi/\sigma\varphi \leq 3$$

$$\chi^2_{relic} \leq 20 \quad .$$

Thus the identification of the τ-signal, when τ decays into a single charged pion plus possible π^0's, could be simply defined by taking a logical OR of the reverse of each one of the conditions listed above. But whereas, in the case of the W → ev events, most of the tracks have a momentum bigger than 10 GeV/c (see Fig. I.3a), in the case of the τ it is just the contrary (see Fig. I.3b).

So a trickier analysis must be done. It consists in distinguishing three regions, according to the p_T of the charged track. In each of these regions, various sets of parameters have to be considered (they are those, among the quantities we have listed before, which may be effectively measured in the particular p_T-range considered). Moreover a strict matching between the three parts of the detector which are involved in this game, must

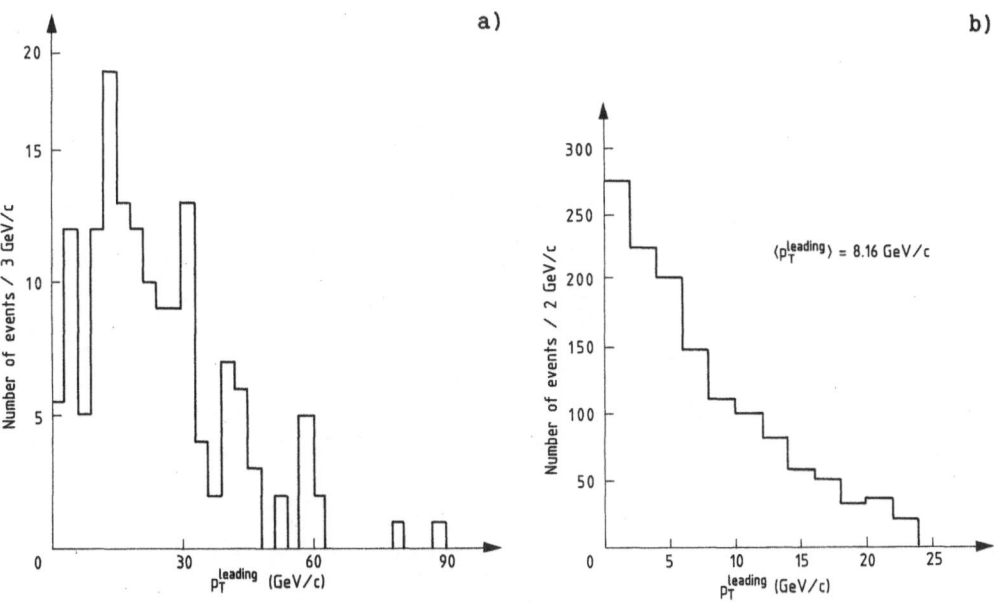

Fig. I.3. Transverse momentum distribution of the leading track for:
a) W → ev events, and b) W → τν$_τ$, τ$^\pm$ → π$^\pm$(nπ^0)ν$_τ$ events.

be defined. The following cases are considered: the first one corresponds to the situation where $p_T^{track} > 10$ GeV/c: the whole set of parameters has to be used. If, in the second case, p_T^{track} is between 5 and 10 GeV/c, the parameter E_{had} is no longer available or its handling becomes too bad. If, in the last case, p_T^{track} is lower than 5 GeV/c and higher than 2 GeV/c, the only quantities which remain reliable are the mismatches in energy and in azimuthal angle between the CD and the calorimetry.

This measurement and the corresponding analysis make wide use of the information provided by the central tracking ($+\vec{B}$), the e.m. and hadronic calorimetry. They rely on an accurate calibration of these three parts of the detector (in particular of the calorimetry).

We have considered here a special case (i.e. the UA1 detector), and therefore have been dealing with the corresponding systematics. However, 'the moral of this tale' is that in a pp collider environment, the search for τ's [even when considering its cleanest signatures: $\tau^{\pm} \to \pi^{\pm}(n\pi^0\text{'s})\nu_{\tau}$] requires in fact to exploit all the resources of a detector to ensure the e/π rejection, and to handle very well the corresponding systematics. Another example of this statement is given in Appendix II.

APPENDIX II: USE OF THE FRAGMENTATION PROPERTIES OF THE JETS,
TO IDENTIFY A $\tau^{\pm} \to 3\pi^{\pm}(n\pi^0\text{'s}) \nu_\tau$ signal.

Based on a study reported in Ref. 13, we want to show here how one may
use the properties of fragmentation of the trigger jet to enhance the signal
provided by the τ_3 events produced by:

$$p\bar{p} \to W^{\pm} \to \tau^{\pm}\nu_\tau \ ,$$

where:

$$\tau^{\pm} \to 3\pi^{\pm} + (n\pi^0\text{'s}) + \nu_\tau \ .$$

It has already been mentioned in Subsection 2.1.2 (iii) that the multiplicity argument in this case (namely $n_{CD} = 3$), is not strong enough to distinguish τ_3 events from standard QCD jets; about 50% of the QCD jets have a
charged multiplicity $\approx$ 3. Moreover, at this energy, a standard QCD jet which
passes the same 'narrowness' requirement and good matching both in space and
in energy between the CD jet and the calorimeter jet as a selected τ_3 candidate, has an effective mass equal to that of a typical τ_3 event.

Therefore, more sophisticated properties have to be emphasized. They
imply mainly properties of the charged part of the jet, given by the tracking device and the $\vec{B}$ field.

We plot the transverse momentum measured for each of the three tracks
of the trigger jet. We compare the corresponding distributions provided by
τ_3 events (as generated by the ISAJET Monte Carlo), by standard QCD jets
(mainly due to light constituents) as given by real data, and by heavy quark
jets (generated by the process: $p\bar{p} \to b\bar{b}$ or $c\bar{c}$, using the ISAJET Monte Carlo).
These distributions are shown in Figs. II-1, II-2, and II-3. We may derive a
set of cuts which allow to enhance the τ-signal. Once the charged multiplicity of the trigger jet is equal to 3, we impose that:

p_T (leading track) $>$ 5 GeV/c

and p_T (next to leading track) $>$ 3 GeV/c

and p_T (next to next to leading track) $>$ 2 GeV/c .

By applying this recipe we obtain that 45% of the τ_3 sample is left,
whereas only 2% of the heavy-flavour jets and 1% of the light-constituent
jets survive these conditions.

If we now consider the τ-sample estimated out of the 56 events found by
UA1 (i.e. 39 τ candidates), this means that only 3 τ_3 events are expected;
but only 0.04 event from light constituents and 0.004 heavy-flavour events

will remain. This gives a signal/background ratio of 3/60, which is quite reasonable.

We thus have a good way to distinguish τ_3 events from standard QCD background (both light and heavy constituents). On the other hand, we cannot, so far, define a precise strategy to distinguish τ_3 from heavy-flavour jets since we have no real experimental knowledge about the fragmentation properties of b-, c- or t-quark jets.

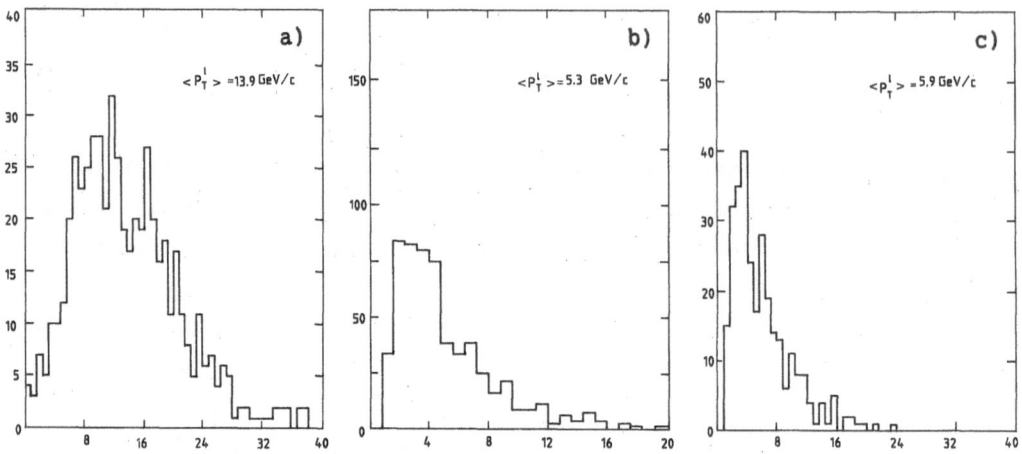

Fig. II.1. Transverse momentum distribution of the leading track for: a) $W \to \tau\nu_\tau$ events; $\tau \to 3\pi^{\pm}(n\pi^0)\nu_\tau$ (ISAJET Monte Carlo); b) QCD jets (real data), mainly light constituents; c) QCD jets, b- and c-quark jets only (ISAJET Monte Carlo).

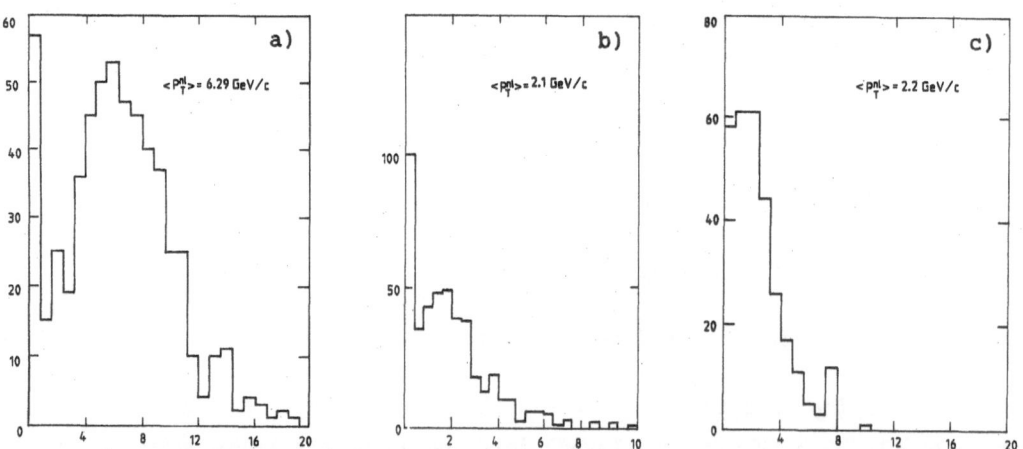

Fig. II.2. Transverse momentum distribution of the next-to-leading particle for: a) $W \to \tau\nu_\tau$ events; $\tau \to 3\pi^{\pm}(n\pi^0)\nu_\tau$ (ISAJET Monte Carlo); b) QCD jets (real data), mainly light constituents; c) QCD jets, b- and c-quark jets only (ISAJET Monte Carlo).

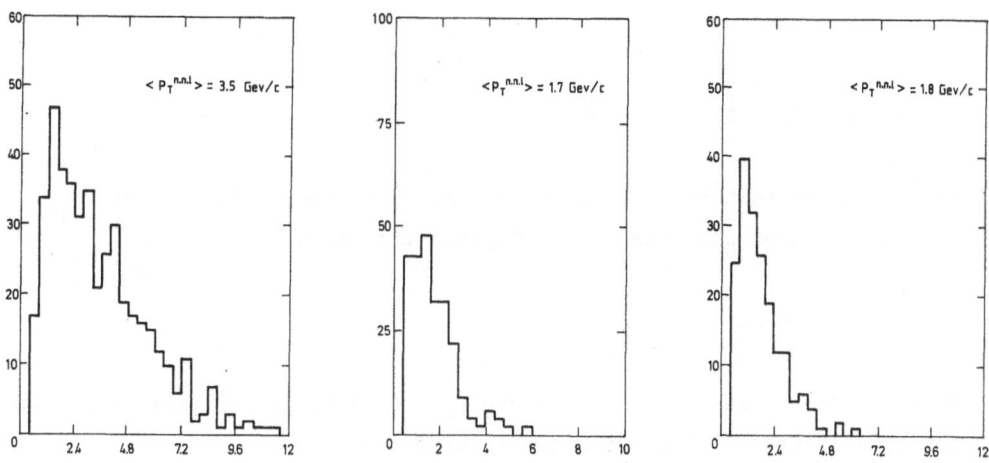

Fig. II.3. Transverse momentum distribution of the next-to-next-to-leading
particle for: a) $W \rightarrow \tau \nu_\tau$ events; $\tau \rightarrow 3\pi^\pm (n\pi^0)\nu_\tau$ (ISAJET Monte
Carlo); b) QCD jets (real data), mainly light constituents;
c) QCD jets, b- and c-quark jets only (ISAJET Monte Carlo).

APPENDIX III: GRAVIPHOTON AND EQUIVALENCE PRINCIPLE

This computation has been done by Scherk[20]. Two atoms of atomic number
A and A' having Z and Z' protons, respectively, and of charge zero, fall
with different accelerations towards the Earth. The force between them is
given by:

$$F = \frac{G}{r^0} [mm' - m^0 m^0 '] .$$

The negative term $m^0 m^0 '$ is due to the exchange of the graviphoton. We get

$$m = Z(m_p + m_e - \Delta) + (A - Z)(m_n - \Delta)$$

$$m^0 = Z(2m_u + m_d + m_e) + (A - Z)(m_u + 2m_d) ,$$

where m_p is the proton mass, m_e is the electron mass and m_n is the neutron
mass in the expression of the usual gravitational mass m; whereas m^0 has
only to do with quarks and electron masses, m_u is the u-quark mass, and m_d
is the d-quark mass.

The acceleration of the atom characterized by (Z, A) is given by

$$a(Z, A) = \frac{G}{r^2} m' \left[1 - \frac{m^0}{m} \frac{m^0{}'}{m'} \right] .$$

If (A', Z') represent the Earth, we can safely replace $m^0{}'/m'$ by $3m_u/m_p$. So the acceleration of the atom (Z, A) towards the Earth is

$$a(Z, A) = \frac{G}{R_E^2} m_E \left[1 - \frac{m^0}{m} \frac{3m_u}{m_p} \right] ,$$

where R_E is the radius of the Earth and m_E its mass. The relative difference in the acceleration of the two atoms is given by

$$\frac{\delta a}{a} = \frac{a(Z, A) - a(Z', A')}{a(Z, A)} - \frac{3m_u}{m_p} \left(\frac{m^0}{m} - \frac{m^0{}'}{m'} \right)$$

$$= \frac{3m_u}{m_p} \frac{Z'A - ZA'}{mm'} \times$$

$$\left[m_e(m_n - \Delta - m_u - 2m_d) + \frac{3}{2}(m_u + m_d)(m_n + m_p) + \frac{1}{2}(m_u + m_d)(m_n + m_p) \right] .$$

In the square bracket, the dominant contribution is given by $m_e m_n$, so that

$$\frac{\delta a}{a} \simeq \frac{3m_u}{m_p} \left(\frac{Z'}{A'} - \frac{Z}{A} \right) \frac{m_e}{m_p} .$$

If we consider, for instance, two atoms such as K and ^{4}He, we get $\delta a/a \simeq 10^{-6}$, which is unacceptable.

The situation is even worse if we consider the fall of K^0 and $\bar{K}^0$ towards the Earth; we get

$$\frac{a(K^0) - a(\bar{K}^0)}{a(K^0)} \simeq 6 \frac{m_u}{m_p} \frac{m_s}{m_{K^0}} \sim 10^{-2} !$$

PHYSICS AT THE CERN pp̄ COLLIDER

AND STATUS OF THE ELECTROWEAK THEORY

L. DiLella

CERN,
Geneva,
Switzerland

1.   INTRODUCTION

These lectures are a review of recent experimental results obtained at the CERN pp̄ Collider on two subjects: the production of high-transverse-momentum jets; and the production and decay of the weak Intermediate Vector Bosons (IVBs), $W^{\pm}$ and Z, which mediate the electroweak interaction.

The data discussed here have been collected during three physics runs which took place between the autumn of 1982 and the end of 1984 at total centre-of-mass energies ($\sqrt{s}$) of 546 GeV (1982 and 1983) and 630 GeV (1984). The total integrated luminosities recorded by the two major experiments, UA1 and UA2, are L = 399 and 452 nb^{-1}, respectively. In a further run at $\sqrt{s}$ = 630 GeV which took place in autumn 1985, an additional integrated luminosity of approximately 450 nb^{-1} has been recorded by each of the two experiments. Some of the results from the analysis of these new data are already available and will be mentioned whenever appropriate.

These lectures begin with a short description of the UA1 and UA2 detectors. We then discuss the results obtained in the study of final states containing high-p_T hadronic jets or isolated single photons, and we compare these results with the predictions of Quantum Chromodynamics (QCD). Finally, we review the measured properties of the IVBs, and we compare them with expectations from the electroweak theory.

2.   THE TWO MAJOR EXPERIMENTS

2.1 The UA1 Experiment

The UA1 experiment is based on a general-purpose magnetic detector,[1] which was designed to achieve an almost complete solid-angle coverage down to polar angles of 0.2° with respect to the beams, using both track detection and calorimetry.

A view of the detector with the two halves of the magnet opened up is shown in Fig. 1a. The *magnet* is a dipole which produces a horizontal field of 0.7 T perpendicular to the beam axis over a volume of $7 \times 3.5 \times 3.5$ m^3.

The magnet contains the *central track detector*, which is a system of drift chambers (maximum drift space 18 cm), filling a cylindrical volume 5.8 m in length and 2.3 m in diameter (see Fig. 1b). There are 6110 sense wires in total, all parallel to the magnetic field, providing an average of ～ 100 points per track.[2] Track coordinates in a plane perpendicular to the magnetic field are determined with a precision of ～ 300 μm by the drift-

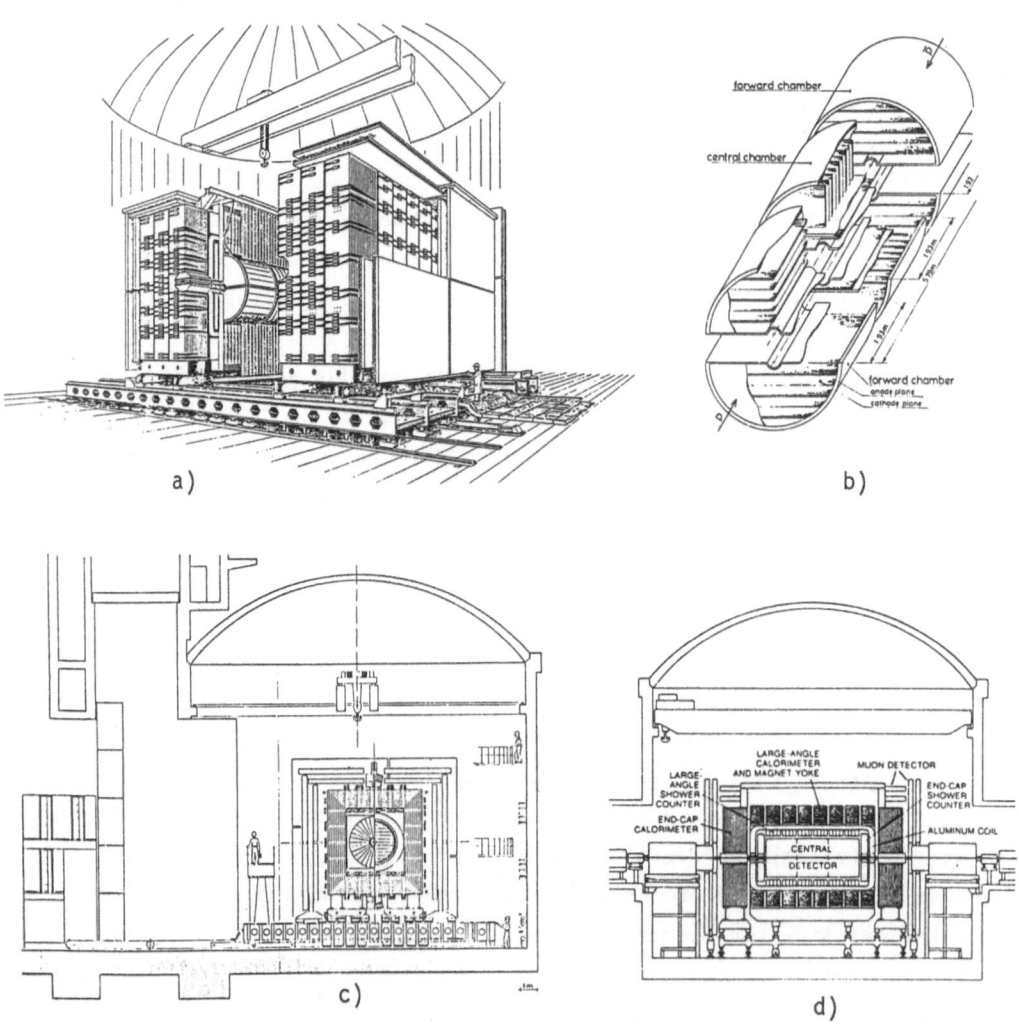

Fig. 1 Views of the UA1 detector: a) a view with the two magnet halves opened up; b) a view of the central tracking detector showing the arrangement of the drift planes; c) a cut of the detector normal to the beam axis; d) a cut of the detector by a vertical plane containing the beam axis.

time, whereas the coordinates along the field are measured with much less accuracy, using the technique of charge division on the sense wires, which also provides a measurement of the energy loss by ionization in the gas.

The magnetic field volume contains also the *large-angle electromagnetic (e.m.) calorimeter*, which covers the polar angle interval 25°-155° and the full azimuth. This calorimeter consists of two half-cylindrical shells on either side of the beam, each made of 24 independent total-absorption counters called 'gondolas'. A gondola subtends an angular interval $\Delta\theta = 5°$, $\Delta\varphi = 180°$. It consists of a multilayer lead-scintillator sandwich with a total thickness equivalent to 26.6 radiation lengths (r.l.) subdivided in depth into four independent segments with thickness 3.3, 6.6, 10.1, and 6.6 r.l., respectively. The scintillation light from each of the four segments is collected by wave-shifting plates on each side of the gondola, and guided to photomultipliers (PMs) located outside the magnet. The energy of an electron is measured with a resolution $\sigma_E/E = 0.15/\sqrt{E}$ (E in GeV). Furthermore, a comparison of the PM pulse heights provides a measurement of the azimuthal (φ) and longitudinal (z) position of the e.m. shower with resolutions σ_φ (radians) $\approx 0.3/\sqrt{E}$ and σ_z (cm) $\approx 6.3/\sqrt{E}$, respectively (E in GeV).

The whole system of gondolas is calibrated periodically with an intense ^{60}Co source. A small number of gondolas has been directly calibrated with electron beams of known energy from the CERN Super Proton Synchrotron (SPS), and the other gondolas are then adjusted to give the same response to the ^{60}Co source of the gondolas whose calibration has been determined in the beam.

The two *end-cap e.m. calorimeters*, also located inside the magnet, cover the polar angle interval 5°-25° with respect to the beams. Each calorimeter consists of 32 azimuthal sectors of lead-scintillator multilayer sandwich at a distance of 3 m from the centre of the beam crossing. These counters have a total thickness of 27 r.l. and are segmented four times in depth. In addition, they contain a position detector (proportional tubes) located at a depth of 11 r.l., which measures the position of the e.m. showers with a precision of ~ 2 mm.

The configuration of both the central and end-cap e.m. calorimeters can be seen in Figs. 1c and 1d.

The yoke of the magnetic spectrometer is laminated, and scintillator is inserted between the iron plates to form the *central hadronic calorimeter* which surrounds both e.m. calorimeters. The hadronic calorimeter (see Figs. 1c and 1d) consists of 450 independent cells, with typical size $\Delta\theta \times \Delta\varphi$ equal to 15° × 18° in the central region and 5° × 10° in the forward regions, respectively. Their thickness is ~ 5 and ~ 7 absorption lengths, respectively. The energy resolution is typically $\sigma_E/E \approx 0.8/\sqrt{E}$ (E in GeV).

Muon detectors surround the magnet yoke (see Fig. 1a). They consist of two planes of drift chambers covering the polar angle interval 5° < θ < 175° and the full azimuth, separated by a distance of 60 cm. Each of these planes is made of 50 drift chambers of ~ 4 × 6 m^2. Each chamber consists of four layers of drift tubes (maximum drift space 7 cm), which define two ortho-gonal coordinates. By knowing the position of the interaction vertex and the muon-track parameters in these chambers, it is possible to obtain an inde-pendent measurement of the muon momentum with a relative precision of ~ 20% through the determination of its deflection in the magnet yoke.

In addition to the detectors just described, more calorimeters and track detectors located along the beam pipe on both sides of the spectro-meter cover the region of polar angles from 5° down to ~ 0.2° with respect to the beam line. A schematic view of the overall detector configuration is shown in Fig. 1d.

2.2 The UA2 Experiment

The UA2 experiment was designed mainly to search for the weak bosons by identifying their decay modes into electrons, and to study final states containing high-p_T jets.

The detector (Fig. 2a) consists of three parts:

- The *vertex detector*, which is a system of cylindrical chambers to re-construct charged-particle tracks in a region without magnetic field.[3] This detector covers the polar angle interval 20° < θ < 160°. It consists of four multiwire proportional chambers (MWPCs) and two drift chambers, all with wires parallel to the beam axis, providing a total of 16 points per track. Reconstruction in three dimensions is achieved in the MWPCs by measuring the charge induced on cathode strips at ±45° to the wires, and in the drift chambers using the technique of charge division. The drift chambers provide also a measurement of the specific ionization for each charged particle.

A 'preshower' counter, consisting of a 1.5 r.l. thick tungsten cylinder followed by a MWPC with cathode strip readout and pulse-height measurement on the wires, is located just behind the last chamber, covering the central region 40° < θ < 140°. As we shall see in subsection 4.3, this device is essential for electron identification.

- The *central calorimeter*,[4] which is a system of 240 independent coun-ters covering the full azimuth (see Fig. 2b) and the polar angle interval 40° < θ < 140°. Each counter has an angular acceptance Δθ × Δφ = 10° × 15° and consists of a first e.m. section (lead/scintillator) 17 r.l. thick, followed by two independent hadronic sections (iron/scintillator), each two absorption lengths thick. Light is collected by wavelength-shifting plates located on two opposite sides of each counter. The energy resolution is

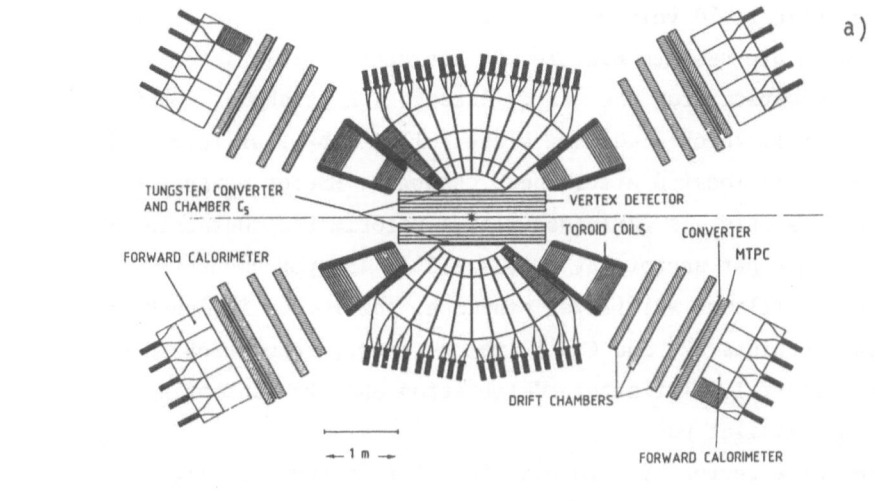

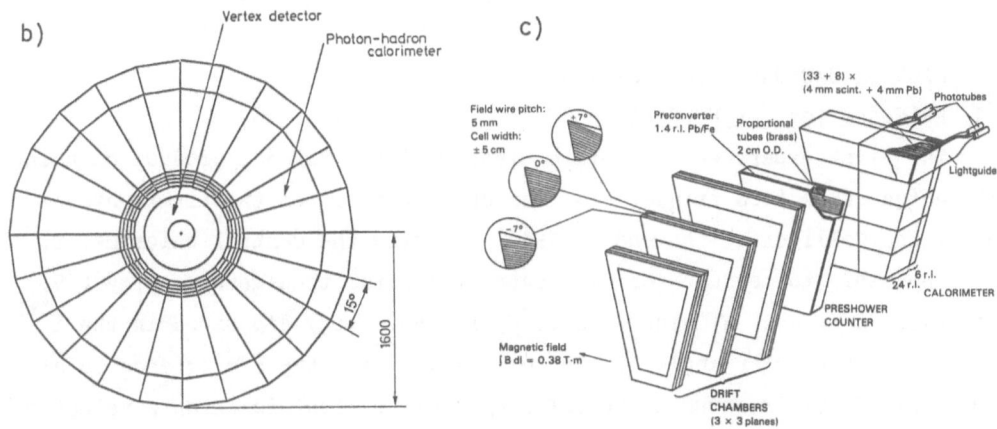

Fig. 2 Views of the UA2 detector: a) cross-section in the vertical plane
containing the beam axis; b) cross-section of the central detector normal to
the beam axis; c) exploded view of a sector in one of the forward detectors.

$\sigma_E/E \approx 0.15/\sqrt{E}$ for e.m. showers, and $\approx 0.32E^{-1/4}$ for hadronic showers (E in
GeV). All counters have been calibrated using 10 GeV electron and muon beams
from the CERN Proton Synchrotron (PS); frequent checks with an intense ^{60}Co
source and light-pulsers are made to monitor the calibration stability. In
addition, a few calorimeter modules are periodically removed from the UA2
detector and transported to an electron and a hadron beam from the CERN SPS
for recalibration.

    - The two *forward detectors*, covering the polar angle intervals 20° <
θ < 37.5° and 142.5° < θ < 160° and the full azimuth. Each detector consists
of twelve sectors in which a toroidal magnetic field is generated by twelve
coils equally spaced in azimuth (the field integral is ~ 0.38 T·m). Follow-

ing the magnetic field volume, each sector contains nine drift chamber planes,[5] which are used to measure the charged-particle momenta together with the information from the vertex detector. A preshower counter, consisting of a 1.5 r.l. thick lead-iron plate followed by four layers of proportional tubes,[6] is located after these chambers. Energy measurement, limited to e.m. showers only, is performed with a calorimeter containing ten independent counters per sector ($\Delta\theta \times \Delta\varphi = 2.5° \times 15°$ per counter). Each counter is a lead-scintillator multilayer sandwich, subdivided in depth into two independent sections, 24 and 6 r.l. thick. Energy resolution is similar to that of the central calorimeter. Calibration and its monitoring are also achieved in a similar way.

A view of a sector of a forward detector is shown in Fig. 2c.

No muon detector is implemented in the UA2 experiment.

3.   HIGH-p_T HADRONIC FINAL STATES

3.1   Evidence for Jet Production

The first experiment to obtain clear evidence for jet production using a technique free from trigger bias was UA2,[7] soon after the first physics run of the Collider at the end of 1981. By using the central calorimeter, it is possible to measure for each event the total transverse energy ΣE_T, defined as $\Sigma E_T = \Sigma_i E_i \sin\theta_i$, where E_i is the energy deposited in the ith cell, θ_i is the polar angle of the cell centre, and the sum extends to all 240 cells. Figure 3 shows a typical ΣE_T distribution[8] for events selected

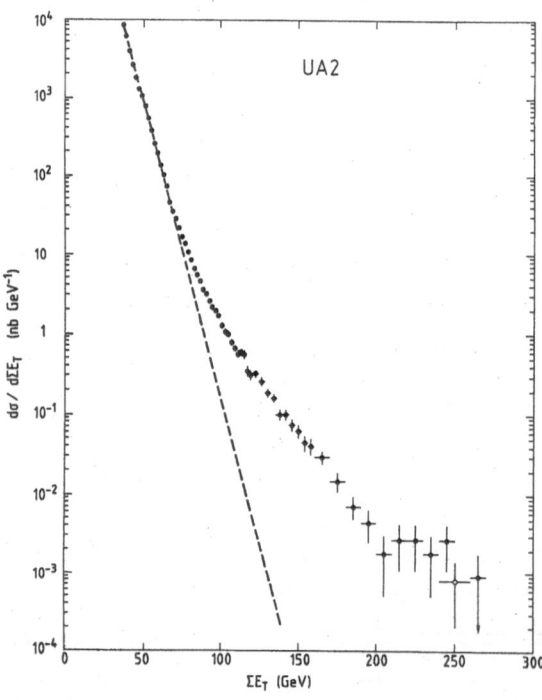

Fig. 3 Distribution of the total transverse energy ΣE_T observed in the UA2 central calorimeter.

by the requirement that $\Sigma\ E_T$ exceeds a given threshold. For $\Sigma\ E_T \gtrsim 60$ GeV there is a clear departure from a single exponential, an effect not seen in lower-energy experiments.[9]

In order to investigate the pattern of energy distribution in the events, energy clusters are constructed by joining all calorimeter cells which share a common side and contain at least 400 MeV. In each event, these clusters are then ranked in order of decreasing transverse energies ($E_T^1 > E_T^2 > E_T^3 > \ldots$). Figure 4a shows the mean values of the fractions $h_1 = E_T^1/\Sigma\ E_T$ and $h_2 = (E_T^1 + E_T^2)/\Sigma\ E_T$ as a function of $\Sigma\ E_T$. Their behaviour reveals that, when $\Sigma\ E_T$ is large enough, a very substantial fraction of $\Sigma\ E_T$ is shared on the average by two clusters with roughly equal transverse energies (an event consisting of only two clusters of equal transverse energies would have $h_1 = 0.5$ and $h_2 = 1$).

The azimuthal separation $\Delta\varphi_{12}$ between the two largest clusters[10] is shown in Fig. 4b for events with $\Sigma\ E_T > 60$ GeV and E_T^1, $E_T^2 > 20$ GeV. A clear peak at $\Delta\varphi_{12} = 180°$ is observed, indicating that the two clusters are co-planar with the beam direction.

The emergence of two-cluster structures in events with large $\Sigma\ E_T$ is even more dramatically illustrated by inspecting the transverse energy dis-tribution over the calorimeter cells. Figure 5 shows such a distribution for

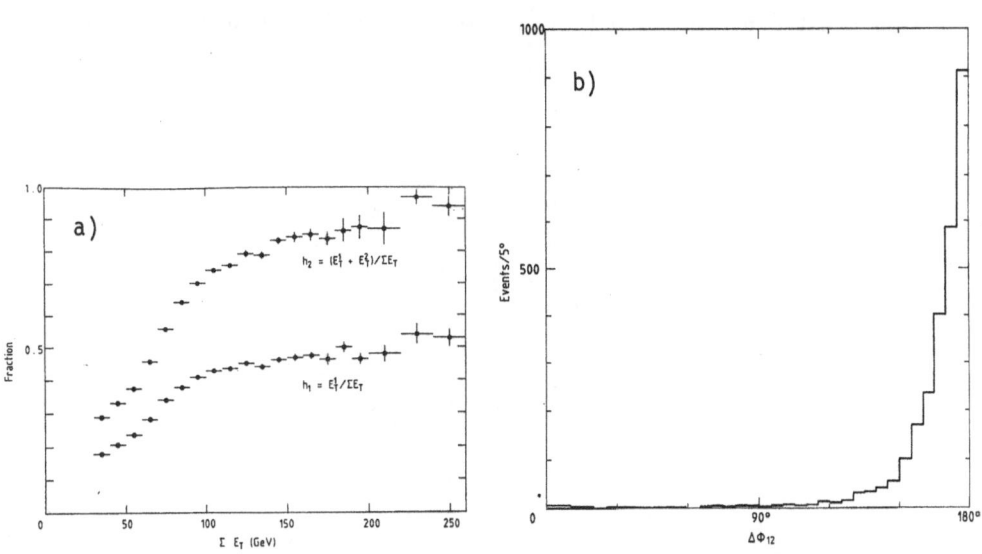

Fig. 4 a) Mean value of the fraction h_1 (h_2) of the total transverse energy $\Sigma\ E_T$ contained in the cluster (in the two clusters) having the largest E_T, as a function of $\Sigma\ E_T$. b) Azimuthal separation between the two largest E_T clusters in events with $\Sigma\ E_T > 60$ GeV and E_T^1, $E_T^2 > 20$ GeV.

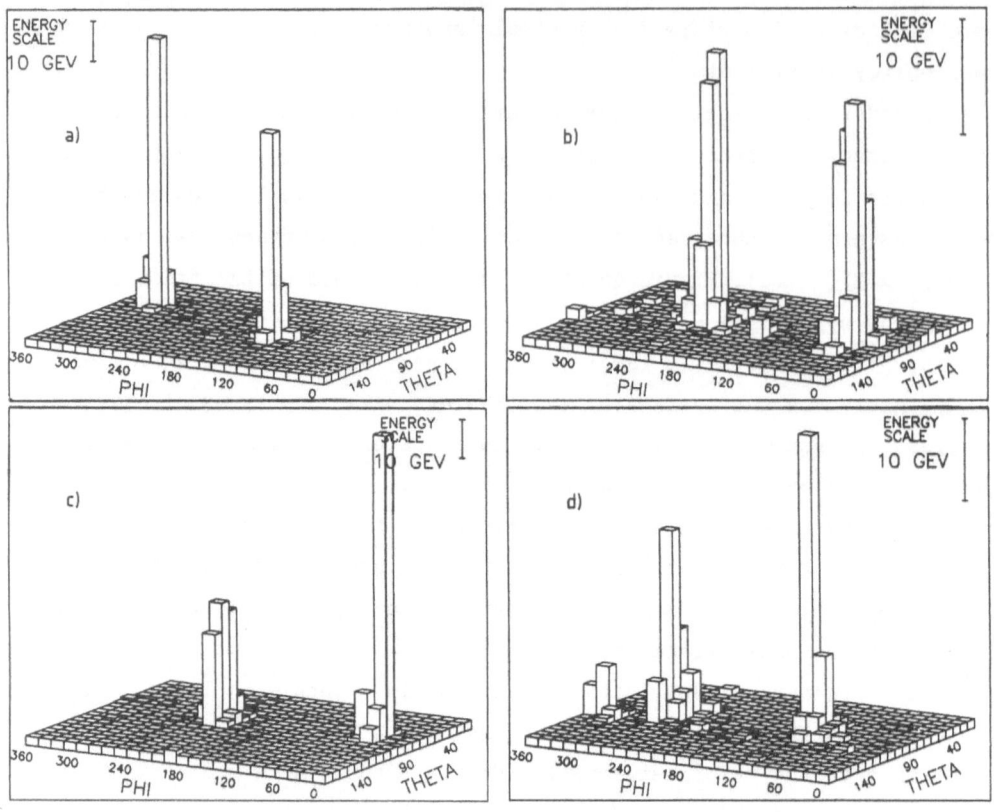

Fig. 5 Four typical transverse energy distributions for events with $\Sigma\,E_T\,>$ 100 GeV in the θ-φ plane. Each bin represents a cell of the UA2 calorimeter.

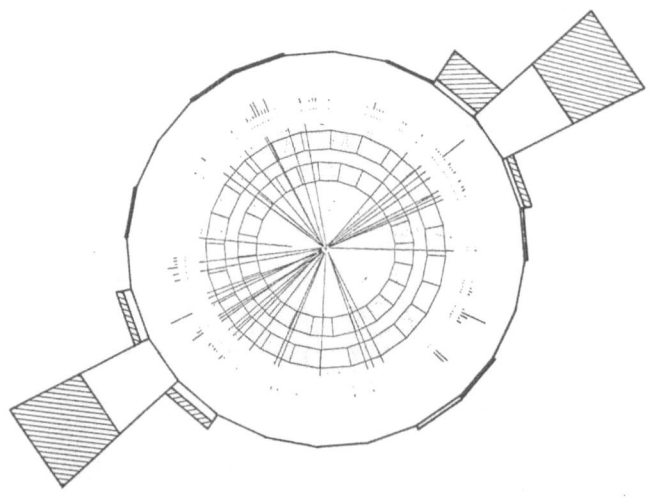

Fig. 6 View of a typical two-jet event perpendicular to the beams in the UA2 detector. The heights of the trapezoids are proportional to transverse energy. The open and shaded areas represent the energy depositions in the electromagnetic and hadronic sections of the calorimeter, respectively.

212

four typical events having $\Sigma\,E_T > 100$ GeV. The transverse energy appears to be concentrated within two (or, more rarely, three) small angular regions. These energy clusters appear to be associated with collimated multiparticle systems (jets), as found by reconstructing the charged-particle tracks in these events (see Fig. 6). Furthermore, longitudinal shower developments, as measured in the calorimeter, are found to be inconsistent, in general, with those of single particles, but consistent with those of jets containing both charged hadrons and photons (from π^0 decay).

High-p_T jets have also been observed by the UA1 experiment[11] and, in pp collisions, by the Axial Field Spectrometer (AFS) Collaboration[12] at the lower $\sqrt{s}$ values available at the CERN Intersecting Storage Rings (ISR).

3.2 Theoretical Interpretation

High-p_T secondaries were first observed[13] in 1972 at the CERN ISR in proton-proton interactions at $\sqrt{s}$ values between 30 and 62 GeV. They were interpreted in the framework of the parton model[14] as being the result of elastic or quasi-elastic scattering of two point-like constituents of the incident protons.[15]

In the parton model the two incident hadrons are considered, at any instant, as being composed of independent point-like constituents (partons), each carrying a fraction x of the incident hadron momentum. Today we know that the partons are quarks (q), antiquarks ($\bar{q}$), and gluons (g), all carrying a new quantum number (colour), and we have a non-Abelian gauge theory, Quantum Chromodynamics (QCD),[16] as the best candidate to explain the strong interaction among these elementary constituents. Large-angle scattering of two high-x partons results in two outgoing partons with high p_T. At this stage the strong forces among partons, which are presumably responsible for colour confinement within the hadrons, induce a final-state interaction among the two high-p_T partons and the other partons, which results in the production of many hadrons (this step is referred to as hadronization, or fragmentation). Since this final-state interaction involves mostly low-momentum transfer mechanisms, the final result is the production of two highly collimated systems of hadrons (jets), each having a total four-momentum approximately equal to that of the parent parton. Because the incident partons have low p_T, the two jets are approximately coplanar with the beam axis; however, their longitudinal momenta are not equal and opposite, in general, since the initial partons may have different x values. All these features are in qualitative agreement with the topology of two-jet events observed at the Collider and described previously.

There are several elementary subprocesses which can contribute to jet production. For each subprocess the differential cross-section, calculated to first order in the strong coupling constant α_s (see the relevant diagrams of Fig. 7), is given by the expression

$$\frac{d\sigma}{d(\cos\theta^*)} = \frac{\pi\alpha_s^2}{2\hat{s}} |M|^2 ,$$ (1)

where θ^* is the scattering angle, and $\hat{s}$ is the square of the total energy in the centre of mass of the two partons. In QCD the coupling constant α_s is a function of Q^2, the square of the four-momentum transfer in the subprocess. In a model with five quark flavours, $\alpha_s(Q^2)$ is expressed as $\alpha_s(Q^2) = 12\pi/[23 \ln (Q^2/\Lambda^2)]$, where Λ is a scale parameter. The property $\alpha_s \to 0$ for $Q^2 \to \infty$, which is called 'asymptotic freedom', allows perturbative calcula- tions of strong processes at high Q^2.[16]

Explicit expressions for $|M|^2$ in Eq. (1) are given in Table 1 for the various subprocesses[17] as a function of the Mandelstam variables $\hat{s}$, $t = -\hat{s}(1 - \cos\theta^*)/2$, and $u = -\hat{s}(1 + \cos\theta^*)/2$ (under the assumption of massless partons). In order to illustrate the relative importance of the subprocesses, Table 1 displays also the numerical values of $|M|^2$ at $\theta^* = 90°$, where $t = u = -\hat{s}/2$. It is evident that terms involving initial gluons, such as gg and qg (or $\bar{q}$g) scattering, are dominant whenever the gluon density in the in- cident hadrons is comparable to that of the quarks.

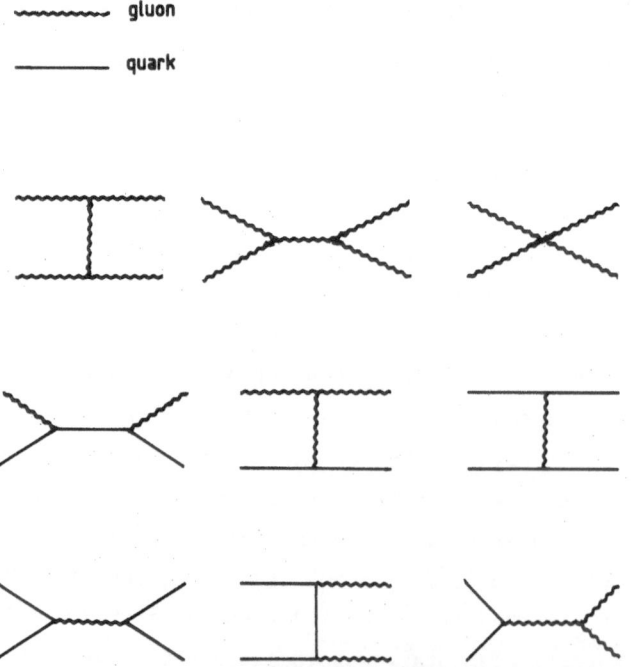

Fig. 7   First-order diagrams for hard parton scattering.

Table 1

Matrix elements for parton scattering

Subprocess	$\|M\|^2$	$\|M\|^2$ at $\theta^* = 90°$
$qq' \rightarrow qq'$ a) $q\bar{q}' \rightarrow q\bar{q}'$ a)	$\dfrac{4}{9} \dfrac{\hat{s}^2 + u^2}{t^2}$	2.22
$qq \rightarrow qq$	$\dfrac{4}{9} \left(\dfrac{\hat{s}^2 + u^2}{t^2} + \dfrac{\hat{s}^2 + t^2}{u^2} \right) - \dfrac{8}{27} \dfrac{\hat{s}^2}{ut}$	3.26
$q\bar{q} \rightarrow q'\bar{q}'$ a)	$\dfrac{4}{9} \dfrac{t^2 + u^2}{\hat{s}^2}$	0.22
$q\bar{q} \rightarrow q\bar{q}$	$\dfrac{4}{9} \left(\dfrac{\hat{s}^2 + u^2}{t^2} + \dfrac{t^2 + u^2}{\hat{s}^2} \right) - \dfrac{8}{27} \dfrac{u^2}{\hat{s}t}$	2.59
$q\bar{q} \rightarrow gg$	$\dfrac{32}{27} \dfrac{u^2 + t^2}{ut} - \dfrac{8}{3} \dfrac{u^2 + t^2}{\hat{s}^2}$	1.04
$gg \rightarrow q\bar{q}$	$\dfrac{1}{6} \dfrac{u^2 + t^2}{ut} - \dfrac{3}{8} \dfrac{u^2 + t^2}{\hat{s}^2}$	0.15
$qg \rightarrow qg$ $\bar{q}g \rightarrow \bar{q}g$	$-\dfrac{4}{9} \dfrac{u^2 + \hat{s}^2}{\hat{u}s} + \dfrac{u^2 + \hat{s}^2}{t^2}$	6.11
$gg \rightarrow gg$	$\dfrac{9}{2} \left(3 - \dfrac{ut}{\hat{s}^2} - \dfrac{\hat{u}s}{t^2} - \dfrac{\hat{s}t}{u^2} \right)$	30.38

a) q and q' denote quarks with different flavours.

3.3 Inclusive Jet Yield

The cross-section for inclusive jet production as a function of the jet p_T and angle of emission θ can be calculated to leading order in α_s as a sum of convolution integrals:[18]

$$\frac{d^2\sigma}{dp_T d(\cos\theta)} = \frac{2\pi p_T}{\sin^2\theta} \sum_{A,B} \int dx_1 \, dx_2 \, F_A(x_1, Q^2) \, F_B(x_2, Q^2)$$

$$\times \, \delta(\hat{s} + t + u) \, \alpha_s^2 \sum_f \frac{\|M\|^2_{AB \rightarrow f}}{\hat{s}} \, , \tag{2}$$

215

where F_A and F_B are structure functions describing the densities of partons A and B in the incident hadrons, and the sum extends over all initial parton types A, B, and all possible final states f.

In the case of incident protons or antiprotons, the structure functions $F(x, Q^2)$ are determined experimentally in deep-inelastic lepton-nucleon scattering experiments ($Q^2 \lesssim 20$ GeV^2) and extrapolated to the Q^2 range of interest (up to ~ 10^4 GeV^2 at Collider energies) according to the predicted QCD evolution.

At Collider energies, jets with p_T around 30 GeV/c produced near 90° arise from hard scattering of partons with relatively small values of x (x $\lesssim$ 0.1). In this region gluon jets are expected to dominate, both because there are many gluons in the nucleon at small x and because the subprocesses involving initial gluons have large cross-sections (see Table 1). This is in contrast with $e^+ e^-$ collisions, where the production of quark jets is the main feature of hadronic final states. Quark jets also dominate at ISR energies ($\sqrt{s} \lesssim 62$ GeV), where high-p_T jets require initial partons with large x values.

There are a number of uncertainties which affect the comparison between the predicted cross-sections and the experimental data. The most obvious one is due to the fact that Eq. (2) predicts the yield of high-p_T massless partons, whereas the experiments measure hadronic jets with a total invariant mass of several GeV. The relation between the parton p_T and the measured cluster transverse energy E_T is usually determined with the help of a QCD-inspired Monte Carlo simulation,[19] in which the outgoing partons evolve into jets according to a specific hadronization model, and the detector response to the hadrons is taken into account.

An important uncertainty in the theoretical predictions arises from the Q^2 extrapolation of the structure functions, especially those describing the gluons.

Another source of theoretical uncertainties is represented by the fact that Eq. (2) does not take into account higher-order effects, such as gluon radiation by the initial and outgoing partons. A complete calculation of these corrections is still missing; their effect is usually described by a multiplicative factor K $\lesssim$ 2.[20]

Finally, in addition to the statistical errors the data are also affected by a number of systematic effects, such as uncertainties in the energy calibration of the calorimeters, in the detector acceptance, and in the knowledge of the integrated luminosity. These effects amount typically to an overall uncertainty of ±50% in the measured jet yields. Altogether, a comparison between the theoretical predictions and the experimental results is only possible, at present, to an accuracy not greater than a factor of 2.

Figure 8 shows the inclusive jet production cross-section around
θ = 90° as a function of the jet p_T, as measured by the UA1[11] and UA2[8] Collaborations at √s = 546 GeV, and by the AFS Collaboration[12] at the ISR. The jet yield at the Collider is much larger than that measured at the ISR. This fact was first pointed out by Horgan and Jacob[18] well before these data were available.

Also shown in Fig. 8 is a band of QCD predictions[18,21] whose width serves to illustrate the theoretical uncertainties. In spite of the experimental and theoretical uncertainties, the agreement between data and theory is remarkable, especially because the theoretical curves are not a fit to the data but represent absolute predictions.

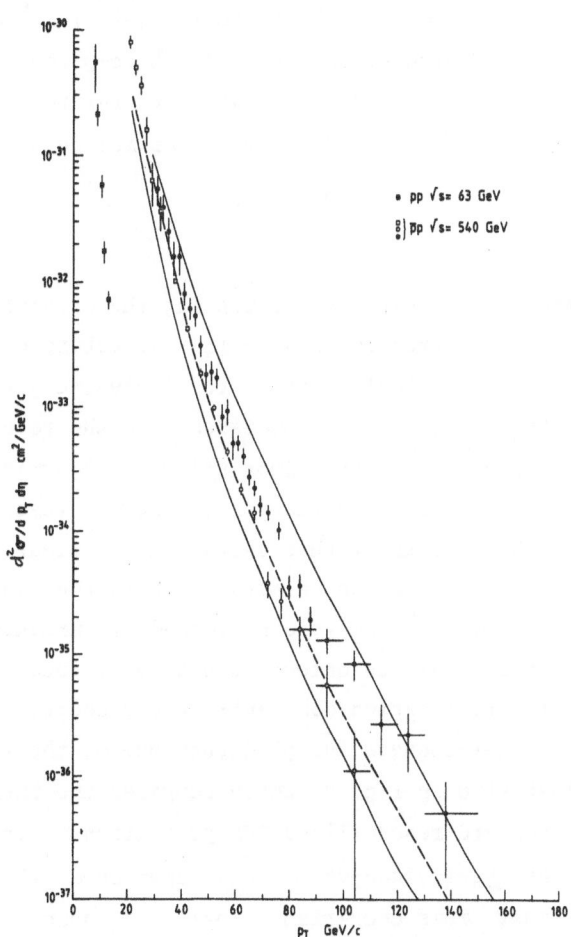

Fig. 8 Cross-sections for inclusive jet production around θ = 90°, as a function of the jet transverse momentum. Collider data (√s = 546 GeV): open circles and squares (UA1);[11] full circles (UA2).[8] ISR data: full squares.[12] The dashed curve represents the original prediction by Horgan & Jacob.[18] The two full curves define a band of QCD predictions.[20,21]

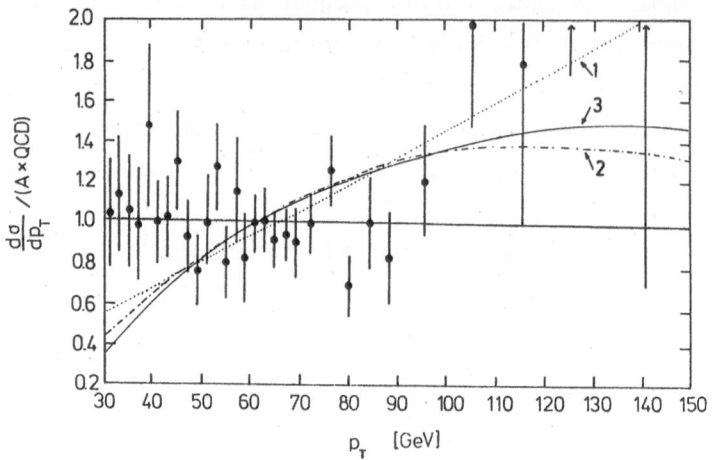

Fig. 9 Full points: ratio between the inclusive jet yield[8] and a renormalized standard QCD prediction. Curve 1: three-gluon vertex in hard-parton scattering removed. Curve 2: three-gluon vertex removed also in the evolution of the structure functions. Curve 3: as for curve 2 with, in addition, α_s = constant.[22]

These data offer the possibility of testing the existence of the three-gluon vertex which results from the non-Abelian structure of QCD. This vertex enters into the QCD calculations of the inclusive jet yield in three different ways: in the elementary subprocesses (see the relevant diagrams of Fig. 7); in the Q^2 dependence of the gluon structure function; and by determining the variation of α_s as a function of Q^2. Such a test can only be performed, however, if the effects of this vertex can be clearly separated from other spurious effects, such as the uncertainties in the theory and the systematic errors in the data. An analysis performed by Furmanski and Kowalski[22] has shown that all these spurious effects can be described to a good approximation by an overall normalization constant; on the contrary, the suppression of the three-gluon vertex changes the p_T dependence of the inclusive jet yield. This is illustrated in Fig. 9, which compares the ratio between the UA2 data[8] and the standard renormalized QCD predictions, with various predictions in which the three-gluon vertex was suppressed, all normalized to the standard QCD result. This comparison provides evidence for the existence of the three-gluon vertex in QCD.

3.4 <u>Energy Dependence</u>

Figure 10 shows a comparison between the cross-section for inclusive jet production measured at $\sqrt{s}$ = 630 GeV and that measured at $\sqrt{s}$ = 546 GeV.[23,24] A clear increase of the jet production cross-section with

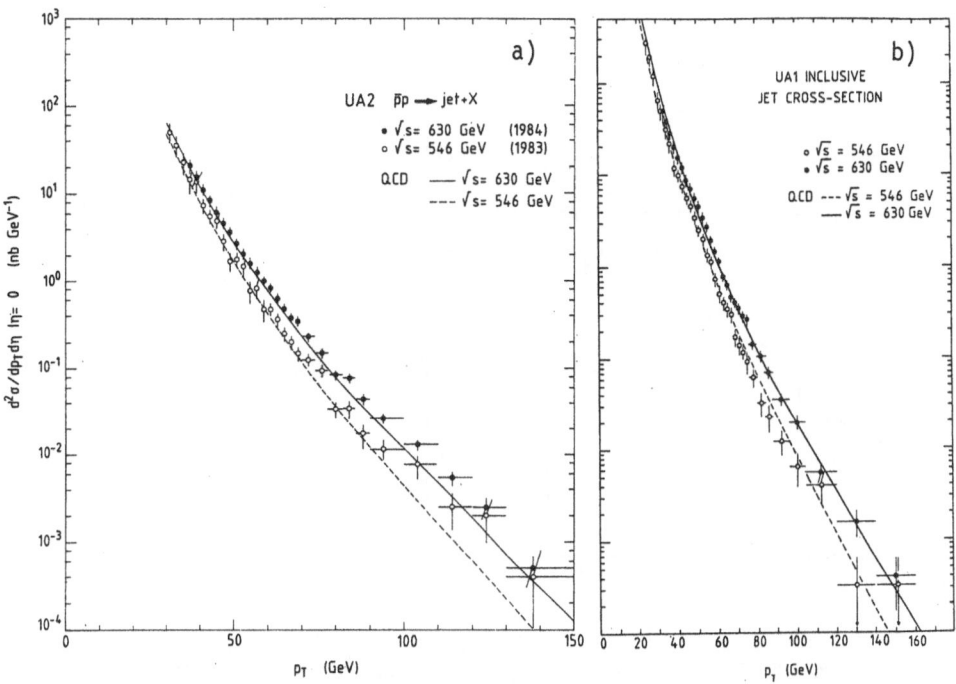

Fig. 10 Inclusive jet production cross-sections at $\sqrt{s}$ = 546 and 630 GeV:
a) UA2 data;[23] b) UA1 data.[24] The curves are QCD predictions.

increasing $\sqrt{s}$ at fixed p_T is visible in Fig. 10. Also shown are QCD predictions based on Eq. (2) and on Table 1, obtained using a parametrization for
the structure functions as given by Eichten et al.[25] In these predictions
the scale is defined to be $Q^2 = p_T^2$.

In order to examine the increase of the jet yield more closely, the
ratio of the cross-sections at the two Collider energies is shown in
Fig. 11. In this case the systematic errors approximately cancel. The residual systematic error, due to a possible energy calibration shift between
the two data sets, is estimated to affect the ratio by ±15% for UA1,[24] and
by less than ±6% for UA2.[23] A p_T-dependent increase of the inclusive jet
cross-section is observed, amounting to a factor of ~ 1.5 at p_T = 40 GeV/c
and to a factor of ~ 2 at p_T = 100 GeV/c when going from $\sqrt{s}$ = 546 GeV to
$\sqrt{s}$ = 630 GeV.

If one ignores the Q^2 dependence of the structure functions and of α_s,
the invariant cross-section for jet production at 90° satisfies the scaling
law $E(d\sigma/dp^3) = A\, p_T^n\, f(x_T)$, where $x_T = 2p_T/\sqrt{s}$, with n = 4.[15] Figure 12 shows
the dimensionless quantity $p_T^4\, E(d\sigma/dp^3)$ plotted as a function of x_T for the
two different beam energies. On this plot the two sets of data overlap, demonstrating that the observed increase of the jet cross-section with $\sqrt{s}$ at
Collider energies is entirely consistent with perfect scaling. However, a

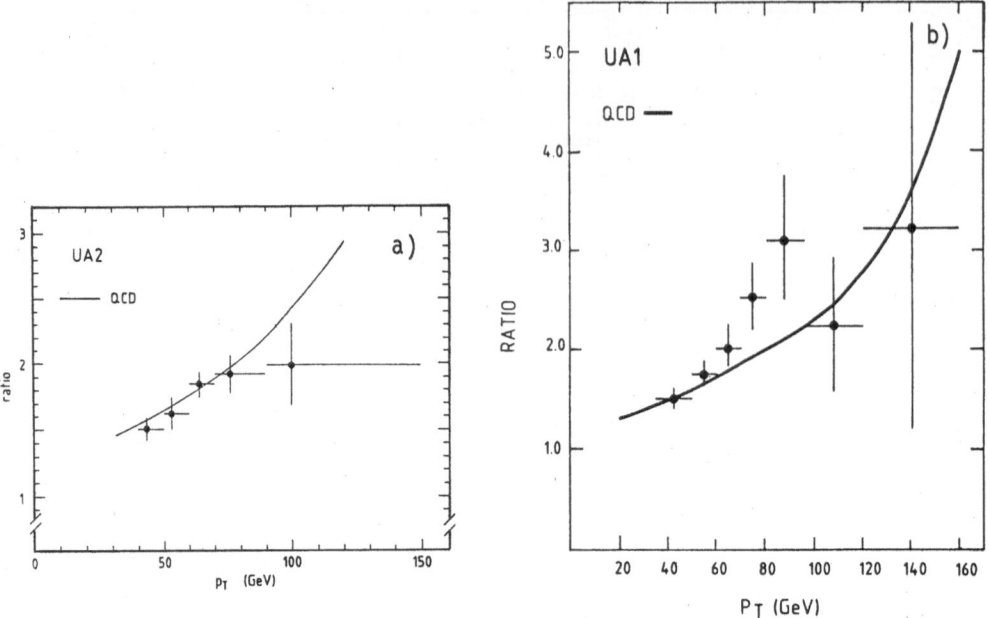

Fig. 11 Ratio of the inclusive jet yields between $\sqrt{s}$ = 630 GeV and $\sqrt{s}$ = 546 GeV: a) UA2 data;[23] b) UA1 data.[24] The curves represent the expected increase with p_T from a QCD calculation.

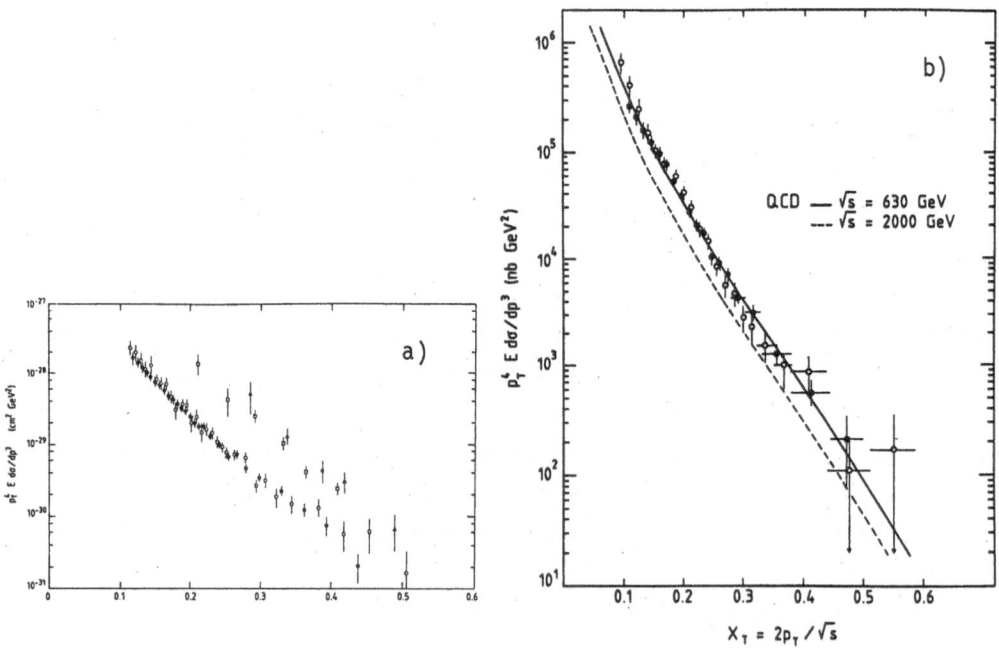

Fig. 12 Scaled invariant jet cross-sections, p_T^4 E$d\sigma/dp^3$, plotted as a function of $x_T = 2p_T/\sqrt{s}$.
a) Open (full) circles: UA2 data at $\sqrt{s}$ = 546 (630) GeV; open triangles (squares): ISR data at $\sqrt{s}$ = 45 (63) GeV;
b) Open (full) circles: UA1 data at $\sqrt{s}$ = 546 (630) GeV.

comparison with measurements at ISR energies in pp collisions[12] shown in Fig. 12a, exhibits the expected scale-breaking effects (we note that, for a given x_T, the value of Q^2 at ISR energies is about two orders of magnitude smaller than the corresponding value at Collider energies). The scale-breaking effects expected in going from Collider energies to the energy of the forthcoming TEV I Collider ($\sqrt{s}$ = 2000 GeV) are shown in Fig. 12b.

In the UA1 experiment, a search for jets has been made in a sample of 'minimum-bias' events recorded during a special Collider run in March 1985.[26] In this run the collision energy was not kept fixed, but varied between two flat-tops at 200 and 900 GeV with a period of 21.6 s. Jets were searched for by means of the same jet-finding algorithm as used for the analysis of hard collisions, but with a lower p_T threshold. The sample of jets found by this method has a p_T spectrum which joins smoothly onto the spectra measured by both UA1 and UA2 at higher p_T. This is illustrated in Fig. 13, which also shows a QCD prediction based on the structure functions of Duke and Owens.[27] The disagreement between theory and experiment in the low-p_T region is not surprising, because of the poor knowledge of the gluon structure function at values of x smaller than 0.04.

Figure 14 shows the inclusive jet cross-sections as a function of the jet p_T, as measured by UA1 at $\sqrt{s}$ = 200 and 900 GeV. The increase of the jet yield with $\sqrt{s}$, clearly shown by these data, can be better illustrated by considering a cross-section $\sigma(jet)$ defined as

$$\sigma(jet) = \int_{-1.5}^{+1.5} d\eta \int_{5 \text{GeV/c}}^{\infty} dp_T (d^2\sigma/dp_T d\eta) \ .$$

Figure 15 shows the behaviour of $\sigma(jet)$ as a function of $\sqrt{s}$. Over the range of $\sqrt{s}$ from 200 to 900 GeV, $\sigma(jet)$ increases by more than a factor of 4,

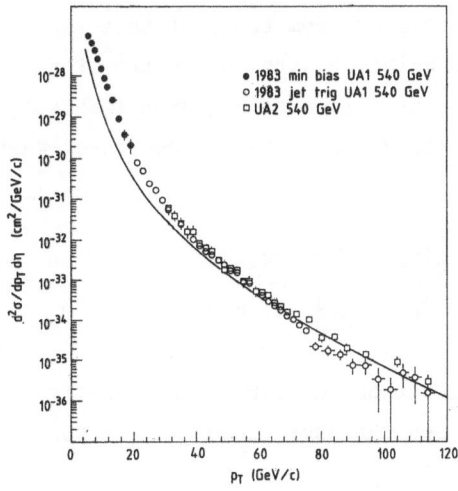

Fig. 13 Comparison of the inclusive jet yield at $\sqrt{s}$ = 546 GeV measured by UA1 in 'minimum-bias' events[26] with previous UA1[11] and UA2[8] measurements at higher p_T. The curve represents a QCD prediction.

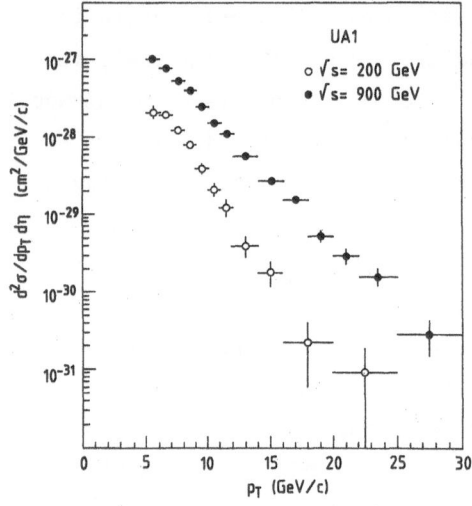

Fig. 14   Inclusive jet yield at
$\sqrt{s}$ = 200 GeV (open circles) and
$\sqrt{s}$ = 900 GeV (full circles).

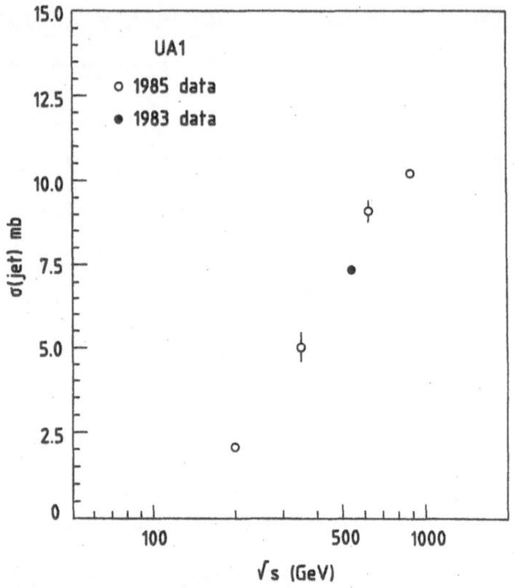

Fig. 15   Integrated jet yield (see
text) as a function of $\sqrt{s}$. Only the
statistical errors are shown.

reaching a value of ~ 10 mb at $\sqrt{s}$ = 900 GeV. This value represents a substantial fraction of the total cross-section, giving support to the rather
old idea[28] that the increase of the total cross-section observed at very
high energy might be entirely due to the increasing rate of hard collisions.

3.5   Angular Distribution of Parton-Parton Scattering

The study of the jet angular distribution in two-jet events provides a
way to measure the angular distribution of parton-parton scattering, and can
therefore be considered as the analogue of Rutherford's experiment in QCD.

In practice, there are complications arising from the fact that the
centre of mass of the two partons does not coincide with the centre of mass
of the colliding hadrons, and, in addition, what is observed experimentally
is a mixture of various subprocesses which occur at different centre-of-mass
energies. We can write

$$\frac{d^3\sigma}{dx_1 dx_2 d(\cos\theta^*)} = \sum_{A,B} \frac{F_A(x_1)}{x_1} \frac{F_B(x_2)}{x_2} \sum_{C,D} \frac{d\sigma_{AB \to CD}}{d(\cos\theta^*)} , \qquad (3)$$

where $F_A(x)/x$ $[F_B(x)/x]$ is the structure function describing the density of
parton A [B] within the incident hadrons, and the sums extend to all possible subprocesses AB → CD.

If the total transverse momentum P_T of the two-jet system is zero, values of θ^*, x_1, and x_2 are obtained for each event using the kinematical relations

$$\theta^* = \sin^{-1}(2p_T/M_{jj}) \ ; \qquad \begin{aligned} x_1 &= [\sqrt{(P_L^2 + M_{jj}^2)} + P_L]/\sqrt{s} \ , \\ x_2 &= [\sqrt{(P_L^2 + M_{jj}^2)} - P_L]/\sqrt{s} \ , \end{aligned}$$

where p_T is the jet transverse momentum, M_{jj} is the invariant mass, and p_L is the total longitudinal momentum of the two-jet system. For events with $P_T \neq 0$ it is not possible to determine θ^* unambiguously. In this case, only events with $P_T \ll p_T$ are used and θ^* is determined according to the convention of Collins and Soper.[29]

Equation (3) may at first sight appear hopeless in view of the many terms involved. However, in the case of $p\bar{p}$ collisions the dominating sub-processes are $gg \to gg$, $gq \to gq$ (or $g\bar{q} \to g\bar{q}$), and $q\bar{q} \to q\bar{q}$, which to a very good approximation have the same $\cos\theta^*$ dependence. Furthermore, for $\cos\theta^* \to 1$ ($t \to 0$) their differential cross-sections are in the ratio $1:4/9:(4/9)^2$ (see Table 1), reflecting the relative strengths of the quark-gluon and gluon-gluon couplings in QCD. Equation (3) can then be approximately factorized as

$$\frac{d^3\sigma}{dx_1 \, dx_2 \, d(\cos\theta^*)} = \left[\frac{1}{x_1} \sum_A F_A(x_1)\right] \left[\frac{1}{x_2} \sum_B F_B(x_2)\right] \frac{d\sigma}{d(\cos\theta^*)} \ .$$

If $d\sigma/d(\cos\theta^*)$ is taken to be the differential cross-section for gluon-gluon elastic scattering, which to leading order in QCD has the form

$$\frac{d\sigma}{d(\cos\theta^*)} = \frac{9}{8} \frac{\pi\alpha_s^2}{2x_1 x_2 s} \frac{(3 + \cos^2\theta^*)^3}{(1 - \cos^2\theta^*)^2} \ , \tag{4}$$

then it becomes possible to write

$$\sum_A F_A(x) = g(x) + \frac{4}{9} [q(x) + \bar{q}(x)] \ , \tag{5}$$

where $g(x)$, $q(x)$, and $\bar{q}(x)$ are respectively the gluon, quark, and anti-quark structure functions of the proton.

The term $d\sigma/d(\cos\theta^*)$ in Eq. (4) contains a singularity at $\theta^* = 0$ with the familiar Rutherford form $\sin^{-4}(\theta^*/2)$, which is typical of gauge vector-boson exchange. In the subprocesses $gg \to gg$ and $gq \to gq$ (or $g\bar{q} \to g\bar{q}$) it arises from the three-gluon vertex. It is also present in the subprocess

$q\bar{q} \rightarrow q\bar{q}$, but in this case it would be present in an Abelian theory as well, as for e^+e^- elastic scattering in QED.

Figures 16 show the $\cos\theta^*$ distributions as measured by UA1[30] (Fig. 16a) and UA2[31] (Fig. 16b). Also shown are QCD predictions, normalized to the total number of events. The UA2 data cover only the range $|\cos\theta^*| < 0.6$ because of the limited polar-angle interval covered by the UA2 calorimeter.

Both sets of data agree with QCD expectations, and they clearly show the increase towards the forward direction expected from the $\sin^{-4}(\theta^*/2)$ singularity. The expectations from a theory with scalar gluons are in strong disagreement with the data (see Fig. 16b).

Figure 17 shows again the angular distribution for two-jet events, as measured in the UA1 experiment.[30] Here the data are plotted as a function of the variable $\chi = (1 + \cos\theta^*)/(1 - \cos\theta^*)$, which has the property that $d\sigma/d\chi$ is constant for Rutherford scattering.[32] In Fig. 17 the data are compared with a QCD prediction which assumes $Q^2 = -\hat{t}$ (full curve), and also with a prediction which assumes exact scaling. We conclude that scale-breaking corrections are required to fit the data, and furthermore that the effective Q^2-scale depends on θ^* and is consistent with $Q^2 = -\hat{t}$.

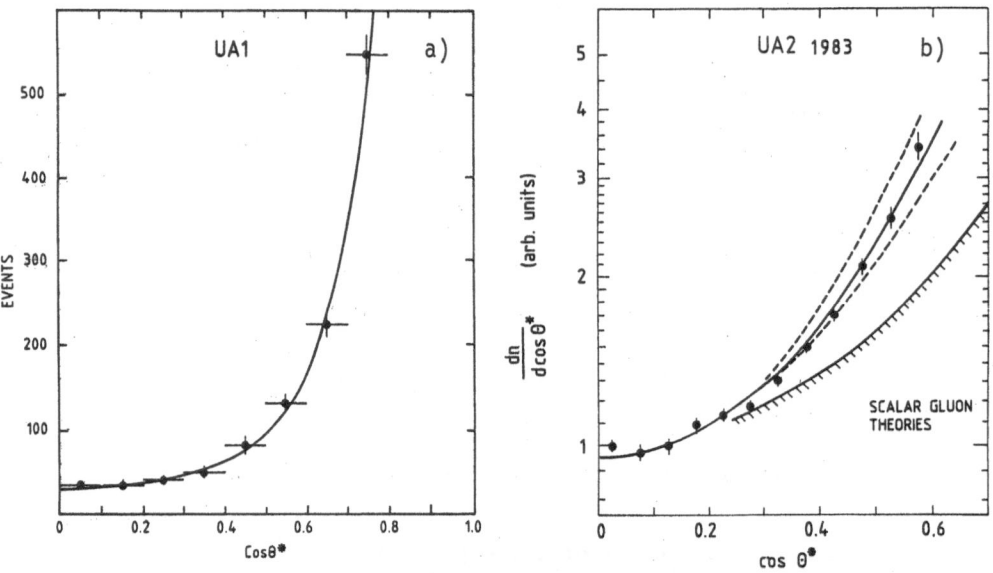

Fig. 16 a) Distribution of $\cos\theta^*$ for hard-parton scattering as measured in the UA1 experiment[30] for two-jet events with M_{jj} between 150 and 250 GeV. b) Distribution of $\cos\theta^*$ for hard-parton scattering as measured in the UA2 experiment.[31] All the different QCD processes (except for $q\bar{q} \rightarrow q'\bar{q}'$), separately normalized to the data, lie in the area between the two dashed curves of Fig. 16b. The full lines are QCD predictions, normalized to the total number of events.

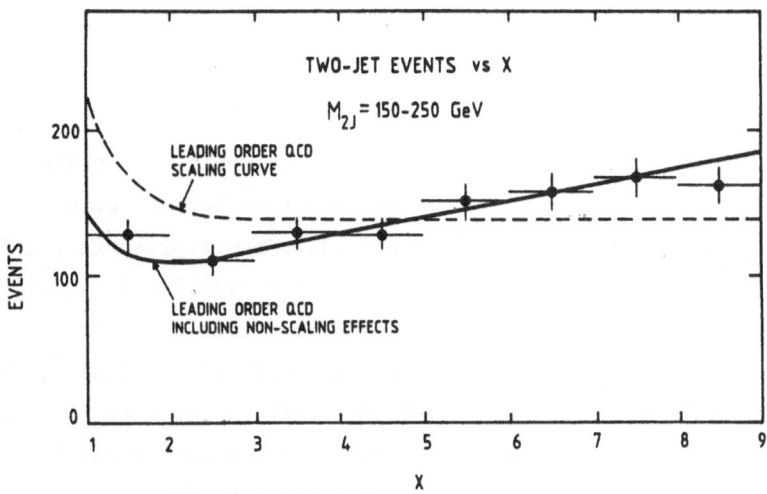

Fig. 17 Two-jet angular distribution[30] plotted as a function of the variable $\chi = (1 + \cos\theta^*)/(1 - \cos\theta^*)$. The curves represent a leading-order QCD prediction with scale-breaking corrections (solid curve) and without them (broken curve).

3.6 <u>Determination of the Proton Structure Function</u>

In order to extract the effective structure function $F(x)$, a quantity $S(x_1, x_2)$ is defined in terms of the measured differential cross-section $d^2\sigma/dx_1 dx_2$:

$$ S(x_1, x_2) = \frac{x_1 x_2 (d^2\sigma/dx_1 dx_2)}{\int_0^{\cos\theta^*_{min}} K[d\sigma/d(\cos\theta^*)]d(\cos\theta^*)} \quad , $$

where θ^*_{min} is the smallest scattering angle for which both jets fall within the detector acceptance (θ^*_{min} is itself a function of x_1 and x_2). The form of $d\sigma/d(\cos\theta^*)$ is taken from Eq. (4), and K is the numerical factor describing higher-order QCD corrections (see subsection 3.3). The approximate factorization property $S(x_1, x_2) = F(x_1)F(x_2)$ is found to be verified by the data within errors.

Figure 18 shows the function $F(x)$ obtained in the UA1[33] and UA2[31] experiments using $\Lambda = 0.2$ GeV and K = 2. In addition to the statistical errors there is a systematic uncertainty of ~ 50% in the overall normalization, which includes one half of the theoretical uncertainty on K [what is actually determined from the data is $F(x)\sqrt{K}$]. Also shown in Fig. 18 are curves representing the function $g(x) + (4/9)[q(x) + \bar{q}(x)]$ as expected from fits to

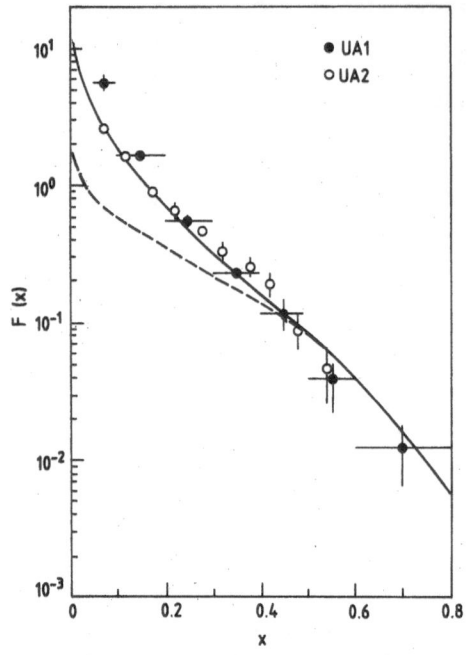

Fig. 18 Effective structure function measured from two-jet events.[31,32] The full curve represents the function $g(x) + (4/9)[q(x) + \bar{q}(x)]$ as expected from fits to ν and $\bar{\nu}$ deep-inelastic scattering data.[34] The broken curve represents the quark contribution alone, equal to $(4/9)[q(x) + \bar{q}(x)]$.

ν and $\bar{\nu}$ deep-inelastic scattering data.[34] The Collider results agree with the behaviour expected at the large Q^2 values appropriate to the Collider experiments ($\langle Q^2 \rangle \approx 2000$ GeV^2), and they show directly the very large gluon density in the proton at small x.

3.7   Total Transverse Momentum of the Two-Jet System

If the two partons which undergo hard scattering have no initial transverse momentum, the total transverse momentum of the final two-jet system, P_T, should be equal to zero. In reality, this does not happen, because the incident partons have a small primordial transverse momentum, and, furthermore, both incident and outgoing partons may radiate gluons.

Experimentally, P_T is determined from the sum of two large and approximately opposite two-dimensional vectors $\vec{p}_{T1}$ and $\vec{p}_{T2}$, and it is therefore sensitive to instrumental effects such as the calorimeter energy resolution and incomplete containment due to edge effects in the detector. These effects can be made small by considering only the component P_η of $\vec{P}_T$ parallel to the bisector of the angle defined by $\vec{p}_{T1}$ and $\vec{p}_{T2}$.

Figure 19 shows the distribution of P_η as measured in the UA2 experiment[31] for two-jet events having $E_T^1 + E_T^2 > 40$ GeV, $\Delta\varphi > 120°$, and with each jet satisfying the conditions $E_T > 10$ GeV, $|\cos\theta| < 0.6$. For $P_\eta < 15$ GeV/c, this distribution corresponds to a P_T distribution of the form $P_T \exp[-AP_T^2]$, with $\langle P_T \rangle = 7 \pm 1$ GeV/c, whilst for $P_\eta > 15$ GeV/c the data lie systematically above this simple parametrization.

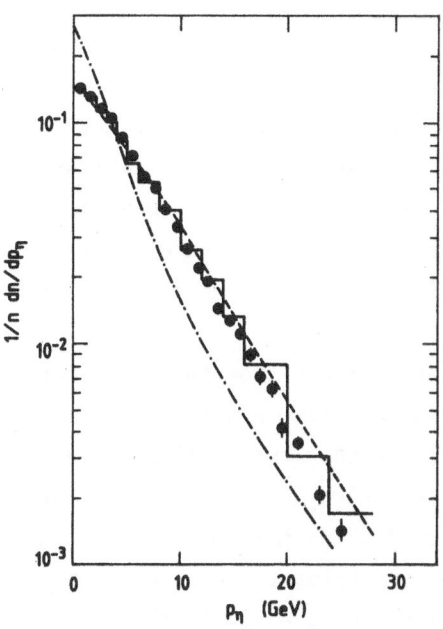

Fig. 19 Distribution of the component P_η of the total transverse
momentum of the two-jet system, as
measured in the UA2 experiment.[31]
The dashed line is a QCD prediction;[35] the dotted-dashed line represents the same prediction, but
assuming that gluons radiate as
quarks. The histogram represents the
standard QCD prediction with the
detector effects taken into account.

These results are in good agreement with a QCD prediction[35] illustrated
by the curve of Fig. 19. In QCD, gluon radiation by a gluon (g → gg) --
which occurs because of the three-gluon vertex -- has a rate 9/4 times
higher than that of q → qg, and predictions based on the assumption that
gluons radiate like quarks disagree with the data (see Fig. 19). Since gluon
jets dominate in the P_T range explored at the Collider, we can conclude that
the good agreement between the data of Fig. 19 and the theoretical predictions can be considered as further evidence in favour of a QCD description
of high-p_T jet production.

3.8 Jet Fragmentation

In the preceding sections we have discussed the properties of jet production independently of the internal structure of the jet itself. In this
section we discuss the properties of the jet internal structure.

High-p_T partons produced far from the mass shell in a hard collision
evolve according to two distinct time-scales. At early times, which correspond to distances much shorter than the typical size of a hadron, gluon
radiation and the creation of $q\bar{q}$ pairs result in the production of a jet
of coloured partons, which can be described using perturbative QCD.[36] The
following step, occurring on a longer time-scale, involves the long-distance
non-perturbative properties of QCD, which lead to the confinement of partons
in colourless bound states, observed as hadrons. Non-perturbative techniques
providing a sufficient understanding of confinement are still missing at
present, and two plausible models are commonly used, one dealing with each

jet independently,[37] the other (referred to as the Lund model) taking into account the configuration of all coloured partons in the final state.[38]

A jet consists in general of many particles (fragments) whose motion with respect to the jet axis can be described by means of two variables: the fractional longitudinal momentum $z = \vec{p} \cdot \vec{P}/P^2$, where $\vec{p}$ is the fragment momentum and $\vec{P}$ is the total jet momentum (obviously $0 < z < 1$); and q_T, the component of $\vec{p}$ perpendicular to $\vec{P}$. The z distribution D(z) is called the jet fragmentation function.

A systematic uncertainty in the scale of z arises from the determination of $\vec{P}$, which depends on the criteria used to define a jet. Furthermore, the function D(z) cannot be measured reliably in the low-z region because the final state contains many low-p_T particles which do not belong to the jet but are associated with the partons which did not take part in the hard collision (these particles, which are called spectators, are absent, of course, in the hadronic final states resulting from e^+e^- collisions).

Figure 20 shows recent UA1 results[39] obtained in a study of two-jet events for which the leading jet has $p_T > 25$ GeV/c. The total jet momentum $\vec{P}$ is measured by the calorimeter, and all charged particles which satisfy the condition $\Delta\varphi^2 + \Delta\eta^2 < 1$ with respect to the jet axis are considered. The band shown in Fig. 20 represents the size of the systematic errors, which at low z are mainly due to the uncertainty in the estimate of the spectator contribution, and at high z to the uncertainty on the calorimeter energy scale.

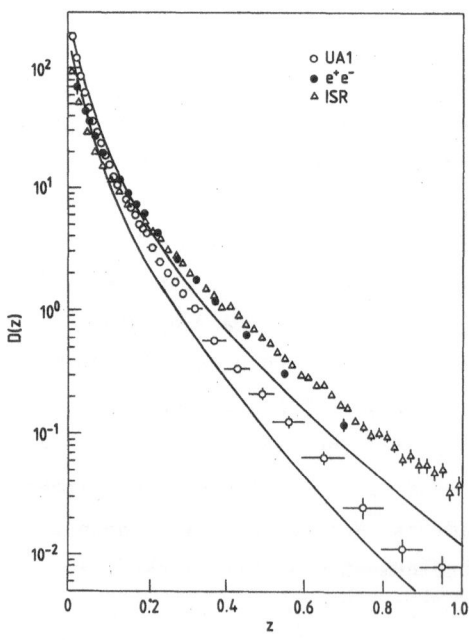

Fig. 20 Jet fragmentation function into charged particles, D(z). Open circles, UA1 data;[39] full circles, e^+e^- data;[40] open triangles, ISR data.[41] The two curves define a band which represents the systematic uncertainties of the UA1 data.

Also shown in Fig. 20 is the fragmentation function D(z) for jets from e^+e^- collisions,[40] and from pp collisions at the ISR.[41] Obviously, the Collider data have a steeper variation with z. This could be the effect of scaling violations ($Q^2 \gtrsim 2000$ GeV^2 for the Collider data, $\lesssim 1000$ GeV^2 for the e^+e^- and ISR data), or a real difference between quark jets (which dominate in e^+e^- and ISR pp data) and gluon jets (which dominate in the p_T range of the UA1 data).

In order to distinguish between these two possibilities, the UA1 Collaboration have developed a method to separate gluon jets from quark jets, which is based on the expected differences in the structure functions of gluons and quarks, and in the matrix elements of the various subprocesses (see Table 1). The probability that the two jets result from subprocess ab → cd, where a, b, c, and d are specific parton types, can then be expressed as

$$\mathcal{P}(ab \to cd) = \frac{\{[F_a(x_1,Q^2)/x_1][F_b(x_2,Q^2)/x_2][d\sigma/d(\cos\theta^*)]\}(ab \to cd)}{\sum_{a,b,c,d} \{[F_a(x_1,Q^2)/x_1][F_b(x_2,Q^2)/x_2][d\sigma/d(\cos\theta^*)]\}(ab \to cd)} ,$$

(6)

where F_a and F_b are structure functions describing the densities of partons a and b; $d\sigma/d(\cos\theta^*)$ is the differential cross-section for the subprocess ab → cd; and the sum extends over all possible subprocesses.

Equation (6) can be used to calculate the probability that parton c is a gluon:

$$\mathcal{P}(g) = \sum_{a,b,d} \mathcal{P}(ab \to gd) ,$$

and the probability that it is a quark (or antiquark):

$$\mathcal{P}(q) = \sum_{a,b,d} \left\{\mathcal{P}(ab \to qd) + \mathcal{P}(ab \to \bar{q}d)\right\} = 1 - \mathcal{P}(g) .$$

Figure 21 shows the distribution of $\mathcal{P}(g)$ for a sample of two-jet events which satisfy the conditions $|\cos\theta^*| > 0.25$ (to enhance the separation between quark and gluon jets) and $1600 < Q^2 < 2600$ GeV^2 (to fix the Q^2-scale). From this distribution two samples are selected: a sample enriched in gluon jets (sample G), defined by the condition $\mathcal{P}(g) > 0.55$; and sample Q, enriched in quark jets, defined by the condition $\mathcal{P}(g) < 0.35$. Sample G has an average value $\langle\mathcal{P}(g)\rangle = 0.65$ and consists, therefore, of gluon jets (65%) and

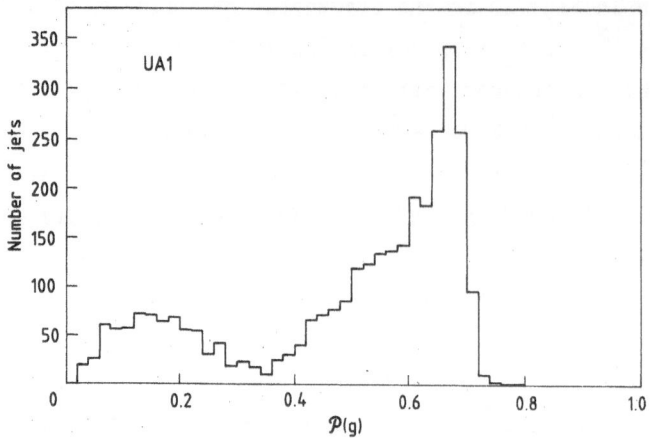

Fig. 21 Distribution of $\mathcal{P}(g)$ (see text) for a sample of two-jet events.

quark jets (35%). Sample Q has $\langle \mathcal{P}(g) \rangle = 0.17$ and consists mostly of quark jets (83%).

The measured fragmentation functions for the two samples, $D_G(z)$ and $D_Q(z)$, can be expressed in terms of the fragmentation functions $D_g(z)$ and $D_q(z)$, which correspond to pure samples of gluon and quark jets, respectively:

$$D_G(z) = 0.65 \, D_g(z) + 0.35 \, D_q(z) \; ,$$

$$D_Q(z) = 0.17 \, D_g(z) + 0.83 \, D_q(z) \; .$$

This system of two linear equations can be solved for each bin of z to obtain $D_g(z)$ and $D_q(z)$.

The result of this procedure is shown in Fig. 22. The softer gluon fragmentation predicted by QCD[36] is indeed observed experimentally. The difference between $D_g(z)$ and $D_q(z)$ cannot be the effect of scaling violations because the average jet transverse momentum is $\langle p_T \rangle = 35$ GeV/c for sample G, and 39 GeV/c for sample Q, indicating that the average value of Q^2 is the same for the two samples.

Scaling violations can be studied by subdividing the events into classes with different $\langle Q^2 \rangle$ values. Figure 23 clearly shows the Q^2 dependence of D(z) for jets with $\mathcal{P}(g) > 0.5$.

Figure 24 shows the average transverse momentum $\langle q_T \rangle$ of jet fragments with respect to the jet axis, as a function of z. For z > 0.1, the jets observed by UA1 are characterized by a value $\langle q_T \rangle = 0.85 \pm 0.02$ GeV/c, which is substantially larger than that of jets observed in e^+e^- collisions[40] or in ISR experiments.[41]

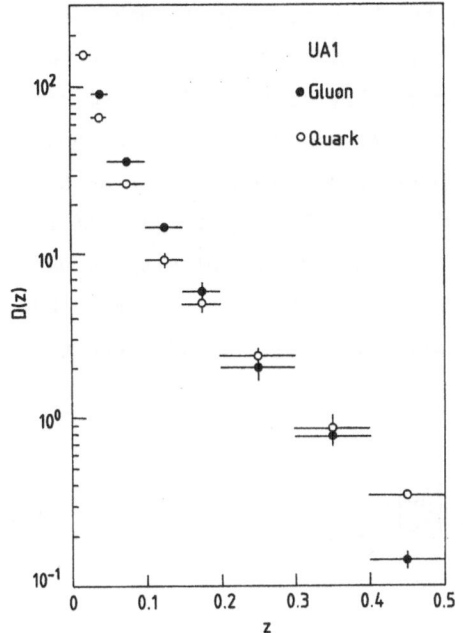

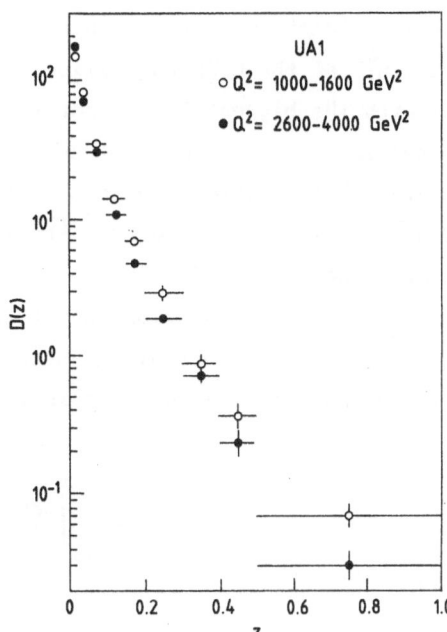

Fig. 22 Fragmentation function D(z) shown separately for gluon and quark jets.

Fig. 23 Q^2 dependence of the fragmentation function D(z) for jets having $\mathcal{P}(g) > 0.5$.

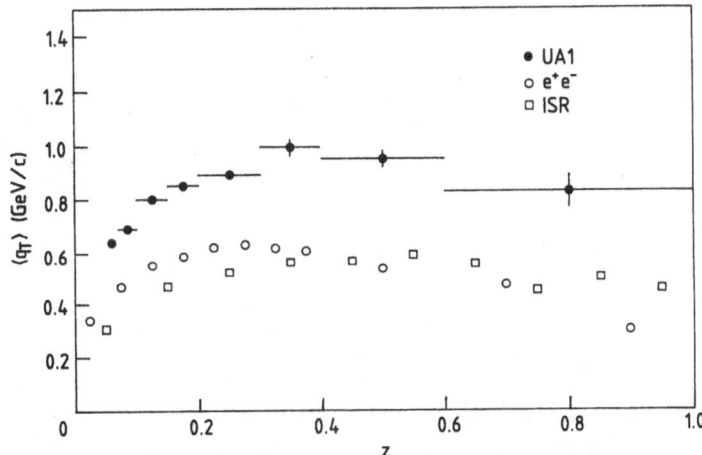

Fig. 24 Average transverse momentum of jet fragments with respect to the jet axis, as a function of z.

3.9 Charged-Particle Multiplicity in a Jet

The average charged-particle multiplicity in a jet, $\langle n_{ch} \rangle$, can be obtained, in principle, by integrating the function D(z) over the range $0 < z < 1$. However, because of the contamination of spectator particles mentioned previously, the uncertainty on D(z) at small z values is too large to make the result meaningful.

An attempt to overcome this difficulty has been made by the UA2 Colla-boration[42] at the CERN $p\bar{p}$ Collider. Events are considered in which the two jets with the highest E_T are separated in azimuth by at least 150° and have an invariant mass M_{jj} of at least 40 GeV. For these events, Fig. 25a shows the distribution $dn/d\varphi$, where φ is the azimuthal separation between each charged particle observed over the range 20° < θ < 160° and the axis of the jet with the highest E_T. A lower limit to the true jet multiplicity is ob-tained under the assumption that there are no jet fragments at $\varphi = \pi/2$, but only charged spectators with a uniform φ distribution. The average charged-particle multiplicity in a jet is then

$$\langle n_{ch} \rangle = \frac{1}{2} \int_0^\pi \frac{dn}{d\varphi} \, d\varphi - \frac{\pi}{2} \left[\frac{dn}{d\varphi} \right]_{\varphi = \pi/2} . \qquad (7)$$

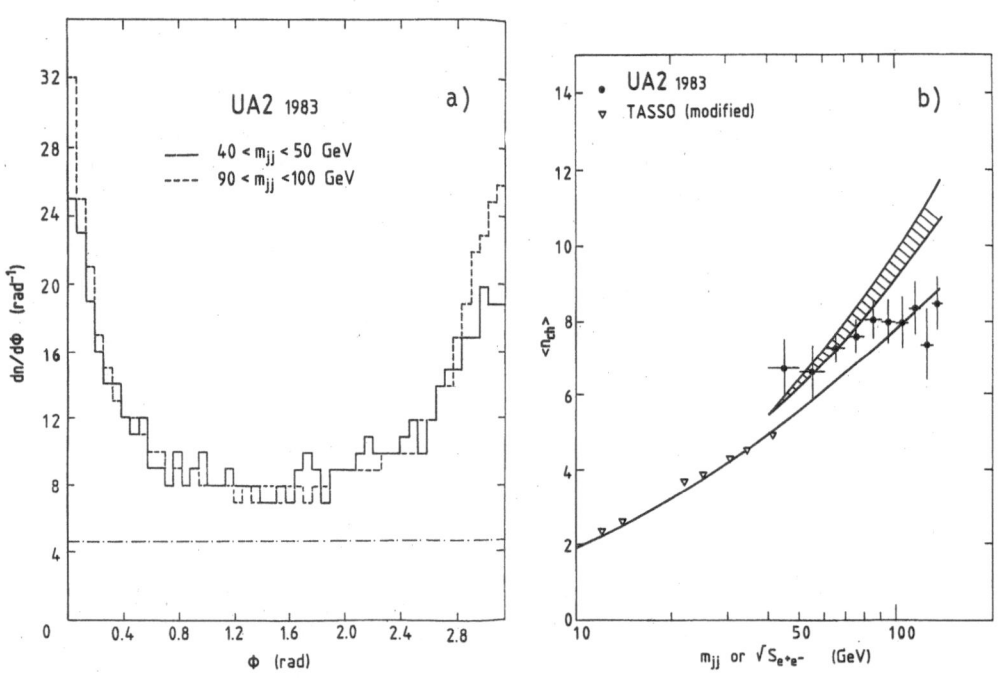

Fig. 25 a) Distribution of the azimuthal separation between the axis of the leading jet and all charged particles, normalized to the total number of events.[42] The dashed-dotted horizontal line represents the azimuthal charged-particle density measured in the same experiment for ordinary soft collisions. b) Mean charged-particle multiplicities in jets derived according to Eq. (7). Open triangles: data from e^+e^- collisions.[40] Full circles: data from UA2.[42] The solid curve is the prediction of the model of Ref. 43 for quark jets; the shaded area represents the predictions for gluon jets.

Values of $\langle n_{ch} \rangle$ thus obtained are shown in Fig. 25b as a function of M_{jj}. Also shown for comparison are values of $\langle n_{ch} \rangle$ as a function of $\sqrt{s}$ for hadronic final states from e^+e^- collisions,[41] modified according to Eq. (7). This procedure allows for a model-independent comparison of jets for $p\bar{p}$ and e^+e^- reactions, suggesting that jets at the $p\bar{p}$ Collider have higher mean multiplicities than would be expected from extrapolation of e^+e^- data. Fragmentation models based on QCD,[43] illustrated by the curves shown in Fig. 25b, predict a higher relative multiplicity of gluon jets with respect to quark jets. At Collider energies one expects that the fraction of gluon jets in the data of Fig. 25b varies from $\sim$ 75% to $\sim$ 30% as M_{jj} varies from 40 to 140 GeV, thus explaining the observed behaviour of $\langle n_{ch} \rangle$ as a function of M_{jj}.

3.10 Three-Jet Final States

Three-jet final states are expected in the framework of QCD as the result of gluon radiation, either by the incident or the outgoing partons, as well as by the virtual parton which is exchanged in the subprocess.

For three final-state (massless) partons, the final-state parton configuration at fixed subprocess centre-of-mass energy $\sqrt{\hat{s}}$ is specified by four independent variables. The leading-order QCD predictions for the subprocess cross-sections are given by

$$d^4\sigma/dx_1\,dx_2\,d(\cos\theta_1^*)d\psi = (1/32\pi^2)|M_3|^2 \ . \tag{8}$$

Explicit expressions for the matrix element M_3 exist in the literature.[44] Neglecting constant and slowly varying factors, $|M_3|^2$ may be written as

$$|M_3|^2 \approx (\alpha_s^3/\hat{s})[x_{1T}^2 x_{2T}^2 x_{3T}^2 (1-x_1)(1-x_2)(1-x_3)]^{-1} \ . \tag{9}$$

In Eqs. (8) and (9) the quantities x_i (i = 1, 2, 3) are the energies (or momenta) of the outgoing partons in the subprocess centre-of-mass, measured in units of $\sqrt{\hat{s}}/2$, and ordered so that $x_1 > x_2 > x_3$; θ_i^* is the angle between parton i and the beam direction ($x_{iT} = x_i \sin\theta_i^*$); and ψ is the angle between the plane defined by partons 2 and 3 and the plane defined by the beam axis and parton 1 (see Fig. 26).

An important feature of Eq. (9) is that for three-parton configurations which approach two-parton configurations ($x_1, x_2 \rightarrow 1$ or $x_{iT} \rightarrow 0$), $|M_3|^2$ diverges as the result of the bremsstrahlung nature of gluon radiation. The comparison of theory with experiment must be restricted, therefore, to a

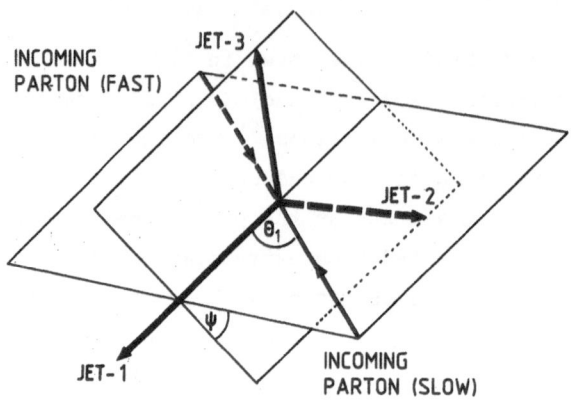

Fig. 26   Three-jet variables defined in the subprocess centre-of-mass frame.

region of phase space where all the jets have reasonably high p_T and are well separated from each other.

In the UA1 experiment,[30] the cuts $\sqrt{\hat{s}} > 150$ GeV, $x_3 < 0.9$, $|\cos\theta_1| <$ 0.6, and $30° < \psi < 150°$ select 173 three-jet events out of a sample corresponding to an integrated luminosity of ~ 100 nb^{-1}. In these events the average p_T values of jets 1 and 3 are ~ 55 and ~ 25 GeV/c, respectively. Figure 27 shows the distribution of these events in the (ψ, $\cos\theta_1$) plane (the sign of $\cos\theta_1$ is defined with respect to the faster of the two incident partons). The projections on the two axes agree with leading-order QCD predictions, and clearly show the expected increase of the cross-section when $|\cos\theta_1| \rightarrow 1$ or $\psi \rightarrow 0°$ (180°).

In the UA2 experiment,[45] the cuts $p_{1T} + p_{2T} + p_{3T} > 70$ GeV/c, $|\vec{p}_{1T} + \vec{p}_{2T} + \vec{p}_{3T}| < 20$ GeV/c, $p_{3T} > 10$ GeV/c, and $|\eta_i| < 0.8$ (i = 1, 2, 3) select 4972 three-jet events corresponding to an integrated luminosity of ~ 310 nb^{-1} (as usual, the jet definition is such that $p_{1T} > p_{2T} > p_{3T}$). Figure 28 shows the distribution of the variable $\cos\omega_{23}^*$, where ω_{23}^* is the angular separation between jets 2 and 3 in the centre of mass of the three-jet system. The increase towards $\cos\omega_{23}^* \rightarrow 1$, in agreement with QCD expectations, can be explained by assuming that the third jet is associated with a hard gluon radiated by parton 2. The drop at $\cos\omega_{23}^* > 0.8$ results from the ~ 30° resolving power of the jet cluster algorithm.

A comparison of the relative yield of three-jet and two-jet events can be used, in principle, to derive a value of α_s which is less affected by the systematic uncertainties on calorimeter energy scale, luminosity, and structure functions than is the case for the inclusive jet yield (see section 3.3). However, there are still large uncertainties related to the choice of the Q^2-scale appropriate to three-jet and two-jet production.

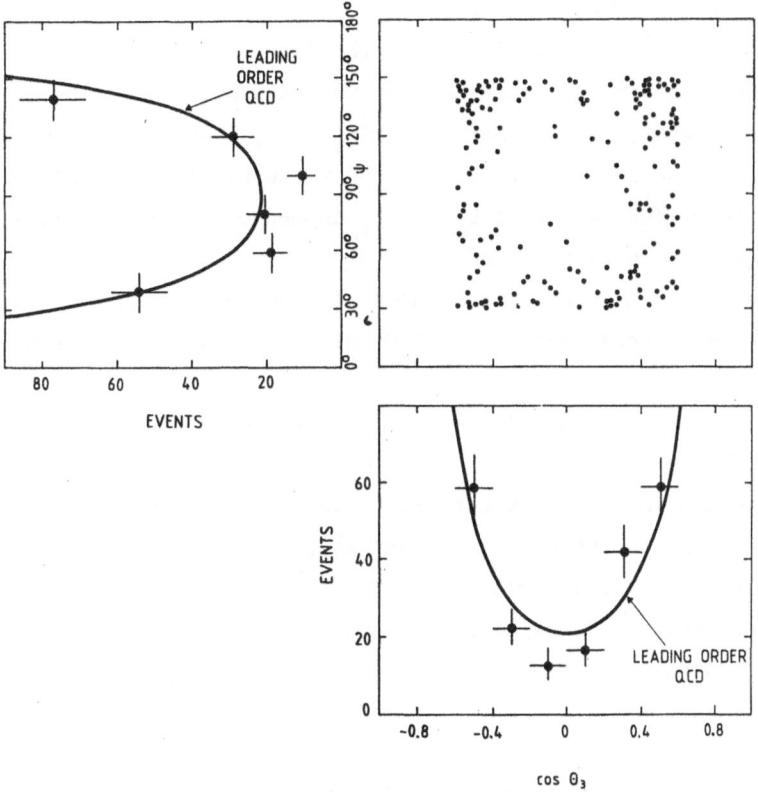

Fig. 27 Distribution of three-jet events, as measured by UA1,[30] in the plane (ψ, $\cos\theta_1$).

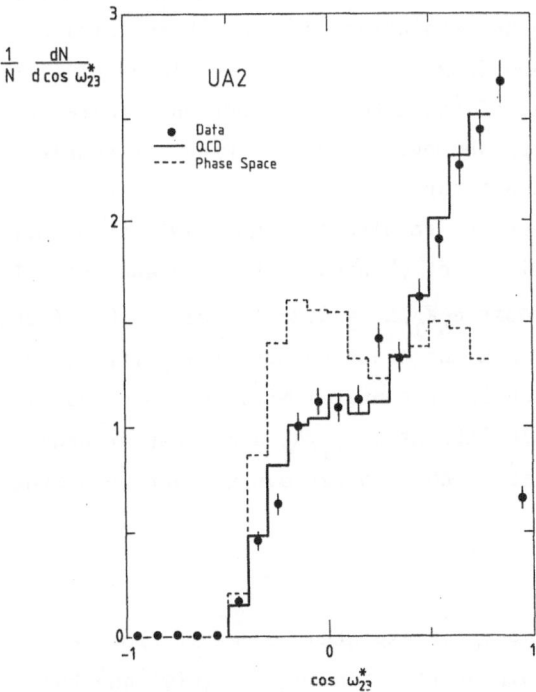

Fig. 28 Distribution of $\cos\omega^*_{23}$, where ω^*_{23} is the angular separation between the axis of jet 2 and jet 3 in the centre of mass of the three-jet system (UA2, Ref. 45).

Furthermore, a quantitative comparison with QCD predictions requires a calculation of the relevant two-jet and three-jet cross-sections beyond leading order. Since such a calculation has not yet been performed, the quantity that one measures is not α_s, but rather $\alpha_s K_3/K_2$, where the parameter K_n (n = 2, 3) represents the effect of the unknown higher-order corrections to the n-jet yield. The values of K_2 and K_3 depend on the kinematical cuts and are expected to be between 1 and 2.

In the UA1 experiment,[30] the three-jet sample is compared with a sample of 1142 two-jet events having $\sqrt{s} > 150$ GeV and $|\cos\hat{\theta}^*| < 0.8$. The result is

$$\alpha_s K_3/K_2 = 0.16 \pm 0.02 \pm 0.03 \qquad (10)$$

under the assumption that the three-jet sample is characterized by a lower Q^2-scale than the corresponding two-jet sample at the same value of $\sqrt{\hat{s}}$ ($\langle Q_{2j}\rangle \approx 0.45\sqrt{\hat{s}}$, $\langle Q_{3j}\rangle \approx 0.3\sqrt{\hat{s}}$, where $Q = \sqrt{Q^2}$).

The UA2 result[45] is based on a comparison of the three-jet sample with a sample of 10566 two-jet events having $p_{1T} + p_{2T} > 70$ GeV/c, $|\vec{p}_{1T} + \vec{p}_{2T}| < 20$ GeV/c and $|\eta_i| < 0.8$ (i = 1, 2). The result is

$$\alpha_s K_3/K_2 = 0.23 \pm 0.01 \pm 0.04 \qquad (10')$$

obtained under the assumption that $Q^2 = P_{1T}^2$ for both two-jet and three-jet events.

In the above results the first error is statistical, whilst the second one results from various systematic uncertainties, such as those arising from the calorimeter energy scale and from the imperfect knowledge of structure and fragmentation functions and of spectator contributions. There is also a possible contamination of four-jet events in the three-jet sample, which is difficult to take into account correctly.

The apparent discrepancy between the results (10) and (10') is almost entirely due to the different definition of Q^2 used in the two analyses. In particular, the UA1 result would become $\alpha_s K_3/K_2 = 0.23 \pm 0.02 \pm 0.04$ if the same Q^2-scale were used for two-jet and three-jet events. This difference indicates that conceptual rather than instrumental limitations preclude a more reliable measurement of α_s. Calculations of K_3/K_2 and a better understanding of the relevant Q^2-scales are needed to improve the determination of α_s by this method.

3.11 Direct Photon Production

Direct photon production at high-p_T is expected to result from two dominant parton subprocesses: the Compton reaction $gq(\bar{q}) \to \gamma q(\bar{q})$ and the

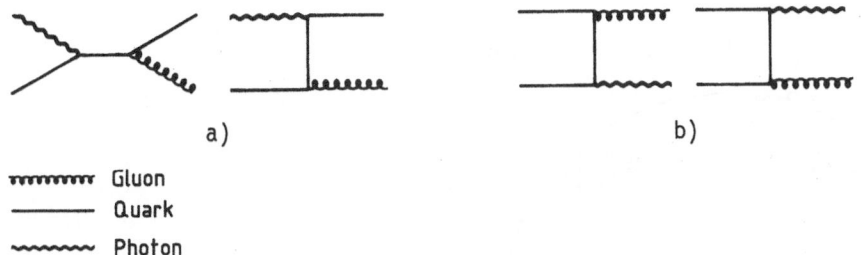

a)

b)

〰〰〰 Gluon

———— Quark

〜〜〜 Photon

Fig. 29 Leading-order diagrams for direct photon production: a) Compton
diagrams; b) annihilation diagrams.

annihilation reaction $q\bar{q} \rightarrow \gamma g$ (see the leading-order diagrams in Fig. 29).
A measurement of the cross-section for direct photon production offers,
therefore, an alternative test of QCD, with the advantages that the photon
energy measurement is not affected by fragmentation effects, and that QCD
calculations have been carried out to the next-to-leading order in α_s.[46]

Naïvely one might expect that the ratio between the yield of direct
photons and that of jets is of the order of $\alpha/\alpha_s \approx 0.05$. In practice, at
Collider energies this ratio is close to $\sim 10^{-4}$. This is a consequence of
the low average quark charge, and especially of the fact that the three-
gluon vertex, which dominates the jet yield at Collider energies (see the
diagrams of Fig. 7 and Table 1), does not contribute to photon production.

Experimentally, the observation of these photons is made difficult by
an overwhelming background of high-p_T π^0's decaying into photon pairs. At
high π^0 energies the $\gamma\gamma$ opening angle is very small (typically $\sim$ 27 mrad at
10 GeV), so that the two photons cannot be resolved. However, high-p_T π^0's
are jet fragments and are accompanied, in general, by other high-p_T particles. On the contrary, single high-p_T photons replace one of the two jets
in the final state, and no other high-p_T particle is expected at small
angles to them. It is possible, therefore, to improve the signal-to-noise
ratio by requiring that the photon candidates be isolated from other particles in the event.

In the UA2 experiment,[47] high-p_T photon candidates having $p_T > 15$ GeV/c
and $|\eta| < 0.8$ are selected by searching for energy clusters in the central
calorimeter which are consistent with an electromagnetic shower (small
lateral dimensions and limited energy leakage in the hadronic calorimeter).
Such clusters are further required to be associated with no charged-particle
track and at most one signal from photon conversion in a cone of 20° half-
aperture around the photon direction. It is estimated that $(55 \pm 6)\%$ of the
photon from the two dominant parton subprocesses satisfy these requirements.
Other sources of single photons, such as quark bremsstrahlung, are strongly
suppressed.

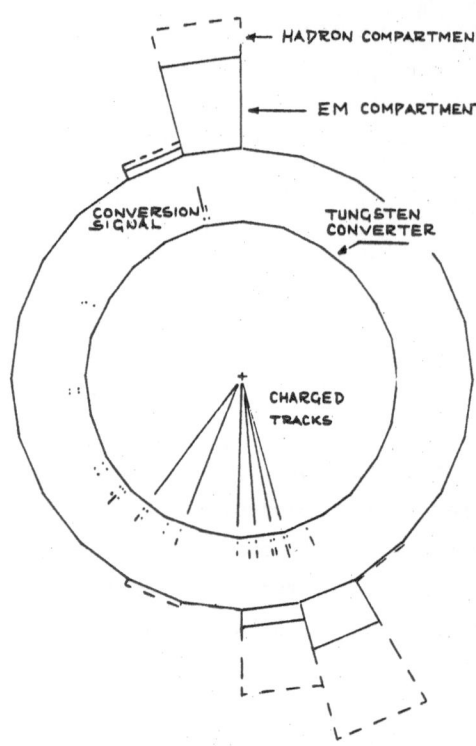

Fig. 30 View of a UA2 event containing a direct photon candidate in a plane perpendicular to the beam axis.

Among the data collected in 1984 ($\sqrt{s}$ = 630 GeV, L = 310 nb^{-1}), a total of 1350 photon candidates are found, of which 995 are associated with a conversion signal in the preshower counter, 1.5 radiation lengths thick (see subsection 2.2). A view of one such event in a plane perpendicular to the beam axis is shown in Fig. 30.

The contribution of π^0 background to this sample is estimated by a method first used by the CCOR Collaboration[48] in an experiment at the CERN ISR. This experiment relies on the different conversion probabilities of single photons and two-photon systems when they cross a thin converter. In the case of UA2, the conversion probability for a single photon is 1 - P = 0.69 $\pm$ 0.02 and 1 - P^2 = 0.90 for a $\pi^0 \to \gamma\gamma$ decay.

If ξ is the fraction of $\pi^0 \to \gamma\gamma$ decays contained in the sample of photon candidates, it can easily be shown that

$$\xi = (P - \lambda)/(P - P^2) , \qquad (11)$$

where λ is the fraction of photon candidates having no conversion signal in the preshower counter. By using Eq. (11), it is found that the π^0 contamination is typically (28 $\pm$ 9)% over the full p_T range.

Figure 31 shows the invariant cross-section for single-photon production at $\sqrt{s}$ = 630 GeV, as a function of the photon transverse momentum. The errors shown are only statistical. There is an additional p_T-independent

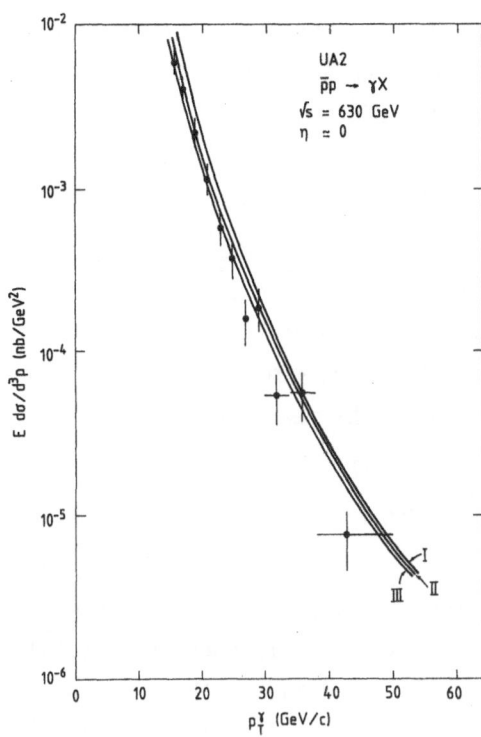

Fig. 31 Invariant cross-section for direct photon production at $\sqrt{s}$ = 630 GeV. Curve I is a QCD prediction for the total direct photon yield[46]. Curves II and III result from excluding bremsstrahlung photons with quark-photon angles less than 20° and 45°, respectively. The error bars do not include the normalization uncertainty.

systematic error of ±25%, which mainly reflects the uncertainties on the luminosity, on the photon energy scale, and on the efficiency of the photon selection criteria.

A QCD prediction which includes next-to-leading-order terms,[46] also shown in Fig. 31, agrees with the data within the statistical and systematic uncertainties.

We finally note that the ratio between the measured photon and jet yields at $\sqrt{s}$ = 630 GeV is ~ 2 × 10^{-4}. Any attempt to extract a γ/π^0 ratio from these data is, of course, a meaningless exercise, because high-p_T π^0's belong to jets, and thus they do not in general satisfy the isolation criteria applied in this analysis.

4. PHYSICS OF THE INTERMEDIATE VECTOR BOSONS

4.1 Properties of the Intermediate Vector Bosons

There are many excellent review papers and textbooks[49] which describe in detail the so-called Standard Model of the unified electroweak theory.[50] Here we recall only the main properties of the Intermediate Vector Bosons (IVBs) which are expected in the framework of this model.

4.1.1 <u>Masses.</u> The unified electroweak theory based on SU(2) x U(1) gauge invariance[50] provides a relation between the Fermi coupling constant G, the fine-structure constant α, and the mass of the W boson, m_W:

$$G = \pi\alpha/(\sqrt{2}\ m_W^2\ \sin^2\theta_W)\ , \qquad (12)$$

where θ_W is the mixing angle between the U(1) gauge field and the third component of the SU(2) gauge field. Using the most recent values,[51] G = (1.16637 ± 0.00002) x 10^{-5} GeV^{-2} and $\alpha^{-1} = 137.03604 \pm 0.00011$, and taking into account radiative corrections, Eq. (12) can be rewritten as

$$m_W = A/[(1 - \Delta r)^{1/2}\ \sin\theta_W]\ , \qquad (12')$$

where A = (37.2810 ± 0.0003) GeV/c^2, and the quantity Δr reflects the effect of the one-loop radiative corrections (to be discussed in subsection 4.15). Its value has been computed to be[52] $\Delta r = 0.0711 \pm 0.0013$. With such a numerical value, Eq. (12') can be simply rewritten as $m_W = 38.681/\sin\theta_W$ GeV/c^2.

The mixing angle θ_W has been measured by a variety of low-energy experiments [deep-inelastic $\nu(\bar{\nu})$-nucleon scattering, elastic $\nu(\bar{\nu})$-e scattering, the scattering of polarized electrons from deuterons, and the measurement of charge asymmetry in the reaction $e^+e^- \rightarrow \mu^+\mu^-$]. Until the middle of 1985, the world average of all available results was[53] $\sin^2\theta_W = 0.220 \pm 0.008$. Recently, two new, more precise results have been obtained from a measurement of the ratio of the neutral-to-charged-current cross-sections for deep-inelastic neutrino-nucleon scattering. The two values are: $\sin^2\theta_W = 0.227 \pm 0.005 \pm 0.005$ (the CDHS Collaboration[54]) and $\sin^2\theta_W = 0.236 \pm 0.005 \pm 0.005$ (the CHARM Collaboration[55]). For both results, the first error represents the statistical uncertainty, and the second one is a theoretical error, common to both experiments, which includes a ± 0.3 GeV/c^2 uncertainty in the mass of the c-quark. The average of these two results is

$$\sin^2\theta_W = 0.232 \pm 0.009\ . \qquad (13)$$

where the theoretical error has been added linearly to the statistical error. This value, together with Eq. (12'), gives

$$m_W = 80.3 \pm 1.5\ \text{GeV/c}^2\ . \qquad (14)$$

The use of the minimal Higgs scheme[50] to generate the IVB masses in the Standard Model provides the basic relation $m_Z = m_W/\cos\theta_W$ between the W and Z masses. With the use of Eqs. (12') and (13) we obtain

$$m_Z = 91.6 \pm 1.2 \text{ GeV/c}^2 . \qquad (14')$$

4.1.2 <u>Decay modes.</u> The main decay modes of the W are also deduced from the Standard Model. The leptonic decay mode $W^- \rightarrow e^- \bar{\nu}_e$ (or $W^+ \rightarrow e^+ \nu_e$) has a partial width given by

$$\Gamma(e\nu_e) = Gm_W^3/(6\pi\sqrt{2}) , \qquad (15)$$

and the partial widths of all other decay modes into a fermion-antifermion pair are simply related to $\Gamma(e\nu_e)$. Neglecting all fermion masses, we find

$$\Gamma(e\nu_e) = \Gamma(\mu\nu_\mu) = \Gamma(\tau\nu_\tau) \qquad (15')$$

and

$$\frac{\Gamma(q\bar{q}')}{\Gamma(e\nu_e)} = 3|U_{qq'}|^2 , \qquad (15'')$$

where the factor of 3 results from the fact that quarks may be emitted in three different colour states, and the quantity $U_{qq'}$ takes into account the mixing between quarks which occurs in the presence of the weak inter-action.[56] For example, in the decay $W^- \rightarrow d\bar{u}$, $|U_{qq'}|$ is just given by $\cos\theta_C$, where θ_C is the Cabibbo angle,[57] whereas in the decay $W^- \rightarrow s\bar{u}$ it is $|U_{qq'}| = \sin\theta_C$. For a quark q belonging to a given fermion family, the sum $\Sigma_{q'}|U_{qq'}|^2$ extended over all possible families is equal to 1. Clearly, W decays into a $q\bar{q}$ pair are expected to appear as two jets of hadrons.

In the case of the decay $W^- \rightarrow \bar{t}b$ (or $W^+ \rightarrow t\bar{b}$) the mass of the t-quark cannot be neglected, and the corresponding phase-space suppression factor is $(1 - m_t^2/m_W^2)^2 (1 + 0.5 \ m_t^2/m_W^2)$, which has the value 0.64 for $m_t = 40$ GeV/c^2.

The numerical value of $\Gamma(e\nu_e)$ from Eq. (15) is 227 MeV/c^2. Assuming three fermion families and a t-quark mass of 40 GeV/c^2, we find

$$\Gamma_W = 10.92 \ \Gamma(e\nu_e) = 2.47 \text{ GeV/c}^2 \qquad (16)$$

for the total width of the W. The branching ratio for the decay mode $W^- \rightarrow e^- \bar{\nu}_e$ is predicted to be $B(e\nu_e) = 1/10.92 \approx 9.2\%$.

As in the case of the W, the Z decays into a fermion-antifermion pair $f\bar{f}$ ($f = \nu$, e^-, μ^-, τ^-, or any quark q). The coupling of the Z to such a pair may be parametrized by an effective Lagrangian given by

$$L_{eff} = -\frac{m_Z}{\sqrt{2}} \left(\frac{G}{\sqrt{2}}\right)^{1/2} \bar{f}\gamma_\mu(V_f - A_f\gamma_5)fZ^\mu \ ,$$

where

$V_f = A_f = 1$ for neutrinos;

$A_f = \pm 1$, $V_f = \pm(1 - 4|Q_f| \sin^2\theta_w)$ for fermions f with charge $Q_f \gtrless 0$.

We note that for charged leptons ($|Q_f| = 1$) the vector coupling V_f is very small because $\sin^2\theta_w$ has a value close to 0.25.

The partial decay widths into a lepton-antilepton pair $\ell\bar{\ell}$ are then given by

$$\Gamma(\ell\bar{\ell}) = \frac{Gm_Z^3}{24\pi\sqrt{2}} (V_\ell^2 + A_\ell^2) \ , \tag{17}$$

and those relative to Z decay into a $q\bar{q}$ pair:

$$\Gamma(q\bar{q}) = 3\, \frac{Gm_Z^3}{24\pi\sqrt{2}} (V_q^2 + A_q^2) \ , \tag{17'}$$

where, as usual, the factor of 3 results from the three possible colour states of the quarks. Quark-antiquark pairs are expected to belong to the same family because flavour changing is suppressed in the weak interaction between neutral currents.

Quantitatively we find

$\Gamma(\bar{\nu}\nu) : \Gamma(e^+e^-) : \Gamma(\bar{u}u) : \Gamma(\bar{d}d) =$

$$= 2 : [1 + (1 - 4\sin^2\theta_w)^2] : 3[1 + (1 - \tfrac{8}{3}\sin^2\theta_w)^2]$$

$$: 3[1 + (1 - \tfrac{4}{3}\sin^2\theta_w)^2] \ , \tag{18}$$

and similar relations for the other two fermion families. It must be noted, however, that in the decay $Z \to t\bar{t}$ the effect of the large top-quark mass m_t introduces a suppression factor which cannot be neglected and is ~ 0.25 for $m_t = 40$ GeV/c^2.

Setting $\sin^2\theta_w = 0.232$ in Eq. (18), we find

$$\Gamma(\nu\bar{\nu}) : \Gamma(e^+e^-) : \Gamma(u\bar{u}) : \Gamma(d\bar{d}) = 2 : 1.01 : 3.44 : 4.43 \tag{18'}$$

and a total width given by

$$\Gamma_Z = 30.1 \; \Gamma(\nu\bar{\nu})/2 = 2.53 \; \text{GeV}/c^2 \tag{19}$$

for $\Gamma(\nu\bar{\nu}) = 168 \; \text{MeV}/c^2$, as obtained from Eq. (17) using the m_Z value given by Eq. (14').

We note that the branching ratio for the decay $Z \rightarrow e^+e^-$ is expected to be $B(e^+e^-) \approx 3.4\%$.

4.1.3 <u>Production of the IVBs</u>. The production of W and Z in hadron collisions is expected to occur as the result of $q\bar{q}$ annihilation, also known as the Drell-Yan mechanism.[58] In the case of $p\bar{p}$ collisions the basic subprocesses are

$$u\bar{d} \rightarrow W^+ \; ,$$
$$d\bar{u} \rightarrow W^- \; ,$$
$$u\bar{u} \rightarrow Z \; , \tag{20}$$
$$d\bar{d} \rightarrow Z \; .$$

Since the transverse momenta of the initial quarks are known to be small, the IVBs are produced with low p_T, and the initial x-values, x_1 and x_2, obey the relation $x_1 x_2 s = m^2$, where m is the IVB mass.

The subprocess $u\bar{d} \rightarrow W^+$ has a cross-section whose leading-order value is given by

$$\hat{\sigma}(x_1, x_2) = \sqrt{2}\pi G m_W^2 \; \cos^2\theta_C \delta(x_1 x_2 s - m_W^2) \; , \tag{21}$$

where we have neglected the W^+ width. To obtain the inclusive cross-section for W^+ production we must perform a convolution integral, which takes into account the x distribution (or structure functions) of the u-quarks and $\bar{d}$-quarks in the incident hadrons, and sum over all relevant subprocesses.

In the case of $p\bar{p}$ collisions, we obtain

$$\sigma(p\bar{p} \rightarrow W^+ + \ldots) = \int_0^1 dx_1 \int_0^1 dx_2 \, \hat{\sigma}(x_1, x_2) W^+(x_1, x_2)$$

$$\tag{22}$$

$$= \sqrt{2}\pi G\tau \; \cos^2\theta_C \int_\tau^1 \frac{dx}{x} \; W^+\left(x, \frac{\tau}{x}\right) \; ,$$

where $\tau = m_W^2/s$ and, neglecting strange and heavier quarks from the sea,

$$W^+(x_1, x_2) = \frac{1}{3} [U(x_1)D(x_2) + D(x_1)U(x_2)] ,$$

U(x) and D(x) being the x distributions of the u- and d-quarks in the incident protons which are identical, of course, to those of the $\bar{u}$- and $\bar{d}$-quarks in the incident antiprotons.

In complete analogy, the cross-section for inclusive Z production in $\bar{p}p$ collisions is given by

$$\sigma(p\bar{p} \to Z + ...) = 2\sqrt{2}\pi G\tau \int_\tau^1 \frac{dx}{x} Z\left(x, \frac{\tau}{x}\right) , \tag{23}$$

where

$$Z(x_1, x_2) = \frac{1}{3} U(x_1)U(x_2) \left(\frac{1}{4} - \frac{2}{3} \sin^2\theta_w + \frac{8}{9} \sin^4\theta_w\right)$$

$$+ \frac{1}{3} D(x_1)D(x_2) \left(\frac{1}{4} - \frac{1}{3} \sin^2\theta_w + \frac{2}{9} \sin^4\theta_w\right) .$$

Estimates of the cross-sections for inclusive production of W and Z, based on Eqs. (22) and (23), have been made since 1977.[59] Next-to-leading order QCD corrections, represented by the graphs of Fig. 32, have recently been included.[60] They have the effect of increasing the cross-sections by ~ 30% at Collider energies, and of giving a non-vanishing transverse momentum to the IVBs. Such a transverse momentum, which on the average is ~ 8 GeV/c at Collider energies, is predicted to increase linearly with $\sqrt{s}$ (it is ~ 15 GeV/c at $\sqrt{s}$ = 2 TeV).

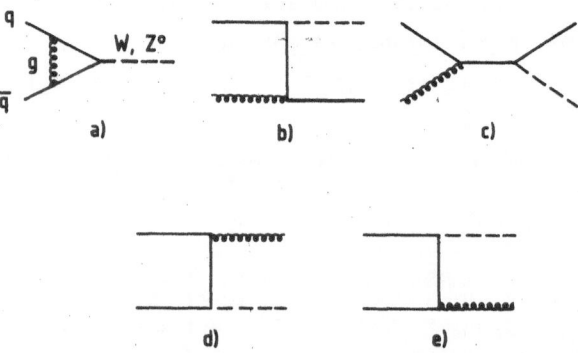

Fig. 32   Next-to-leading order diagrams for IVB production.

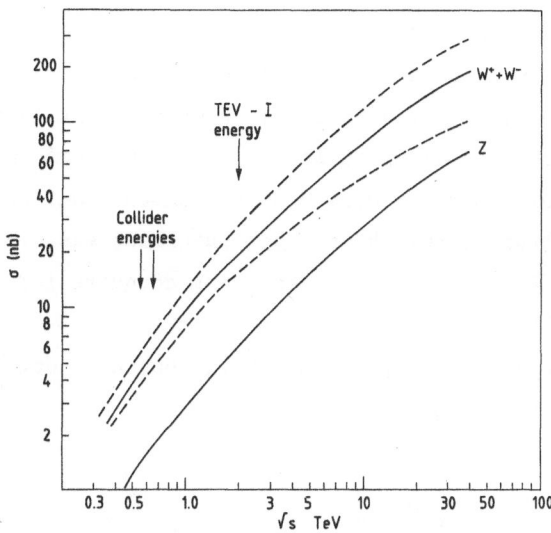

Fig. 33   QCD predictions for inclusive production of W and Z as a function of $\sqrt{s}$.[60] The two broken lines illustrate the theoretical uncertainty on σ_W.

Figure 33 shows the predicted W and Z inclusive cross-sections as a function of $\sqrt{s}$,[60] together with their theoretical uncertainties.

We note that at Collider energies the production of IVBs is dominated by subprocesses involving at least one valence quark, since $\sqrt{x_1 x_2} = m/\sqrt{s} \approx$ 0.13. This is no longer the case at higher energies, such as that of the forthcoming TEV I Collider, where annihilation of $q\bar{q}$ pairs with both partons belonging to the sea play an important role in W and Z production.

4.2   Detection of the IVBs

We have seen in the previous subsection that the IVBs are produced, in general, with a rather low transverse momentum. Their decay products, however, if emitted not very close to the beam direction, will have high p_T because of the large amount of energy released in the decay.

For both W and Z a large fraction of the decay modes (about 75%) consist of $q\bar{q}$ pairs which appear as two high-p_T hadronic jets. In searching for such configurations we expect a background from the continuum of high-p_T two-jet events discussed in Section 3. To estimate this background we integrate the measured two-jet cross-section[23] over the two-jet mass interval $65 < m_{jj} < 100$ GeV/c^2, for kinematical configurations in which both jets are produced centrally ($|\eta| < 0.85$). The result, $\sigma(2\text{-jets}) \approx 250$ nb, is approximately two orders of magnitude larger than the cross-section for production of the IVBs, followed by their decay into two jets with the same kinematical configuration.

Such an unfavourable signal-to-noise ratio cannot be easily reduced experimentally because there is no way to distinguish jets from W and Z decay from those which result from hard-parton scattering. A measurement of the two-jet invariant mass with excellent resolution (of the order of the natural width of the IVBs) would certainly help to identify a peak super-imposed on the smooth mass distribution of the background. However, in both UA1 and UA2, jet energies are measured using conventional calorimetry, and the resolution so obtained for the two-jet invariant mass in the region of the IVB masses is ∼ 8 GeV/c^2, a value far from adequate for the event sample available at present.

For these reasons, both experiments have chosen to detect the IVBs by identifying their leptonic decays:

$$W \rightarrow e^{\pm} \nu_e (\bar{\nu}_e) \left. \right\} \text{ in both UA1 and UA2 ,}$$
$$Z \rightarrow e^+ e^-$$

$$W \rightarrow \mu^{\pm} \nu_\mu (\bar{\nu}_\mu) \left. \right\} \text{ in UA1 only ,}$$
$$Z \rightarrow \mu^+ \mu^-$$

for which the backgrounds are much lower, as we shall see later, in spite of the smaller branching ratios.

The decision to detect electrons instead of muons in the UA2 experiment was based on the possibility of measuring electron energies with good preci-sion using compact e.m. calorimeters (typically only ∼ 25 cm thick), whereas muon detection implies large magnetic field volumes to measure its momentum, and the use of thick iron absorbers.

When searching for the decay $Z \rightarrow e^+ e^-$, or $Z \rightarrow \mu^+ \mu^-$, the Z mass can be directly determined by measuring the energies (or momenta) of the two leptons and the angle α between their directions:

$$m^2 = 4E_1 E_2 \sin^2\frac{\alpha}{2} , \tag{24}$$

and a good mass resolution Δm (smaller than the natural width Γ_Z) is essen-tial for determining Γ_Z.

The mass resolution Δm can be estimated from a knowledge of the measur-ing errors ΔE_1, ΔE_2, and $\Delta \alpha$:

$$\frac{\Delta m}{m} = \frac{1}{2} \sqrt{\left(\frac{\Delta E_1}{E_1}\right)^2 + \left(\frac{\Delta E_2}{E_2}\right)^2 + \left(\frac{\Delta \alpha}{\tan\frac{\alpha}{2}}\right)^2} . \tag{25}$$

The opening angle α is measured using tracking devices, and its error Δα can easily be made as low as 0.01 rad, or less. Its effect on Δm/m is then negligible, because at Collider energies α is generally larger than 90°.

Electron energies are measured in calorimeters with a resolution $\Delta E/E \approx 0.15/\sqrt{E}$ (E in GeV, see Section 2), giving

$$\frac{\Delta m}{m} \approx \frac{0.15}{\sqrt{m}} , \qquad (26)$$

where m is in GeV/c^2, for a symmetric decay configuration. For m ≈ 90 GeV/c^2 the mass resolution is Δm ≈ 1.4 GeV/c^2, an adequate value for measuring Γ_Z.

For the decay $Z \rightarrow \mu^+\mu^-$, muon momenta p_1 and p_2 replace the electron energies in Eqs. (24) and (25). They are determined by measuring the muon deflection in the magnetic field with a resolution which is typically Δp/p ≈ 0.5% p (p in GeV/c) in the UA1 detector (there is only a 0.6 mm sagitta in a 1 m long track from a 45 GeV/c particle). In this case we obtain

$$\frac{\Delta m}{m} \approx 1.8 \times 10^{-3} \, m ,$$

where m is in GeV/c^2, which corresponds to a mass resolution Δm ≈ 14 GeV/c^2 at m ≈ 90 GeV/c^2 and is a much worse value than in the e^+e^- case. Furthermore, Δm/m increases linearly with m, whereas for e^+e^- pairs it decreases as $m^{-1/2}$, showing the superiority of the e^+e^- channel in searching for possible new particles with masses larger than those of the IVBs.

In the case of the decay $Z \rightarrow e^+e^-$ the actual mass resolution is worse than that given by Eq. (26) because it is difficult to keep the calorimeter calibration under control over the long period of time during which data are collected. At the $p\bar{p}$ Collider there is no heavy particle of known mass decaying into e^+e^- pairs whose production rate is high enough to be used as a calibration standard. The calibration is therefore monitored by a variety of methods, such as the use of radioactive sources, light pulsers, etc., and by periodically recalibrating as many calorimeter modules as possible with electron beams of known energy.

The calibration uncertainty increases Δm/m by a systematic contribution of the order of 1.5% to be added in quadrature to Eq. (26).

The detection of the decay mode $W \rightarrow \ell\nu(\bar{\nu})$, where ℓ is either an electron or a muon, is based on a completely different method, because the neutrino cannot be detected. In the W rest frame the lepton energy is just $m_W/2$. If we neglect the W transverse momentum p_T^W, the lepton transverse momentum $p_T = (m_W/2) \sin\theta^*$ (where $\sin\theta^*$ is the decay angle with respect to the beam direction in the W rest frame) is a Lorentz invariant, and the p_T spectrum can be obtained from the decay angular distribution $f(\theta^*)$ by a simple change of variables:

$$\frac{dn}{dp_T} \propto \frac{1}{\sqrt{1 - (2p_T/m_W)^2}} \; f\left[\sin^{-1}\left(\frac{2p_T}{m_W}\right)\right] \; .$$

The singularity at $p_T = m_W/2$ arises only from the change of variables and it is often called the Jacobian peak. The W transverse motion and the W natural width have the effect of smearing this peak. However, if the mean value of p_T^W is much smaller than $m_W/2$, the Jacobian peak is still the most distinctive feature of the charged-lepton p_T distribution in the decay $W \rightarrow \ell\nu$.

Experimentally, the problem consists in detecting high-p_T charged leptons (either $e^\pm$ or $\mu^\pm$) over the largest possible solid angle, and in searching for a peak structure in their p_T spectrum. A fit to such a structure using the form expected for the Jacobian peak from $W \rightarrow \ell\nu$ is a method of determining the W mass.

In a search for high-p_T leptons the main background is represented by high-p_T hadrons, or jets of hadrons, which are misidentified as leptons. To

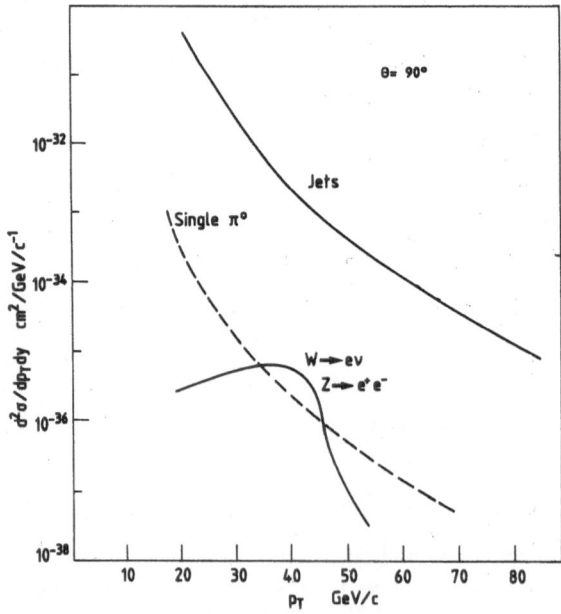

Fig. 34 Comparison of the inclusive jet yield at the Collider with the yield of single high-p_T hadrons and of high-p_T leptons expected from the decay of the IVBs.

248

illustrate the relative level of the signal and of the main background source, in Fig. 34 the inclusive p_T spectrum of hadronic jets is compared with that expected from $W \rightarrow \ell\nu$ decay. Also shown in Fig. 34 is the expected spectrum of high-p_T single hadrons. We see that near the Jacobian peak the relative rates are approximately in the ratios jets:single hadrons:leptons $\approx$ 10^3:1:1. This is a very favourable experimental situation because rejection factors against multihadrons of the order of 10^4, or larger, are not difficult to achieve.

4.3 Electron Identification

In a search for high-p_T electrons at the Collider, the main background results from hadronic jets consisting of one or more high-p_T π^0's and one charged particle, which could be either a charged jet fragment or an electron from photon conversion. Such a configuration is hard to distinguish from a genuine electron in the UA1 and UA2 detectors.

As a consequence, both experiments use somewhat similar electron identification criteria, which aim at a high rejection against jets while maintaining a reasonably high efficiency for electrons. These criteria require the energy deposition in the calorimeter to match that expected from an isolated electron, which implies that a very large fraction of events containing electrons inside or near a jet are rejected. In the following, we describe the main cuts used to select an inclusive electron sample.

i) The transverse energy E_T associated with the energy cluster in the calorimeter is required to exceed a given threshold. For the data discussed here, this threshold is set at 15 GeV for both experiments.

ii) The shower energy leakage into the hadronic compartment of the calorimeter is required to be small.

iii) In the UA1 experiment the longitudinal development of the electromagnetic shower, which is measured over four samples, is required to be compatible with that expected for an electron. In the UA2 experiment, the lateral shower profile is required to be small, using the small cell size of the calorimeters.

iv) The presence of a charged-particle track pointing to the energy cluster in the calorimeter is required.

v) In the UA1 experiment the momentum measurement is used to ensure that the track transverse momentum is larger than 7 GeV/c (or compatible with 15 GeV/c within 3σ). In UA2, a magnetic field exists only in the two forward regions $20° < \theta < 37.5°$ with respect to the beams. In these regions the track momentum is required to match the particle energy, as measured in the calorimeter, within 4σ.

vi) A distinctive feature of the UA2 detector is the presence of a 'pre-shower' counter (see subsection 2.2) in front of the calorimeters. This counter is used to verify that the electromagnetic shower is initiated in the converter and is also associated with the measured charged-particle track, as expected for electrons. This is done in practice by requiring that the observed signal exceeds a threshold corresponding to several minimum-ionizing particles, and, furthermore, that its position in space matches the track impact point within the space resolution of the counter itself (a few millimetres).

vii) Finally, an explicit isolation criterion is applied. In UA1 this is done by requiring only a limited amount of transverse energy (typically less than 3 GeV) associated with charged particles and calorimeter cells contained in a cone of ∽ 40° half-angle around the electron track. In UA2 this cone has a half-angle of typically ∽ 15°.

The combination of all these cuts is estimated to be ∽ 75% efficient for isolated electrons in both experiments. Figure 35a shows the distribution of the transverse momentum p_T^e for the electron candidates in the UA1 experiment,[61] selected from the 1984 data sample ($\sqrt{s}$ = 630 GeV, L = 263 nb^{-1}). The UA2 distribution, corresponding to the full data sample,[62] is shown in Fig. 35b.

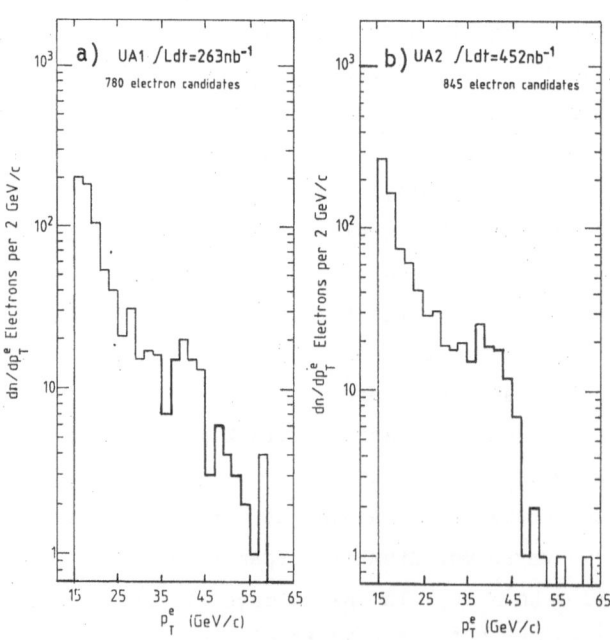

Fig. 35 Inclusive transverse momentum spectrum of electron candidates:
a) UA1, 1984 sample; b) UA2, full sample.

At this stage of the analysis, most of the electron candidates are mis-identified high-p_T hadrons, as suggested by the rapidly falling p_T^e distributions at low values of p_T^e. However, a shoulder in the region $p_T^e \approx 40$ GeV/c is clearly visible in both distributions of Fig. 35. Such a structure is expected from the Jacobian peak which results from the kinematics of W → ev decay.

4.4 Neutrino Identification

The presence of a non-interacting high-p_T neutrino in the final state is characteristic of W → ev decays. Since a large fraction of the total collision energy is carried by particles at very small angles, which cannot be detected because they remain inside the machine vacuum pipe, only the missing transverse momentum $\vec{p}_T^{miss}$ can be reliably measured. For events containing an electron candidate, $\vec{p}_T^{miss}$ is identified with the neutrino transverse momentum $\vec{p}_T^v$.

In the UA1 experiment, for events containing an electron candidate of transverse momentum $\vec{p}_T^e$, one defines $\vec{p}_T^v$ as

$$\vec{p}_T^v = -\vec{p}_T^e - \sum_i \vec{p}_T^i ,$$

where $\vec{p}_T^i$ is a vector with magnitude equal to the energy deposited in the ith cell of the calorimeter, and directed from the event vertex to the estimated impact point on the cell. The sum is extended over all calorimeter cells.

However, the central part of the UA1 detector ($|\eta| < 1.5$, where η is the pseudorapidity) has imperfect calorimetry in two azimuthal sectors at $\pm 15°$ to the vertical axis. For this reason, the measurement of $\vec{p}_T^v$ is unreliable whenever $\vec{p}_T^v$ is within these regions. After rejecting such events, the $\vec{p}_T^v$ resolution becomes almost Gaussian.

In the case of the UA2 detector, there is no particle detection at angles $\theta < 20°$ to the beams. Furthermore, the two forward regions ($20° < \theta < 40°$ to the beams) provide only partial detection because of incomplete azimuthal coverage (due to 12 toroid coils) and incomplete hadronic calorimetry. This results in non-Gaussian tails in the $\vec{p}_T^{miss}$ resolution. The probability of losing one of the jets in a two-jet event varies between ~ 10% at $p_T = 15$ GeV/c and ~ 2% at $p_T = 40$ GeV/c.

For each event containing an electron candidate, the UA2 definition of $\vec{p}_T^v$ is

$$\vec{p}_T^v = -\vec{p}_T^e - \sum_j \vec{p}_T^j - \lambda \vec{p} ,$$

where the sum extends to all reconstructed jets with $p_T^j > 3$ GeV/c, and the vector $\vec{P}$ is the total transverse momentum carried by the system of all other particles not belonging to the jet. The factor λ, of the order of 1.5, is an empirical correction factor which takes into account the non-linearity of the calorimeter response to low-energy particles. Its value is determined by applying the condition $\langle \vec{p}_T^v \rangle = 0$ to the sample of $Z \rightarrow e^+e^-$ events observed in UA2.

4.5 The Final W → ev Event Samples

Figure 36 shows, for UA1 and UA2, the distribution of the events containing at least one electron candidate with $p_T^e > 15$ GeV/c in the (p_T^e, p_T^v) plane. In the high-p_T^e region ($p_T^e \gtrsim 25$ GeV/c), signals from W → ev ($p_T^v \approx p_T^e$) and $Z \rightarrow e^+e^-$ ($p_T^v \approx 0$) are clearly visible above the background of mis-identified hadrons, which is dominant at low p_T^e. It must be noted that the UA1 sample corresponds to the 1984 data only, while the UA2 distribution represents the full sample.

The projection of the two distributions onto the p_T^v axis (Fig. 37) demonstrates clearly that the rejection power against background of a cut on p_T^v is much larger in UA1 than in UA2. The UA1 data (Fig. 37a) show two well-separated classes of events: those with $p_T^v > 15$ GeV/c, which show the characteristic Jacobian structure expected from W → ev decay; and those with

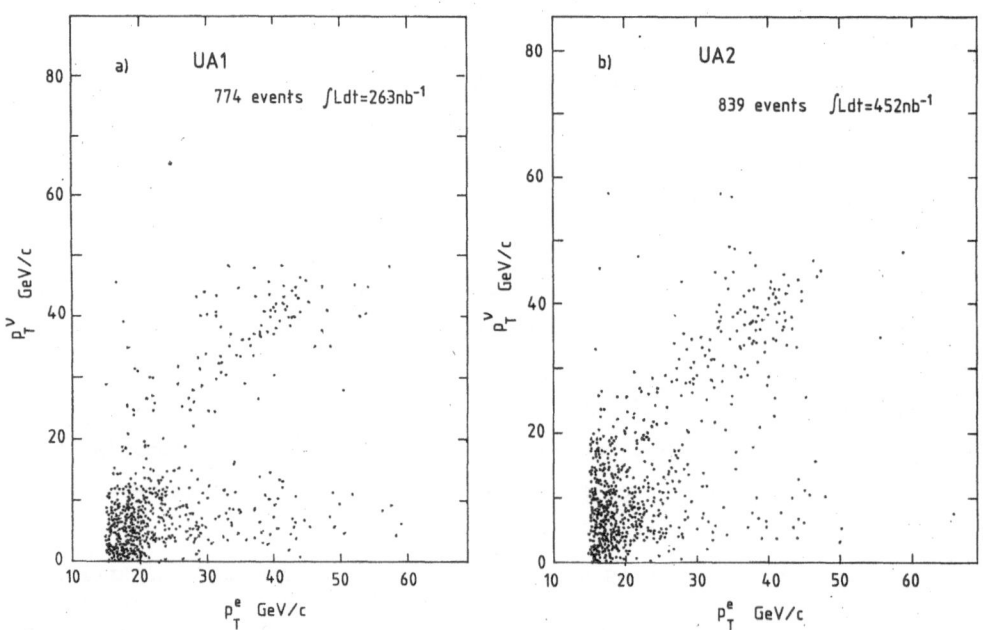

Fig. 36 Distribution of the events containing at least one electron candidate in the p_T^e, p_T^v plane: UA1, 1984 sample; b) UA2, full sample.

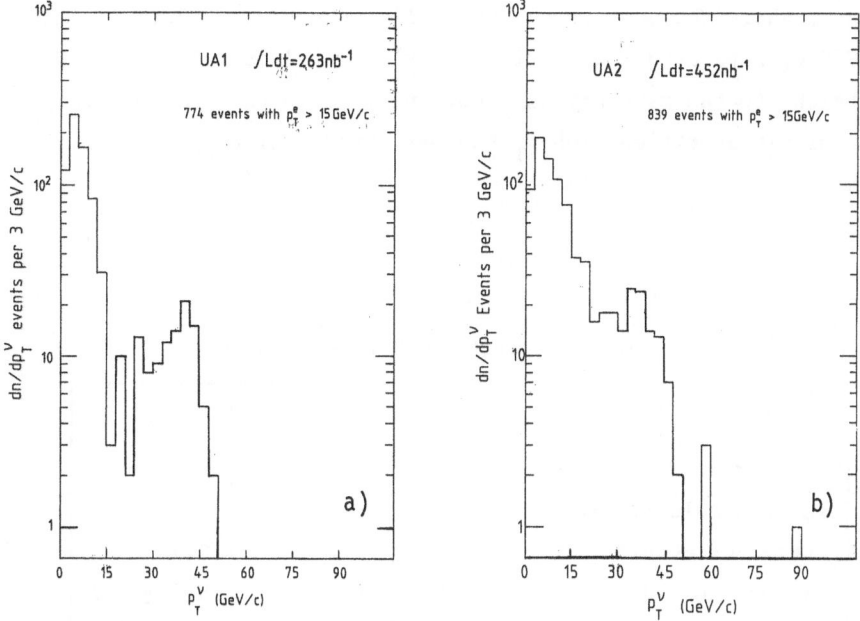

Fig. 37 Distribution of p_T^v for events containing at least one electron candidate with p_T^e > 15 GeV/c: a) UA1, 1984 sample; b) UA2, full sample.

p_T^v < 15 GeV/c, which are mostly misidentified hadronic final states. In the latter class the non-zero value of p_T^v is the effect of the p_T^v resolution. The separation between the two classes of events is much less clear in the UA2 data (Fig. 37b), as a result of the non-Gaussian tails of the p_T^v resolution in the UA2 experiment.

The final UA1 W → ev sample is defined by requiring p_T^v > 15 GeV/c. In the full data sample this condition is satisfied by 172 events. Background contributions to this sample are listed in Table 2. The background from misidentified hadrons is estimated from the shape of the p_T^v resolution. After rejecting events for which either $\vec{p}_T^e$ or $\vec{p}_T^v$ points in the direction of the vertical axis within ±15°, 148 events are left. The p_T^e distribution of these events is shown in Fig. 38a.

The final UA2 W → ev sample is defined by applying a topological cut which minimizes the background from misidentified hadrons accompanied by a jet at opposite azimuth. As a measure of the fraction of the electron transverse momentum balanced by jets at opposite azimuth, the quantity

$$\varrho_{opp} = -\vec{p}_T^e \cdot \sum_j \vec{p}_T^j / |\vec{p}_T^e|^2$$

is defined, where the sum extends over all reconstructed jets (if any) with $p_T^j > 3$ GeV/c, separated in azimuth from $\vec{p}_T^e$ by at least 120°. Most $W \to e\nu$ decays belong to the category of events with large p_T imbalance ($\varrho_{opp} \approx 0$), whereas for misidentified high-p_T hadrons, values of ϱ_{opp} near unity are expected.

Table 2

W → ev event samples and backgrounds

	UA1	UA2
p_T^e threshold (GeV/c)	15	25
Number of events	172	119
Hadronic background	5.3 ± 1.9	5.6 ± 1.7
Z → e (detected) e (undetected)	–	4.4 ± 0.8
W → τν_τ (τ → e$\nu_e\nu_\tau$)	9.1 ± 0.5	1.7 ± 0.2
W → τν_τ (τ → $\nu_\tau\pi^0$ + hadrons)	2.7 ± 0.4	–
W → ev signal	155 ± 13	107 ± 11

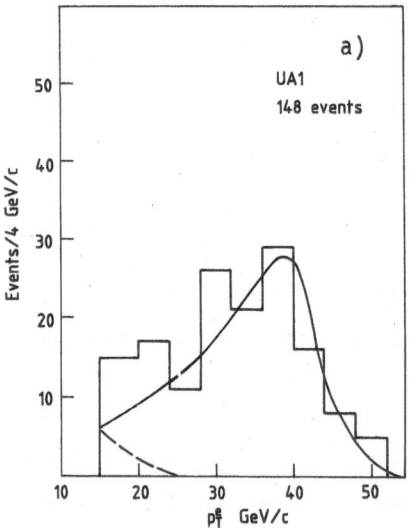 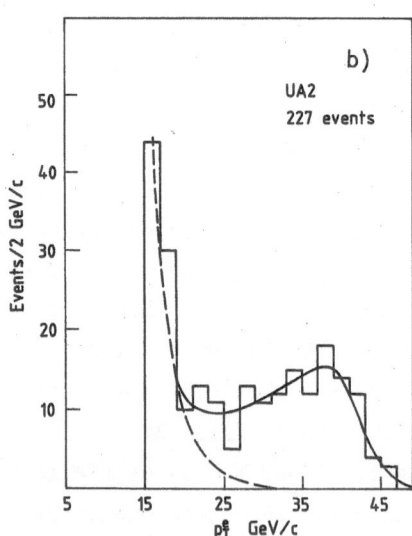

Fig. 38 a) The p_T^e distribution for electron candidates in events having $p_T^\nu > 15$ GeV/c (UA1); b) the p_T^e distribution for electron candidates satisfying the requirement $\varrho_{opp} < 0.2$ (UA2). Broken curves: estimated hadronic background. Full curves: expected distributions for values of m_W as given by Eqs. (27) and (27').

Figure 38b shows the p_T^e distribution of all electron candidates satisfying $\varrho_{opp} < 0.2$. Although the Jacobian peak structure has been strongly enhanced with respect to the inclusive p_T^e spectrum of Fig. 35b, the hadronic background is still dominant for $p_T^e < 20$ GeV/c. For this reason the final UA2 $W \rightarrow ev$ sample consists only of the 119 events having $p_T^e > 25$ GeV/c. Background contributions to this sample are listed in Table 2. The background from misidentified hadrons is estimated from a sample of high-p_T π^0's which have the opposite jet outside the detector acceptance.

We note (see Table 2) that the contribution from $W \rightarrow \tau\nu_\tau$, followed by by $\tau \rightarrow ev_e\nu_\tau$, is larger in UA1 than in UA2 because of the lower p_T^e threshold used to define the final sample. However, the contribution from $Z \rightarrow e^+e^-$ decays with one electron outside the detector acceptance is larger in UA2, because the probability of detecting both electrons is only ~ 60% in UA2, while it is nearly 100% in UA1.

4.6 Cross-Sections for Inclusive W Production

The cross-sections for inclusive W production, followed by the decay $W \rightarrow ev$, σ_W^e, are computed in a straightforward way from the number of events in the samples, after subtracting the various background contributions.

The results obtained by the UA1 and UA2 experiments, including the data taken during the 1985 run, are listed separately in Table 3 for $\sqrt{s} = 546$ and 630 GeV. The quoted systematic uncertainties arise mainly from the uncertainties on the total luminosity ($\pm15\%$ in UA1, and $\pm8\%$ in UA2 which benefits from the measurement of the total cross-section by UA4[63] in the same intersection).

These results are consistent with (but systematically higher than) the corresponding theoretical predictions,[60] also given in Table 3. These pre-

Table 3

W $\rightarrow$ ev cross-sections

	σ_W^e (nb)		$r = \dfrac{\sigma_W^e \ (630 \text{ GeV})}{\sigma_W^e \ (546 \text{ GeV})}$
	$\sqrt{s} = 546$ GeV	$\sqrt{s} = 630$ GeV	
UA1	$0.55 \pm 0.08 \pm 0.09$	$0.60 \pm 0.05 \pm 0.09$	1.09 ± 0.18
UA2	$0.59 \pm 0.10 \pm 0.07$	$0.59 \pm 0.05 \pm 0.07$	1.00 ± 0.19
Theory[60]	$0.36 \begin{array}{l} + 0.11 \\ - 0.05 \end{array}$	$0.45 \begin{array}{l} + 0.14 \\ - 0.08 \end{array}$	1.26 ± 0.02

(The second error is the systematic uncertainty.)

dictions have large uncertainties arising from uncertainties in the structure functions and higher-order QCD corrections.

In contrast, the ratio $r = \sigma_W^e(630\ \text{GeV})/\sigma_W^e(546\ \text{GeV})$ has negligible systematic errors and small theoretical uncertainties. The average of the two measurements, $r = 1.05 \pm 0.13$ is lower than the theoretical prediction,[60] $r = 1.26 \pm 0.02$, but not statistically inconsistent with it.

4.7 Determination of the W Mass

To extract a value of m_W from the $W \to e\nu$ event samples, both experiments define for each event a transverse mass m_T:

$$m_T^2 = 2p_T^e p_T^\nu (1 - \cos\Delta\varphi)\ ,$$

with the property that $m_T \leq m_W$, where $\Delta\varphi$ is the azimuthal separation between $\vec{p}_T^{\,e}$ and $\vec{p}_T^{\,\nu}$. A Monte Carlo simulation is then used to generate m_T distributions for different values of m_W, and the most probable value of m_W is found by a maximum-likelihood fit to the experimental distributions. This technique has the advantage that m_T is rather insensitive to the W transverse motion, contrary to other variables such as p_T^e or p_T^ν.

In the UA1 experiment, a sample of 86 events for which both p_T^e and p_T^ν exceed 30 GeV/c is selected from the 148 events of Fig. 38. The m_T distribution for this sample, which is virtually background-free, is shown in Fig. 39a.

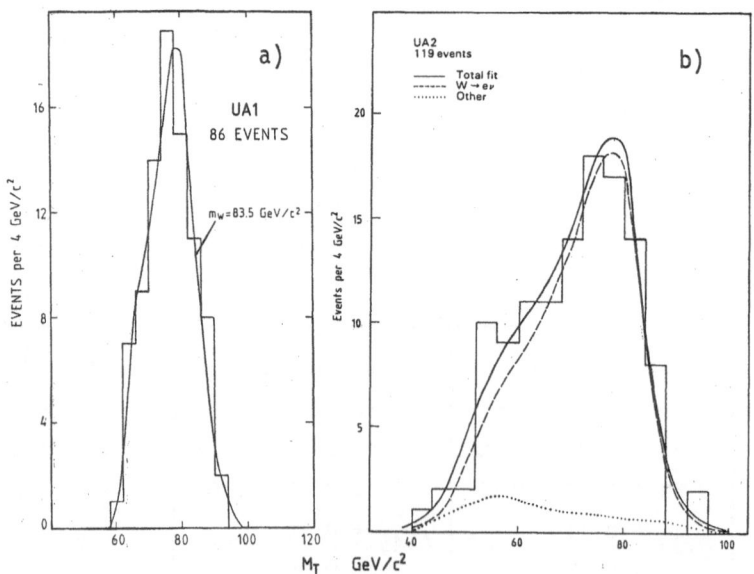

Fig. 39 Electron-neutrino transverse mass distribution for the UA1 (a) and UA2 (b) $W \to e\nu$ event samples. The curves represent best fits to the data.

In UA2, the m_T distribution of the 1982-84 W → ev sample (119 events) is shown in Fig. 39b. The backgrounds discussed in subsection 4.5 have a negligible effect on the mass determination, because they are predominantly at small m_T, and the best-fit value of m_W depends mainly on the upper edge of the distribution.

The results of the fit are

$$\text{UA1:} \quad m_W = 83.5^{+1.1}_{-1.0} \text{ (stat.)} \pm 2.8 \text{ (syst.) GeV/c}^2, \tag{27}$$

$$\text{UA2:} \quad m_W = 80.1 \pm 0.6 \text{ (stat.)} \pm 0.5 \text{ (syst}_1) \pm 1.3 \text{ (syst}_2) \text{ GeV/c}^2. \tag{27'}$$

The UA2 result includes the data taken during the 1985 run, and corresponds to a total integrated luminosity of 880 nb^{-1}.

The systematic error in Eq. (27) and the second systematic error in Eq. (27') reflect the uncertainty on the absolute energy scale of the calorimeters, which is $\pm 3\%$ in UA1 and $\pm 1.6\%$ in UA2. These errors are quoted separately because they cancel in the ratio m_W/m_Z. The first systematic error of ± 0.5 GeV/c^2 in the UA2 result is mainly due to uncertainties in the measurement of p_T^ν.

The smaller systematic uncertainty of the UA2 experiment [see Eqs. (27) and (27')] results from a better control of the calorimeter calibration, which was performed for all calorimeter cells, using beams of known energies, before the start of the experiment. This calibration has then been repeated periodically for a fraction of the calorimeter modules. A recalibration of 39 calorimeter cells performed in July 1986 is consistent with the quoted uncertainty.

A fit to the m_T distributions, using the W width Γ_W as a second free parameter, provides a way to obtain an upper limit on Γ_W. The results from the two experiments are $\Gamma_W <$ 6.5 GeV/c^2 (UA1) and $<$ 7 GeV/c^2 (UA2) at the 90% confidence level.

4.8 Charge Asymmetry in the Decay W → ev

At the energies of the CERN p$\bar{\text{p}}$ Collider, W production is dominated by q$\bar{\text{q}}$ annihilation involving at least one valence quark or antiquark. As a consequence of the V - A coupling, the helicity of the quarks (antiquarks) is -1 (+1) and the W is almost fully polarized along the $\bar{\text{p}}$ beam. At higher energies (e.g. at the TEV I Collider) the contribution from q$\bar{\text{q}}$ annihilation with both partons belonging to the sea is important, and the W polarization is greatly reduced.

Similar helicity arguments applied to W → ev decay predict that the leptons (e$^-$ or ν_e) should be preferentially emitted opposite to the direction of the W polarization, and antileptons (e$^+$ or $\bar{\nu}_e$) along it. More precisely, the angular distribution of the charged lepton in the W rest frame has the form $dn/d(\cos\theta^*) \propto (1 + \cos\theta^*)^2$, where θ^* is the e^+ (e^-) angle in the W rest frame, measured with respect to a direction parallel (antiparallel) to the W polarization. This axis coincides with the direction of the incident $\bar{p}$ (p) beam only if the W transverse momentum p_T^W is zero. For $p_T^W \neq 0$ the initial parton directions are not known, and the Collins-Soper convention[29] is used to define θ^*.

A further complication arises from the fact that p_L^ν is not measured, and the condition that the invariant mass of the $e\nu$ pair be equal to m_W gives two solutions for p_L^ν or p_L^W. The UA1 analysis[61] retains only those events for which one solution is unphysical and the charge sign is unambiguously determined (75 events). In UA2,[62] the solution corresponding to the smaller value of $|p_L^W|$ is chosen.

Figure 40a shows the $\cos\theta^*$ distribution from UA1, corrected for the detector acceptance. The expected $(1 + \cos\theta^*)^2$ form agrees well with the data.

In the UA2 experiment only the charge-averaged $\cos\theta^*$ distribution can be measured, because there is no magnetic field over most of the solid angle. The experimental data agree well with the form $1 + \cos^2\theta^*$ expected in this case (see Fig. 40b).

It has been shown[64] that for a particle of arbitrary spin J, one expects

$$\langle \cos\theta^* \rangle = \langle \lambda \rangle \langle \mu \rangle / J(J + 1) ,$$

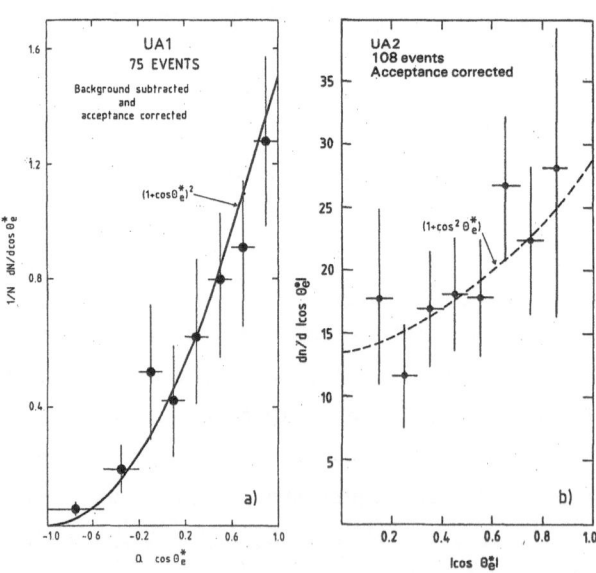

Fig. 40 a) The UA1 distribution of the product $Q \cos\theta^*$ for electrons from $W \to e\nu$ decay, where $Q = +1$ (-1) for e^+ (e^-), and $\theta^* = 0$ along the $\bar{p}$ beam. b) The $|\cos\theta^*|$ distribution, as measured by UA2.

where $\langle\mu\rangle$ and $\langle\lambda\rangle$ are, respectively, the global helicity of the production system ($u\bar{d}$) and of the decay system ($e\bar{v}$). For V - A one has $\langle\mu\rangle = \langle\lambda\rangle = -1$, which, together with the assignment J = 1, leads to a maximal value $\langle\cos\theta^*\rangle$ = 0.5. The UA1 result[61] is $\cos\theta^* = 0.43 \pm 0.07$, in agreement with maximal helicity states at production and decay, and with the assignment J = 1. For J = 0 one obviously expects $\langle\cos\theta^*\rangle = 0$; for any other spin value $J \geq 2$, $|\langle\cos\theta^*\rangle| \leq 1/6$. Hence the UA1 result demonstrates that the W boson has indeed spin 1.

It must be noted that no measurement done so far can distinguish between V - A and V + A. In the latter case, all helicities change sign and the angular distribution remains the same. In order to separate the two alternatives, a direct measurement of the lepton helicity would be needed.

In the UA2 experiment,[62] 28 electrons with $p_T^e >$ 20 GeV/c and $\varrho_{opp} <$ 0.2 are observed in the two forward regions where a magnetic field is present. Of these, 20 are observed in the region favoured by the V - A coupling and 8 in the opposite region, giving an asymmetry A = 0.43 $\pm$ 0.17. This value is in good agreement with the result of a Monte Carlo calculation, A = 0.53, which assumes a V - A decay matrix element.

More generally, if x is the ratio between the A and V couplings, the distribution of the charged lepton with respect to the direction of the proton beam has the form

$$dn/d(\cos\theta^*) \propto (1 - q\cos\theta^*)^2 + 2q\alpha\cos\theta^*,$$

where q = -1 (+1) for electrons (positrons), and $\alpha = [(1 - x^2)/(1 + x^2)]^2$. The distribution $d^2n/dp_T^e\,d\theta_e$ for the 28 W $\rightarrow$ ev events observed in UA2 is consistent with α = 0, as expected for V - A coupling. A maximum likelihood fit gives $\alpha <$ 0.39 (68% confidence level), corresponding to 0.48 $< |x| <$ 2.1. We note that the value of α does not provide any information about the sign of x, or about the choice of x or 1/x.

4.9 Longitudinal Momentum of the W

Figure 41 shows the distribution of the fractional beam momentum carried by the W boson, $x_W = 2p_L^W/\sqrt{s}$, for the total UA1 sample (here the smaller $|p_L^W|$ value is chosen in events for which both solutions are physical). This distribution, shown separately for $\sqrt{s}$ = 546 and 630 GeV, is expected to reflect the structure functions of the annihilating partons. From the well-known relations

$$x_W = x_p - x_{\bar{p}},$$

$$m_W^2/s = x_p x_{\bar{p}} \qquad \text{(if } p_T^W \ll m_W\text{)},$$

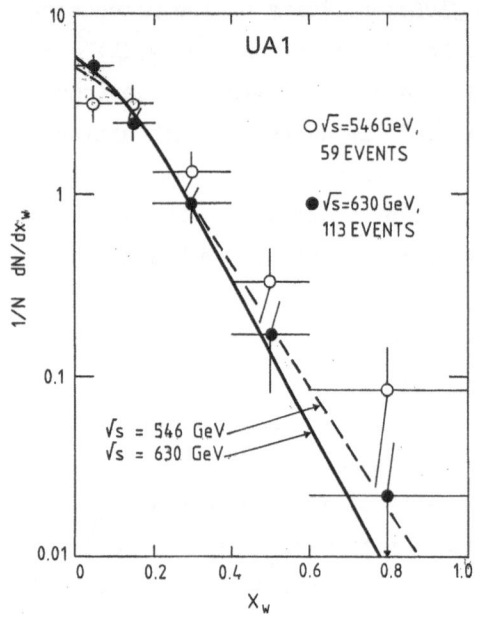

Fig. 41 The UA1 x_W distribution. The curves are expectations from $q\bar{q}$ annihilation, using the structure functions of Ref. 25.

one can extract the fractional momentum x_p $(x_{\bar{p}})$ of the parton contained in the incident p $(\bar{p})$. Using the events with an unambiguous determination of the charge sign (118 events in the UA1 sample), $x_p(x_{\bar{p}})$ can be identified with the fractional momentum of a u $(\bar{d})$-quark for a W^+, and a d $(\bar{u})$-quark for a W^-.

Figures 42a and 42b show the resulting u- and d-quark x-distributions, which agree with the expectation from the structure functions (the parametrization of Eichten et al.[25] is used in Figs. 42).

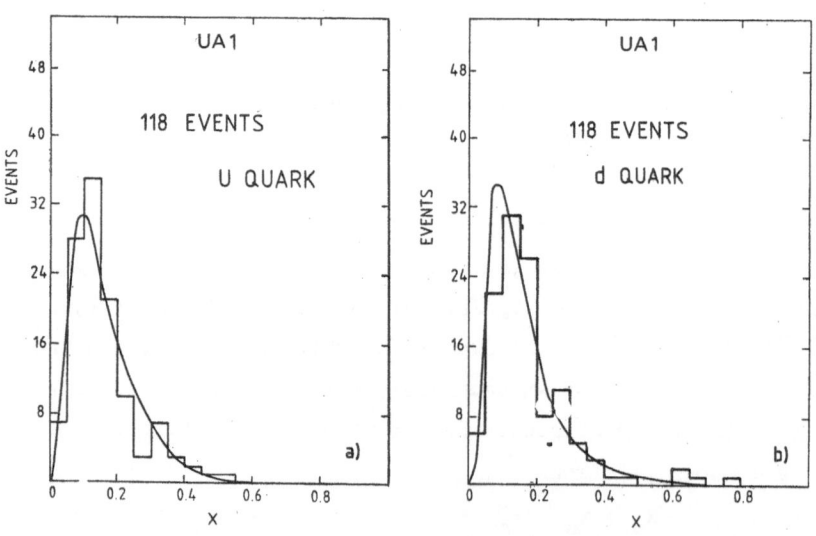

Fig. 42 The x-distributions for a) u-quarks, and b) d-quarks, as measured by UA1. The curves are expectations from the structure functions of Ref. 25.

4.10 <u>Transverse Momentum of the W</u>

The W transverse momentum p_T^W is obtained by adding the measured vectors $\vec{p}_T^e$ and $\vec{p}_T^\nu$. In the case of the UA2 sample, the topological cut discussed in subsection 4.5 may reject high-p_T^W events if they contain jets emitted opposite to the decay electron in a plane normal to the beams. For this reason the cuts $p_T^e > 25$ GeV/c and $\varrho_{opp} < 0.2$ (see subsection 4.5) used to define the W → eν sample are replaced by the cuts $p_T^e > 15$ GeV/c, $p_T^\nu > 25$ GeV/c, and $m_T > 20$ GeV/c^2. Figures 43 show the p_T^W distributions for UA1 and UA2. The two sets of data agree well with each other and with QCD predictions[60].

The events containing at least one jet with $E_T > 5$ GeV are shown as dashed areas in Figs. 43. As expected, they correspond to W bosons produced at high values of p_T^W. The E_T distributions of these jets are shown in Fig. 44, where they are compared with the results of a QCD calculation to second order in α_s.[65] In UA2, 28% of the W's are observed in association with at least one jet, whereas this fraction is 38% in UA1. This difference reflects the larger acceptance of the UA1 detector for jets. As shown in Fig. 45, the jets associated with the UA1 sample are seen to display an angular distribution of the form dn/d|cosθ^*| ∝ (1 − |cosθ^*|)$^{-1}$, where θ^* is the angle between the jet and the beam direction in the W-jet(s) rest frame. Such a form is expected from higher-order QCD diagrams.[65]

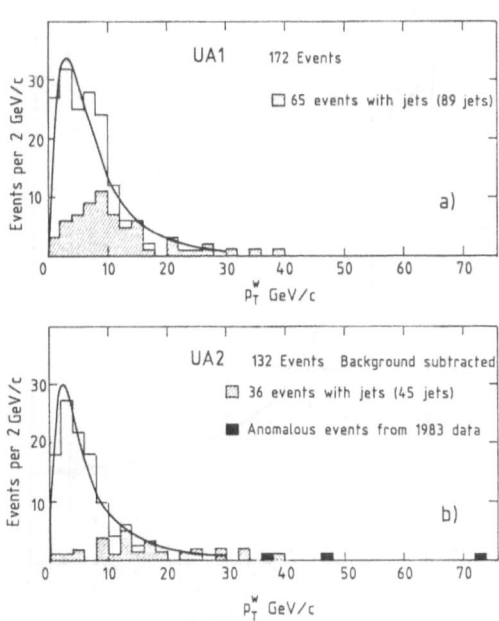

Fig. 43 The p_T^W distribution, as measured by UA1 (a) and UA2 (b). The shaded regions represent events which contain at least one jet with $E_T > 5$ GeV. The three events shown as dark areas in (b) are the anomalous UA2 events discussed in Ref. 66. The curves are QCD predictions from Ref. 60.

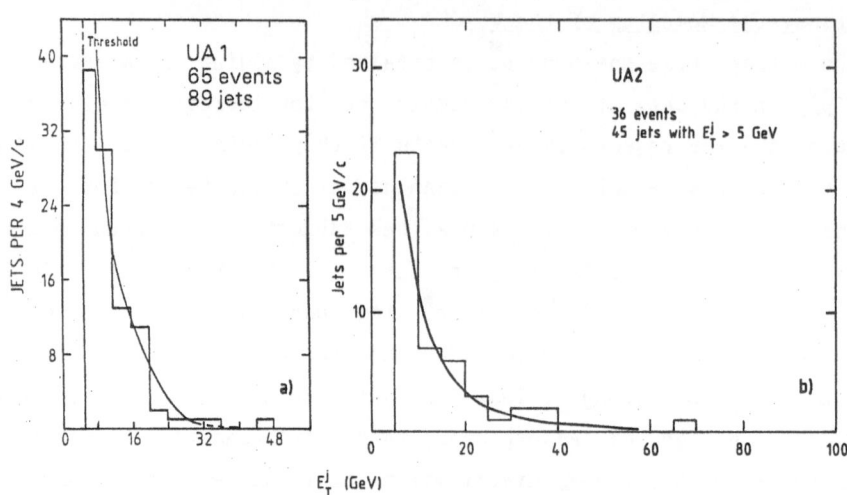

Fig. 44   The E_T distribution of jets produced in association with W → ev events for UA1 (a) and UA2 (b). Only jets with E_T > 5 GeV are considered. The curves are QCD predictions from Ref. 65.

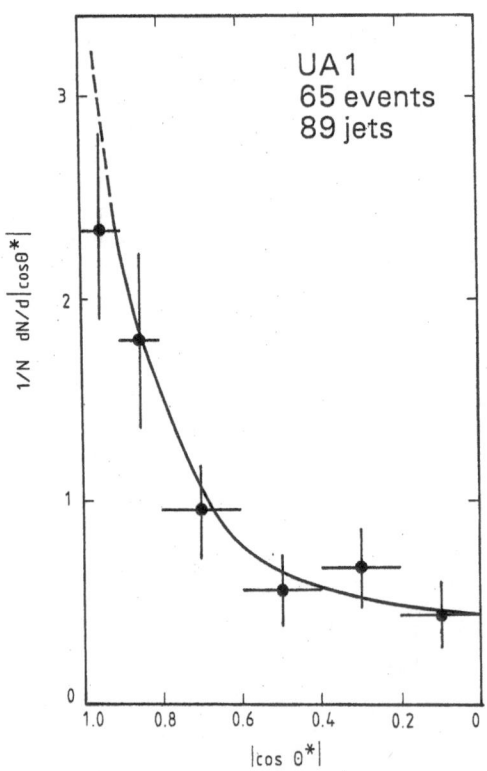

Fig. 45   Distribution of $|\cos\theta^*|$ for jets produced in association with W → ev events. The angle θ^* is the jet angle in the W-jet(s) rest frame with respect to the beam axis. Only jets with E_T > 5 GeV are considered. The curve is a QCD prediction.[65]

It must be recalled that the UA2 Collaboration published the observation of three anomalous events from the 1983 data sample.[66] These events could be interpreted as W → ev decays in association with very hard jets. Two of these events, emphasized in Fig. 43b, could not be easily interpreted as originating from higher-order QCD processes because of the very high values of p_T^W and of the jet transverse energies. No additional event as extreme as these has been observed in the 1984 sample, in spite of a more than threefold increase of the integrated luminosity, thus making the interpretation of these events in terms of higher-order QCD processes much more likely.

The QCD prediction[60] shown in Fig. 43b, normalized to the observed number of events having p_T^W < 30 GeV/c, gives 2.7 events for p_T^W > 30 GeV/c, to be compared with 6 observed events and an estimated background of 0.7 events. This QCD prediction is affected by an uncertainty of ±50% which results mainly from uncertainties in the structure functions. In the curve of Fig. 43b the structure functions given by Glück et al.[67] are used.

4.11 The Z → e⁺e⁻ Event Samples

As observed in subsection 4.3, the efficiency of the electron identification criteria is about 75% in both experiments, and their application to both electrons from Z decay would result in the loss of about 50% of the Z → e⁺e⁻ event samples. The selection of electron-pair candidates is therefore performed by using less selective but more efficient criteria which require that both energy clusters be compatible with an electron from calorimeter information alone, and that at least one cluster satisfy the full electron identification criteria. Such a selection leads to samples of electron pair candidates whose invariant mass m_{ee} has the distributions shown in Fig. 46. Above a threshold of 20 GeV/c², both distributions show a rapidly falling continuum at mass values of less than 50 GeV/c², and a well-separated peak near $m_{ee} \approx 90$ GeV/c². The events in this peak are interpreted as Z → e⁺e⁻ decays (18 events in UA1,[61] 16 in UA2[62]).

The low-mass continuum is mostly due to background from two-jet events. Such a background is negligible under the Z peak, as demonstrated by the fact that no event is observed in a wide interval of m_{ee} between the low-mass continuum and the peak. Background estimates give less than 0.3 events under the Z peak for both distributions of Fig. 46.

A value of m_Z can be obtained from the m_{ee} distributions in the high-mass region by a maximum likelihood fit of a Breit-Wigner shape distorted by the experimental mass resolution. The UA1 result, based on 14 well-measured e⁺e⁻ pairs, is

$$m_Z = 93.0 \pm 1.4 \text{ (stat.)} \pm 3.2 \text{ (syst.)} \text{ GeV/c}^2 \ . \tag{28}$$

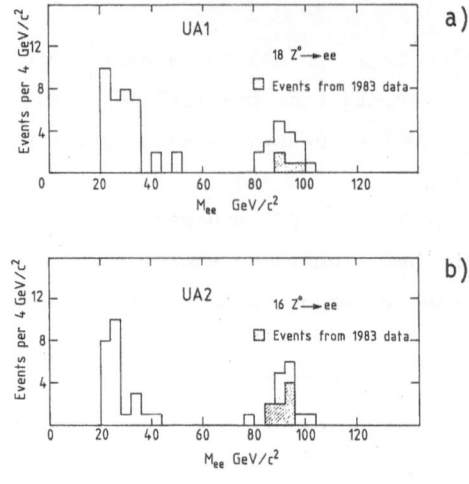

Fig. 46 Invariant mass distribution of electron pairs measured in the UA1
(a) and UA2 (b) experiments (1985 data not included). The shaded events are
$Z \to e^+e^-$ decays observed in the 1982-83 data samples.

The corresponding UA2 result, which includes the 1985 data and is based on
27 well-measured e^+e^- pairs, is

$$m_Z = 92.1 \pm 1.1 \text{ (stat.)} \pm 1.5 \text{ (syst.) GeV/c}^2 \ . \tag{28'}$$

In both cases the systematic error reflects the uncertainty in the absolute
calibration of the calorimeter energy scale.

Cross-sections for inclusive Z production followed by the decay $Z \to e^+e^-$, σ_Z^e, are listed in Table 4. Within the rather large statistical and systematic errors, the results of the two experiments are consistent with each other and also with the theoretical predictions.[60]

Table 4

$Z \to e^+e^-$ cross-sections

	σ_Z^e (pb)	
	$\sqrt{s}$ = 546 GeV	$\sqrt{s}$ = 630 GeV
UA1	$42 {\,}^{+\,25}_{-\,18} \pm 6$	$73 \pm 14 \pm 11$
UA2	$115 \pm 40 \pm 10$	$73 \pm 15 \pm 10$
Theory[60]	$42 {\,}^{+\,13}_{-\,6}$	$51 {\,}^{+\,16}_{-\,10}$

(The second error is the systematic uncertainty.)

4.12 The Z Width and the Number of Neutrino Species

Within the context of the Standard Model, the value of the Z width Γ_Z is related to the number of fermion doublets for which the decay $Z \to f\bar{f}$ is kinematically allowed. Under the assumption that for any additional fermion family only the neutrino is significantly less massive than $m_Z/2$, we can write

$$\Gamma_Z = \sum_\ell \Gamma_Z(\ell\bar{\ell}) + \sum_q \Gamma_Z(q\bar{q}) + N_\nu \Gamma_Z(\nu\bar{\nu}) \tag{29}$$

where $\ell \equiv$ e, μ, τ, $q \equiv$ u, d, c, s, t, b, and N_ν is the number of neutrino species. We recall that $\Gamma_Z(\nu\bar{\nu}) \approx 0.17$ GeV/c^2.

A value of Γ_Z can be obtained, in principle, by the same maximum likelihood fit as used to determine m_Z. However, the expected width for three fermion families ($\Gamma_Z \approx 2.5$ GeV/c^2) is comparable with the experimental mass resolution achieved in UA1 and UA2. Under these circumstances the determination of Γ_Z depends critically on a precise knowledge of both the measurement errors and the shape of the experimental resolution. In the present experiments, given also the small event samples available, this method does not lead to reliable results.

A model-dependent method,[68] which does not depend on the mass resolution, consists in measuring the ratio $R = \sigma_W^e/\sigma_Z^e$, which is related to the ratio of total widths, Γ_Z/Γ_W, by the equation:

$$(\Gamma_Z/\Gamma_W) = R(\sigma_Z/\sigma_W)\,[\Gamma_Z(e^+e^-)/\Gamma_W(e\nu)] \tag{30}$$

where σ_Z (σ_W) is the inclusive cross-section for Z (W) production. This equation can be easily derived from the definition of σ_W^e and σ_Z^e.

The measured values of R are

$$\text{UA1:} \quad R = 8.9\,{}^{+\,1.6}_{-\,1.3} \tag{31}$$

and

$$\text{UA2:} \quad R = 7.4\,{}^{+\,1.5}_{-\,1.2} \tag{31'}$$

where in both cases the errors are dominated by statistics because the value of the integrated luminosity cancels out. These two values can be statistically combined to give

$$R = 8.3\,{}^{+\,1.1}_{-\,1.0} \; ; \tag{31''}$$

$$R < 10.2 \qquad (95\% \text{ confidence level}) .$$

In Eq. (30) both σ_Z/σ_W and the ratio of the leptonic partial widths depend on $\sin^2\theta_W$, explicitly through the neutral current couplings and implicitly via the W and Z masses. However their product is constant to within 1% over the range $\sin^2\theta_W = 0.232 \pm 0.009$ [see Eq. (13)]. A further, more serious uncertainty results from the use of different sets of structure functions in the theoretical calculation of σ_Z/σ_W. Recent estimates[60] give values of σ_Z/σ_W ranging between the two extreme values 0.285 and 0.325. Using R as in Eq. (31") and the value $\sigma_Z/\sigma_W = 0.305 \pm 0.020$, we evaluate

$$\Gamma_Z/\Gamma_W = 0.94 \begin{array}{c} + 0.13 \\ - 0.11 \end{array} \text{ (stat.) } \pm 0.06 \text{ (theor.)}$$

$$\Gamma_Z/\Gamma_W < 1.16 \pm 0.08 \text{ (theor.) (95\% confidence level)}$$

(32)

where the theoretical error reflects the maximum uncertainty on σ_Z/σ_W as quoted above.

The ratio of total widths is sensitive to both the number of neutrino types, N_ν, and to the mass of the top quark, m_t. The dependence of $\Gamma_W(t\bar{b})$ on m_t has been given in Subsection 4.1.2; that of $\Gamma_Z(t\bar{t})$ is discussed in a recent paper by Kühn et al.[69]. The results are summarized in Fig. 47, which

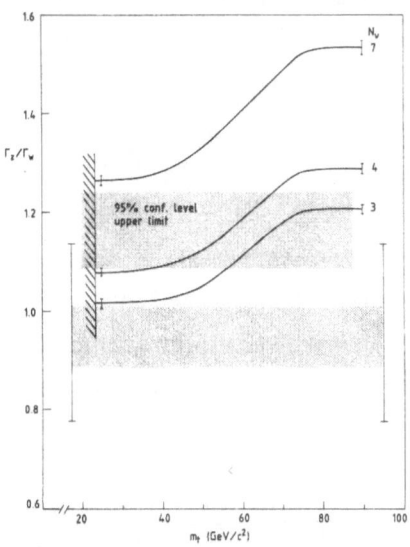

Fig. 47 The value of Γ_Z/Γ_W, as obtained by combining the UA1 and UA2 measurements, displayed with both statistical and theoretical uncertainties. Also shown is the 95% confidence level upper limit with its theoretical uncertainty. The solid lines are the expected variations of Γ_Z/Γ_W with m_t for 3, 4 and 7 neutrino species. The superimposed error bars represent the $\sin^2\theta_W$ uncertainty (see text). The shaded region marks the lower limit on m_t ($m_t > 23$ GeV) as determined by e^+e^- experiments at PETRA.

shows the expected variation of Γ_Z/Γ_W with m_t if the number of light neutrino species is respectively 3, 4 or 7. The error bars at the end of each line show the effect of varying $\sin^2\theta_W$ in the range 0.232 ± 0.009. Also shown in Fig. 47 are the experimental evaluations of Γ_Z/Γ_W as given in Eq. (32). From the 95% confidence level upper limit using the conservative upper estimate of $\sigma_Z/\sigma_W = 0.325$, the data exclude more than six neutrino species, independently of m_t. This limit decreases to three neutrino species in the case $m_t > 68$ GeV. Hence it is not possible to have at the same time a heavy t-quark and more than three neutrino species.

We finally note that the method just discussed will become obsolete as soon as the forthcoming Z factories (SLC, LEP) begin operation, because the Z width, Γ_Z, will be directly measured with a precision of ± 50 MeV/c^2 at these machines. At that moment, Eq. (30) will be used to determine Γ_W with a precision far superior to any other possible method.

4.13 The Decay W → μν

The observation of high-p_T muons is possible only in the UA1 experiment, since there is no muon detector in UA2. The requirements of an isolated track consistent with a high-p_T muon ($p_T^\mu > 15$ GeV/c) and of a missing transverse momentum in excess of 15 GeV/c select a sample of 47 events.[61] The p_T^μ distribution of this sample is shown in Fig. 48, where it is compared with the spectrum expected from W → μν decay for $m_W = 83$ GeV/c^2. The Jacobian peak structure, clearly observed in the decay W → eν, is not visible in this case because of the distortions due to the poor momentum resolution of the magnetic spectrometer for high-momentum tracks. We recall that in UA1 the momentum resolution for a 40 GeV/c muon is ~ ± 20%, whereas it is ~ ± 2.5% for a 40 GeV/c electron as measured in the calorimeter.

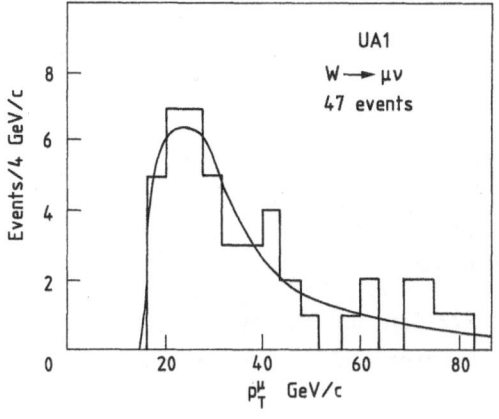

Fig. 48 Transverse momentum spectrum for muons from W → μν decay (UA1). The expectation from $m_W = 83$ GeV/c^2 is also shown.

It must also be noted that the number of events in the W → μν sample is ∼ 1/3 of that of the W → eν sample. This is due to the incomplete angular coverage of the muon trigger, which was not operational over the whole solid angle during data taking.

Whilst the W → μν sample cannot be used to obtain a precise measurement of m_W, it provides a measurement of σ_W^μ, the cross-section for W production followed by the decay W → μν. The measured ratio[61] $\sigma_W^\mu/\sigma_W^e = 1.07 \pm 0.17$ (stat.) $\pm$ 0.13 (syst.) represents a test of μ-e universality.

4.14 The Decay Z → μ⁺μ⁻

By requiring the simultaneous presence of two muons, each having $p_T^\mu >$ 3 GeV/c and such that their invariant mass $m_{\mu\mu}$ exceeds 6 GeV/c², a total of 222 events are selected in the UA1 experiment.[61] The $m_{\mu\mu}$ distribution of these events (Fig. 49) shows a rapidly falling continuum at low masses, well separated from 10 events which all have $m_{\mu\mu} > 70$ GeV/c². These high-mass events, which all consist of μ⁺μ⁻ pairs, are identified as Z → μ⁺μ⁻ decays. The errors affecting the individual mass values are also shown in Fig. 49. These errors are asymmetric because momentum measurement by magnetic deflection gives errors which are Gaussian in 1/p and not in p. The event with m $m_{\mu\mu} > 140$ GeV/c² in Fig. 49 is consistent with a Z → μ⁺μ⁻ decay within the large measurement error.

A best fit of a Breit-Wigner shape distorted by the mass resolution

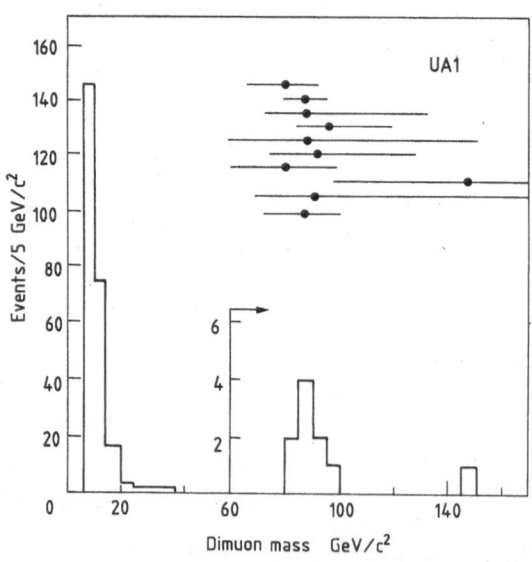

Fig. 49 Invariant mass distribution for 222 dimuons observed in the UA1 experiment. Also shown are the measurement errors of the 10 events interpreted as Z → μ⁺μ⁻ decays.

gives the value[61] $m_Z = 88.8^{+5.5}_{-4.6}$ GeV/c^2, in good agreement with the value

obtained from the e^+e^- channel [see Eqs. (28) and (28')]. These events correspond to a measured cross-section $\sigma^\mu_Z = 51.4 \pm 17.1$ (stat.) $\pm$ 8.6 (syst.) pb, which agrees with the measured σ^e_Z values, and also with theoretical predictions based on μ-e universality (see Table 4).

4.15 Comparison with the Standard Model

In order to compare the measurements of m_W and m_Z with the predictions of the Standard Model, we must use suitably renormalized and radiatively corrected quantities.[70] We shall use the scheme where $\sin^2\theta_W$ is defined as[52]

$$\sin^2\theta_W = 1 - (m_W/m_Z)^2 , \qquad (33)$$

which leads to Eq. (12') for the W mass prediction, and, by the assumption of the minimal Higgs scheme, to the prediction

$$m^2_Z = 4A^2/[(1 - \Delta r) \sin^2 2\theta_W] \qquad (34)$$

for the Z mass (see subsection 4.1.1).

Using Eqs. (12') and (34) we can extract two values of $\sin^2\theta_W$ from the values of m_W and m_Z measured in each experiment. We then combine them to obtain

$$\text{UA1:} \quad \sin^2\theta_W = 0.216^{+0.004}_{-0.005} \text{ (stat.)} \pm 0.014 \text{ (syst.)} , \qquad (35)$$

and

$$\text{UA2:} \quad \sin^2\theta_W = 0.232 \pm 0.004 \text{ (stat.)} \pm 0.008 \text{ (syst.)} . \qquad (35')$$

In both cases the systematic error reflects the uncertainty on the mass scale from the uncertainty of the calorimeter calibration.

Because of the systematic errors, a weighted average of the two results is not possible. Within errors, the two values of $\sin^2\theta_W$ agree with each other, and also with the value of $\sin^2\theta_W$ obtained from low-energy experiments [see Eq. (13)].

By using Eq. (33) it is possible to measure $\sin^2\theta_W$ with no systematic error from the mass scale. The results are

$$\text{UA1:} \quad \sin^2\theta_W = 0.194 \pm 0.031 ,$$

and

$$\text{UA2:} \quad \sin^2\theta_W = 0.242 \pm 0.025 \ .$$

The weighted average of these two results is

$$\sin^2\theta_W = 0.223 \pm 0.019 \ ,$$

which represents a less precise measurement than the results given by Eqs. (35) and (35'), in view of the limited event samples available at present.

By using the $\sin^2\theta_W$ definition given by Eq. (33) we have implicitly assumed that the ϱ parameter, defined as[71]

$$\varrho = m_W^2 / (m_Z^2 \cos^2\theta_W) \tag{36}$$

is equal to 1. We can test this assumption by combining Eqs. (12') and (36) to obtain

$$\varrho = m_W^2 / [m_Z^2 \ (1 - B^2/m_W^2)] \ ,$$

where $B^2 = A^2/(1 - \Delta r)$. The UA1 result is

$$\varrho = 1.028 \pm 0.037 \ (\text{stat.}) \pm 0.019 \ (\text{syst.}) \ ,$$

and the corresponding UA2 result is

$$\varrho = 0.986 \pm 0.026 \ (\text{stat.}) \pm 0.006 \ (\text{syst.}) \ .$$

These two results agree with each other and with the value $\varrho = 1.02 \pm 0.02$ obtained by the low-energy experiments.[52,72] We recall that the value $\varrho = 1$ corresponds to the minimal Standard Model with only one isodoublet of complex Higgs fields.

The relations (12') and (34) may also be used to measure the radiative correction parameter Δr. Eliminating $\sin^2\theta_W$ from Eqs. (12') and (34) we obtain

$$1 - \Delta r = (A^2/m_W^2)/(1 - m_W^2/m_Z^2)$$

from which, using the UA2 mass values, we deduce

$$\Delta r = 0.073 \pm 0.022 \ (\text{stat.}) \pm 0.030 \ (\text{syst.})$$

in agreement with the theoretical prediction, $\Delta r = 0.0711 \pm 0.0013$ (see Subsection 4.1.1).

The interest in a precise test of the radiative corrections to the IVB masses arises from the fact that these corrections are quite sensitive to the t-quark mass, as well as to the existence of further fermion families with large mass splittings between the two members of the same isodoublet. For example in the standard three-family case, for $m_t \gg m_W$, Δr is modified by a contribution δ which is asymptotically given by:[52]

$$\delta = -\frac{3\alpha}{16\pi} \frac{\cos^2\theta_W}{\sin^4\theta_W} \frac{m_t^2}{m_W^2} .$$

(37)

The sensitivity of Δr to the mass of the Higgs boson m_H is much weaker (the standard value $\Delta r \approx 0.07$ assumes $m_H = m_Z$). For $m_H/m_Z < 10$ an increase of only < 0.009 is expected for Δr.[52]

From the dependence of Δr on m_t given by Eq. (37), and from the UA2 lower limit $\Delta r > 0.0124$ (90% confidence level), we deduce the upper limit $m_t < 246$ GeV/c^2 (90% confidence level).

5. CONCLUSIONS

The main feature of the results reviewed in these lectures is the remarkable agreement between theoretical expectations and experimental observations.

In the study of final states containing high-p_T jets, the present statistical and systematic errors are comparable with the theoretical uncertainties, which result partly from inadequate knowledge of structure functions and partly from the lack of higher-order calculations. In this field, any further progress must come not only from increased experimental precision but also from improvements in the theory. In this respect measurements of high-p_T photons are particularly promising because they are free from systematic uncertainties related to the definition of a jet, and the process involves a smaller number of diagrams.

These considerations do not apply to the physics of W and Z bosons. In this field there are measurements for which the present experimental uncertainties are still large and must be improved. An important example is represented by the measurement of the W and Z masses, whose present accuracy is barely sufficient to demonstrate the need for radiative corrections in the Standard Model. The Z mass will probably be measured with an absolute precision of ± 50 MeV/c^2 by the next generation of e^+e^- machines (SLC, LEP). However, a meaningful test of the Standard Model requires also a measurement of the W mass with comparable accuracy. Such a measurement could be done at $\bar{p}p$ colliders, given a sufficiently large event sample and electromagnetic calorimetry of excellent quality. Here the method would consist in measuring the ratio m_W/m_Z, which is not affected by the systematic uncertainties on the mass scale, and using the value of m_Z from e^+e^- experiments.

At CERN, the Antiproton Collector (ACOL) now under construction will become operational in the second half of 1987, providing approximately tenfold increase of the Collider luminosity. By that time, the two main experiments, UA1 and UA2, will have undergone major upgrading.

By the end of 1989 the data samples should contain ~ 5000 W → ev and ~ 500 Z → e^+e^- decays in each experiment, corresponding to an integrated luminosity of about 10 pb^{-1}. It will then be possible to measure m_W/m_Z with a precision of ~ $\pm 2.5 \times 10^{-3}$, which would then give an error of ~ ± 200 MeV/c^2 on m_W. This is comparable with the precision expected from the study of the process $e^+e^- \rightarrow W^+W^-$ in the second phase of the LEP project. This phase, which requires the use of additional superconductive RF cavities to increase the machine energy beyond the threshold for production of W pairs, will not be operational before 1993 at the earliest.

Acknowledgements

I warmly thank Professor T. Ferbel, Director of ASI-1986, for organizing a very interesting and lively School in the beautiful environment of the Virgin Islands.

I am deeply indebted to the Scientific Reports Editing and the Text Processing Sections at CERN for their competence and patience in preparing the final version of this paper.

REFERENCES

1. A. Astbury et al., Phys. Scri. 23:397 (1981).
2. M. Barranco Luque et al., Nucl. Instrum. Methods 176:175 (1980).
   M. Calvetti et al., Nucl. Instrum. Methods 176:255 (1980).
3. M. Dialinas et al., report LAL-RT/83-14 (1983).
4. A. Beer et al., Nucl. Instrum. Methods 224:360 (1984).
5. C. Conta et al., Nucl. Instrum. Methods 224:65 (1984).
6. K. Borer et al., Nucl. Instrum. Methods 227:29 (1984).
7. M. Banner et al., Phys. Lett. 118B:203 (1982).
8. P. Bagnaia et al., Phys. Lett. 138B:430 (1984).
9. C. DeMarzo et al., Nucl. Phys. B211:375 (1983).
   B. Brown et al., Phys. Rev. Lett. 49:711 (1982).

10. P. Bagnaia et al., Z. Phys. C20:117 (1983).
11. G. Arnison et al., Phys. Lett. 123B:115 (1983) and 132B:214 (1983).
12. T. Akesson et al., Phys. Lett. 118B:185 (1982) and 123B:133 (1983).
13. B. Alper et al., Phys. Lett. 44B:521 (1973).
    M. Banner et al., Phys. Lett. 44B:537 (1973).
    F. W. Büsser et al., Phys. Lett. 46B:471 (1973).
14. R. P. Feynman, Phys. Rev. Lett. 23:1415 (1969).
15. S. M. Berman et al., Phys. Rev. D4:3388 (1971).
16. For a review see: F. Wilczek, Annu. Rev. Nucl. Part. Sci. 32:177
    (1982).
17. B. L. Combridge et al., Phys. Lett. 70B:234 (1977).
18. R. Horgan and M. Jacob, Nucl. Phys. B179:441 (1981).
19. F. E. Paige and S.D. Protopopescu, Proc. 1982 DPF Summer Study on
    Elementary Particle Physics and Future Facilities, Snowmass,
    Colorado, 1982 (AIP, New York, 1983), p. 471.
20. M. Furman, Nucl. Phys. B197:413 (1982).
21. B. Humpert, preprint CERN-TH 3934 (1984).
    N. G. Antoniou et al., Phys. Lett. 128B:257 (1983).
    Z. Kunszt and E. Pietarinen, Phys. Lett. 132B:453 (1983).
22. W. Furmanski and H. Kowalski, Phys. Lett. 148B:247 (1984).
23. J. A. Appel et al., Phys. Lett. 160B:349 (1985).
24. G. Arnison et al., preprint CERN-EP/86-29 (1986) (to be published in
    Phys. Lett. B).
25. E. Eichten et al., Rev. Mod. Phys. 56:579 (1984).
26. F. Ceradini (UA1 Collaboration) Proc. Int. Europhysics Conf. on High-
    Energy Physics, Bari, 1985 (Laterza, Bari, Italy, 1985), p. 363.
27. D. W. Duke and J. F. Owens, Phys. Rev. D30:49 (1984).
28. D. Cline et al., Phys. Rev. Lett. 31:491 (1973).
29. J. C. Collins and D. E. Soper, Phys. Rev. D16:2219 (1977).
30. G. Arnison et al., Phys. Lett. 158B:494 (1985).
31. P. Bagnaia et al., Phys. Lett. 144B:283 (1984).
32. G. Arnison et al., Phys. Lett. 136B:294 (1984).
33. B. L. Combridge and C. J. Maxwell, Nucl. Phys. B239:428 (1984).
34. F. Bergsma et al., Phys. Lett. 153B:111 (1985).
35. M. Greco, Z. Phys. C26:567 (1985).
36. G. Sterman and S. Weinberg, Phys. Rev. Lett. 39:1436 (1977).
    M. B. Einhorn and B. G. Weeks, Nucl. Phys. B146:445 (1978).
    K. Shizuya and S.-H. H. Tye, Phys. Rev. D20:1101 (1979).
37. A. Bassetto et al., Phys. Rep. 100:201 (1983).
38. B. Andersson et al., Phys. Rep. 97:31 (1983).
39. P. Ghez (UA1 Collaboration), Proc. 5th Topical Workshop on Proton-
    Antiproton Collider Physics, Saint-Vincent, 1985 (World Scientific,
    Singapore, 1985), p. 112.
    G. Arnison et al., preprint CERN-EP/86-55 (submitted to Nucl. Phys. B).
40. M. Althoff et al., Z. Phys. C22:307 (1984).
41. T. Akesson et al., Z. Phys. C30:27 (1986).
42. P. Bagnaia et al., Phys. Lett. B144:291 (1984).
43. G. Marchesini and B. R. Webber, Nucl. Phys. B238:1 (1984).
    B. R. Webber, Nucl. Phys. B238:492 (1984).
44. Z. Kunszt and E. Pietarinen, Nucl. Phys. B164:45 (1980).
    T. Gottschalk and D. Sivers, Phys. Rev. D21:102 (1980).
    F. A. Berends et al., Phys. Lett. 118B:124 (1981).
45. J.A. Appel et al., Z. Phys. C30:341 (1986).
46. P. Aurenche et al., Phys. Lett. 140B:97 (1984).
47. M. Fraternali (UA2 Collaboration), same Proceedings as Ref. 26, p. 470.
48. A. L. S. Angelis et al., Phys. Lett. 94B:106 (1980) and 98B:115 (1981).
49. See, for example:
    J. Ellis et al., Annu. Rev. Nucl. Part. Sci. 32:443 (1982).
    L. B. Okun', Leptons and quarks (North-Holland, Publ. Co., Amsterdam,
    1982).

50. S. Glashow, Nucl. Phys. $\underline{22}$:579 (1961).
    S. Weinberg, Phys. Rev. Lett. $\underline{19}$:1264 (1967).
    A. Salam, Proc. 8th Nobel Symposium, Aspenäsgården, 1968, ed. N. Svartholm (Almqvist and Wiksell, Stockholm, 1968), p. 367.
51. M. Aguilar-Benitez et al. (Particle Data Group), Phys. Lett. $\underline{170B}$:36 (1986).
52. W. J. Marciano and A. Sirlin, Phys. Rev. $\underline{D29}$:945 (1984);
    F. Jegerlehner, Bielefeld preprint BI-TP 1986/8;
    W. Hollick, preprint DESY 86-049 (1986).
53. W. J. Marciano and A. Sirlin, Nucl. Phys. $\underline{B189}$:442 (1981).
    For a more recent review, see J. Panman, preprint CERN-EP/85-35 (1985).
54. H. Abramowicz et al., preprint CERN-EP/86-38 (1986), submitted to Phys. Rev. Letters.
55. F. Bergsma, A precise determination of the electroweak mixing angle from semileptonic neutrino scattering (in preparation).
56. M. Kobayashi and K. Maskawa, Prog. Theor. Phys. $\underline{49}$:652 (1973).
57. N. Cabibbo, Phys. Rev. Lett. $\underline{10}$:531 (1963).
58. S. D. Drell and T. M. Yan, Ann. Phys. (USA) $\underline{66}$:578 (1971).
59. L. B. Okun' and M. B. Voloshin, Nucl. Phys. $\underline{B120}$:459 (1977).
    C. Quigg, Rev. Mod. Phys. $\underline{94}$:297 (1977).
    J. Kogut and J. Shigemitsu, Nucl. Phys. $\underline{B129}$:461 (1977).
    F. E. Paige, report BNL-27066 (1979).
60. G. Altarelli et al., Nucl. Phys. $\underline{B246}$:12 (1984).
    G. Altarelli, R. K. Ellis and G. Martinelli, Z. Phys. $\underline{C27}$:617 (1985).
    M. Diemoz and G. Martinelli, private communication.
61. M. Levi (UA1 Collaboration), same Proceedings as Ref. 26, p. 346.
    G. Arnison et al., Nuovo Cimento Lett. $\underline{44}$:1 (1985).
    G. Arnison et al., Phys. Lett. $\underline{166B}$:484 (1986).
    G. Arnison et al., Europhys. Lett. $\underline{1}$:327 (1986).
    Preliminary UA1 results from the 1985 runs are in: UA1 Collaboration, preprint CERN-EP/86-115 (1986).
62. J. A. Appel et al., Z. Phys. $\underline{C30}$:1 (1986).
    UA2 Collaboration, Measurement of the Standard Model parameters from a study of W and Z bosons (submitted to Phys. Lett. B).
63. M. Bozzo et al., Phys. Lett. $\underline{147B}$:392 (1984).
64. M. Jacob, Nuovo Cimento $\underline{9}$:826 (1958).
65. S. D. Ellis, R. Kleiss and J. Stirling, Phys. Lett. $\underline{154B}$:435 (1985).
66. P. Bagnaia et al., Phys. Lett. $\underline{139B}$:105 (1984).
67. M. Glück et al., Z. Phys. $\underline{C13}$:119 (1982).
68. F. Halzen and K. Mursula, Phys. Rev. Lett. $\underline{51}$:857 (1983).
    K. Hikasa, Phys. Rev. $\underline{D29}$:1939 (1984).
69. J.H. Kühn, A. Reiter and P.M. Zerwas, Nucl. Phys. $\underline{B272}$:560 (1986).
70. A. Sirlin, Phys. Rev. $\underline{D22}$:971 (1980).
    W. J. Marciano, Phys. Rev. $\underline{D20}$:274 (1979).
    M. Veltman, Phys. Lett. $\underline{91B}$:95 (1980).
    F. Antonelli et al., Phys. Lett. $\underline{91B}$:90 (1980).
71. D. Ross and M. Veltman, Nucl. Phys. $\underline{B95}$:135 (1975).
    P. Q. Hung and J. J. Sakurai, Nucl. Phys. $\underline{B143}$:81 (1978).
72. J. Kim et al., Rev. Mod. Phys. $\underline{53}$:211 (1980).
    P. Langacker, Proc. 22nd Int. Conf. on High-Energy Physics, Leipzig, 1984 (Akad. Wissenschaften der DDR, Berlin-Zeuthen, 1984), p. 215.

WEAK DECAYS OF HEAVY QUARKS

David G. Hitlin

California Institute of Technology
Pasadena, CA 91125

INTRODUCTION

We now have a great deal of experimental information on the weak decays of heavy quarks, and a considerable body of calculations which serve to organize and explain the data. These lectures will treat the subject of weak decays of heavy quarks in some detail, summarizing the current experimental and theoretical situation and briefly looking at prospects for future progress in this area. As the literature in this area has now become quite vast, no attempt will be made to treat the subject historically.

STATES CONTAINING HEAVY QUARKS

We will be concerned primarily with hadrons containing charmed and bottom quarks; there is, as of this writing, little experimental information on top quark-containing hadrons.

There are three pseudoscalar charmed mesons, the $(D^0(c\bar{u}), D^+(c\bar{d}))$ isodoublet and the $D_s^+(c\bar{s})$ isosinglet. There are four pseudoscalar B mesons, the $(B_d^0, B_u^-$ isodoublet and the B_s^0, B_c^- isosinglets. All of these are $^1S_0$ combinations of light and heavy quarks; all have weak decays. The vector partners $(^3S_1)$ decay through electromagnetic transitions or single pion emission. The low Q value of the $D^* \to D\pi$ transition has proven particularly valuable in isolating weak hadronic D decays.

Figure 1 shows the lowest-lying $J^P = 0^-$ mesons of the $\underline{35} + \underline{1}$ multiplets in the six quark model.

In systems with one heavy and one light quark, the $^3S_1 - {}^1S_0$ mass splitting is governed by the long distance behavior of the potential. For a linear potential, the

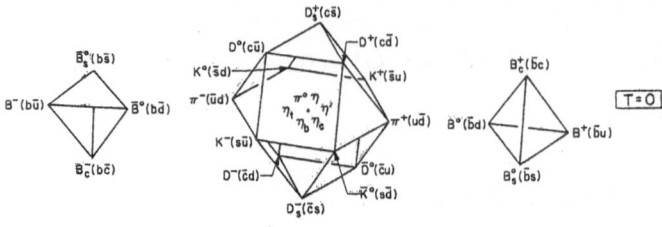

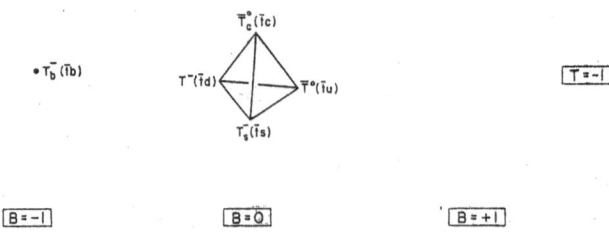

Fig. 1 The $J^P = 0^-$ mesons of the $\underline{35} + \underline{1}$ multiplets in the six quark scheme.

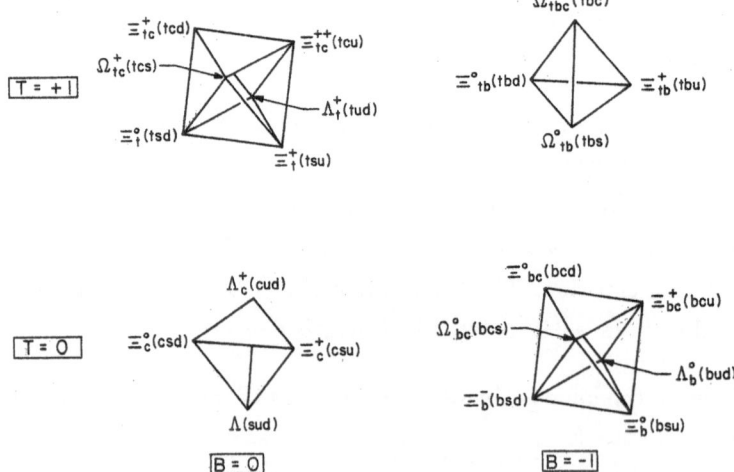

Fig. 2 The $J^P = \frac{1}{2}^-$ baryons of the $\underline{20}_A$ multiplet.

276

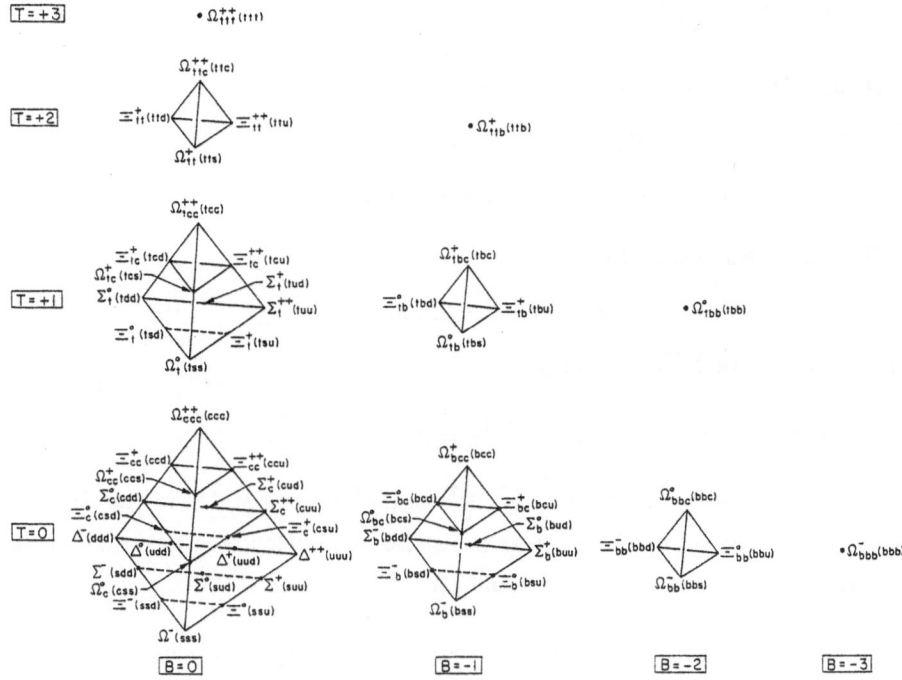

Fig. 3   The $J^P = \frac{3}{2}^+$ baryons of the $\underline{56}_S$ multiplet.

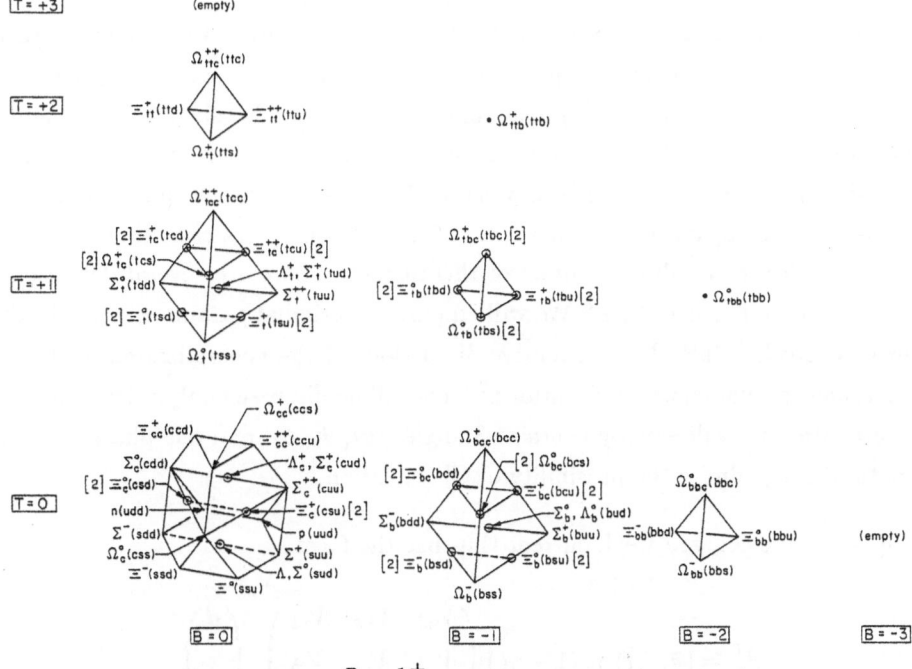

Fig. 4   The $J^P = \frac{1}{2}^+$ baryons of the $\underline{70}_M$ multiplet.

splitting should be essentially independent of the meson mass.[1] This is indeed the case, as the mass-squared differences of the $\rho - \pi$, $K^* - K$, $D^* - D$, $D_s^+ - D_s$ and $B^* - B$ systems all fall in the range[2,3] 0.55-0.57 GeV^2.

The D^0 and D^+ masses[2] are well established:

$$M_{D^0} = 1864.6 \pm 0.6 \text{ MeV}$$

$$M_{D^+} = 1863.3 \pm 0.6 \text{ MeV} \ .$$

The D_s^+ mass[2] has now settled at

$$M_{D_s^+} = 1970.5 \pm 2.5 \text{ MeV} \ ,$$

after residing above 2 GeV initially.

Measurements of B meson masses will be discussed below.

Figures 2, 3, and 4 show the $J^P = \frac{1}{2}^-$, $\frac{3}{2}^+$ and $\frac{1}{2}^+$ baryon states in the six quark model. The heavy baryon mass splittings have also been calculated in detail[4] in bag and potential models. Four charmed baryons (Λ_c^+, Ξ_c^+, Ξ_c^0 and Ω_c^0) decay weakly.

The experimental status of each of these areas of investigation is in constant flux. Heavy quark production and decay remains an active area of study at fixed target machines and with both $\bar{p}p$ and e^+e^- colliders.

The Hadronic Weak Current

Within the context of the Standard Model, an anomaly-free, renormalizable theory is ensured by placing the charged leptons in doublets with their associated neutrinos and the quarks in left-handed weak isospin doublets and right-handed singlets. The hadronic current is not then described by these left-handed doublets, but by a generalization of the Cabibbo prescription due to Kobayashi and Maskawa.[5] For N quark generations, mixing is described by an $N \times N$ unitary matrix. In the six quark case, three angles and one phase suffice for a full description of this matrix. There are several definitions of these parameters in general use.[6] We will employ the original parametrization in these lectures. There have also been extensive discussions of the generalization of the K-M matrix to four or more quark generations.[7] This will be discussed only briefly here. The parametrization we will employ contains 3 angles: θ_1, θ_2, θ_3, and one phase: δ, which, in a natural way, admits the possibility of CP violation.

The hadronic charged weak current thus has the form

$$J_u^T = (\bar{u}, \bar{c}, \bar{t}) \, \gamma_\mu \, (1 - \gamma_5) \begin{pmatrix} V_{ud} & V_{us} & V_{ub} \\ V_{cd} & V_{cs} & V_{cb} \\ V_{td} & V_{ts} & V_{tb} \end{pmatrix} \begin{pmatrix} d \\ s \\ b \end{pmatrix} \ ,$$

with the K-M matrix completely described as:

$$
\begin{pmatrix} V_{ud} & V_{us} & V_{ub} \\ V_{cd} & V_{cs} & V_{cb} \\ V_{td} & V_{ts} & V_{tb} \end{pmatrix} = \begin{pmatrix} c_1 & s_1 c_3 & s_1 s_3 \\ -s_1 c_2 & c_1 c_2 c_3 - s_2 s_3 e^{i\delta} & c_1 c_2 c_3 + s_2 c_3 e^{i\delta} \\ -s_1 s_3 & c_1 s_2 c_3 + c_2 s_3 e^{i\delta} & c_1 s_2 s_3 - c_2 c_3 e^{i\delta} \end{pmatrix}.
$$

Here, $c_i \equiv \cos\theta_i$, $s_i \equiv \sin\theta_i$.

The values of the K-M angles and phase are determined by a fitting to a number of experimental inputs.

We will use the recent evaluation of K-M matrix elements by Kleinknecht and Renk.[8] The couplings are obtained from the following types of measurements:

V_{ud}: comparison of muon decay with nuclear beta decay, incorporating radiative corrections.

V_{us}: rates of K_{e_3} decay or semileptonic hyperon decay. The V_{us} obtained by these two methods differs by more than the experimental error, perhaps due to SU(3) breaking effects, which are thought to be larger in hyperon decays. The K_{e_3} data are therefore used.

V_{cd}: comparison of ratios of single and di-muon production in ν and $\bar{\nu}$ reactions, which can be analyzed to yield the rate of single charm quark production by the charged weak current. Extraction of V_{cd} then requires a knowledge of $B(c \to \mu X)$, which is obtained from the measurements of $B(D \to eX)$ on the $\psi(3770)$ resonance in $e^+ e^-$ annihilation. Note that the recent Mark III measurement[9] $B(D^0 \to \pi^- e^+ \nu_e) = (0.4^{+0.4}_{-0.3} \pm 0.1)\%$ provides a determination of $|V_{cd}/V_{cs}|^2 = 0.06 \pm 0.03 \pm 0.01$, with many fewer assumptions than the technique presently employed, but that the statistics are, at present, quite limited.

V_{cs}: rate for the exclusive decays $D \to Ke\nu$, with the assumption that $f_+(0)$, which differs from 1 due to SU(4) breaking, can be calculated by sum rules, and neglecting the q^2 dependence of f_+.

V_{ub}: study of the end point spectrum for electrons produced in semileptonic B meson decay. As no evidence for $b \to u$ transitions has been found, only a limit on $|V_{ub}|/|V_{cb}|$ can be placed. The analysis of the endpoint spectrum has been controversial; the limit set is model dependent. Kleinknecht uses $|V_{ub}|/|V_{cb}| < 0.19$ at 90% confidence level. See the discussion on semileptonic B decay for more details.

V_{cb}: uses the measured b quark lifetime (a differing mixture of B containing hadrons in different experiments) together with the average B meson semileptonic branching ratio.

With these experimental inputs, the K-M matrix elements can be found by a least-squares adjustment. On the assumption of the existence of only three quark generations, unitarity allows a reasonably precise extraction of the matrix elements, despite our paucity of knowledge on V_{td}, V_{ts} or V_{tb}. Kleinknecht obtains:

$$
\begin{array}{c}
\quad\quad d \quad\quad\quad\quad s \quad\quad\quad\quad b \\
\begin{array}{c} u \\ c \\ t \end{array}
\left(
\begin{array}{ccc}
0.9743 - 0.9757 & 0.219 - 0.225 & 0.000 - 0.008 \\
0.219 - 0.225 & 0.9733 - 0.9748 & 0.039 - 0.050 \\
0.002 - 0.017 & 0.037 - 0.048 & 0.9987 - 0.9993
\end{array}
\right),
\end{array}
$$

where the numbers quoted are the 90% confidence level allowed range.

A similar analysis, allowing the possibility of four quark generations, provides a possible window for new physics. Using the Bose-Paschos[7] parametrization, Kleinknecht finds:

$$
\begin{array}{c}
\quad\quad d \quad\quad\quad\quad\quad s \quad\quad\quad\quad\quad b \quad\quad\quad\quad\quad b' \\
\begin{array}{c} u \\ c \\ t \\ t' \end{array}
\left(
\begin{array}{cccc}
0.9708 - 0.9750 & 0.218 - 0.224 & 0.000 - 0.008 & 0.028 - 0.092 \\
0.169 - 0.237 & 0.66 - 0.98 & 0.039 - 0.050 & 0.00 - 0.78 \\
0.00 - 0.17 & 0.00 - 0.77 & 0.030 - 0.999 & 0.000 - 0.997 \\
0.00 - 0.15 & 0.00 - 0.35 & 0.000 - 0.999 & 0.05 - 0.9993
\end{array}
\right),
\end{array}
$$

using the constraint $f_+(0) < 1$ for D semileptonic decay. Clearly, we have little evidence for or against a fourth generation at the current level of precision. Observation of a t quark and measurement of its couplings will provide the next major step in this area of analysis.

Wolfenstein[10] has pointed out that there appears to be a hierarchy of the coupling constants between generations. That is, ignoring CP violation, one can parametrize the K-M matrix as:

$$
\left(
\begin{array}{ccc}
1 & \lambda & \lambda^3 \\
-\lambda & 1 & \lambda^2 \\
-\lambda^3 & -\lambda^2 & 1
\end{array}
\right),
$$

with $\lambda \sim \sin\theta_1$. We will later employ this observation to gain some insight into the mixing of neutral heavy quark systems.

Transitions within a quark doublet, e.g., $c \rightarrow s$ transitions, will be referred to as "Cabibbo-allowed", for obvious historical reasons, while those between doublets, *e.g.*, $c \to d$, $b \to c$, will be called "Cabibbo-suppressed". Suppressed transitions are characterized by couplings smaller than those for allowed transitions by one or more orders in the parameter λ.

Heavy Quark Decay Modes

In the current-current picture, we can isolate three distinct classes of heavy quark decay:

Purely Leptonic Decay

For pseudoscalar mesons, purely leptonic decays $P \to l\bar{\nu}_l$ (Figure 5) provide a clean measure of the axial-vector coupling of the hadronic current to vacuum, i.e., of the matrix element:

$$\langle 0|J_\mu|P \rangle = -if_p q_{p_\mu} .$$

The decay rate is

$$\Gamma(P(Q\bar{q}) \to l\bar{\nu}_l) = \frac{G_F^2}{8\pi}|V_{Qq}|^2 f_p^2 m_p m_l^2 \left[1 - \frac{m_l^2}{m_p^2} \right]^2 .$$

While f_π and f_K are, of course, well measured, no such measurements exist for heavy-quark-containing mesons. The reason is that the helicity suppression is so great that purely leptonic branching ratios are quite small. Thus, leptonic decays contribute in only a negligible way to heavy meson lifetimes. Table I shows estimates of D^+, D_s^+ and B^+ meson purely leptonic branching ratios, using current lifetimes (see below), for $f_P \equiv f_K = .128$ GeV. Direct observation of purely leptonic decay is clearly difficult. Theoretical predictions for the size of f_D, f_F and f_B have varied over the years, but most potential-model-based estimates have now converged to the region of .1–.25 GeV.[11] The only available experimental information comes from Mark III,[12] which has analyzed its large sample of $\psi(3770)$ to set a limit on $B(D^+ \to \mu^+ \nu_\mu)$. This is done by searching a sample of events in which D^+ hadronic decays have been reconstructed for a muon in the recoil spectrum and zero missing mass. No such events have been found, leading to a limit of $B(D^+ \to \mu^+ \nu_\mu) < 8.4 \times 10^{-4}$ at 90% confidence level. When theoretical uncertainties in the D^+ lifetime and V_{cd} are included, this leads to a limit

$$f_D < 340 \text{ MeV at 90\% Confidence Level.}$$

There is, as yet, no comparable information in the b quark sector, although $B \to \tau\nu_\tau$ may eventually be experimentally accessible.

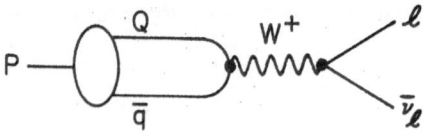

Fig. 5 Purely leptonic decay of a
heavy $Q\bar{q}$ meson.

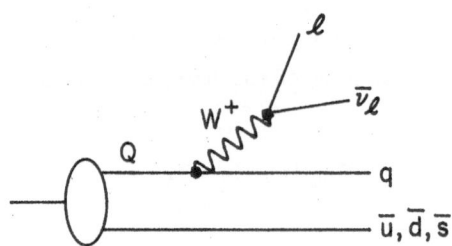

Fig. 6 Semileptonic decay of a heavy $Q\bar{q}$
meson.

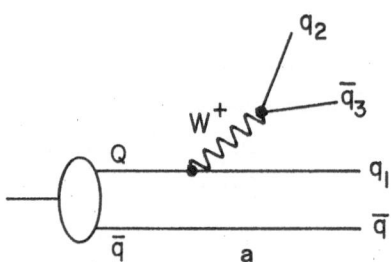

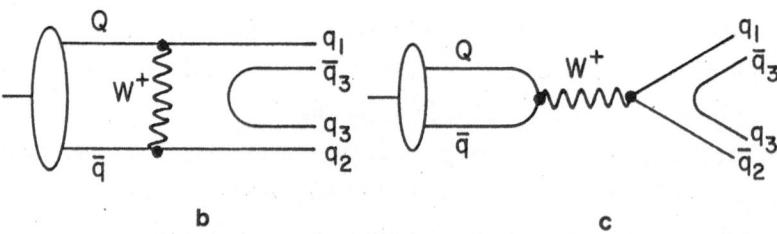

Fig. 7 Hadronic decays of a heavy $Q\bar{q}$ meson in the valence quark
approximation. a) light quark spectator decay; b) W exchange; c) annihilation. The exchange and annihilation amplitudes (generically "non-spectator amplitudes") are helicity-
suppressed at the light quark vertex.

Table I Estimates of Purely Leptonic Branching Fractions

l \ P $B(P \to l\nu_l)$	D^+	D_s^+	B_u^+	B_c^+
$e^+\nu_e$	3.3×10^{-9}	3.7×10^{-8}	4.9×10^{-12}	4.4×10^{-10}
$\mu^+\nu_\mu$	1.4×10^{-4}	1.6×10^{-3}	2.1×10^{-7}	2.6×10^{-5}
$\tau^+\nu_\tau$	3.3×10^{-4}	1.5×10^{-2}	4.5×10^{-5}	5.6×10^{-3}

Semileptonic Decays

Heavy meson semileptonic decays occur primarily through beta decay of the heavy quark (Figure 6). In this so-called spectator approximation, the rate for semileptonic decay can be scaled from muon decay:

$$\Gamma(P \to Xl\bar{\nu}_l) = \Gamma(Q \to ql\bar{\nu}_l) = \frac{G_F^2}{192\pi^3}|V_{Qq}|^2 m_Q^5 \,.$$

The decay matrix element is a product of terms involving the weak hadronic current and the leptonic current. A more detailed treatment of semileptonic decays, taking into account final state hadrons and form factors, will be given below.

Hadronic Decays

In the valence quark approximation, hadronic decays of heavy mesons can occur through three types of processes, shown in Figure 7 for the case of charm decay. The dominant diagram is flavor decay, the W exchange and annihilation diagrams being suppressed by helicity conservation at the light quark vertex. As the matrix element here is a product of terms each involving the weak hadronic current, calculations are non-trivially model dependent. Within the valence quark picture, it is consistent to retain only the flavor decay contribution, leading to the spectator model. This naïve picture, as we shall see, must be considerably refined, but we shall employ it for the time being, in order to estimate the lifetimes of heavy quark states.

Figure 8 illustrates all the spectator diagrams for charmed meson decay. Clearly, the expected lifetime for charmed mesons can be scaled from muon decay:

$$\tau_0(D^0, D^+, D_s^+) = \frac{1}{\Gamma_0} \simeq \frac{1}{5}\left(\frac{m_\mu}{m_c}\right)^5 [\Im|V_{cs}|^2 + |V_{cd}|^2]$$

$$\simeq 3 \times 10^{-13}\text{sec for } m_c = 1.5 \text{ GeV} \,,$$

with a phase space factor $\Im \simeq 0.5$.

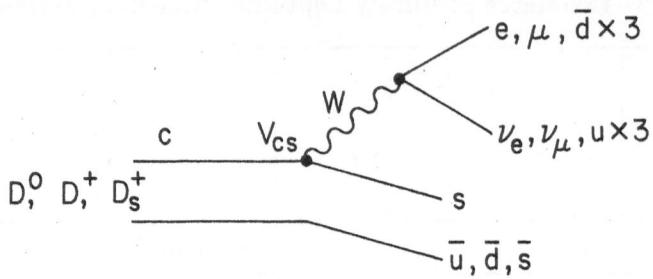

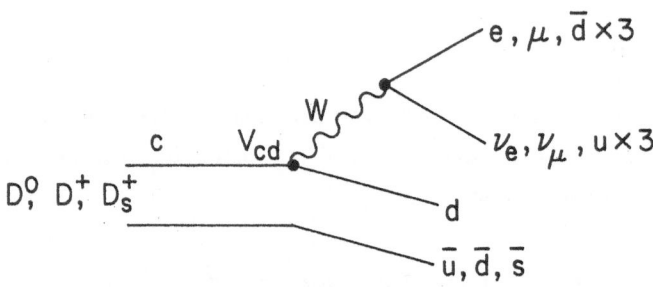

Fig. 8 Cabibbo-allowed (a) and -suppressed (b) semileptonic and spectator hadronic decays of D mesons,
leading to naïve estimate of $\sim 20\%$ for semileptonic
branching ratio.

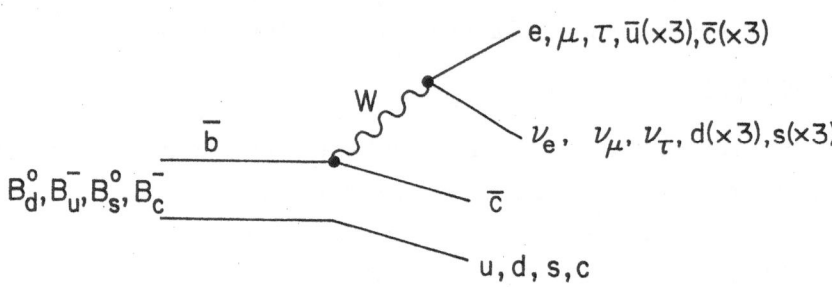

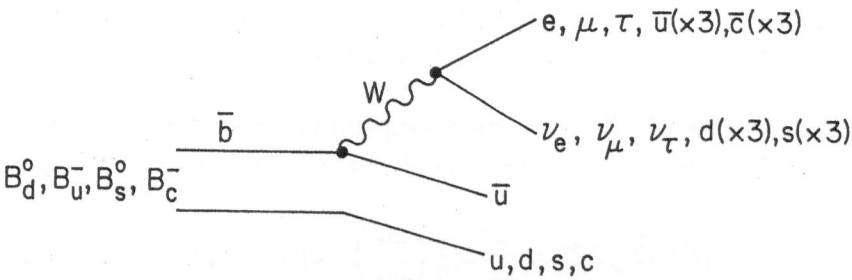

Fig. 9 Cabibbo-allowed (a) and -suppressed (b) semileptonic and spectator hadronic decays of B mesons, leading to naïve estimate of
$\sim 13\%$ for semileptonic branching ratio.

The B meson lifetime can be estimated in a similar way (Figure 9):

$$\tau_0(B_u^-, B_d^0, B_s^0, B_c^-) \simeq \frac{1}{9}\left(\frac{m_\mu}{m_b}\right)^5 [\Im|V_{bc}|^2 + |V_{bu}|^2]$$

$$\simeq 2 \times 10^{-12} \sec \text{ for } m_b = 5 \text{ GeV},$$

again, with a phase space factor which, fortuitously, is about 0.5. The first measurements of τ_b evoked surprise at the long lifetime, due, as we now see, to the small value of V_{bc}.

The important observation, however, is that in this simple picture, the lifetimes of all charmed mesons are *identical*, as are the lifetimes of all B mesons. While the lifetimes of individual species of B mesons have not been measured, there is now clear evidence that the lifetimes of charmed mesons are not identical. Before we try to understand the problem, let us briefly review the status of heavy quark lifetime measurements.

LIFETIMES

There have been a large number of experiments on the lifetimes of charmed and b quark-containing states. In the case of charm, measurements of lifetimes of individual states abound. In the b case, only the lifetime of mixtures of b-containing hadrons has been measured. Many compilations of these measurements have been made. Tables II, III and IV, taken from the recent review of Lüth,[13] summarize the current situation.

A clear pattern has emerged in the charm sector. The naïve extrapolation from muon decay gives a surprisingly good average estimate of charm lifetimes, but individual charmed particle lifetimes are far from identical. In particular,

$$\tau(D^+) > \tau(D^0) \simeq \tau(D_s^+) > \tau(\Lambda_c^+) \ .$$

An attempt at understanding these lifetimes will require, as we shall see, an investigation of purely hadronic charm decays which takes us beyond the naïve spectator model.

The average lifetime of b-containing hadrons is quite long. There are bounds on $\tau(B_u^+)/\tau(B_d^0)$ obtained at the $\Upsilon(4S)$ resonance, but these are not yet of sufficient precision to shed much light on the detailed mechanism of b quark decay.

While the naïve picture of heavy quark decay does describe, crudely, the lifetimes of heavy mesons, its failure to account for differing charmed particle lifetimes is sufficient motivation for us to examine our overly simplified initial premises.

Table II Measurements of Lifetimes of Charmed Mesons

Experiment	D^+ Decays	$\tau(10^{-13}s)$	D^0 Decays	$\tau(10^{-13}s)$	D_s^+ Decays	$\tau(10^{-13}s)$
E-531[14]	23	$11.1\pm^{4.4}_{2.9}$	58	$4.3\pm^{0.7}_{0.5}\pm^{0.1}_{0.2}$	6	$2.6\pm^{1.6}_{1.1}$
WA-58[15]	27	$5.0\pm^{1.5}_{1.0}\pm1.9$	44	$3.6\pm^{1.2}_{0.8}\pm0.7$		
SHF[16]	48	$8.6\pm1.3\pm^{0.7}_{0.3}$	50	$6.1\pm0.9\pm0.3$		
NA-16[17]	15	$8.4\pm^{3.5}_{2.2}$	16	$4.1\pm^{1.3}_{1.0}$		
NA-18[18]	7	$6.3\pm^{4.9}_{2.3}\pm1.5$	9	$4.1\pm^{2.6}_{1.3}\pm0.5$		
NA-27[19]	40	$10.7\pm^{2.8}_{1.8}$	60	$4.1\pm^{0.7}_{0.6}$		
NA-1[20]	98	$9.5\pm^{3.1}_{1.9}$	51	$4.3\pm^{1.4}_{0.9}$		
NA-11[21]	28	$10.6\pm^{3.6}_{2.4}\pm1.6$	26	$3.7\pm^{1.0}_{0.7}\pm0.5$	9	$3.2\pm^{1.5}_{1.0}$
NA-11[22]	69	$11.2\pm^{1.6}_{1.3}\pm0.8$				
NA-32[23]	42	$9.8\pm^{1.9}_{1.5}$	42	$3.9\pm^{0.6}_{0.5}$	21	$3.3\pm^{1.0}_{0.6}$
E-691[24]	969	$10.6\pm0.5\pm0.3$	1360	$4.35\pm0.15\pm0.10$	99	$4.8\pm^{0.6}_{0.5}\pm0.2$
DELCO[25]				$4.6\pm1.5\pm^{0.7}_{0.6}$		
MKII[26]	16	$8.9\pm^{3.8}_{2.7}\pm1.3$	66	$4.7\pm^{0.9}_{0.8}\pm0.5$		
HRS[27]	114	$8.1\pm1.2\pm1.6$	53	$4.2\pm0.9\pm0.6$	13	$3.5\pm^{2.4}_{1.8}\pm0.9$
TASSO[28]			13	$4.3\pm^{2.0}_{1.4}\pm0.8$	7	$3.4\pm^{2.9}_{1.6}\pm0.7$
CLEO[29]	247	$11.4\pm1.6\pm0.7$	317	$5.0\pm0.7\pm0.4$	87	$4.6\pm2.2\pm0.5$
ARGUS[30]			249	$4.89\pm0.65\pm0.60$	86	$4.4\pm1.4\pm1.0$
Total	1743	$10.29\pm^{0.44}_{0.40}$	2414	$4.42\pm^{0.15}_{0.14}$	319	$4.37\pm^{0.48}_{0.38}$

Table III Measurements of Lifetimes of Charmed Baryons

Experiment	Λ_c^+ Decays	$\tau(10^{-13}s)$	Ξ_c^+ Decays	$\tau(10^{-13}s)$	Ω_c^0 Decays	$\tau(10^{-13}s)$
E-531[14]	13	$2.0\pm^{0.7}_{0.6}$				
WA-58[15]	11	$2.3\pm^{0.9}_{0.6}\pm0.4$				
NA-27[20]	9	$1.2\pm^{0.5}_{0.3}$				
NA-32[31]	14	$1.4\pm^{0.5}_{0.3}\pm0.3$				
E-691[32]	65	$2.0\pm^{0.5}_{0.4}\pm0.3$				
E-400[33]			59	$6.2\pm^{1.8}_{1.6}$		
WA-62[34]			82	$4.8\pm^{2.1}_{1.0}\pm^{2.0}_{1.0}$	3	$7.9\pm2.8\pm2.0$
Total	112	$1.8\pm^{0.3}_{0.2}$	141	$5.7\pm^{1.6}_{1.1}$	3	7.9 ± 3.4

Table IV Measurements of the Average Lifetimes of Beauty Particles

	MARK II[35]	MAC[36]	DELCO[37]	HRS[38]	JADE[39]	TASSO[40]
Luminosity pb^{-1}	200	220	214	200	63	25
Cuts p (GeV/c)	2.0	2.0	1.0	2.0	2.0	
p_t (GeV/c)	1.5	1.5	1.0	1.5	1.8	
No. Leptons	284	561	113	301	113	
Signal Fraction	0.64	0.70	0.79	0.53	0.75	0.30
Resolution (μm)	90	50	150	100	450	140
Impact Parameter (μm)	80 ± 17	129 ± 19	249 ± 49	80 ± 27	282 ± 66	92 ± 17
Lifetime (10^{-12}s)	0.85 ± 0.17	1.29 ± 0.20	$1.17^{+0.27}_{-0.22}$	$1.02^{+0.41}_{-0.37}$	$1.80^{+0.51}_{-0.30}$	1.57 ± 0.32
Systematic Error	± 0.21	± 0.17	$\pm^{0.17}_{0.16}$	(included)	± 0.40	$\pm^{0.37}_{0.34}$

The Heavy Quark Weak Decay Hamiltonian

Summing hadronic and semileptonic decay processes, the bare spectator quark-parton Hamiltonian for weak hadronic decays of heavy mesons has the following form:

For charmed mesons:

$$H_W^{(0)}(\Delta c = -1) = \frac{G_F}{\sqrt{2}}[V_{cs}(\bar{s}c)_L + V_{cd}(\bar{d}c)_L]$$
$$\times [V_{ud}^*(\bar{d}u)_L + V_{us}^*(\bar{s}u)_L$$
$$+ \bar{e}\nu_e + \bar{\mu}\nu_\mu + \bar{\tau}\nu_\tau].$$

For bottom mesons:

$$H_W^{(0)}(\Delta b = -1) = \frac{G_F}{\sqrt{2}}[V_{cb}(\bar{c}b)_L + V_{ub}(\bar{u}b)_L]$$
$$\times [V_{ud}^*(\bar{d}u)_L + V_{us}^*(\bar{s}u)_L$$
$$+ V_{cs}^*(\bar{s}c)_L + V_{cd}^*(\bar{d}c)_L$$
$$+ \bar{e}\nu_e + \bar{\mu}\nu_\mu + \bar{\tau}\nu_\tau].$$

For top mesons:

$$H_W^{(0)}(\Delta t = -1) = \frac{G_F}{\sqrt{2}}[V_{tb}(\bar{b}t)_L + V_{ts}(\bar{s}t)_L + V_{ts}(\bar{s}t)_L + V_{td}(\bar{d}t)_L]$$
$$\times [V_{ud}^*(\bar{d}u)_L + V_{us}^*(\bar{s}u)_L + V_{cs}^*(\bar{s}c)_L$$
$$+ V_{cd}^*(\bar{d}c)_L + V_{cb}^*(\bar{b}c)_L + V_{ub}^*(\bar{b}u)_L$$
$$+ \bar{e}\nu_e + \bar{\mu}\nu_\mu + \bar{\tau}\nu_\tau],$$

where, e.g., $(\bar{s}c)_L = \sum_{i=1}^{3} \bar{s}_i\gamma^\mu(1-\gamma_5)c_i$. For both c and b decays, the point approximation is valid, as $q^2 \ll M_W^2$; for t decay this may not obtain.

These bare Hamiltonia generate a hierarchy of decays characterized by the size of the relevant K-M elements. For example, in hadronic charmed particle decays, terms involving $V_{cs}V_{ud}^*$ are dominant (Cabibbo-allowed). Strong interaction effects modify this simple picture in several significant ways:

- weak couplings are renormalized
- the color structure of the interaction may be distorted by octet currents
- new chiral structures, e.g., $V + A$ currents may appear
- there may be induced S and T interactions

In practice, the last two effects are small, but the first two are quite significant. QCD provides a framework in which these strong interaction effects can be estimated.[41] The lowest order corrections, i.e., $O(\alpha_s)$ hard gluon contributions, fall into two classes. Those which involve self energy or vertex corrections to the four quark operators are absorbed in the renormalization of G_F, while those involving gluon exchange along with the $W^\pm$ are logarithmically divergent. Cutting the latter off at M_W yields a first order correction to $H_W^{(0)}$:

$$H_W^{(1)} = -\frac{G_F V_{cs} V_{ud}^*}{\sqrt{2}} \frac{3\alpha_s}{8\pi} \log\left(\frac{M_W^2}{\mu^2}\right) (\bar{s}\lambda_a c)_L (\bar{d}\lambda^a u)_L$$

for Cabibbo-allowed hadronic charm decays, with analogous corrections for the other terms. Thus the hard gluon corrections induce a new operator, which has the same chiral and flavor structure, but involves color octet currents, λ_a. Semileptonic decays clearly receive no such corrections.

The physical effects of hard gluon exchanges are most easily seen if we make a Fierz transformation:

$$(\bar{s}\lambda_a c)_L (\bar{d}\lambda_a u)_L = -\frac{2}{3}(\bar{s}c)_L(\bar{d}u)_L + 2(\bar{s}d)_L(\bar{u}c)_L\,.$$

If we then sum the contributions to the Hamiltonian:

$$\begin{aligned}
H_W &= H_W^{(0)} + H_W^{(1)} + \cdots \\
&\simeq \frac{G_F}{\sqrt{2}} V_{cs}V_{ud}^* \left[\left(1 + \frac{\alpha_s}{4\pi}\ln\frac{M_W^2}{\mu^2}\right)(\bar{s}c)_L(\bar{d}u)_L \right. \\
&\quad \left. - \frac{3\alpha_s}{4\pi}\ln\frac{M_W^2}{\mu^2}(\bar{s}d)_L(\bar{u}c)_L\right],
\end{aligned}$$

we see that the hard gluon exchange has two effects:

- the strength of the charged current interaction is renormalized
- there is an induced effective neutral current interaction.

Similar effects are expected, of course, for the Cabibbo-suppressed charm terms as well as for the $\Delta b = -1$ and $\Delta t = -1$ Hamiltonians.

It is useful to define the symmetric and antisymmetric linear combinations

$$O_\pm(\Delta c = -1) = \frac{1}{2}[(\bar{s}c)_L(\bar{u}d)_L \pm (\bar{s}d)_L(\bar{u}c)_L]\,.$$

The bare Hamiltonian for Cabibbo-allowed hadronic charm decays then has the form:

$$H_W^{(0)} = \frac{G_F}{\sqrt{2}} V_{cs} V_{ud}^* [O_+ + O_-],$$

while the first order corrected version is:

$$H_W = \frac{G_F}{\sqrt{2}} V_{cs} V_{ud}^* [c_+^{(1)} O_+ + c_-^{(1)} O_-],$$

where

$$c_+^{(1)} = 1 - \frac{\alpha_s}{2\pi} \ln \frac{M_W^2}{\mu^2}$$

$$c_-^{(1)} = 1 + \frac{\alpha_s}{2\pi} \ln \frac{M_W^2}{\mu^2}.$$

Clearly, the bare light quark spectator model is recovered in the limit $c_+^{(1)} = c_-^{(1)} = 1$.

In a similar way, the dominant term in the hadronic decay of B mesons can be written as

$$H_W = \frac{G_F}{\sqrt{2}} V_{cb} V_{ud}^* [c_+^{(1)} O_+ + c_-^{(1)} O^-],$$

where

$$O_\pm (\Delta b = -1) = [(\bar{c}b)_L (\bar{d}u)_L \pm (\bar{d}b)_L (\bar{c}u)_L],$$

where again the sum over quark colors is understood.

The effective four quark operators $O_\pm$ may usefully be classified by their color SU(3) transformation properties. The current-current structure is $\underline{3} \times \underline{3} = \underline{3}^* + \underline{6}$. The $\underline{3}^*$ representation of SU(3) is antisymmetric, while the $\underline{6}$ is symmetric. Hard gluon corrections thus enhance the $\underline{3}^*$ and suppress the $\underline{6}$. In the case of the $\Delta s = -1$ Hamiltonian which governs K decay, a similar decomposition obtains. Since O_- transitions are $\Delta I = \frac{1}{2}$, while O_+ transitions can be either $\Delta I = \frac{1}{2}$ or $\frac{3}{2}$, the origin of the $\Delta I = \frac{1}{2}$ rule is likely to be found in QCD corrections. Thus, the $\Delta I = \frac{1}{2}$ rule is often referred to as "$\underline{6}$ dominance". These QCD modifications to the structure of the current-current Hamiltonian governing inclusive weak hadronic decays can be calculated to all orders in α_s using renormalization group methods.[42] In terms of the anomalous dimensions d_+ and d_-:[43]

$$c_\pm = \left[\frac{\alpha_s(\mu^2)}{\alpha_s(M_W^2)} \right]^{\frac{d_\pm}{2b}} = \left[1 + \alpha_s(\mu^2) \frac{b}{4\pi} \ln \left(\frac{M_W^2}{\mu^2} \right) \right]^{\frac{d_\pm}{2b}}.$$

Here, $b = 11 - \frac{2N_f}{3}$, where N_f is the number of fermionic degrees of freedom.

In the leading log (LL) approximation, the value of $c_\pm$ depends on the effective coupling constant g_s, defined by:

$$c_\pm\left(g_s, \frac{M_W}{\mu}\right) = c_\pm\left(0, \frac{M_W}{\mu}\right)\left[\frac{\ln(M_W^2/\Lambda_{QCD}^2)}{\ln(\mu^2/\Lambda_{QCD}^2)}\right]^{\frac{d_\pm}{2b}}.$$

Note that since $d_- = -2d_+$, we have the relation $c_- = 1/c_+^2$.

The calculation has, in fact, been carried out including two loop corrections to the β function.[44] These "NLL" corrections are small, lending confidence to the prescription:

$$c_\pm\left(g_s, \frac{M_W}{\mu}\right) = \left[\frac{\alpha_s(\mu^2)}{\alpha_s(M_W^2)}\right]^{\frac{d_\pm}{b}}\left[1 + \frac{\alpha_s(\mu^2) - \alpha_s(M_W^2)}{\pi}\rho_\pm\right],$$

where

$$\rho_+ = \frac{1}{b}\left[\frac{-221}{24} + \frac{5}{9}N_f\right] + \frac{1}{b^2}\left[51 - \frac{19}{3}N_f\right]$$

$$\rho_- = \frac{1}{b}\left[\frac{263}{12} - \frac{10}{9}N_f\right] + \frac{1}{b^2}\left[-102 + \frac{38}{3}N_f\right].$$

The c_+ and c_- coefficients for $N_f = 6$ are shown in Fig. 10 as a function of the scale μ. Clearly, hard gluon exchanges modify the charm Hamiltonian to a greater degree than they do the bottom and top sectors.

We can most clearly visualize the affect of QCD corrections on heavy quark weak processes by rewriting the Hamiltonian once more. Defining

$$c_1 = \frac{c_+ + c_-}{2} \quad , \quad c_2 = \frac{c_+ - c_-}{2},$$

we have, for the $\Delta c = -1$ Hamiltonian:

$$H_W(\Delta c = -1) = \frac{G_F}{\sqrt{2}}V_{cs}V_{ud}^*[c_1(\bar{s}c)_L(\bar{u}d)_L + c_2(\bar{s}d)_L(\bar{c}u)_L],$$

and for the $\Delta b = -1$ Hamiltonian:

$$H_W(\Delta b = -1) = \frac{G_F}{\sqrt{2}}V_{cb}V_{ud}^*[c_1(\bar{c}b)_L(\bar{u}d)_L + c_2(\bar{d}b)_L(\bar{c}u)_L],$$

where c_1, c_2 are evaluated at the appropriate scale μ. In the absence of QCD corrections, $c_1 = 1$, $c_2 = 0$. Thus, hard gluon exchange is responsible for:

• Shortening heavy quark lifetimes, as the non-leptonic decay width is enhanced:

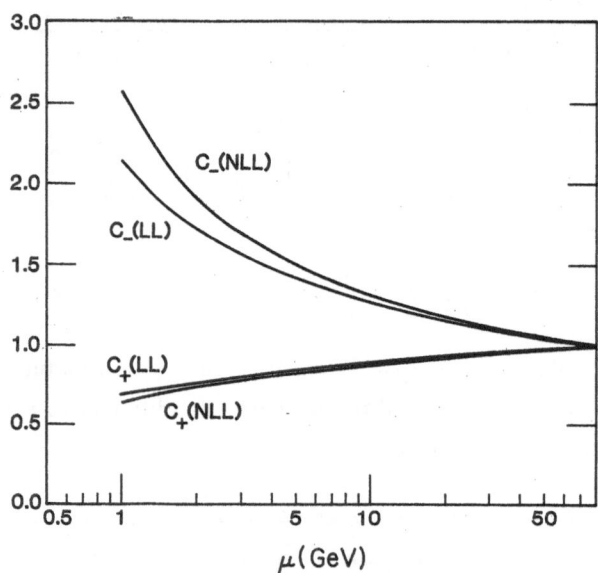

Fig. 10 Scale dependence of the operator coefficients c_- and c_+ as calculated in the leading log (LL) approximation, and to the next-to-leading-log (NLL) order.

$$\Gamma_{NL} = (2c_+^2 + c_-^2)\Gamma_0 \,.$$

- Inducing an "effective neutral current" interaction, proportional to c_2^2. This term also often referred to as a "color-suppressed" interaction.

- Reducing semileptonic branching ratios, the so-called "non-leptonic enhancement", since, e.g.,

$$\Gamma(c \to su\bar{d}) = (2c_+^2 + c_-^2)\Gamma(c \to sl\nu_l) \,,$$

so that

$$B(c \to eX) = \frac{1}{2 + 2c_+^2 + c_-^2} \simeq 15\% \,,$$

rather than the naïve value of 20%. The reduction in phase space due to non-zero quark masses, also, of course, slightly reduces the naïve estimate. The effect on the B semileptonic branching ratio is correspondingly smaller, as the renormalization scale is 5 GeV, rather than ~ 1.5 GeV, as for charm.

- The $\Delta I = \frac{1}{2}$ rule in non-leptonic K decay and its generalizations to the heavy quark sector.

These modifications are all born out by experiment, providing striking support for Quantum Chromodynamics. In one important area, however, we must go farther. The framework we have developed thus far ensures, up to small phase space corrections, that the lifetimes and semileptonic branching ratios of all charmed mesons are identical, as are those of all B mesons. As D^0, D^+ and D_s^+ lifetimes are manifestly not identical, we must extend our investigation to a more detailed study of non-leptonic decays. As experimental data improve, measurements of the separate lifetimes of the individual B mesons will undoubtedly be made. With appropriate scaling, the results discussed below will apply equally to the b quark sector.

Comparison of the lifetimes and semileptonic branching ratios of the D^0 and D^+ mesons provides the first as to the origin of the differing lifetimes. Neglecting small Cabibbo-suppressed semileptonic decays of the D^+, isospin invariance tells us that $\Gamma(D^0 \to e^+X) = \Gamma(D^+ \to e^+X)$, as both are purely spectator decays: $c \to se\nu_e$. We can then relate the ratios of lifetimes and semileptonic branching fractions:

$$\frac{B(D^+ \to e^+X)}{B(D^0 \to e^+X)} = \frac{\Gamma(D^+ \to e^+X)}{\Gamma(D^+ \to \text{all})} \cdot \frac{\Gamma(D^0 \to \text{all})}{\Gamma(D^0 \to e^+X)}$$

$$\simeq \frac{\Gamma(D^0 \to \text{all})}{\Gamma(D^+ \to \text{all})} = \frac{\tau(D^+)}{\tau(D^0)} \,.$$

Mark III has measured[45] the separate D^0 and D^+ semileptonic branching fractions:

$$B(D^+ \to e^+ X) = (17.0 \pm 1.9 \pm 0.7)\%,$$

$$B(D^0 \to e^+ X) = (7.5 \pm 1.1 \pm 0.4)\%.$$

The ratio is

$$\frac{B(D^+ \to e^+ X)}{B(D^0 \to e^+ X)} = 2.3^{+0.5+0.1}_{-0.4-0.1},$$

surprisingly close to the world average[13] directly measured lifetime ratio:

$$\frac{\tau(D^+)}{\tau(D^0)} = 2.27 \pm 0.13.$$

The lifetime and semileptonic branching ratios can be combined to extract the partial widths for semileptonic decay:

$$\Gamma_S L(D^0) = 1.70 \pm 0.28 \times 10^{11} \text{ sec}^{-1}$$

$$\Gamma_S L(D^+) = 1.65 \pm 0.21 \times 10^{11} \text{ sec}^{-1}.$$

Thus, we see that the source of the differing charmed meson lifetimes must be sought in the hadronic decay sector.

As $B(D^+ \to e^+ X)$ is slightly larger than the QCD corrected estimate, we are still, however, faced with a puzzle. Are the hadronic decays of the D^0 enhanced by a factor of two, or are the D^+ decays inhibited? Scaling from muon decay is not helpful due to the fifth power dependence of the lifetime on the (unknown) charmed quark mass. There is a direct experimental approach, however. It so happens, perhaps *a fortiori*, that the hadronic decays of the D meson are dominated by two body and quasi-two body decay modes. It has been possible to reconstruct a large number, in fact a large majority of such exclusive decays, thus allowing us to approach the matter of the charmed meson lifetime, a manifestly *inclusive* matter, by summing over a large number of measured *exclusive* channels.

CHARMED QUARK DECAYS

D Hadronic Decays

A full understanding of the mechanism of weak decays of charmed mesons requires a detailed study of exclusive decay modes. We now have a great deal of information on the exclusive decays of the D^0 and D^+, and are beginning to learn some important facts about D_s^+ decays. For clarity, we will discuss Cabibbo-allowed and -suppressed decays separately.

The classic technique for reconstruction of exclusive states is of course, the invariant mass technique. Charmed particles are produced in hadronic reactions, photoproduction, neutrino reactions and high energy e^+e^- annihilation.

Charm production cross sections range from $\sim 10^{-3}$ of the total cross section in hadronic reactions, to $\sim 10^{-2}$ in photoproduction, to $\sim 10^{-1}$ in neutrino reactions to as much as 30% in e^+e^- annihilation. Much progress has been made in recent years in sophisticated triggering schemes to enrich event samples, and recently, in the use of high precision vertex detectors, which greatly enhance signal-to-background ratios by allowing reconstruction of detached secondary vertices.

In order to reduce combinatoric background, several methods are employed. The GIM mechanism ensures that Cabibbo-allowed charm hadronic decays produce either K_s^0 or charged kaons of a specific sign (i.e., $D^0 \rightarrow K^-$, etc.). Semileptonic decays similarly produce leptons of a specific sign ($D^0 \rightarrow e^+$). Such correlations are important both in isolating a signal and demonstrating its validity. Given sufficient spatial resolution, the relatively long D lifetime allows isolation of a separated vertex, substantially enhancing signal-to-noise ratios (see Fig. 11). Since at high energy, $D^*(1^-)$ production is three times larger than $D(0^-)$ production, a requirement that the mass of a reconstructed D and a slow pion form a D^* mass can be a powerful method of suppressing background.

Most information on D^0 and D^+ exclusive decay modes, however, comes from low energy e^+e^- annihilation at SPEAR. Due to the existence of the $\psi(3770)$ resonance, which decays predominantly to $D\bar{D}$ pairs, the experimental situation is greatly simplified, making it possible to study exclusive decays in great detail. The $\psi(3770)$ resonance (Fig. 12) is a very prominent feature in the e^+e^- total cross section. The $\psi(3770)$ decays almost entirely to $D\bar{D}$, making it possible to measure absolute branching ratios for exclusive D decays by simultaneously measuring the $\psi(3770)$ cross section. The only direct measurement which bears on this point[47] shows that $B(\psi(3770) \rightarrow D\bar{D}) = 1.09 \pm 0.23$. As will be discussed further below, it is in fact possible to measure absolute branching ratios at the $\psi(3770)$ without knowing the cross section. Two other $\psi(3770)$ decay modes are expected from conventional mechanisms: a) the OZI violating decay $\psi(3770) \rightarrow J/\psi\pi\pi$ and b) decays to the χ states, the $^3P_{0,1,2}$ charmonium levels. Preliminary evidence for both of these decays, at a small level, has been found.[48] Lipkin has advanced the idea that nearby resonances can enhance OZI violation in $\psi(3770)$ decay,[49] although no direct evidence for this exists.

The final states produced in $\psi(3770) \rightarrow D\bar{D}$ decay are particularly simple. Decays to other charmed states or decays involving even so much as an additional slow pion, are kinematically forbidden. Further, the D's have a unique momentum: $P_{D^0} = 274$ MeV/c and $P_{D^+} = 237$ MeV/c. This will prove particularly useful in the reconstruction of

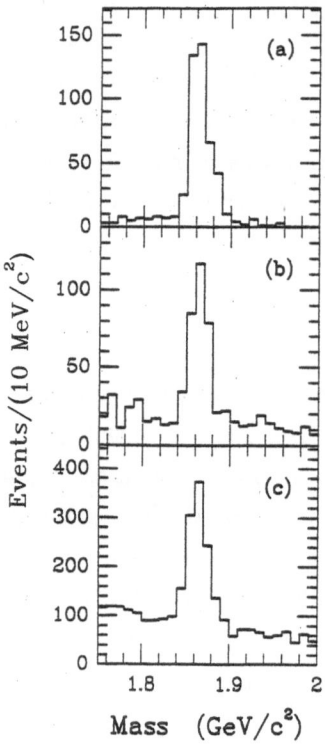

Fig. 11 Three D^0 mass spectra from experiment E691. a) $D^0 \to K^-\pi^+$, with the requirement that there be an additional π^+ in the event which fits the reaction $D^{*+} \to \pi^+ D^0$; b) $D^0 \to K^-\pi^+\pi^+\pi^-$, also with the D^* requirement, and c) $D^0 \to K^-\pi^+$ without the D^* requirement.

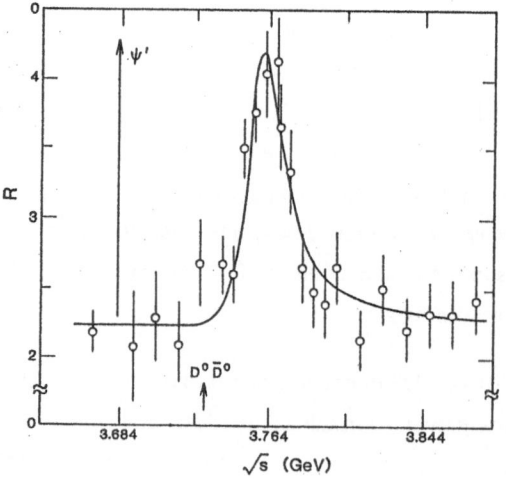

Fig. 12 The Mark II measurement[46] of the resonance parameters of the $\psi(3770)$ resonance in e^+e^- annihilation (radiatively corrected).

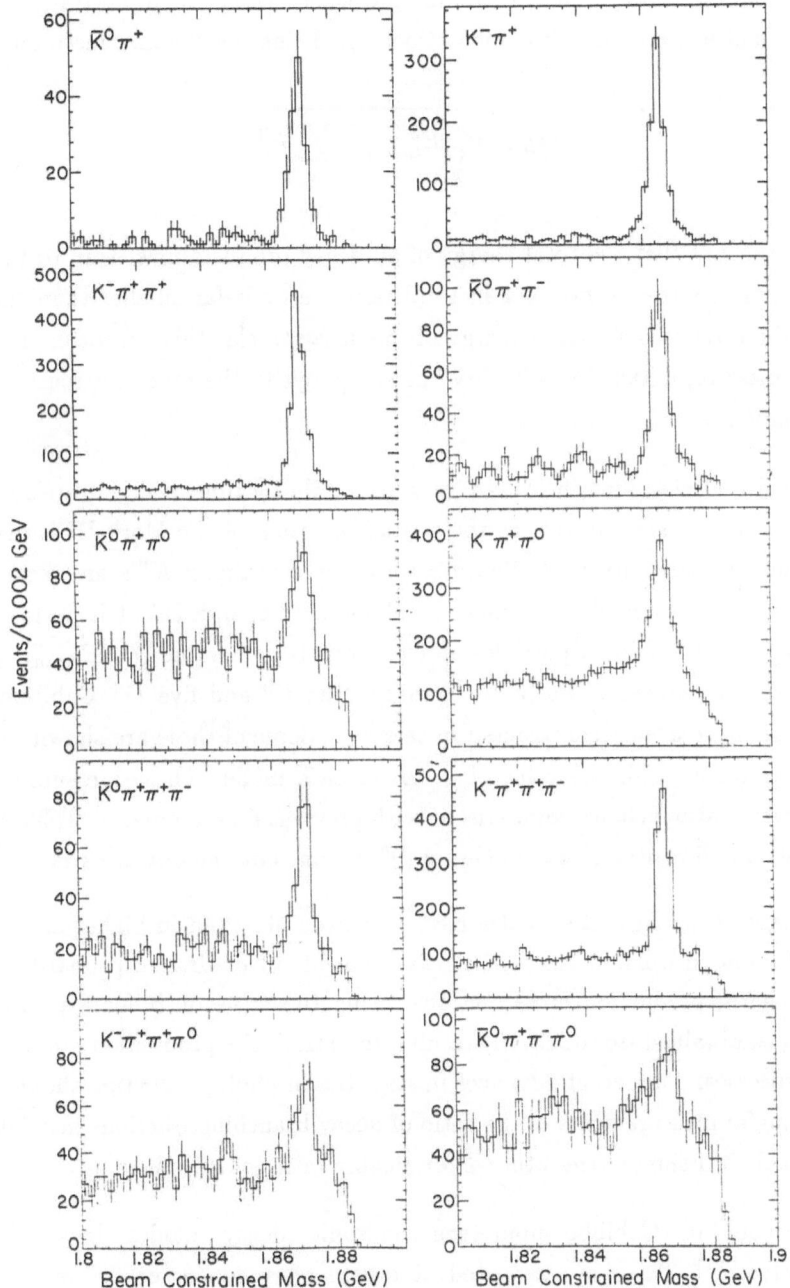

Fig. 13 Beam constrained mass plots from Mark III for five D^+ and five D^0 Cabibbo-allowed decays.

Cabibbo-suppressed hadronic decays and of semileptonic decays.

Another unique feature of D production at the $\psi(3770)$ allows reconstruction of masses for hadronic decay modes with high precision. Rather than plotting the conventional invariant mass of particle combinations, we define the "beam-constrained mass":

$$M_{BC} = \sqrt{E_{\text{beam}}^2 - \sum_i P_i^2} \, ,$$

exploiting the fact that the total energy of all decay products must sum to the energy of each beam. As the uncertainty in the beam energy is far smaller than the uncertainty in the total reconstructed energy of the decay tracks, this approach yields much improved mass resolution ($\sigma \sim 3$ MeV) as compared to the invariant mass technique ($\sigma \sim 12$ MeV).

Charged particles, identified as K or π primarily by time-of-flight, are required to originate at the primary vertex, as the spatial precision of the Mark III is insufficient to allow direct sensitivity to D lifetimes. Decays containing K^0's are reconstructed through the $\pi^+\pi^-$ channel. For modes involving a π^0 or η, a 2C fit is performed: the total energy of the charged particles and photons is constrained to E_{beam}, while the photons are constrained to the π^0 or η mass. Ten D^0 and five D^+ Cabibbo-allowed decay modes have been reconstructed in this way. Some of these are shown in Fig. 13. The measurement yields the value of $\sigma \cdot B$ for each mode. These are summarized in Table V, which also includes comparisons with previous measurements. Table VI shows some limits on all neutral decay modes of D^0 mesons obtained by the Crystal Ball.[60]

Hadronic D meson decay modes have also been observed in higher energy experiments. Except in isolated cases, the measurements of $\sigma \cdot B$, are primarily useful in studying charm production cross sections, using the value of B for a particular decay mode, as obtained from $\psi(3770)$ results, to extract the production cross section in various processes. The recent advances in statistics in photoproduction should soon allow meaningful measurements of the ratio of decay branching fractions into interesting modes. Table VII summarizes other D branching ratio measurements.

A number of Cabibbo-suppressed hadronic decay modes have also been reconstructed.[61,69] These are expected at a rate of $\sim tan^2\theta_c$ or 5% of comparable allowed modes. Since the signals are smaller, the problem of $K - \pi$ misidentification assumes greater significance. The unique momentum of D's at the $\psi(3770)$ can be exploited to display explicitly background due to incorrect particle assignment. For most Cabibbo-suppressed modes, this involves plotting the invariant mass for events in which the reconstructed D momentum is within 50 MeV of the expected value. Background is measured using sidebands of reconstructed "D" momentum. Figures 14 and

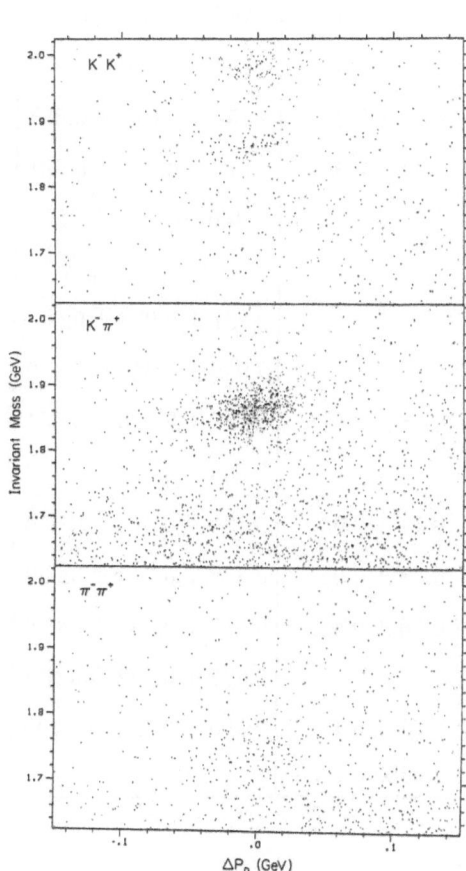

Fig. 14 Scatter plots of invariant mass *vs.* difference from expected D^0 momentum for the Cabibbo-allowed decay $D^0 \to K^-\pi^+$ and two Cabibbo-suppressed channels, $D^0 \to K^-K^+$ and $\pi^-\pi^+$ (Mark III).

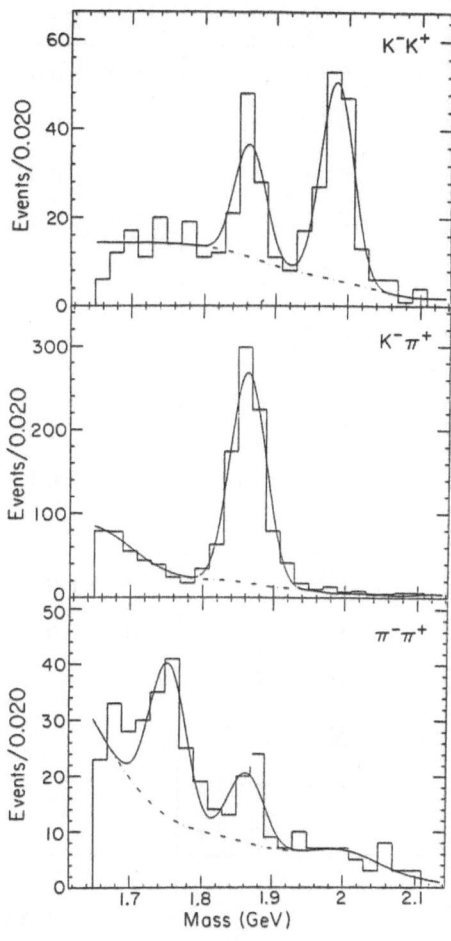

Fig. 15 Invariant mass spectra for the Cabibbo-allowed decay $D^0 \to K^-\pi^+$ and the two Cabibbo-suppressed decays $D^0 \to K^-K^+$ and $D^0 \to \pi^-\pi^+$, for events within ± 50 MeV of the expected D^0 momentum (Mark III). Note that peaks due to $\pi - K$ misidentification are clearly separated from the actual signal.

Table V

Comparison of $\sigma \cdot B$ Measurements for Cabibbo-Allowed

D Meson Decays at the $\psi(3770)$

Decay Mode	LGW[50] $\sqrt{s} = 3.774$ GeV	Mark II[51] $\sqrt{s} = 3.771$ GeV	Mark III[52,69] $\sqrt{s} = 3.768$ GeV
$D^0 \to K^-\pi^+$	0.25 ± 0.05	0.24 ± 0.02	$0.248 \pm 0.009 \pm 0.014$
$\bar{K}^0\pi^0$		0.18 ± 0.08	$0.108 \pm 0.020 \pm 0.010$
$K^-\pi^+\pi^0$	1.4 ± 0.6	0.68 ± 0.23	$0.759 \pm 0.044 \pm 0.083$
$\bar{K}^0\pi^+\pi^-$	0.46 ± 0.12	0.30 ± 0.08	$0.372 \pm 0.030 \pm 0.031$
$\bar{K}^0 K^+ K^-$			$0.05 \pm^{0.02}_{0.01} \pm 0.01$
$\bar{K}^0\phi$			$0.05 \pm^{0.03}_{0.02} \pm^{0.02}_{0.01}$
$\bar{K}^0\omega$			$0.187 \pm 0.073 \pm 0.047$
$\bar{K}^0\eta$			$0.088 \pm 0.039 \pm 0.012$
$K^-\pi^+\pi^+\pi^-$	0.36 ± 0.11	0.68 ± 0.11	$0.525 \pm 0.026 \pm 0.054$
$\bar{K}^0\pi^+\pi^-\pi^0$			$0.666 \pm 0.113 \pm 0.153$
$D^+ \to \bar{K}^0\pi^+$	0.14 ± 0.05	0.14 ± 0.03	$0.135 \pm 0.012 \pm 0.010$
$K^-\pi^+\pi^+$	0.36 ± 0.06	0.38 ± 0.05	$0.388 \pm 0.013 \pm 0.029$
$\bar{K}^0\pi^+\pi^0$		0.78 ± 0.48	$0.417 \pm 0.081 \pm 0.075$
$K^-\pi^+\pi^+\pi^0$			$0.177 \pm 0.042 \pm 0.042$
$\bar{K}^0\pi^+\pi^+\pi^-$		0.51 ± 0.18	$0.305 \pm 0.031 \pm 0.030$
$K^-\pi^+\pi^+\pi^+\pi^-$		$< 0.23@90\%$ CL	

Table VI

Crystal Ball

Limits on All-Neutral

Decay Modes of D^0 Mesons

Mode	$\sigma_{D^0} \cdot B$ (nb) @ 90% CL
$\pi^0\pi^0$	< 0.019
$\eta^0\pi^0$	< 0.05
$\eta^0\eta^0$	< 0.07
$\pi^0\pi^0\pi^0$	< 0.11
$\eta^0\pi^0\pi^0$	< 0.45

Table VII

Additional D Meson Hadronic Branching Ratios

Mode	Experiment	Measurement	Branching Ratio (%)
$D^0 \to K^-\pi^+\pi^-\pi^-$	Aguilar et al.[53]	topological	7.1 ± 2.5
	Bailey et al.[54]	$\dfrac{B(D^0 \to K^-\pi^+\pi^+\pi^-)}{B(D^0 \to K^-\pi^+)} = 2.0 \pm 1.0$	8.4 ± 4.3
	ARGUS[55]	$\dfrac{B(D^0 \to K^-\pi^+\pi^+\pi^-)}{B(D^0 \to K^-\pi^+)} = 2.17 \pm 0.36$	9.1 ± 1.7
$D^0 \to \bar{K}^0\pi^+\pi^-$	Avery et al.[56]	$\dfrac{B(D^0 \to \bar{K}^0\pi^+\pi^-)}{B(D^0 \to K^-\pi^+)} = 1.7 \pm 0.8$	7.1 ± 3.4
$D^+ \to K^-\pi^+\pi^+$	Aguilar et al.[57]	topological	$14 \pm 6 \pm 5$
$D^+ \to K^-\pi^+\pi^+\pi^0$	Aguilar et al.[58]	$\dfrac{B(D^+ \to K^-\pi^+\pi^+\pi^0)}{B(D^+ \to K^-\pi^+\pi^+)} = 0.57 \pm 0.65 \pm .17$	5.2 ± 6.0
$D^+ \to \bar{K}^{*0}K^+$	E691 [59]	$\dfrac{B(D^+ \to \bar{K}^{*0}K^+)}{B(D^+ \to K^-\pi^+\pi^+)} = 0.075 \pm 0.016 \pm 0.007$	0.68 ± 0.19
$D^+ \to K^-K^+\pi^+_{nonres}$	E691[59]	$\dfrac{B(D^+ \to K^-K^+\pi^+)}{B(D^+ \to K^-\pi^+\pi^+)} = 0.055 \pm 0.014 \pm 0.006$	0.50 ± 0.16
$D^+ \to \phi\pi^+$	E691[59]	$\dfrac{B(D^+ \to \phi\pi^+)}{B(D^+ \to K^-\pi^+\pi^+)} = 0.094 \pm 0.016 \pm 0.006$	0.86 ± 0.20

Table VIII

Mark III Results on Cabibbo-Suppressed Decays of D Mesons

Decay Channel	Ratio	Decay Channel	Ratio
D^0 Decays		D^+ Decays	
$\dfrac{B(\pi^-\pi^+)}{B(K^-\pi^+)}$	$0.033 \pm 0.010 \pm 0.006$	$\dfrac{B(\bar{K}^0 K^+)}{B(\bar{K}^0\pi^+)}$	$0.317 \pm 0.086 \pm 0.048$
$\dfrac{B(K^-K^+)}{B(K^-\pi^+)}$	$0.122 \pm 0.018 \pm 0.012$	$\dfrac{B(\pi^+\pi^0)}{B(\bar{K}^0\pi^+)}$	$< 0.15@90\%$ CL
$\dfrac{B(\bar{K}^0 K^0)}{B(K^-\pi^+)}$	$< 0.11@90\%$ CL	$\dfrac{B(\pi^-\pi^+\pi^+)}{B(K^-\pi^+\pi^+)}$	$0.042 \pm 0.016 \pm 0.010$
$\dfrac{B(\bar{K}^{*0} K^0 + cc)}{B(K^{*-}\pi^+) + B(K^-\rho^+)}$	$< 0.034@90\%$ CL	$\dfrac{B(K^- K^+\pi^+)}{B(K^-\pi^+\pi^+)}$	$0.059 \pm 0.026 \pm 0.009$
$\dfrac{B(\bar{K}^{*-} K^+ + cc)}{B(K^{*-}\pi^+) + B(K^-\rho^+)}$	0.05 ± 0.03	$\dfrac{B(\phi\pi^+)}{B(K^-\pi^+\pi^+)}$	$0.084 \pm 0.021 \pm 0.011$
$\dfrac{B(\pi^-\pi^+\pi^0)}{B(D^0 \to \text{all})}$	$0.011 \pm 0.004 \pm 0.002$	$\dfrac{B(\bar{K}^{*0} K^+)}{B(K^-\pi^+\pi^+)}$	$0.048 \pm 0.021 \pm 0.011$
$\dfrac{B(\pi^-\pi^+\pi^+\pi^-)}{B(D^0 \to \text{all})}$	$0.015 \pm 0.006 \pm 0.002$		

15 show an example of this technique. The invariant mass projections shown in Fig. 15 for events within 50 MeV of the correct D momentum, display clear D signals, as well as cleanly separated misidentification peaks. Branching ratios are expressed as ratios to closely related Cabibbo-allowed modes, to minimize systematic errors and facilitate comparison with models. These are summarized in Table VIII. The decays $D^+ \to \pi^+\pi^0$, $D^0 \to \pi^+\pi^-\pi^0$ and $D^0 \to \pi^+\pi^-\pi^+\pi^-$ are subject to large combinatorial backgrounds when analyzed by this technique. These modes, also shown in Table VIII, are, therefore, reconstructed by the beam-constrained method in the recoil spectrum against a sample of reconstructed Cabibbo-allowed decays: branching ratios so obtained are thus absolutely normalized.

Two methods are available to extract branching fractions at the $\psi(3770)$ from the primary $\sigma \cdot B$ measurements. The most common method has been to measure the cross section at the peak of the $\psi(3770)$ and normalize the results on the assumption that the $\psi(3770)$ decays solely to $D\bar{D}$ pairs. The various measurements of the $\psi(3770)$ resonance are shown in Fig. 16 and Table IX. Unfortunately, these do not agree very well, although there is a tendency for later measurements to yield smaller values of the resonance height.

This has led the Mark III group to develop a new method which provides an absolute normalization of D meson branching ratios, independent of the production cross section. The technique consists of comparing the number of events in which a single D hadronic decay is reconstructed ("single tags") with the number in which a pair of D's is reconstructed ("double tags"). If the appropriate reconstruction efficiencies for single and double tag combinations are known, the individual branching ratios can be extracted without reference to the production cross section.[64] The original publication[65] of absolutely normalized branching ratios using this procedure suffered from a contamination due to two backgrounds, which have now been removed.[66] As these branching ratios are used to extract B inclusive and exclusive branching ration, as well as charm production cross sections, care should be exercised to ensure that the corrected normalization is henceforth employed. The previous normalization has been corrected downward by slightly more than twenty percent, now yielding absolute branching ratios which are consistent with branching ratios extracted using the conventional technique, when the most recent $\psi(3770)$ cross section measurements are employed.

Table X shows the results of the revised fit, which allows extraction of individual absolutely normalized D branching ratios, and since the number of produced D's is an output of the fit, an indirect measurement of the $\psi(3770)$ cross section, on the assumption that all $\psi(3770)$ decays produce $D\bar{D}$. Note that the fit yields $\sigma_{D^0}/\sigma_{D^+} = 1.36 \pm 0.23 \pm 0.14$, consistent with that expected from the small D^0/D^+ mass difference.

Table XI summarizes additional Mark III hadronic branching ratios with the revised

Table IX

Direct Measurements of $\psi(3770)$ Cross Section

	Lead Glass Wall[62]	Mark II[46]	Crystal Ball[63]
$\sqrt{s}$ (GeV)	3.774	3.771	3.771
σ (nb)	10.3 ± 1.6	$7.0 \pm 0.8 \pm 0.7$	6.4 ± 0.9

Table X

Summary of σ_D (nb), $\sigma \cdot B$ (nb) and $B(\%)$ for the Revised Mark III

Determination of Absolute D Branching Ratios, Compared with Earlier Results

Channel	Lead-Glass Wall	Mark II	Crystal Ball	This Experiment
σ_{D^0}	11.5 ± 2.5	$8.0 \pm 1.0 \pm 1.2$	6.8 ± 1.2	$5.8 \pm 0.5 \pm 0.5$
$\sigma \cdot B(K^-\pi^+)$	0.25 ± 0.05	0.24 ± 0.02		$0.248 \pm 0.009 \pm 0.014$
$B(K^-\pi^+)$	2.2 ± 0.6	3.0 ± 0.6		$4.2 \pm 0.4 \pm 0.3$
$\sigma \cdot B(K^-\pi^+\pi^+\pi^-)$	0.36 ± 0.10	0.68 ± 0.11		$0.525 \pm 0.026 \pm 0.054$
$B(K^-\pi^+\pi^+\pi^-)$	3.2 ± 1.1	8.5 ± 2.1		$9.1 \pm 0.8 \pm 0.8$
$\sigma \cdot B(K^-\pi^+\pi^0)$	1.4 ± 0.6	0.68 ± 0.23		$0.759 \pm 0.044 \pm 0.083$
$B(K^-\pi^+\pi^0)$	12 ± 6	8.5 ± 3.2		$13.3 \pm 1.2 \pm 1.3$
σ_{D^+}	9.1 ± 2.0	$6.0 \pm 0.7 \pm 1.0$	6.0 ± 1.1	$4.2 \pm 0.6 \pm 0.3$
$\sigma \cdot B(K^-\pi^+\pi^+)$	0.36 ± 0.06	0.38 ± 0.05		$0.388 \pm 0.013 \pm 0.029$
$B(K^-\pi^+\pi^+)$	3.9 ± 1.0	6.3 ± 1.5		$9.1 \pm 1.3 \pm 0.5$
$\sigma \cdot B(\bar{K}^0\pi^+)$	0.14 ± 0.05	0.14 ± 0.03		$0.135 \pm 0.012 \pm 0.010$
$B(\bar{K}^0\pi^+)$	1.5 ± 0.6	2.3 ± 0.7		$3.2 \pm 0.5 \pm 0.2$
$\sigma \cdot B(\bar{K}^0\pi^+\pi^0)$		0.78 ± 0.48		$0.417 \pm 0.081 \pm 0.075$
$B(\bar{K}^0\pi^+\pi^0)$		12.9 ± 8.4		$10.2 \pm 2.5 \pm 1.5$
$\sigma \cdot B(\bar{K}^0\pi^+\pi^+\pi^-)$		0.51 ± 0.8		$0.31 \pm 0.03 \pm 0.03$
$B(\bar{K}^0\pi^+\pi^+\pi^-)$		8.4 ± 3.5		$6.6 \pm 1.4 \pm 0.5$

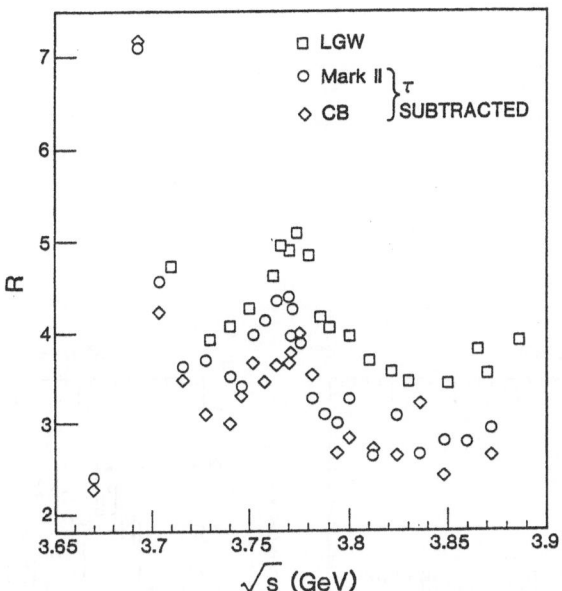

Fig. 16 Comparison of three measurements of the $\psi(3770)$ resonance of SPEAR. Note that the Mark II and Crystal Ball (CB) results are τ-subtracted, while the earlier Lead Glass Wall (LGW) measurement is not.

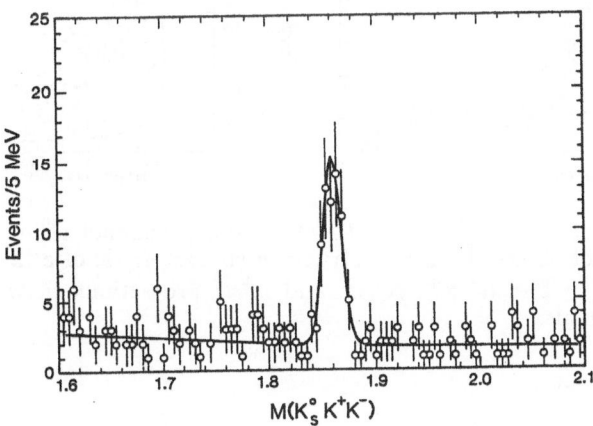

Fig. 17 ARGUS plot of $M(K_s^0 K^+ K^-)$ with the $K^+ K^-$ required to be a ϕ, showing evidence for the decay $D^0 \to \bar{K}^0 \phi$.

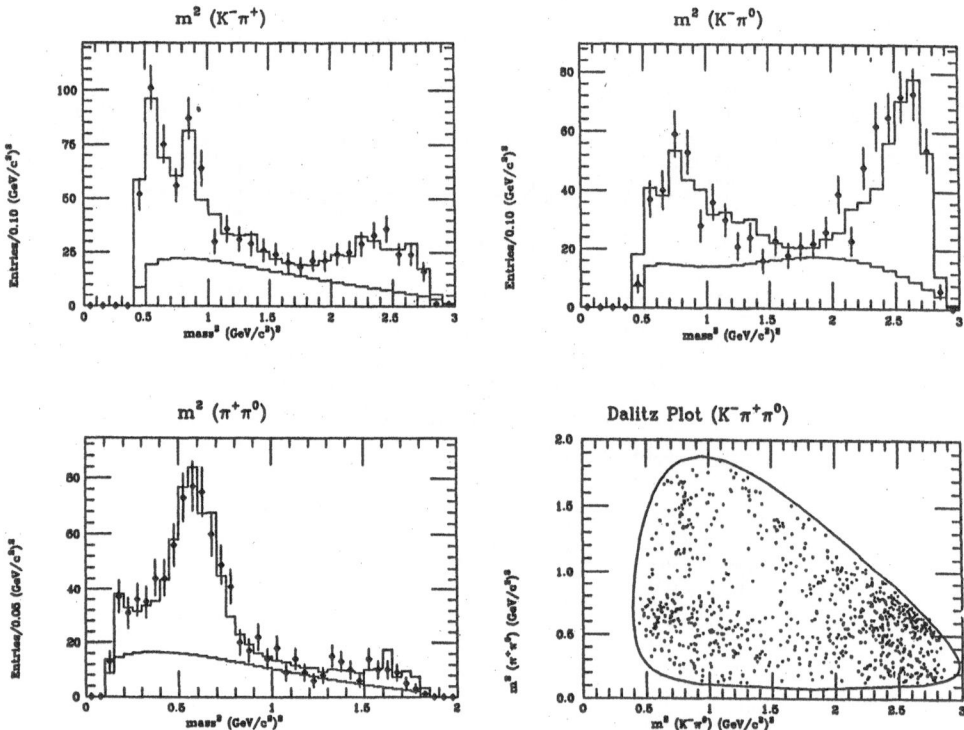

Fig. 18 Dalitz plot analysis by Mark III of the decay channel $D^0 \to K^-\pi^+\pi^0$. The Dalitz plot shows the $L = 1$ structure characteristic of a decay of the type $P \to PV$. The $K^-\pi^+$, $K^-\pi^0$ and $\pi^+\pi^0$ projections show prominent K^* and ρ signals.

normalization. Implications for other measurements will be discussed below.

The decay $D^0 \to \bar{K}^0 \phi$ has sparked special interest as a possible way of obtaining direct evidence for non-spectator hadronic decays. The ARGUS $\bar{K}^0 \phi$ signal is shown in Fig. 17. The three observations of the $\bar{K}^0 \phi$ mode are summarized in Table XII. This "smoking gun" mode will be discussed in further detail below.

The four $D \to K\pi\pi$ decays provide an interesting opportunity to study D meson couplings to pseudoscalar-vector (PV) channels. By fitting the Dalitz plots, using a maximum likelihood method, to functions parametrizing possible resonant substructure through relativistic Breit-Wigner functions, together with non-resonant and background contributions, Mark III[70] has isolated the PV content of the $K\pi\pi$ decays. The fits allowed for complex phases between various amplitudes. The background is determined using a sample of events from a region in beam-constrained $K\pi\pi$ mass below the D mass.

Figure 18 shows the Dalitz plot of the $K^-\pi^+\pi^0$ mode, together with the three projections and the result of the fit. Both ρ^+ and K^{*-} resonant structure is apparent. Table XIII summarizes the results of the fits. The study of the $D^+ \to K^-\pi^+\pi^+$ Dalitz plot is complicated by the existence of two identical final state pions. A non-resonant Bose-symmetric term is required in addition to the ordinary non-resonant phase space term in order to provide an adequate representation of the data. Both $K^{*-}\pi^+$ and $\bar{K}^{*0}\pi^+$ can be measured in two final states. For the former, the agreement is excellent, while for the latter, wherein the comparison involves the $K^-\pi^+\pi^+$ channel, the agreement is poor. Comparison of these results with theoretical expectations will be discussed below.

D Semileptonic Decays

As we have seen, semileptonic processes play an important role in the phenomenology of charmed and bottom meson decay. The ratio of inclusive semileptonic branching ratios of the D^+ and D^0 mesons provided, for a time, the most precise indication of the nonequality of D^+ and D^0 lifetimes. This relationship is based on the fact that these semileptonic decays are spectator processes, as they predominantly are, in both the Cabibbo-allowed and -suppressed sector. There are two semileptonic decay processes (see Fig. 19), Cabibbo-allowed for the D_s^+ and Cabibbo-suppressed for the D^+, which could originate from annihilation diagrams. Both are expected to be small, not only because of wave-function suppression, but because final state hadrons must arise from flavor singlet gluon emission.

Inclusive semileptonic branching ratios, interesting in their own right, are needed to extract information on the charmed sea from dimuon events produced in neutrino reac-

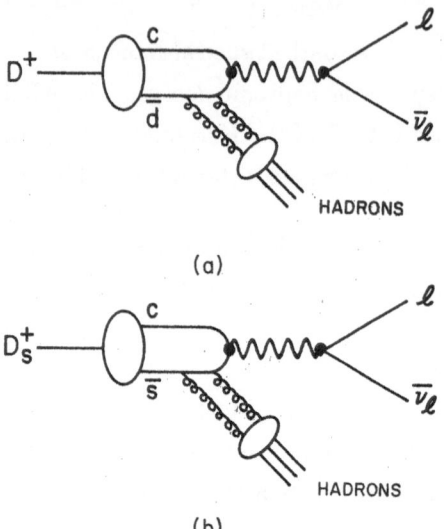

(a)

(b)

Fig. 19 Semileptonic decays arising from non-spectator amplitudes.
a) $D^+ \rightarrow l\nu_l +$hadrons, which is Cabibbo-suppressed, and
b) $D_s^+ \rightarrow l\nu_l +$ hadrons, which is Cabibbo-allowed.

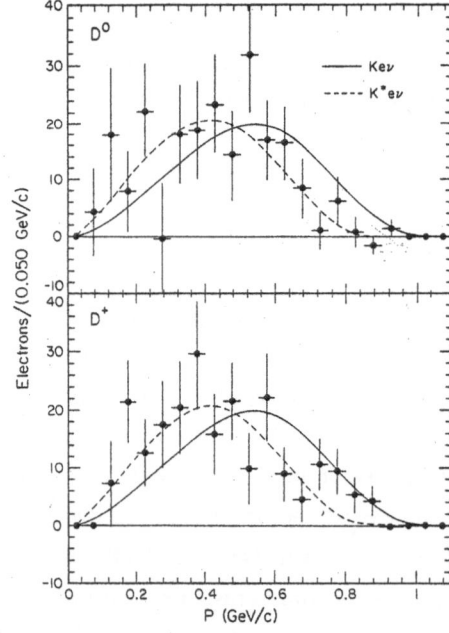

Fig. 20 Inclusive electron spectra for D^0 and D^+ decay at the $\psi(3770)$ from Mark III, with the expected shape for $Ke\nu$ and $K^*e\nu$ superimposed.

Table XI

Other D Meson Branching Ratios Using
the Revised Mark III Absolute Normalization

Decay Mode	Branching Ratio (%)
$D^0 \rightarrow K^- K^+$	$0.51 \pm 0.09 \pm 0.06$
$D^0 \rightarrow \pi^- \pi^+$	$0.14 \pm 0.04 \pm 0.03$
$D^+ \rightarrow K^+ \bar{K}^0$	$1.01 \pm 0.32 \pm 0.18$
$D^+ \rightarrow \pi^+ \pi^- \pi^+$	$0.38 \pm 0.15 \pm 0.09$
$D^+ \rightarrow K^- K^+ \pi^+_{nonres}$	$0.54 \pm 0.25 \pm 0.09$
$D^+ \rightarrow \phi \pi^+$	$0.77 \pm 0.22 \pm 0.11$
$D^+ \rightarrow K^+ \bar{K}^{*0}$	$0.44 \pm 0.20 \pm 0.11$
$D^0 \rightarrow \bar{K}^0 \phi$	$0.86^{+0.50+0.31}_{-0.41-0.17}$
$D^0 \rightarrow \bar{K}^0 K^+ K^-_{nonres}$	$0.85^{+0.27+0.19}_{-0.24-0.18}$
$D^0 \rightarrow \bar{K}^0 K^0$	≤ 0.46 at 90% CL

Table XII

Measurements of $B(D^0 \rightarrow \bar{K}^0 \phi)$

Experiment	Measurement	$B(D^0 \rightarrow \bar{K}^0 \phi)^\dagger$ (%)
ARGUS[67]	$\dfrac{B(D^0 \rightarrow \bar{K}^0 \phi)}{B(D^0 \rightarrow \bar{K}^0 \pi^+ \pi^-)} = 0.155 \pm 0.033$	$0.99 \pm 0.24 \pm 0.14$
CLEO[68]	$\dfrac{B(D^0 \rightarrow \bar{K}^0 \phi)}{B(D^0 \rightarrow \bar{K}^0 \pi^+ \pi^-)} = 0.142 \pm 0.049$	$0.91 \pm 0.33 \pm 0.13$
Mark III[69]		$0.86 \pm^{0.50}_{0.41} \pm^{0.31}_{0.17}$

$\dagger$Using revised Mark III measurements of $\sigma \cdot B(D^0 \rightarrow \bar{K}^0 \pi^+ \pi^-) = 0.37 \pm 0.030 \pm 0.31$
and revised derived value of $\sigma_D = 5.8 \pm 0.5 \pm 0.6$ nb at the $\psi(3770)$.

Table XIII

Mark III $D \to K\pi\pi$ Pseudoscalar-Vector Branching Fractions

Decay Mode	Fraction (%)	Phase (degrees)	BR (%)
$D^0 \to K^-\pi^+\pi^0$			$13.3 \pm 1.2 \pm 1.3$
$K^-\rho^+$	$81 \pm 3 \pm 6$	0.0	$10.8 \pm 0.4 \pm 1.7$
$K^{*-}\pi^+$	$12 \pm 2 \pm 3$	154 ± 11	$4.9 \pm 0.7 \pm 1.5$
$\bar{K}^{*0}\pi^0$	$13 \pm 2 \pm 3$	7 ± 7	$2.6 \pm 0.3 \pm 0.7$
non-resonant	$9 \pm 2 \pm 4$	52 ± 9	$1.2 \pm 0.2 \pm 0.6$
$D^0 \to \bar{K}^0\pi^+\pi^-$			$6.4 \pm 0.5 \pm 1.0$
$\bar{K}^0\rho^0$	$12 \pm 1 \pm 7$	93 ± 30	$0.8 \pm 0.1 \pm 0.5$
$K^{*-}\pi^+$	$56 \pm 4 \pm 5$	0.0	$5.3 \pm 0.4 \pm 1.0$
non-resonant	$33 \pm 5 \pm 10$	—	$2.1 \pm 0.3 \pm 0.7$
$D^+ \to \bar{K}^0\pi^+\pi^0$			$10.2 \pm 2.5 \pm 1.6$
$\bar{K}^0\rho^+$	$68 \pm 8 \pm 12$	0.0	$6.9 \pm 0.8 \pm 2.3$
$\bar{K}^{*0}\pi^+$	$19 \pm 6 \pm 6$	43 ± 23	$5.9 \pm 1.9 \pm 2.5$
non-resonant	$13 \pm 7 \pm 8$	250 ± 19	$1.3 \pm 0.7 \pm 0.9$
$D^+ \to K^-\pi^+\pi^+$			$9.1 \pm 1.3 \pm 0.4$
$\bar{K}^{*0}\pi^+$	$13 \pm 1 \pm 7$	105 ± 8	$1.8 \pm 0.2 \pm 1.0$
non-resonant	$79 \pm 7 \pm 15$	0.0	$7.2 \pm 0.6 \pm 1.8$

tions, are used to normalize charm production cross sections in hadronic reactions and are central to many limits placed on neutral meson mixing. Continuum measurements in e^+e^- annihilation have produced values of the average semileptonic branching ratio for c quarks (and b quarks, see below), without distinguishing the individual parent charmed hadron. It has been possible to use the unique characteristics of $D\bar{D}$ production at the $\psi(3770)$, however, to obtain the separate D^0 and D^+ inclusive semileptonic branching ratios, as well as to measure particular exclusive channels and, in the case of $D \to Ke\nu$, to extract the q^2 dependence of the vector form factor.

The key to these studies is the Mark III collection of reconstructed hadronic D decays at the $\psi(3770)$, which provide a large tag sample of 3435 ± 39 $D^0 \to K^-\pi^+$, $K^-\pi^+\pi^0$, $K^-\pi^+\pi^+\pi^-$ and 1729 ± 20 $D^+ \to \bar{K}^0\pi^+$, $K^-\pi^+\pi^+$ events. The rest of the event consists of decay products of the accompanying $\bar{D}^0$ or D^-. (In the above, all references to a particle or decay imply the charge conjugate particle or decay as well.) Thus, for example, the fraction of electrons in the recoil spectrum against hadronic tags provides a direct measurement of the semileptonic branching fraction. The astute reader will have noticed, by the way, that the uncertainty in the number of tags quoted above is less than $\sqrt{N_{\text{tags}}}$. This is so because each tag provides a single independent experiment in which an electron may be identified. Were there no background to the tags, there would be no uncertainty as to the number of such experiments. Thus, the uncertainty in the number of tag trials is just the uncertainty in the number of background events.

In order to measure the inclusive semileptonic branching ratio, it is necessary to know rather precisely the probability of misidentifying a hadron in the recoil spectrum as an electron. As the GIM current produces, e.g., e^+ from D^0 and D^+, it is possible to directly measure this misidentification probability by counting the number of "wrong sign" electrons. One must also apply small corrections for γ conversions and Dalitz decays, as well as for background under the tag sample. Table XIV summarizes measurements of individual and average D inclusive semileptonic branching fractions.

The inclusive lepton spectra obtained separately for D^0 and D^+ by the Mark III[45] are shown in Fig. 20, with expected spectra from $Ke\nu$ and $K^*e\nu$, each separately normalized, superimposed. It is clear that the actual spectra are from not pure $D \to Ke\nu$ decay, but contain heavier hadron components which soften the electron momentum distribution.

Using the hadronic tag sample, the Mark III has now been able to reconstruct a significant number of exclusive semileptonic decays.[9] This has allowed explicit measurement of the hadron content of the final state. In this case, the tag sample consists of 3130 ± 39 D^0 hadronic decays in the channels $K^-\pi^+$, $K^-\pi^+\pi^+\pi^-$ and $K^-\pi^+\pi^0$, over a background of 1512 ± 39 events, and 1695 ± 28 D^+ decays in the channels $K^-\pi^+\pi^+$,

Table XIV

Inclusive D Semileptonic Branching Ratios

Mode	Experiment	Branching Ratio (%)
$B(D^0 \to e^+ X)$	Mark III[45]	$7.5 \pm 1.1 \pm 0.4$
	Mark II[71]	5.5 ± 3.7
	DELCO[72]	< 4.0 @ 95% CL
$B(D^+ \to e^+ X)$	Mark III	$17.0 \pm 1.9 \pm 0.7$
	Mark II	16.8 ± 6.0
	DELCO	$22.0 \pm^{4.4}_{2.2}$
$B(D \to e^+ X)$	Mark III	$11.7 \pm 1.0 \pm 0.5$
(assuming	Mark II	10.0 ± 3.2
$\frac{\sigma(D^0)}{\sigma(D^+)} = 1.27$)	DELCO	8.0 ± 1.5
	LGW[47]	7.6 ± 2.8

$\bar{K}^0\pi^+$, $\bar{K}^0\pi^+\pi^+\pi^-$ and $\bar{K}^0\pi^+\pi^0$, with a background of 765 ± 28. The $\psi(3770)$ decay to $D\bar{D}$ characteristics are particularly valuable in this case, since both the charm and momentum of the D recoiling against the hadronic tag are known. Thus, both the kaon (in the case of charged K's) and the lepton charge of the recoil can be required to be consistent with the GIM current. Since $\vec{p}_D$ is known, there is no kinematic ambiguity of the type usually encountered in the study of semileptonic K decay; rather than balancing only the transverse momentum of the missing neutrino, we know its total momentum.

The many possible sources of background to semileptonic channels will not be discussed in detail here. The most obvious one will, however, serve to illustrate the method of analysis. $D^0 \to K^-\pi^+\pi^0$, in which the π^+ is misidentified as an e^+ and both photons from the π^0 are missed appear as candidates for $D^0 \to K^-e^+\nu_e$. We can define $U \equiv E_{\text{miss}} - |\vec{P}_{\text{miss}}|$, which is 0 for a missing neutrino and positive for a missing π^0. Figure 21 shows the distribution in U vs. the beam-constrained mass of the tag for D^0 events, showing that it is possible to isolate $K^-e^+\nu_e$ quite clearly. Other backgrounds, such as those from $K^-\pi^+$, $K^-\pi^0e^+\nu_e$ and $K^0_L\pi^-e^+\nu_e$ are removed by cuts on the invariant mass of the Ke system and on the total missing energy. After these cuts and kinematic fitting of the complete event, 47 events are reconstructed in the $D^0 \to K^-e^+\nu_e$ channel, with a background of 2.1 events, yielding a branching ratio $B(D^0 \to K^-e^+\nu_e) = (3.9 \pm 0.6 \pm 0.6)\%$.

The analysis of channels such as $D^0 \to K^-\pi^0e^+\nu_e$, $D^0 \to \bar{K}^0\pi^-e^+\nu_e$, $D^+ \to \bar{K}^0e^+\nu_e$ and $D^+ \to K^-\pi^+e^+\nu_e$ are very similar. Semimuonic channels have also been studied. The problem is again very similar, although here the problem is to identify muons clearly without using the muon detection system, as the muon system lower threshold of 600 MeV/c admits only $\sim 20\%$ of muons from D decay, and the solid angle coverage is only 60% of 4π.

Figure 21 and 22 show the distribution of U vs. beam constrained tag mass for all reconstructed semileptonic modes. Table XV shows the exclusive branching ratios obtained. The D^0 modes sum to a value close to the inclusive result, while the D^+ exclusive modes are $\sim 2\sigma$ lower than the inclusive result. Table XVI compares the Mark III measurements of inclusive and exclusive semileptonic branching ratios.

E691 has reported a preliminary result,[73] based on 30% of their data, on the branching ratio $B(D^0 \to K^-e^+\nu_e)$. With a signal of 57 ± 9 events, the average of two analysis methods is

$$\frac{B(D^0 \to K^-e^+\nu_e)}{B(D^0 \to K^-\pi^+)} = 0.73 \pm 0.12 \pm 0.13 \ .$$

When combined with $B(D^0 \to K^-\pi^+) = (4.2 \pm 0.4 \pm 0.3)\%$, this yields

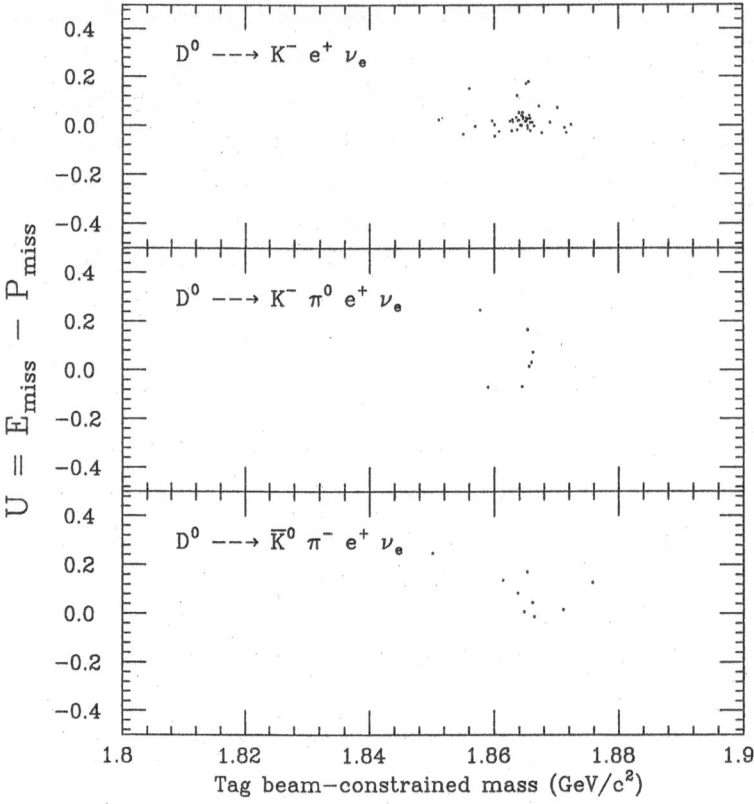

Fig. 21 Scatter plot of $E_{\text{missing}} - |P_{\text{missing}}|$ *vs.* beam-constrained mass of the reconstructed hadronic decay (tag) for three D^0 exclusive semileptonic decay modes (Mark III).

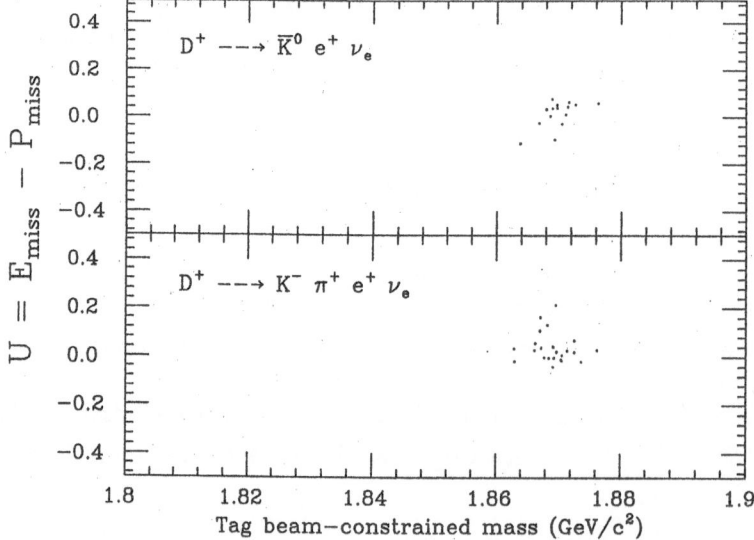

Fig. 22 Scatter plot of $E_{\text{missing}} - |P_{\text{missing}}|$ *vs.* beam-constrained mass of the reconstructed hadronic decay (tag) for two D^+ exclusive semileptonic decay modes (Mark III).

Table XV

Mark III Measurements of Exclusive Semileptonic
D Meson Branching Ratios

Decay Mode	Events Signal (Bkgnd)	Branching Ratio (%)
$D^0 \to K^- e^+ \nu_e$	47 (2.1)	$3.9 \pm 0.6 \pm 0.6$
$K^- \pi^0 e^+ \nu_e$	7 (1.1)	$1.7 \pm^{0.9}_{0.7} \pm 0.6$
$\bar{K}^0 \pi^- e^+ \nu_e$	9 (0.7)	$2.2 \pm^{0.9}_{0.7} \pm 0.4$
$\pi^- e^+ \nu_e$	3 (0.9)	$0.4 \pm^{0.4}_{0.3} \pm 0.1$
$D^0 \to K^- \mu^+ \nu_\mu$	56 (9.4)	$4.1 \pm 0.7 \pm 1.2$
$\bar{K}^0 \pi^- \mu^+ \nu_\mu$	20 (8.5)	$2.7 \pm^{1.1}_{1.0} \pm 1.6$
$D^+ \to \bar{K}^0 e^+ \nu_e$	15 (1.1)	$6.3 \pm^{2.0}_{1.6} \pm 1.1$
$K^- \pi^+ e^+ \nu_e$	24 (1.2)	$3.9 \pm^{0.9}_{0.8} \pm 0.7$
$D^+ \to \bar{K}^0 \mu^+ \nu_\mu$	37 (8.9)	$10.2 \pm^{2.2}_{2.1} \pm 3.6$

Table XVI

Comparison of Mark III Inclusive and Exclusive
D Meson Semileptonic Branching Ratios

Decay Mode	Branching Ratio (%)
$D^0 \to (K^- + (K\pi)^-)l^+ \nu_l$	7.9 ± 1.5
$e^+ X$	$7.5 \pm 1.1 \pm 0.4$
$D^+ \to (\bar{K}^0 + (K\pi)^0)l^+ \nu_l$	13.3 ± 2.5
$e^+ X$	$17.0 \pm 1.9 \pm 0.7$

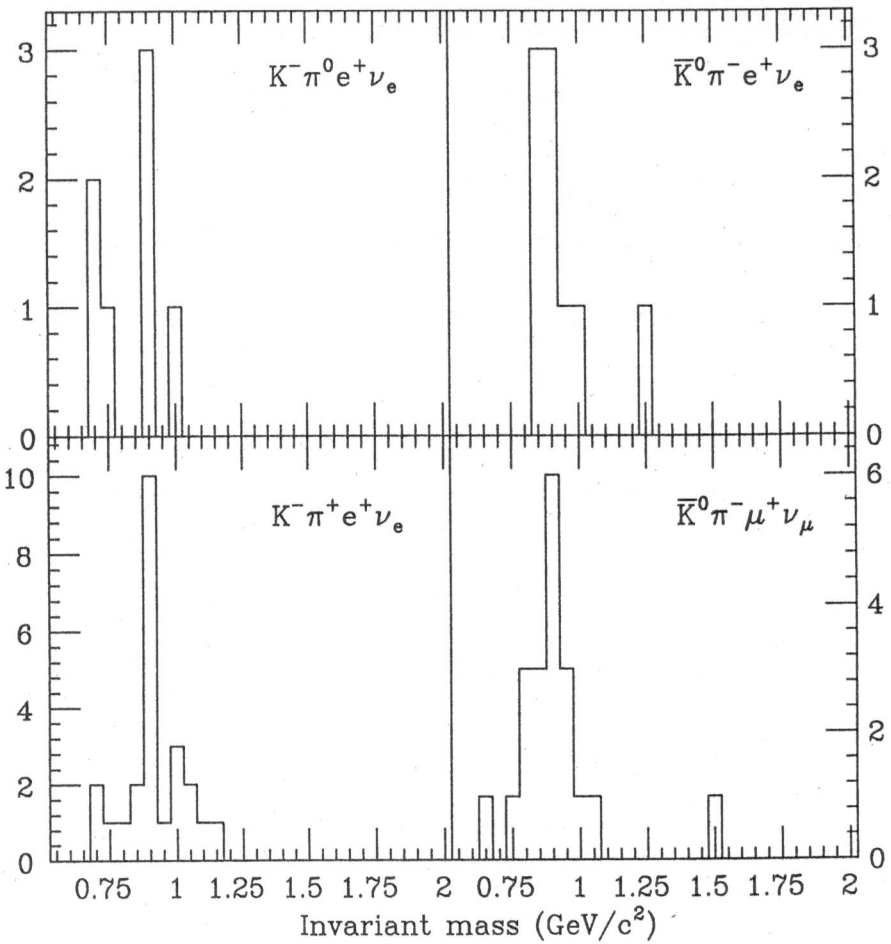

Fig. 23 Distribution of $K\pi$ invariant mass for four D_{l4} exclusive semileptonic decay modes (Mark III).

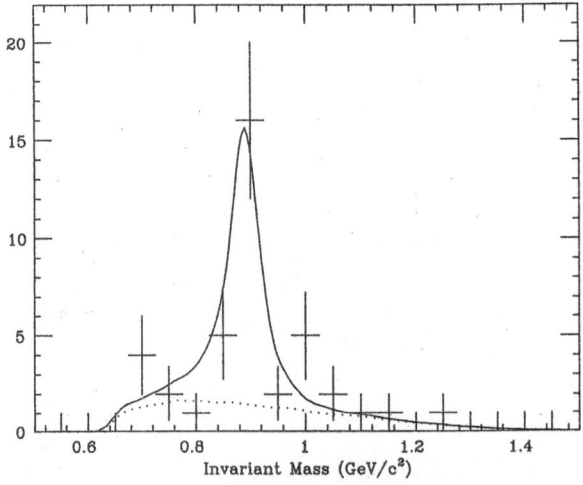

Fig. 24 Fit of a $K^*(890)$ Breit-Wigner plus s-wave phase space for the sum of the four D_{l4} decay modes shown above.

$$B(D^0 \to K^- e^+ \nu_e) = 3.1 \pm 0.8 \ ,$$

when statistical and systematic errors are combined in quadrature. The agreement with the Mark III result is thus satisfactory.

It is interesting to plot the $K\pi$ mass in the various $D \to K\pi l\nu_l$ channels. These are shown individually in Figure 23; the sum is plotted in Figure 24, together with a fit to the sum of a K^* Breit-Wigner and s-wave $K\pi$ phase space. The $K\pi$ final state is not entirely K^*: there is a significant non-resonant contribution. The fit with s-wave $K\pi$ phase space yields

$$\frac{B(D \to K^* e^+ \nu_e)}{B(D \to K\pi e^+ \nu_e)} = 0.55 \pm 0.13 .$$

The result using p-wave phase space is 0.57 ± 0.13. The fact that there is a non-resonant component cannot be accommodated in the quark model treatment of semileptonic decay of, for example, Grinstein *et al.*[74] The implications of this problem on using B semileptonic decay spectra to extract the $b \to u/b \to c$ fraction will be discussed below.

In the case of the $D^0 \to K^- e^+ \nu_e$ channel, the Dalitz plot distribution has also been analyzed to extract the t dependence of the vector form factor.[9] The matrix element for Cabibbo-allowed semileptonic D decay may be written as

$$M = \frac{G_F}{\sqrt{2}} J_\mu^{\text{lepton}} J_{\text{hadron}}^\mu \, ,$$

where

$$J_\mu^{\text{lepton}} = \nu_e \gamma_\mu (1 - \gamma_5) e^+ \, ,$$

$$\begin{aligned} J_{\text{hadron}}^\mu = \sqrt{2} |V_{cs}| [f_+(t)(P_D^\mu + P_K^\mu) \\ + f_-(t)(P_D^\mu - P_K^\mu)] \, , \end{aligned}$$

and the four momentum transfer $t \equiv (P_D - P_K)^2 = M_D^2 + M_K^2 - 2M_D E_K$.

Since $(P_D - P_K)$ is proportional to M_l, the second term in J_{hadron}^μ may be neglected for any heavy quark semileptonic decay at the present level of precision. Thus, in the D center-of-mass, we can write the Dalitz plot distribution as

$$\frac{d\Gamma}{dx_e dx_k} \propto |f_+|^2 \left(2x_e - x_e x_k + x_e^2 + x_k - \frac{M_K^2}{M_D^2} - 1 \right) \, ,$$

where $x_e = E_e/E_{e_{max}}$, $x_K = E_K/E_{K_{max}}$.

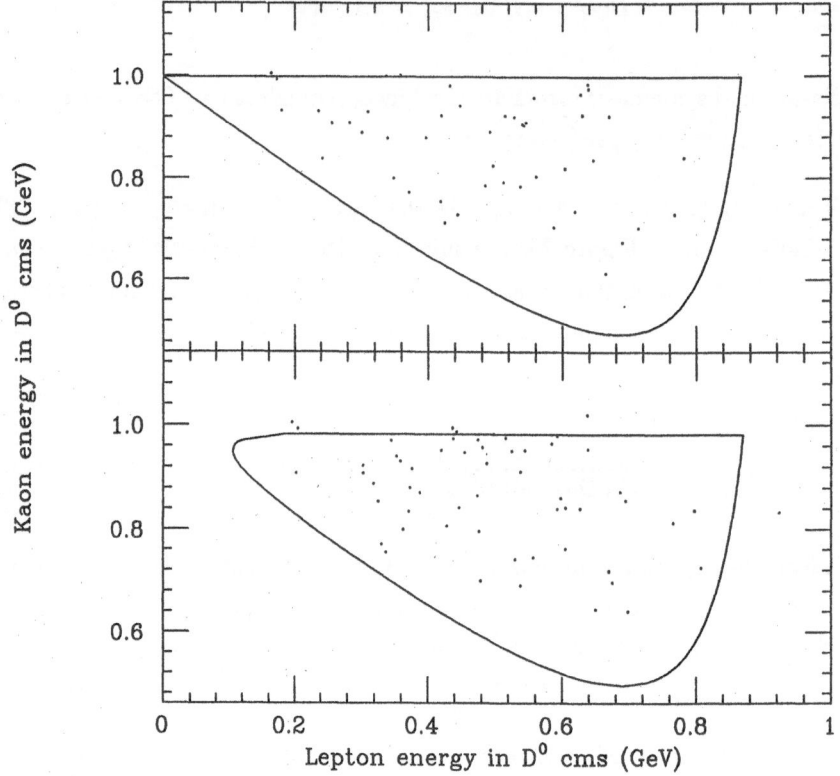

Fig. 25   Dalitz plot for D^0_{e3} and $D^0_{\mu3}$ exclusive events (Mark III).

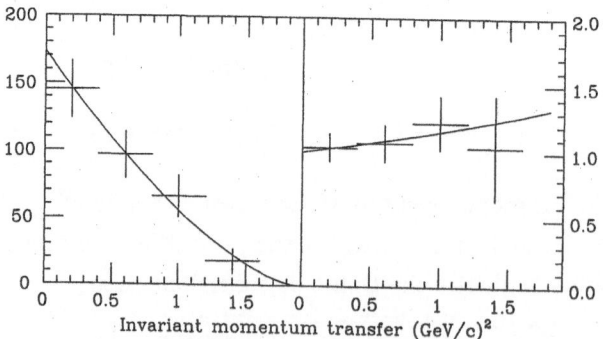

Fig. 26   a) Kaon momentum spectrum and b) $f_+(t)/f_+(0)$ vs. t for the Mark III D_{l3} events. The best fit curve is superimposed.

The Dalitz plot distribution for the 47 $D^0 \to K^- e^+ \nu_e$ events is shown in Figure 25. The t dependence of f_+ is given by a once-subtracted dispersion relation, dominated by the lowest $1^- c\bar{s}$ resonance, the D_s^{+*}. We thus expect

$$f_+(t) = \frac{f_+(0)}{1 - \dfrac{t}{M_{D_s^{+*}}^2}} .$$

Since t depends only on E_K, a fit to the kaon spectrum suffices to extract $M_{D_s^{+*}}$. Figure 26 shows this spectrum together with a curve for the best fit value of $M_{D_s^{+*}}$:

$$M_{D_s^{+*}} = 2.1^{+1.7}_{-0.4} \text{ GeV} .$$

Thus, within rather large errors, the t dependence of the f_+ form factor follows the naïve expectation.

In the case of semileptonic D decays to final states containing K^*, $K\pi$, etc., there may be five t dependent form factors. The data is not yet sufficiently precise to allow an attempt to extract them.

Semileptonic decays of the D_s^+ are also of interest, although as yet there is no experimental information in this area. Two points should be noted. First, now that there are reliable lifetime measurements for all three 0^- charmed mesons, it would be very interesting to be able to compare their inclusive semileptonic branching ratios to ascertain whether the relationship seen for D^0 and D^+ decays is maintained. Second, the study of the hadronic component in exclusive decays could be especially rewarding. In the case of the $c\bar{s}$ annihilation amplitude (Fig. 19), the source of hadrons is singlet multiple gluon emission. Thus, along with ϕ and η, which are expected to dominate spectator D_s^+ semileptonic decay, one might expect some amount of glueball production.

Rare Decays

Decays which violate family number, such as $K_L \to \mu e$ or $D^0 \to \mu e$ are of particular interest in the context of unified models which contain, e.g., SU(5) scalar leptoquarks. The existing limit

$$B(K_L^0 \to \mu e) < 6 \times 10^{-6} \text{ at } 90\% \text{ CL} ,$$

does not exclude a measurable rate for $D^0 \to \mu e$ in, for example, the model of Buchmüller and Wyler.[75] Mark III has also been able to place an upper limit on this mode:

$$B(D^0 \to \mu e) < 1.2 \times 10^{-4} \text{ at } 90\% \text{ CL} .$$

This translates into a (model-dependent) lower bound on an SU(5) lepto-quark mass of more than about 1.5 TeV/c^2. Two other limits on this branching ratio, at approximately

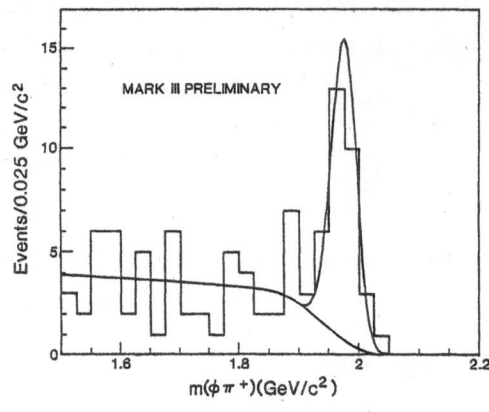

Fig. 27 Three Mark III D_s^+ hadronic decays:

a) $D_s^+ \to \phi\pi^+$

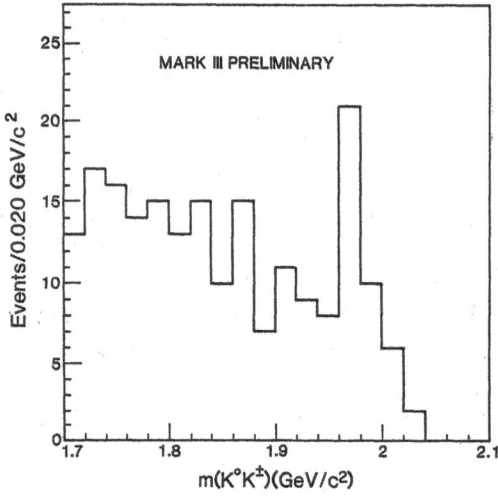

b) $D_s^+ \to K^0 K^+$

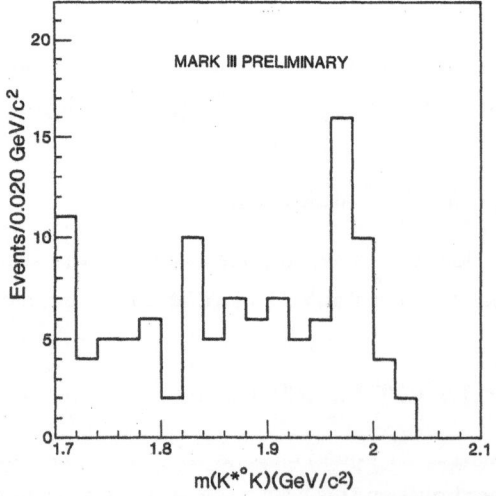

c) $D_s^+ \to K^{*0} K^+$.

one order of magnitude less sensitivity have also been set.[76] The Mark II finds $B(D^0 \to \mu^\pm e^\mp) < 2.1 \times 10^{-3}$, while ACCMOR sets a limit of 9×10^{-4}, both at the 90% confidence level.

D_s^+ Hadronic Decays

Much less is known experimentally about the hadronic decays of the D_s^+, although this is an area in which rapid progress can be expected. Observations of the D_s^+ have had a checkered history, with several early experiments[77,78,79,80] finding signals in $\eta\pi^+$, $\eta\pi^+\pi^+\pi^-$, etc. at masses above 2 GeV, while current measurements clearly place the mass in the 1970 MeV region.

Modern observations of D_s^+ decays are summarized in Table XVII. In the absence of a well-defined feature in the e^+e^- cross section analogous to the $\psi(3770)$, extraction of D_s^+ branching ratios requires a number of assumptions. The clearest set of assumptions applies to production in high energy e^+e^- annihilation far above threshold. A typical set of such assumptions is:

a) charmed quark production accounts for 4/10 of the total hadronic cross section below $b\bar{b}$ threshold and 4/11 above.

b) the probability of producing an $s\bar{s}$ pair from the vacuum is 15%, which thereby specifies the fraction of D_s^+ production compared to total charm production.

c) the D_s^+ production cross section is typically measured only above a given D_s^+ momentum. Thus, the measured cross section $\frac{s}{\beta}\frac{d\sigma}{dx_{D_s^+}}$ must be extra extrapolated to $x_{D_s^+} = 0$.

The observations of $D_s^+ \to \phi\pi^+$ can be used with this *ansatz* to extract the $\phi\pi^+$ branching ratio. The ratio of other D_s^+ branching fractions, which of course can be obtained indirectly, are also shown in Table XVII. The uncertainties quoted are statistical only. Figure 27 shows three D_s^+ hadronic decays observed by Mark III.

There is, as yet, no experimental information on D_s^+ semileptonic decay. Such an inclusive measurement would provide an interesting comparison with D^0 and D^+ rates: a small branching ratio would support the speculation that D^0 and D_s^+ lifetimes are shorter than that of the D^+ due to non-leptonic enhancement associated with non-spectator diagrams. As Cabibbo-allowed D_s^+ semileptonic decay can proceed via $c\bar{s}$ annihilation as well as by a spectator process, a measurement of exclusive semileptonic modes would be particularly interesting.

Theory of Charmed Meson Hadronic Decays

The wealth of precise data on charmed particle lifetimes and hadronic and semileptonic branching ratios has led to a number of detailed theoretical treatments. In order

Table XVII

Hadronic Decays of D_s^+

Decay Mode	Experiment	Measurement	$\sqrt{s}$	Branching Ratio (%)
$D_s^+ \to \phi\pi^+$	CLEO[81]	$R(D_s^+)B(D_s^+ \to \phi\pi^+) = 2.2 \pm 0.5$	10	4.4 ± 1.1
	TASSO[82]		35-44	$3.3 \pm 1.6 \pm 0.4$
	ARGUS[83]	$R(D_s^+)B(D_s^+ \to \phi\pi^+) = 1.47 \pm 0.32 \pm 0.20$	10	$3.2 \pm 0.7 \pm 0.5$
	HRS[84]	$R(D_s^+)B(D_s^+ \to \phi\pi^+) = 1.35 \pm 0.34$	29	3.3 ± 1.0
	Mark III*[85]	$\sigma \cdot B = (34 \pm 6 \pm 14)\text{pb}$	4.14	
	E691**[59]			4.0 ± 0.9
$K^-K^+\pi^+$	TPC[86]	$\dfrac{\sigma_{D_s}(z > 0.4)B(D_s^+ \to \text{"}K^+K^-\pi^+\text{"})}{\sigma_h}$ $= (3.34 \pm 0.95 \pm 1.11) \times 10^{-2}$		
	E691[59]	$\dfrac{B(D_s^+ \to K^+K^-\pi^+)}{B(D_s^+ \to \phi\pi^+)} = 0.3 \pm 0.1 \pm 0.1$		1.2 ± 0.6
	ACCMOR[87]	5 events hadronically produced		
$\phi\pi^+\pi^-\pi^+$	ARGUS[83]	$\dfrac{B(D_s^+ \to \phi\pi^+\pi^-\pi^+)}{B(D_s^+ \to \phi\pi^+)} = 1.11 \pm 0.37 \pm 0.28$	10	
	E691[59]	$\dfrac{B(D_s^+ \to \phi\pi^+\pi^-\pi^+)}{B(D_s^+ \to \phi\pi^+)} = 1.0 \pm 0.4 \pm 0.2$		
K^0K^+	Mark III*†[85]	$\sigma \cdot B = (32 \pm 6 \pm 10)\text{pb}$	4.14	
K^*K^+	ARGUS[88]	$\dfrac{B(D_s^+ \to K^{*0}K^+)}{B(D_s^+ \to \phi\pi^+)} = 1.44 \pm 0.37$	10	
	Mark III*†[85]	$\sigma \cdot B = (31 \pm 6 \pm 11)\text{pb}$	4.14	
	E691[59]	$\dfrac{B(D_s^+ \to K^{*0}K^+)}{B(D_s^+ \to \phi\pi^+)} = 1.0 \pm 0.2 \pm 0.1$		4.0 ± 0.9

* $\sigma \cdot B \equiv \sigma(e^+e^- \to D_s^\pm D_s^{*\mp})B(D_s^+ \to \text{mode})$
** $B(D_s^+ \to \phi\pi^+)$ assumed to be 4.0% with no error
† Preliminary

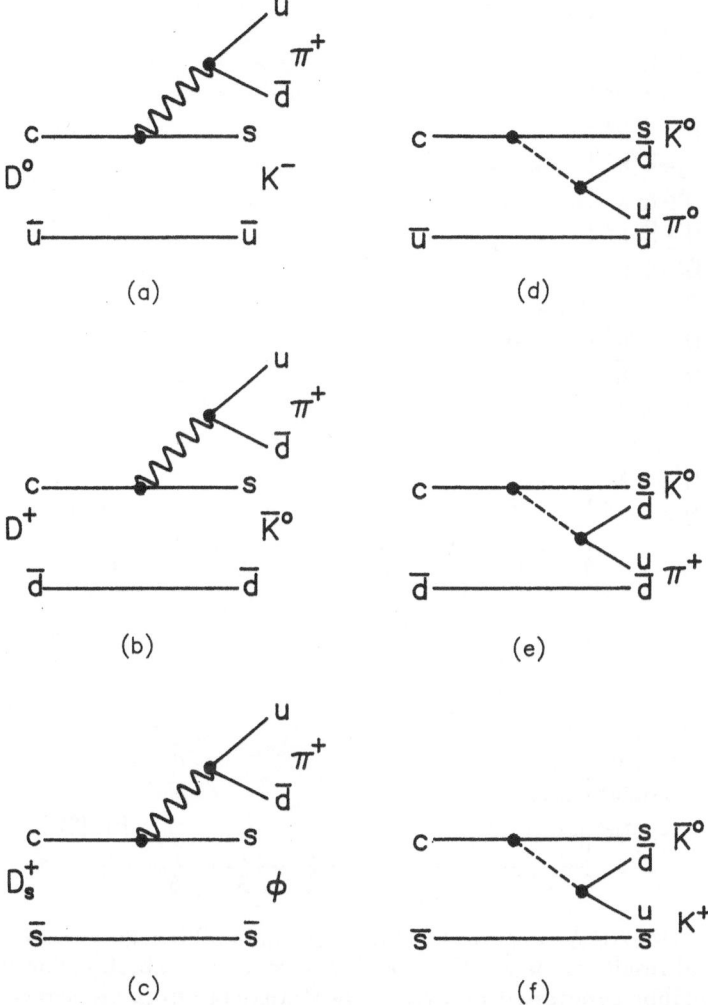

Fig. 28 Spectator amplitudes for charmed meson hadronic decay arising from QCD-motivated renormalization of operator coefficients.
Diagrams a), b) and c) have the same operator structure as in
the simple valence quark picture, although they now enter with
coefficient $c_1 = (c_+ + c_-)/2$. Diagrams d), e) and f) are "effective
neutral current" or "color suppressed" amplitudes, which have
coefficient $c_2 = (c_+ - c_-)/2$. Note that in the D^+ case, the two
amplitudes lead to the same final state, and may thus interfere.

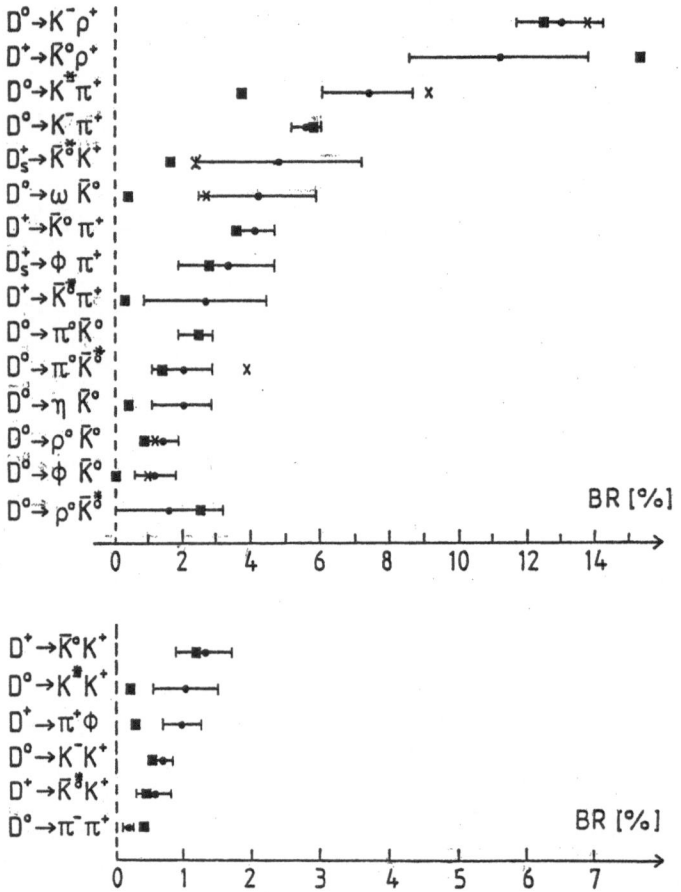

Fig. 29 Comparison of the experimental charmed meson branching ratios with the theoretical results of Bauer, Stech and Wirbel for a) Cabibbo-allowed decays and b) Cabibbo-suppressed decays. Only statistical errors are shown. Calculated values indicated by black squares are derived from factorized amplitudes with $a_1 \simeq 1.3$ and $a_2 \simeq -0.55$. Calculated values indicated with an "x" include annihilation contributions. Note that the normalization of D^0 and D^+ branching ratios is that derived from the Mark III double tag method *before* the recent downward revision of $\sim 20\%$. The new normalization thus has the effect of changing the branching ratio scale by this amount for the D^0 and D^+ modes. As the normalization of the D_s^+ decays is in any case somewhat arbitrary, and the experimental errors are large, the revised scale has essentially no effect on the quality of the fit. The revised values of a_1 and a_2 will shift by 10%.

to orient ourselves, let us begin with a few simple observations using, primarily, the Mark III data.

If we assume the validity of the QCD-motivated discussion above, we are justified in using the current-current Hamiltonian derived therein, with a range of c_-/c_+ values close to that calculated with the leading-log approximation and its extensions. For the case of Cabibbo-allowed charmed mesons, the amplitudes at the quark level for the light quark spectator Hamiltonian are shown in Fig. 29. The four quark operators generate two distinct amplitudes for each charmed meson. The first, with coefficient $c_1 = \frac{1}{2}(c_+ + c_-)$ is called "color-allowed". The second, with coefficient $c_2 = \frac{1}{2}(c_+ - c_-)$, is due to the effective neutral current generated by the Fierz transformation; it is usually referred to as the "color-suppressed" amplitude due to the color mismatch which impedes final state hadron formation.

There are other amplitudes which should also be considered. Figure 7 shows these two "non-spectator" processes, which can, in principle, contribute to Cabibbo-allowed D^0 and D_s^+ decays. These amplitudes are helicity-suppressed at the light quark vertex: their neglect yields the light quark spectator model.

In treating exclusive hadronic decays, it is important to recognize from the outset the complex relation between the quark level operators and the actual hadrons; the explicit structure of hadrons is certain to play an important role in understanding exclusive decays. Final state interactions are also likely to be important, as charmed meson masses lie in a region where resonance rescattering effects cannot be neglected. We will return to this question presently.

Bauer, Stech and and Wirbel[89] have dealt with the hadron problem by appealing to factorization to replace the product of quark currents with products of effective hadron currents:

$$H_{\text{eff}}(\Delta c = -1) = \frac{G_F}{\sqrt{2}} V_{cs} V_{ud}^* [a_1 (\bar{u}d')_H (\bar{s}'c)_H$$
$$+ a_2 (\bar{s}'d')_H (\bar{u}c)_H] \quad,$$

where the subscript H now denotes hadron field operators. The QCD coefficients c_1 and c_2 are replaced by new effective coefficients a_1 and a_2:

$$a_1 = (c_1 + \xi c_2)_{\mu = m_c} \, , \quad a_2 = (c_2 + \xi c_1)_{\mu = m_c} \, ,$$

where the color-mismatch factor ξ, which has the value $1/N$ (i.e., $1/3$) at the quark level, is now a free parameter. This is justified by the fact that the effect of color octet currents within explicit hadrons is not, a priori, understood.

The factorization hypothesis then allows us to write the amplitude for specific exclusive processes in terms of meson decay constants and hadron current matrix elements.

For example:

$$A(D^0 \to K^-\pi^+) = \frac{G_F}{\sqrt{2}} V_{cs} V_{ud}^* a_1 (-if_\pi) p_\lambda^\pi \langle K^-|(\bar{s}c)^\lambda|D^0\rangle_{q^2=m_\pi^2} \ ,$$

$$A(D^0 \to \bar{K}^0\pi^0) = \frac{G_F}{\sqrt{2}} V_{cs} V_{ud}^* a_2 \left(\frac{-if_\pi}{\sqrt{2}}\right) p_\lambda^\pi \langle \bar{K}^0|(\bar{s}c)^\lambda|D^0\rangle_{q^2=m_\pi^2} \ .$$

The hadron matrix elements may be evaluated, using, for example, relativistic harmonic oscillator wave functions.

At the quark level, with $\xi = \frac{1}{3}$, we can see that the ratio of these two $D \to K\pi$ rates would be given by

$$R = \frac{\Gamma(D^0 \to \bar{K}^0\pi^0)}{\Gamma(D^0 \to K^-\pi^+)} = \frac{1}{2}\left(\frac{2c_+ - c_-}{2c_+ + c_-}\right)^2 \ ,$$

for the nominal values of the QCD coefficients.

Table XVIII shows the experimental value of this ratio: $R = 0.45 \pm 0.08 \pm 0.05$. This quark-level estimate cannot account for the data. Two D^0 decay modes into pseudoscalar-vector channels; $D^0 \to K^*\pi$ and $K\rho$ can also be estimated in the same way (see Table XVIII). In the ratio $\Gamma(00)/\Gamma(-+)$ the hadron matrix elements, which are, in general, different for PP and PV channels, drop out; the dependence on the operator coefficients is identical. Clearly, this naïve estimate is incorrect. If we recognize that it is the *effective* coefficients which contribute to exclusive decays, however, and that the assumption $\xi = \frac{1}{3}$ is not justified in this picture, we can do better. If there are no color-suppressed amplitudes at all ($\xi = 0$), then we have

$$R = \frac{1}{2}\left(\frac{c_+ - c_-}{c_+ + c_-}\right)^2$$

$$\simeq .17 \quad for \quad c_-/c_+ = 2.4 \ ,$$

with larger values possible for larger ratios of a_2/a_1. This is dramatic evidence for the lack of color-suppressed amplitudes in these decays. Two other experimental ratios which involve color-mismatched decays, one for D^+ and one for D_s^+, also shown in Table XVIII, support the idea that there is an inhibition of color-suppressed amplitudes.

Bauer *et al.* have extended the factorization idea to a large number of hadronic D (and B) decays. With appropriate pole masses parametrizing the q^2 dependence of the current matrix elements calculated at $q^2 = 0$ with oscillator wave functions, they are able to account for a large number of measurements (see Fig. 30). Their model includes primarily spectator amplitudes. Annihilation amplitudes are included by fitting to the $D^0 \to \bar{K}^0\phi$ rate (see below). Final state interactions are also included.

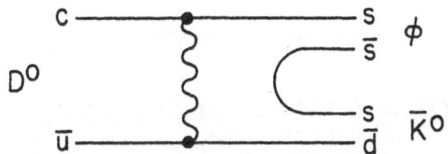

Fig. 30 Exchange amplitude for $D^0 \to \bar{K}^0 \phi$ decay, the "smoking gun" test, as this decay cannot arise from a simple spectator diagram.

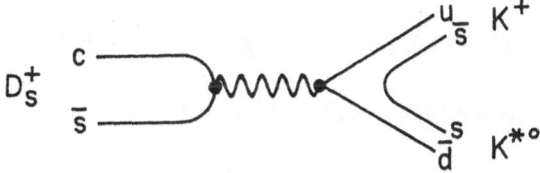

Fig. 31 Annihilation amplitude for $D_s^+ \to K^{*0} K^+$ decay. This decay can also arise from a color-suppressed spectator amplitude.

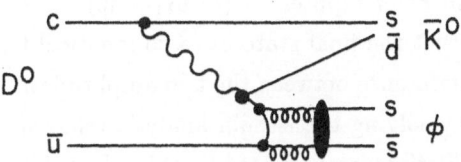

Fig. 32 Rescattering amplitude for $D^0 \to \bar{K}^0 \phi$ decay which casts doubt on the simple interpretation of this mode as clear evidence for the existence of non-spectator processes.

A more detailed comparison of $D \to PP$ and $D \to PV$ decay modes related by isospin allows study of the role of final state interactions, which do turn out to play a role. Such an analysis has been done by Chau and Cheng,[90] and Kamal,[91] as well as by Bauer, Stech and Wirbel. An isospin decomposition of related modes yields:

$$A(D^0 \to K^- \pi^+) = \frac{1}{\sqrt{3}}(\sqrt{2}A_{1/2} + A_{3/2})$$

$$A(D^0 \to \bar{K}^0 \pi^0) = \frac{1}{\sqrt{3}}(-A_{1/2} + \sqrt{2}A_{3/2})$$

$$A(D^+ \to \bar{K}^0 \pi^+) = \sqrt{3}A_{3/2} \ ,$$

where the isospin $\frac{1}{2}$ and $\frac{3}{2}$ amplitudes are, in general, complex:

$$A_I = |A_I| \exp(i\delta_I) \ .$$

These amplitudes obey the sum rule

$$A(D^0 \to K^- \pi^+) + \sqrt{2}A(D^0 \to \bar{K}^0 \pi^0) = A(D^+ \to \bar{K}^0 \pi^+) \ .$$

Similar isospin relations hold for the $D \to K^* \pi$ and $D \to K\rho$ channels.

No solutions are possible for relatively real amplitudes. The Mark III data can, however, be fit with complex ratios of the isospin $\frac{1}{2}$ and $\frac{3}{2}$ amplitudes (see Table XIX). It is clear that final state interactions play a role in these two-body decays. The $\bar{K}\rho$ and $\bar{K}^* \pi$ channels can themselves communicate via final state interactions, and can also mix with the (as yet unmeasured) $\bar{K}^* \rho$ channel and (for $I = \frac{1}{2}$ amplitudes) with $\bar{K}^0 \phi$, $\bar{K}^0 \omega$ and $\bar{K}^{*0} \omega$.

There has been much discussion in the literature of two additional effects which could, in principle, play a role in charmed meson lifetime differences. The first is Pauli interference,[92,93] which can occur only in D^+ decay. Note (see Fig. 29) that both the color-allowed ($\sim c_1$) and color-suppressed ($\sim c_2$) amplitudes produce the same final state in D^+ decay and that the final state has two identical fermions. It is, therefore, reasonable to expect interference between the two amplitudes. This interference can be seen to be destructive by solving the isospin analysis referred to above for the c_1 and c_2 coefficients. The D^0 lifetime is unaffected by this effect. The amount of interference can be accounted for by a parameter α (neglecting Cabibbo-suppressed decays):

$$\frac{\tau(D^+)}{\tau(D^0)} = \frac{2 + 2c_+^2 + c_-^2}{2 + 2(1 + \alpha)c_+^2 + (1 - \alpha)c_-^2} \ ,$$

where α can be evaluated as an overlap integral:

$$\alpha = 192\pi^2 \frac{|\psi(0)|^2}{M_D} \ .$$

Table XVIII

Ratios of Charmed Meson Branching Fractions

Involving Color-Suppressed Amplitudes

Ratio	Mark III Measurement
$\dfrac{B(D^0 \to \bar{K}^0 \pi^0)}{B(D^0 \to K^- \pi^+)}$	$0.45 \pm 0.08 \pm 0.05$
$\dfrac{B(D^0 \to \bar{K}^0 \rho^0)}{B(D^0 \to K^- \rho^+)}$	$0.11 \pm 0.04 \pm 0.02$
$\dfrac{B(D^0 \to \bar{K}^{*0} \pi^0)}{B(D^0 \to K^{*-} \pi^+)}$	$0.29 \pm 0.14 \pm 0.09$
$\dfrac{B(D^+ \to \phi \pi^+)}{B(D^+ \to K^- \pi^+ \pi^+)}$	$0.08 \pm 0.02 \pm 0.01$
$\dfrac{B(D_s^+ \to \bar{K}_s^0 K^+)}{B(D_s^+ \to \phi \pi^+)}$	$0.44 \pm 0.12 \pm 0.21$

In potential models

$$f_D^2 = \frac{12|\psi(0)|^2}{M_D} \ .$$

Thus,

$$\alpha = 16\pi^2 \left(\frac{f_D}{M_D}\right)^2 \ .$$

Maximal destructive interference can occur for $\alpha = 1$, which occurs for the reasonable value $f_D = 150$ MeV. With maximal interference, however, the ratio $\tau(D^+)/(\tau(D^0)$ does not exceed 1.5. If, however, we neglect color-suppressed terms, we have

$$\frac{\tau(D^+)}{\tau(D^0)} = \frac{2 + c_1^2 + c_2^2}{2 + c_1^2 + c_2^2 + 2\alpha c_1 c_2} \ ,$$

which admits larger values of the lifetime ratio.

Table XX compares D^0 and D^+ lifetimes and semileptonic branching ratios calculated by Buras *et al.* assuming large Pauli interference in D^+ decay with and without the inclusion of color-suppressed amplitudes. The neglect of color-suppressed amplitudes clearly improves the agreement with experiment.

The underlying reasons for the success of the phenomenological approach of Bauer, Stech and Wirbel can perhaps be found in a study of the relation of the $1/N$ expansion to the D hadronic decay matrix elements recently undertaken by Buras, Gérard and Rückl. We have already seen that neglect of color-suppressed amplitudes dramatically improves our understanding of the Mark III D branching ratio measurements. How may we justify this procedure? The work of Buras *et al.* provides a motivation. They first classify the amplitudes into three classes, then further investigate the order of the various contributions in a $1/N$ expansion. Buras *et al.* also define a diagrammatic language which facilitates such a classification. I will reproduce only sufficient detail here to make the basic points; the reader is referred to the original paper for more detail.

Consider the leading $(O(\sqrt{N}))$ contributions to the decays $D \to K\pi$. These are of three types:

I: $\quad \Gamma(D^0 \to K^-\pi^+) \sim |a_1|^2$

II: $\quad \Gamma(D^0 \to \bar{K}^0\pi^0) \sim |a_2|^2$

III: $\quad \Gamma(D^+ \to \bar{K}^0\pi^+) \sim |a_1 + xa_2|^2 \ .$

The factor x is process dependent; in the SU(3) limit, $x = 1$.

Table XIX

Isospin Amplitudes and

Phase Shifts Derived from

$$D \to PP \ \& \ D \to PV$$

Channel	$\mid A_{1/2}/A_{3/2} \mid$	$\delta_{1/2} - \delta_{3/2}$
$D \to \bar{K}\pi$	3.67 ± 0.27	$(77 \pm 11)^0$
$D \to \bar{K}^*\pi$	3.22 ± 0.97	$(84 \pm 13)^0$
$D \to \bar{K}\rho$	3.12 ± 0.40	$(0 \pm 26)^0$

Table XX

Comparison of Inclusive D^0 and D^+
Measurements with Estimates Including
Pauli Interference, with and without
Inclusion of Color-Suppressed Amplitudes

Measurement	Color-suppressed terms neglected	Color-suppressed terms included	Experiment
$\tau(D^0)$ $(\times 10^{-13}$ sec)	5.5	7.0	$4.42\pm^{0.15}_{0.14}$
$B(D^0 \to e^+X)$ (%)	11.5	14.4	7.5 ± 1.2
$\tau(D^+)$ $(\times 10^{-13}$ sec)	9.5	9.8	$10.29\pm^{0.44}_{0.40}$
$B(D^+ \to e^+X)$ (%)	19.7	20.4	17.0 ± 2.0
$\tau(D^+)/\tau(D^0)$	1.7	1.4	$2.25\pm^{0.15}_{0.14}$

In addition to the leading terms, there are others. The effective neutral current or color-suppressed terms, produced by the Fierz transformation and color algebra, generate an amplitude proportional to $a_2/\sqrt{N}$ from the term $\sim a_1\sqrt{N}$. Annihilation amplitudes are also proportional to $a_1/\sqrt{N}$, as are final state interactions. Soft gluon interactions are responsible for non-planar diagrams $\sim a_1/N^{3/2}$. The terms proportional to $1/\sqrt{N}$ can be further divided into two classes. There is a factorizable part, which is **retained** in the standard formulation, and a non-factorizable part due to soft gluon exchange, which is dropped. The Buras *et al.* ansatz can thus be summarized as follows. The amplitude for the channel $D \to K\pi$ can be written as an expansion in $1/N$:

$$A(D \to K\pi) = \sqrt{N}\left[b_0 + \frac{1}{N}(b_1 + b'_1) + O\left(\frac{1}{N^2}\right) + \cdots\right].$$

The b_0 and b_1 terms are factorizable; both are retained in the standard approach. The b'_1 and higher order terms are non-factorizable, as they are due to multiple soft gluon exchange and final state interactions, and are usually not retained. Thus, the rate for $D^0 \to K^-\pi^+$ in the standard approach is:

$$\Gamma(D^0 \to K^-\pi^+) \sim |a_1 + \frac{1}{3}a_2|^2.$$

Buras *et al.* point out that since the color-suppressed amplitude, b_1, is higher order in $1/N$, it, too, should be dropped. Thus,

$$\Gamma(D^0 \to K^-\pi^+) \sim |a_1|^2$$

to leading order in $1/N$. Thus, the phenomenological fit of Bauer *et al.*, which finds $\xi = 0$, is, in fact, just the neglect of factorizable terms which are higher order in $1/N$.

With this observation, one can just take over the fit to Mark III data of Bauer *et al.*, obtaining, thereby, both its successes and failures. Since annihilation diagrams and final state interactions are both higher order in $1/N$, this approach cannot cope with either the large $D^0 \to \bar{K}^0\phi$ rate nor with differences in the Cabibbo-suppressed regime such as $B(D^0 \to K^-K^+) \neq B(D^0 \to \pi^-\pi^+)$. The contribution of the annihilation process is, in fact, accounted for by fitting to the $\bar{K}^0\phi$ rate. Nonetheless, the $1/N$ expansion does serve to clarify major areas in the phenomenology of hadronic D decays.

Non-Spectator Decays

The lifetime differences among D^0, D^+ and D_s^+ mesons are due to a combination of effects in the hadronic decay sector. Final state interference effects are significant in D^+ decay, while non-spectator diagrams could be important in D^0 and D_s^+ decay.

A search for the process $D^0 \to \bar{K}^0 \phi$ was originally proposed by Bigi and Fukugita[95] as a "smoking gun" test of the existence of exclusive non-spectator processes (see Fig. 31). They predicted a rate of .2 − 1%, for the non-spectator decay. Spectator processes can produce this mode only through OZI violation, leading to a much smaller predicted rate. Another non-spectator process, e.g., $D_s^+ \to \bar{K}^{*0} K^+$ can also arise from color-suppressed spectator processes (Fig. 32), so this mode does not rate smoking gun status. The $\bar{K}^0 \phi$ mode was first found by ARGUS,[67] and has now also been seen by Mark III[69] and CLEO.[68] The branching ratios are shown in Table XII, normalized to the Mark III value for $B(D^0 \to \bar{K}^0 \pi^+ \pi^-)$. The rate is indeed large. There has been some debate over the extraction of the $\bar{K}^0 \phi$ signal from background, but the three measurements are in good agreement.

Do these observations then provide the "smoking gun" for exclusive non-spectator processes? This has engendered a lively debate. The problem is that it is difficult to understand how f_D can be sufficiently large to provide a wave function overlap which produces a 1% branching ratio. Donoghue[96] has pointed out that rescattering effects required by unitarity can generate $D^0 \to \bar{K}^0 \phi$ in the absence of annihilation amplitudes, via diagrams shown in Figure 33. Hussain and Kamal[97] contend, however, that the inelasticity required to generate $\bar{K}^0 \phi$ at the 1% level would also produce a large rate for $D^0 \to \bar{K}^0 \eta$, which is not observed. Baur et al.[98] have studied annihilation processes in the $1/N$ expansion approach. The rate is

$$\Gamma(D^0 \to \bar{K}^0 \phi) = |\frac{G_F}{\sqrt{2}} V_{cs}^* V_{ud} a_2 \langle \bar{K}^0 \phi | (\bar{s}d)_L | 0 \rangle \langle 0 | (\bar{u}c)_L | D^0 \rangle |^2 .$$

Writing $a_2(\xi) = c_2 + \xi c_1$, they observe that varying c_-/c_+ in the range $2.4 \leq \frac{c_-}{c_+} \leq 3.2$ produces

$$-0.27 \leq a_2 \leq -0.10 \quad , \quad \text{for} \quad \xi = \frac{1}{3} ,$$

and

$$-0.75 \leq a_2 \leq -0.52 \quad , \quad \text{for} \quad \xi = 0 .$$

Since, in the standard approach,

$$\Gamma(D^0 \to \bar{K}^0 \phi) = \Gamma_{\text{ann}} |c_2 + \frac{1}{3} c_1|^2 ,$$

while only the c_2 term is leading order in $1/N$, retaining only the leading term enhances the $D^0 \to \bar{K}^0 \phi$ rate by an order of magnitude. The problem is that Γ_{ann} itself is still small. Γ_{ann} can be parametrized as

$$\Gamma_{\text{ann}} \sim |\langle \bar{K}^0 \phi | (\bar{s}d)_L | 0 \rangle \langle 0 | (\bar{u}c)_L | D^0 \rangle |^2$$
$$\sim |F(m_D^2, 0^-) f_D|^2 .$$

$F(m_D^2, 0^-)$ can be approximated by the appropriate pole, but its evaluation at $q^2 \simeq$

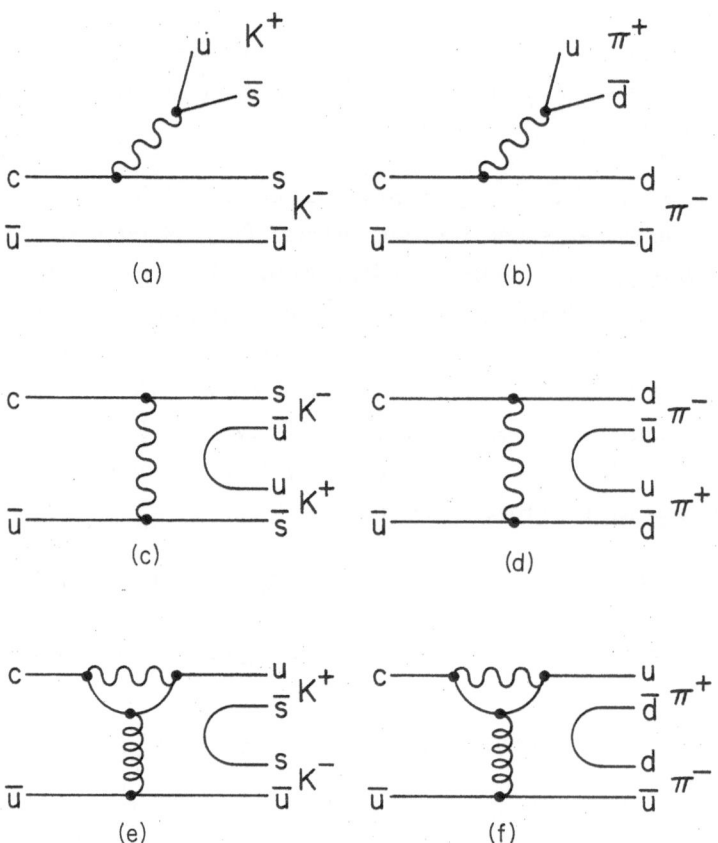

Fig. 33 Amplitudes contributing to the Cabibbo-suppressed decays, $D^0 \rightarrow K^- K^+$ and $D^0 \rightarrow \pi^+ \pi^-$. Diagrams a) and b) are the ordinary spectator amplitudes; c) and d) are exchange amplitudes; e) and f) are Penguin diagrams, which explicitly involve gluon exchange.

m_D^2 is difficult. The result is extremely sensitive to the degree of PCAC violation and $SU(3) \otimes SU(3)$ breaking at large q^2. Baur *et al.* parametrize this violation by a parameter v, with $v = 1$ in the PCAC limit. Then,

$$F(m_D^2, 0^-) \simeq a \left[\frac{m_D^2}{m_D^2 - m_K^2 - v} \right]$$

$$= \begin{cases} \dfrac{a m_D^2}{m_D^2 - m_K^2} & \text{for} \quad v = 0 \\[3mm] \dfrac{a m_K^2}{m_D^2 - m_K^2} & \text{for} \quad v = 1 . \end{cases}$$

Within this model, they then find:

	$v = 0$	$v = 1$	
$B(D^0 \to \bar{K}^0 \phi)$	7×10^{-4}	10^{-5}	standard
	7×10^{-3}	10^{-4}	leading $1/N_c$

We are thus led to the conclusion that the observation of $D^0 \to \bar{K}^0 \phi$ at the 1% level could be understood in the $1/N$ picture, if there were surprisingly large PCAC violation at large q^2, as evidence for non-spectator processes. It is more likely, however, that rescattering effects play at least some part. A test would be provided by a measurement of the branching ratio for $D_s^+ \to \rho^0 \pi^+$. This decay also proceeds via a non-spectator diagram. If it is large, there would be good evidence for a primary weak hadronic non-spectator process; if it is small, then the origin of $D^0 \to \bar{K}^0 \phi$ is likely to be in rescattering effects.

Cabibbo-Suppressed Decays

Hadronic D meson decays involving a single Cabibbo-suppressed vertex should naively occur at a rate of $tan^2\theta_c \sim 5.5\%$ compared to related allowed decays. Inspection of Table VIII, however, shows that while this may be approximately the case in the large, their are several anomalies in detail.

In particular, the rates for $D^0 \to K^- K^+$ and $D^0 \to \pi^- \pi^+$ differ by a factor of 3.7. Even allowing for phase space differences, this result is surprising. It is not clear at this time whether this difference arises from final state interactions or from the contribution of Penguin diagrams[99] (see Fig. 33). The very large ratio $B(D^+ \to \bar{K}^0 K^+)/B(D^+ \to \bar{K}^0 \pi^+)$ is likely to arise from the effect of destructive Pauli interference, which is operative in the case of $D^+ \to \bar{K}^0 \pi^+$ but not for the Cabibbo-suppressed channel.

Further progress in understanding Cabibbo-suppressed channels likely awaits a measurement of $B(D^+ \rightarrow \pi^+\pi^0)$. This exotic final state should not be subject to final state effects and thus is fairly unambiguously predicted to have the value $\frac{1}{2}tan^2\theta_c \times B(D^+ \rightarrow \bar{K}^0\pi^+)$.

Charmed Baryons

We are primarily concerned here with the weak decays of charmed baryons. Four charmed baryons (Λ_c^+, Ξ_c^+, Ξ_c^0 and Ω_c^0) decay weakly. The others will decay to these states via radiative or strong decays. In contrast to the case of charmed mesons, there is no particular inhibition of non-spectator amplitudes in charmed baryon decay. We might, therefore, expect the charmed baryons which decay weakly to have shorter lifetimes than charmed mesons. This is indeed the case for the Λ_c^+. The small number of events seen in Ξ_c^+ and Ω_c^0 decay indicate somewhat longer lifetimes.

There have been quite a number of measurements of the Λ_c^+ mass, as well as of the $\Lambda_c^+ - \Sigma_c$ mass difference. The Λ_c^+ mass also underwent an unusual transformation, in that early measurements which placed it at 2260 MeV were supplanted by improved measurements which raised the mass to 2281 MeV.[2]

The normalization of weak decay branching ratios for the charmed baryons depends on early measurements from Mark II,[100] together with a string of assumptions. All other branching ratio determinations are typically normalized to the Mark II results. The first hint of charmed baryon production in e^+e^- annihilation came from the observation of a step in $p+\bar{p}$ production at SPEAR energies.[101] The Mark II then remeasured this with increased precision, and also found a step in $\Lambda+\bar{\Lambda}$ production ($\Delta R(\Delta+\bar{\Lambda}) = 0.10\pm0.03$). Steps in the normalization procedure are:

1) the step $\Delta R(p + \bar{p}) = 0.31 \pm 0.06$ is entirely due to the onset of charmed baryon production

2) all charmed baryons so produced cascade to the Λ_c^+

3) $\dfrac{\Lambda_c^+ \rightarrow p}{\Lambda_c^+ \rightarrow all} = 0.6 \pm 0.1.$

Thus,
$$\sigma(\Lambda_c^+ + \bar{\Lambda}_c^+) = \frac{\Delta R(p + p) \times \sigma_{\mu\mu}}{\sigma(\Lambda_c^+ \rightarrow p)/\sigma(\Lambda_c^+ \rightarrow all)}$$
$$= 1.7 \pm 0.4 \text{ nb} \quad at \quad \sqrt{s} = 5.2 \text{ GeV}.$$

Table XXI summarizes the various measurements of Λ_c^+ branching ratios normalized to this result.

The Ξ_c^+ has been observed in the charged hyperon beam[109] at CERN in the channel $\Lambda K^-\pi^+\pi^+$ at $m(\Xi_c^+) = 2460 \pm 15$ MeV. The observed width is compatible with the experimental resolution.

Table XXI

Hadronic Decays of Λ_c^+

Decay Mode	Experiment	Measurement	Branching Ratio (%)	Comments
$\Lambda_c^+ \to pK^-\pi^+$	Mark II[100]	$\sigma(\Lambda_c + \bar{\Lambda}_c)B(\Lambda_c^+ \to pK\pi) = (37 \pm 12)\text{pb}$	2.2 ± 1.0	a
	ARGUS[102]	$\sigma \cdot B = (15.6 \pm 4.3)\text{pb}$	—	
pK^0	Mark II[103]	$\sigma(\Lambda_c + \bar{\Lambda}_c)B(\Lambda_c^+ \to p\bar{K}^0) = (18 \pm 8)\text{pb}$	1.1 ± 0.7	a
	ARGUS[102]	$\sigma \cdot B = (15.6 \pm 4.3)\text{pb}$	2.2 ± 0.6	b
pK^{*0}	Mark II	$\dfrac{B(\Lambda_c^+ \to pK^{*0})}{B(\Lambda_c^+ \to pK^-\pi^+)} = 0.18 \pm 0.10$	0.4 ± 0.2	a
	Basile et $al.$[104]	$\dfrac{B(\Lambda_c^+ \to pK^{*0})}{B(\Lambda_c^+ \to pK^-\pi^+)} = 0.42 \pm 0.24$	0.9 ± 0.5	b
	ARGUS[102]	$\dfrac{B(\Lambda_c^+ \to pK^{*0})}{B(\Lambda_c^+ \to pK^-\pi^+)} < 0.32$ @ 90% CL	< 0.7 @ 90% CL	b
$\Delta^{++}K^-$	Mark II[103]	$\dfrac{B(\Lambda_c^+ \to \Delta^{++}K^-)}{B(\Lambda_c^+ \to pK^-\pi^+)} = 0.17 \pm 0.07$	0.4 ± 0.2	a
	Basile et $al.$[104]	$\dfrac{B(\Lambda_c^+ \to \Delta^{++}K^-)}{B(\Lambda_c^+ \to pK^-\pi^+)} = 0.40 \pm 0.17$	0.9 ± 0.4	b
	ARGUS[102]	$\dfrac{B(\Lambda_c^+ \to \Delta^{++}K^-)}{B(\Lambda_c^+ \to pK^-\pi^+)} < 0.15$ @ 90% CL	< 0.33 @ 90% CL	b
$p\bar{K}_0\pi^+\pi^-$	ARGUS[102]	$\sigma \cdot B = (24.0 \pm 7.5)\text{pb}$	3.4 ± 1.1	b
	BIS-2[105]	$\dfrac{B(\Lambda_c^+ \to p\bar{K}^0\pi^+\pi^-)}{B(\Lambda_c^+ \to \Lambda\pi^+\pi^-\pi^+)} = 4.3 \pm 1.2$	3.9 ± 1.1	c
$\Lambda\pi^+$	Baltay et $al.$[106]	$\dfrac{B(\Lambda_c^+ \to \Lambda\pi^+)}{B(\Lambda_c^+ \to p\bar{K}^0)} = 0.67 \pm 0.78 \pm 0.35$	0.74 ± 0.86	d
	Kitagaki et $al.$[107]	$\dfrac{B(\Lambda_c^+ \to \Lambda\pi^+)}{B(\Lambda_c^+ \to p\bar{K}^0)} = 0.51 \pm 0.62 \pm 0.27$	0.56 ± 0.68	d
$\Lambda\pi^+\pi^+\pi^-$	ARGUS[102]	$\sigma \cdot B = (6.34 \pm 1.16 \pm 1.20)\text{pb}$	0.9 ± 0.2	b
	CLEO[108]		2.8 ± 1.3	

a) See Text

b) Normalized to Mark II $B(\Lambda_c^+ \to pK^-\pi^+)$

c) Normalized to ARGUS $\sigma \cdot B(\Lambda_c^+ \to \Lambda\pi^+\pi^+\pi^-)/\sigma \cdot B(\Lambda_c^+ \to pK^-\pi^+)$

d) Normalized to Mark II $B(\Lambda_c^+ \to p\bar{K}^0)$

The Ξ_c^+ has also recently been seen in 800 GeV neutron beam production at FNAL.[33] Two peaks are seen in the $\Lambda K \pi \pi$ invariant mass spectrum, which are interpreted as

$$\Xi_c^+ \to \Lambda K^- \pi^+ \pi^+ \qquad \text{and} \qquad \Xi_c^+ \to \Sigma^0 K^- \pi^+ \pi^+$$
$$\hookrightarrow \Lambda \gamma.$$

The mass is $m(\Xi_c^+) = 2448 \pm 5 \pm 30$ MeV, with $\sigma \cdot B(\Sigma^0 K \pi \pi)/\sigma \cdot B(\Lambda K \pi \pi) = 0.83 \pm 0.4$.

Three events interpreted at $\Omega_c^0 \to \Xi^- K^- \pi^+ \pi^+$ have also been reported,[110] with a mass $m(\Omega_c^0) = 2740$ MeV and a lifetime $\tau(\Omega_c^0) = (7.9 \pm 2.8 \pm 2.0) \times 10^{-13}$ sec.

The Σ_c^{++} was found in the first bubble chamber event at BNL,[111] which also found the Λ_c^+, via the cascade

$$\nu_\mu p \to \mu^- \Sigma_c^{++}(2426)$$
$$\hookrightarrow \Lambda_c^+ \pi^+$$
$$\hookrightarrow \Lambda \pi^+ \pi^- \pi^-.$$

Twenty events involving Σ_c^{++} cascades to Λ_c^+ in the $\Lambda \pi^+$, $\Lambda 3\pi$, $\bar{K}^0 p$ and $\bar{K}^0 p \pi^+ \pi^-$ channels were found in the 15 foot bubble chamber at FNAL,[112] allowing a measurement of the $\Sigma_c^{++} - \Lambda_c^+$ mass difference: $m(\Sigma_c^{++} - m(\Lambda_c^+) = 168 \pm 3$ MeV. The Σ_c^+ was observed in BEBC[113] with a track-sensitive target in the cascade

$$\nu_\mu p \to \mu^- \Sigma_c^+ \pi^+$$
$$\hookrightarrow \Lambda_c^+ \pi^0$$
$$\hookrightarrow K^- p \pi^+,$$

resulting in $m(\Sigma_c^+) - m(\Lambda_c^+) = 168 \pm 3$ MeV.

ARGUS[102] has recently reported new measurements of the Σ_c^0 and Σ_c^{++}, finding $m(\Sigma_c^{++}) - m(\Lambda_c^+) = 168.4 \pm 0.6$ MeV with 89 ± 20 events and $m(\Sigma_c^0) - m(\Lambda_c^+) = 167.0 \pm 0.7$ MeV. It thus appears that the $\Sigma_c^{++}, \Sigma_c^+, \Sigma_c^0$ multiplet is nearly degenerate, as expected.

Of the four charmed baryons which are expected to decay by weak interactions, three, the Λ_c^+, Ξ_c^+ and Ω_c^0 have thus been seen, the Ξ_c^0 has yet to be found. There are fundamental differences between the phenomenology of weak decays of charmed baryons and that of charmed mesons. We will use the Λ_c^+ as an example to explore the differences.

The $\Lambda_c^+(cdu)$ can decay weakly via a Cabibbo-allowed spectator process (Fig. 35a), or a W exchange process (Fig. 35b). The latter is not helicity-suppressed as in the charmed meson case, and, in fact, exchange diagrams are expected to dominate Λ_c^+

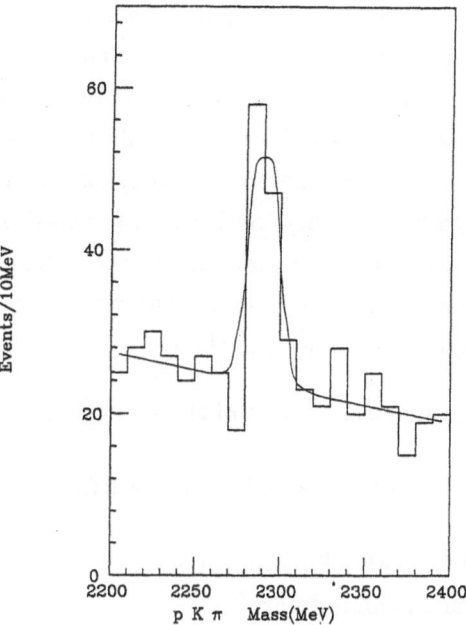

Fig. 34 Invariant mass of $pK\pi$ from experiment E691, showing Λ_c^+ photoproduction.

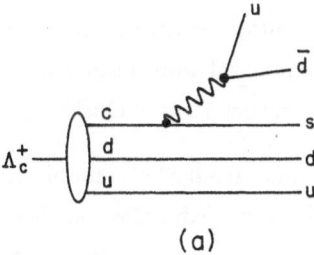

(a)

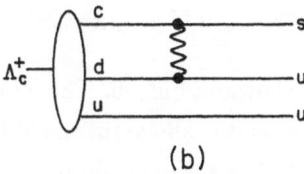

(b)

Fig. 35 Typical amplitudes for the weak decay of the Λ_c^+: a) Spectator diagram; b) exchange diagram, which in the case of baryons is not helicity-suppressed.

decay. In the naïve quark model, the (ud) quarks are in a $J = 0$, $I = 0$ state. Thus, the (cd) quarks are in $J = 1$ or $J = 0$ configurations with statistical weights $3 : 1$. The decay of the spin 1 configuration to two fermions in an s state is helicity-suppressed, but such a decay of the spin 0 system is allowed. As the spectator diagram results in two u quarks in the final state, Pauli interference effects will also be important. Since the baryon wavefunction is antisymmetric in color, only matrix elements of O_- are non-zero. The most significant distinction between charmed baryon and meson decay is thus that non-spectator effects enter at the valence quark level in baryon decay,[41] and are therefore relatively more important. There have been a number of detailed calculations of the non-leptonic decay branching ratios of charmed baryons.[114−118]

There has been a single measurement of the Λ_c^+ semileptonic branching ratio:[119]

$$B(\Lambda_c^+ \to e^+ X) = 0.045 \pm 0.017 \ ,$$

which, together with the short Λ_c^+ lifetime, points up the importance of non-leptonic enhancements in charmed hadron decay.

B QUARK DECAYS

The decay phenomenology of b quark containing hadrons is even richer than that for the charm sector. Since the dominant processes proceed through charm quark containing states, the experimental problem is, however, more difficult. Nonetheless, we have a great deal of experimental information on B meson decays, although the emphasis is more on inclusive, rather than exclusive, processes.

Most of this information comes from e^+e^- annihilation. The existence of the $\Upsilon(4S)$ resonance, which decays to B meson pairs close to threshold has been as important to the b sector as has been the $\psi(3770)$ to studies in the charm sector. Continuum studies at PEP and PETRA have yielded lifetime measurements, semileptonic branching ratios and the observation of $B^0 \bar{B}^0$ mixing. A single, very striking emulsion event[120] has been found by WA75 at CERN in 350 GeV/c π^- interactions.

The Υ family spectrum is shown in Fig. 36. The $\Upsilon(4S)$ at $10 \cdot 577$ GeV, is noticeably wider than the three lower Υ's, as it is above threshold for explicit $B\bar{B}$ production. The $\Upsilon(4S)$ has a width of $20 \pm 2 \pm 4$ MeV and a height of $\sim .6$ nb at the peak. CLEO, CUSB, Crystal Ball and ARGUS have exploited the unique features of the $\Upsilon(4S)$ to provide a great deal of information on the properties of B mesons. There is good evidence that the $\Upsilon(4S)$ decays entirely to $B\bar{B}$ mesons:

$$B(\Upsilon(4S) \to B\bar{B}) > 0.83 \,.$$

At PEP and PETRA energies, b quark production represents 1/11 of the total hadronic

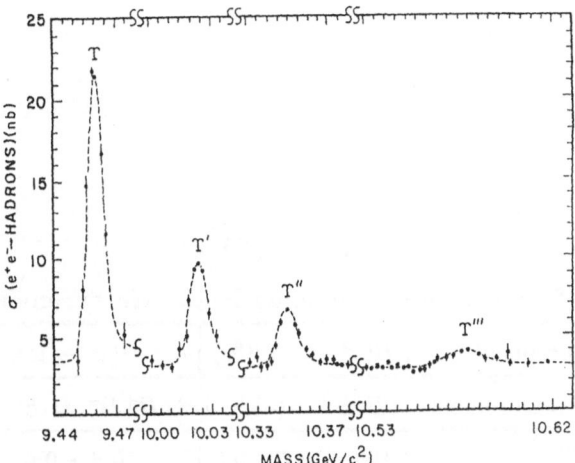

Fig. 36 Four lowest members of the Υ family (CUSB). The Υ^{111} or $\Upsilon(4S)$ lies above $B\bar{B}$ meson threshold.

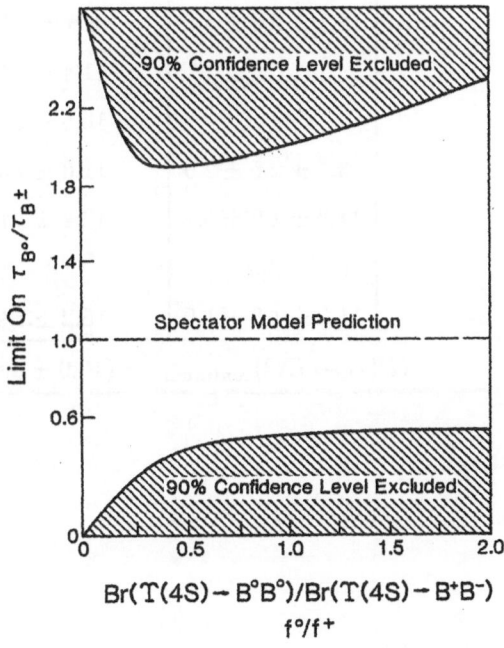

Fig. 37 CLEO results on the ratio of B^0 to $B^{\pm}$ lifetimes.

Table XXII

B Semileptonic Branching Ratio Measurements

Experiment	$B(B \to eX)(\%)$	$B(B \to \mu X)(\%)$
$\Upsilon(4S)$ Measurements $\;(B^0\bar{B}^0 = (45 \pm 5)\%, B^+B^- = (55 \pm 5)\%)$		
CLEO[121]	$12.0 \pm 0.7 \pm 0.5$	$10.8 \pm 0.6 \pm 1.0$
CUSB[122,123]	$13.2 \pm 0.8 \pm 1.4$	$11.2 \pm 0.9 \pm 0.1$
ARGUS[124]	$12.0 \pm 0.9 \pm 0.8$	
$\langle B(B \to lX)\rangle_{4S} = (11.8 \pm 0.3 \pm 0.6)\%$		
Continuum Measurements		
TASSO[125]	$11.1 \pm 3.4 \pm 4.0$	$11.7 \pm 2.8 \pm 1.0$
CELLO[126]	$14.1 \pm 5.8 \pm 3.0$	$8.8 \pm 3.4 \pm 3.5$
JADE[127]		$11.4 \pm 1.8 \pm 2.5$
Mark J[128]		$10.5 \pm 1.5 \pm 1.3$
Mark II[129]	$13.5 \pm 2.6 \pm 2.0$	$11.6 \pm 2.1 \pm 1.7$
MAC[130]	$11.3 \pm 1.9 \pm 3.1$	$12.4 \pm 1.8 \pm 2.2$
DELCO[131]	14.6 ± 2.8	
TPC[132]	$11.0 \pm 1.8 \pm 1.0$	$15.2 \pm 1.9 \pm 1.2$
$\langle B(B \to lX)\rangle_{continuum} = (12.0 \pm 0.6 \pm 1.5)\%$		

cross section. Little is known about hadronic production cross sections, apart from an upper limit established on the b production cross section from the single WA75 event.

Naive counting of available channels (Figure 9) yields an estimate of the semileptonic branching ratio $B(b \to cl\nu) \simeq 13\%$ when the phase space reduction for $\bar{c}s$ pairs is taken into account. As the renormalization scale is now 5 GeV, hard gluon effects produce a smaller non-leptonic enhancement than in D decay: $(2c_+^2 + c_-^2)/3 = 1.14$ for $c_+ = 0.85$, $c_- = 1.40$. Thus, we expect a B semileptonic branching ratio of 11-12%. This is nicely confirmed by the data, shown in Table XXII. On the $\Upsilon(4S)$, only B_d^0 and B_u^- are produced, while, in the continuum, all types of B mesons and baryons contribute. The two types of measurements, therefore, need not agree. Nonetheless, within the precision of the experiments, the results are identical. The continuum measurements indicate that the muonic branching ratio may be higher than the electronic branching ratio, although this is, of course, unlikely.

Since semileptonic decay is predominantly a spectator process, it is possible to use the $\Upsilon(4S)$ data to place bounds on the ratio of B^0 and B^+ lifetimes, since

$$\frac{\tau(B^0)}{\tau(B^+)} = \frac{B(B^0 \to lX)}{B(B^+ \to lX)}.$$

If the fraction of $B^0\bar{B}^0(B^+B^-)$ produced at the $\Upsilon(4S)$ is denoted by $f_0(f_+)$, then

$$\frac{N_{B\bar{B}} \, N_{l+l-}}{\epsilon N_l} = \frac{f_0[B(B^0 \to lX)]^2 + f_+[B(B^+ \to lX)]^2}{f_0 B(B^0 \to lX) + f_+ B(B^+ \to lX)}.$$

CLEO[133] has used this relation (Fig. 37) to place the bound

$$0.43 < \frac{\tau(B^0)}{\tau(B^+)} < 2.03,$$

for $f_0 = (1 - f_+) = 0.4$. As can be seen from the figure, this method is not very sensitive near equal lifetimes.

Limits on $|V_{bu}|/|V_{bc}|$

The endpoint spectrum in semileptonic B meson decay provides the most sensitive method to evaluate $|V_{bu}|/|V_{bc}|$. As the lepton spectrum in $b \to ul\nu_l$ extends beyond that for $b \to cl\nu_l$, the endpoint region around 2.5 GeV/c allows extraction of the allow amount of $b \to u$ even though the $b \to c$ leptons dominate the bulk of the spectrum. The $\Upsilon(4S)$ resonance is the optimal place for this investigation, as B mesons are produced copiously at low momentum. While the first CLEO and CUSB results were based on the spectator model of Altarelli et al.,[134] which used an order α_s QCD calculation, analogous to radiative corrections to muon decay, difficulties in fitting improved data have led to a reexamination of the technique and a realization of the model-dependence of the extracted limits.

The newest limits come from ARGUS[135] and CLEO[136] data. The problem is best illustrated in Fig. 38, which shows the lepton spectra for $b \to X_c e \nu_e$ and $b \to X_u e \nu_e$ calculated using a constituent quark model with harmonic oscillator wave functions by Grinstein, Isgur and Wise.[74] The softening of both spectra due to use of explicit hadronic final states, compared to the free quark model, is clearly evident. Since the experimental spectra fit quite satisfactorily to $b \to c l \nu_l$ alone, the problem is to determine how much $b \to u l \nu_l$ can be tolerated.

ARGUS sets a limit for $b \to u/b \to c$ of $< 12\%$ at 90% confidence using the spectator model and a fit to the Grinstein $et\ al.$ model above 1.6 GeV/c. CLEO has done an elaborate study using six models and differing portions of the lepton spectrum. The most sensitive result comes from a fit to the 2.2-2.6 GeV/c region: $b \to u/b \to c < 2.9\%$ for the Altarelli $et\ al.$ model and $< 4.5\%$ for the Grinstein $et\ al.$ model. Other models yield limits as high as 6%. There is thus a substantial model dependence to the limit. Two further caveats should be mentioned. First, the Grinstein $et\ al.$ model is subject to a $b \to u$ normalization uncertainty of a factor of two, and second, this model has not successfully reproduced the Mark III data on exclusive D semileptonic decays, as described above. Thus, substantial uncertainty still attends the extraction of this K-M matrix element. The branching ratio limits can be used to extract a limit on $|V_{bu}/V_{bc}| \lesssim 0.2$.

Exclusive B Meson Decays

Much less is known about exclusive B meson decays than about those of charmed mesons. Both ARGUS and CLEO have reconstructed a small sample of exclusive decays, however, which allow us a preliminary glimpse into the details of the weak decays of the b quark system.

It is primarily the existence of the $\Upsilon(4S)$ resonance, which decays exclusively to $B\bar{B}$ pairs, which has made exclusive decay studies possible. Exclusive state reconstruction is difficult in that the final state multiplicity is large ($\langle n_{\mathrm{charged}} \rangle \simeq 6/B_{\mathrm{meson}}$) and that is in general necessary to also reconstruct a charmed meson in order to find a B meson.

Charmed channels thus far employed in such reconstruction are $D^0 \to K^- \pi^+$, $K^- \pi^+ \pi^+ \pi^-$, $D^+ \to K^- \pi^+ \pi^+$ in addition to $\psi \to e^+ e^-$ and $\mu^+ \mu^-$. B meson branching ratios so derived are thus scaled by the appropriate D meson branching ratios. The use of the beam-constraint technique here, as in similar work at the $\psi(3770)$, results in improved mass resolution, in this case of the order of 5 MeV. A danger of this technique is that low momentum pions (~ 100 MeV/c), which are abundantly produced, can be missed without influencing the resultant beam-constrained mass by more than a few MeV. This was apparently the case with early CLEO exclusive results.[137] This effect, together with the new Mark III branching ratios, accounts for the large changes between

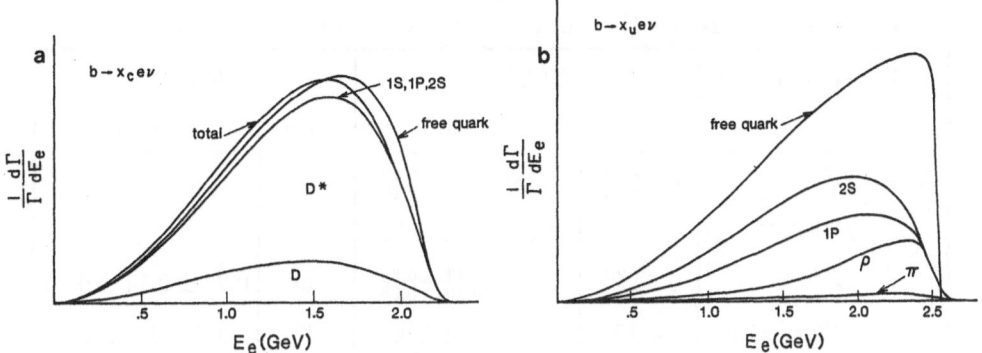

Fig. 38 Electron spectra arising from a) $b \rightarrow ce\nu$ and b) $b \rightarrow ue\nu$ decay in the quark model calculation of Grinstein *et al.*

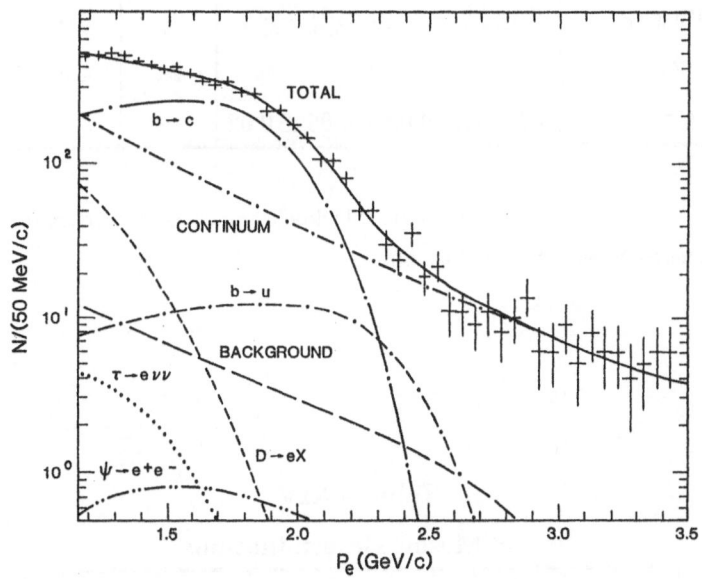

Fig. 39 ARGUS electron momentum spectra from B meson decay on the $\Upsilon(4S)$, with various calculated components of the spectrum.

Table XXIII

Exclusive B Meson Hadronic Branching Ratio Measurements[†]

Decay Mode	CLEO		ARGUS	
	Events	$B(\%)$	Events	$B(\%)$
$\bar{B}^0 \rightarrow D^+\pi^-$	$4.3\pm^{2.4}_{2.1}$	$0.48 \pm^{0.26}_{0.23} \pm^{0.12}_{0.10}$		
$D^0\pi^+\pi^-$	< 10	< 3.0 @ 90% CL		
$D^{*+}\pi^-$	$5.3\pm^{2.8}_{2.2}$	$0.25 \pm^{0.13}_{0.10} \pm^{0.08}_{0.05}$	5	$0.27 \pm 0.14 \pm 0.10$
$D^{*+}\pi^+\pi^-\pi^-$	< 15	< 3.7 @ 90% CL	27	$3.3 \pm 0.9 \pm 1.6$
$D^{*+}\pi^-\pi^0$			8	$1.5 \pm 0.8 \pm 0.8$
$\psi\bar{K}^{*0}$	5.0 ± 2.2	$0.41 \pm 0.19 \pm 0.03$		
$B^- \rightarrow D^0\pi^-$	$14.0\pm^{4.6}_{3.9}$	$0.36 \pm^{0.12}_{0.10} \pm^{0.08}_{0.05}$		
$D^+\pi^-\pi^-$	$1.2\pm^{2.0}_{1.1}$	$0.20 \pm^{0.33}_{0.18} \pm^{0.19}_{0.06}$		
$D^{*0}\pi^-$		0.51 ± 0.38		
$D^{*+}\pi^-\pi^-$	$2.7\pm^{1.9}_{1.7}$	$0.16 \pm^{0.11}_{0.10} \pm^{0.06}_{0.03}$	7	$0.5 \pm 0.2 \pm 0.3$
$D^{*+}\pi^-\pi^-\pi^0$			24	$4.3 \pm 1.3 \pm 2.6$
ψK^-	3.0 ± 1.7	$0.09 \pm 0.06 \pm 0.02$		

[†]Note that these measurements use the Mark III D hadronic branching ratio before the recent revision. They should therefore be scaled upward by 20%.

Table XXIV

B Meson Determinations

	CLEO	ARGUS
$M_{\bar{B}^0}$ (MeV)	$5281.3 \pm 0.8 \pm 2.0$	$5278.2 \pm 1.0 \pm 3.0$
M_{B^-} (MeV)	$5279.3 \pm 0.8 \pm 2.0$	$5275.8 \pm 1.3 \pm 3.0$
$\Delta M(\bar{B}^0_d - B^-_u)$	$3.1 \pm 1.8 \pm 2.0$	$2.4 \pm 1.6 \pm 1.0$

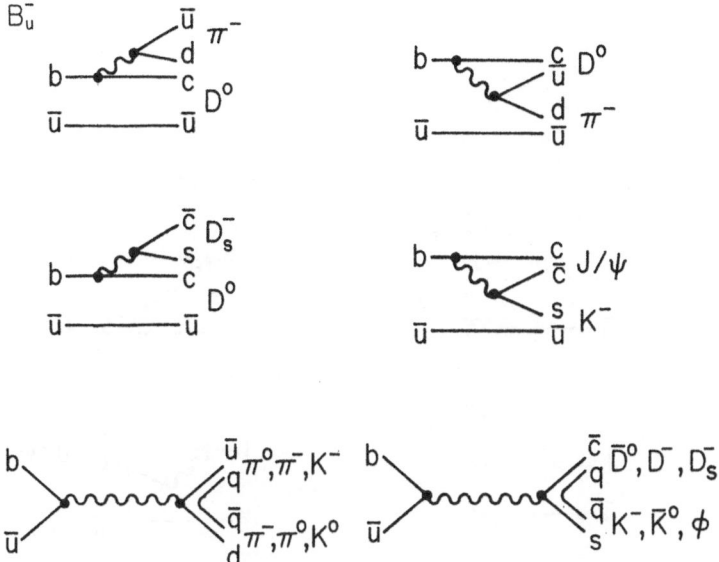

Fig. 40 Amplitudes contributing to B_u^- hadronic decay. a) and b) are the K-M-allowed spectator processes; c) and d) the allowed color-suppressed spectator processes; e) and f) are K-M-suppressed annihilation amplitudes.

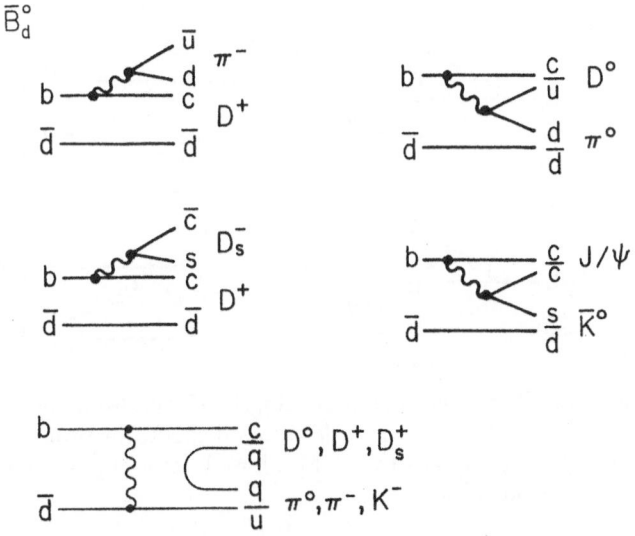

Fig. 41 Amplitudes contributing to $\bar{B}_d^0$ hadronic decay. a) and b) are the K-M-allowed spectator processes; c) and d) the allowed color-suppressed spectator processes; e) is the K-M-suppressed exchange amplitude.

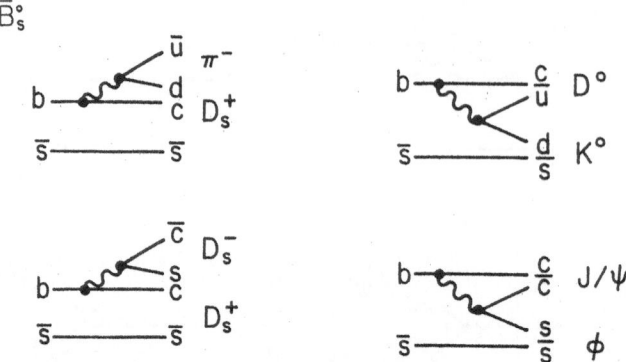

Fig. 42 Amplitudes contributing to $\bar{B}_s^0$ hadronic decay. a) and b) are the K-M-allowed spectator processes; c) and d) are the K-M-allowed color-suppressed amplitudes.

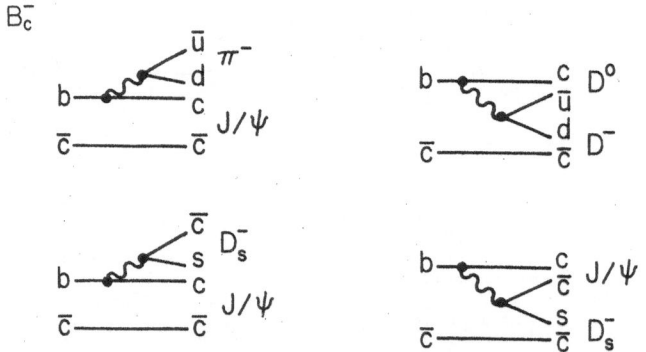

Fig. 43 Amplitudes contributing to B_c^- hadronic decay. a) and b) are the K-M-allowed spectator processes; c) and d) are the K-M-allowed color-suppressed amplitudes.

the original results and the CLEO[138] results, shown together with recent ARGUS[139] branching ratios in Table XXIII. The total number of reconstructed exclusive B decays in the two experiments is of the order of one hundred.

Figure 44 shows the beam-constrained mass distribution for the new CLEO results, while Fig. 45 shows the ARGUS data. Comparison of neutral and charged state exclusive masses allows a determination of the individual B_d^0 and B_u^- masses. These are summarized in Table XXIV. Note that the CESR and DORIS II mass scales, as seen in the measurements of the various Υ masses, are slightly different. The average $B_0 - B^-$ mass difference:

$$\Delta m(B^0 - B^-) = 2.8 \pm 1.1 \pm 2.0 \text{ MeV}$$

is, however, unaffected by this discrepancy. Note that the quoted exclusive branching ratios are also sensitive to assumed B^0/B^- production ratio on the $\Upsilon(4S)$ and that the two groups use slightly different values.

Limits have also been placed on a number of other exclusive B decay modes. These are summarized in Table XXV. These limits are already sufficient to exclude the possibility of a large contribution of penguin diagrams in exclusive B decays.

Inclusive B Meson Decays

In addition to studies of inclusive semileptonic processes, there have been studies of inclusive production of the ψ and of charmed mesons in B decay.

J/ψ production in B decay occurs via the color-mismatched diagrams shown in Figs. 40-43. Isolation of a ψ signal at the $\Upsilon(4S)$ via the $\mu^+\mu^-$ and e^+e^- decay channels is facilitated by a selection of events with, for example, the Fox-Wolfram parameter $H_2 < 0.3$ to discriminate against continuum background. Figure 46 shows the ARGUS signal,[144] which also provides a hint of ψ' production. Table XXVI summarizes the experimental results. Note that the ARGUS result implies that about 20% of the inclusive J/ψ production comes from ψ' decay.

The momentum distribution of the ψ in $B \to \psi X$ for the two experiments is shown in Fig. 47. While statistics are low, it appears that the CLEO momentum spectrum[142] peaks at higher values than that of ARGUS. This would lead to differing conclusions as to the mass of the system accompanying the ψ in B decay.

Since no direct evidence of $b \to u$ transitions has been obtained, it is of interest to compare the various measurements of charm production in B meson decay with models of $b \to c$ transitions. In the limit $\Gamma(b \to u) = 0$, the spectator model predicts 1.15 charmed quarks to be produced per b decay, since both vertices involving a W can produce charm.

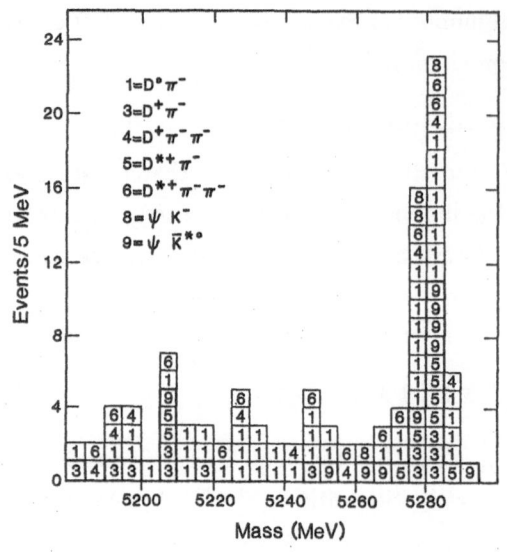

Fig. 44   CLEO results on exclusive
B meson hadronic decay.

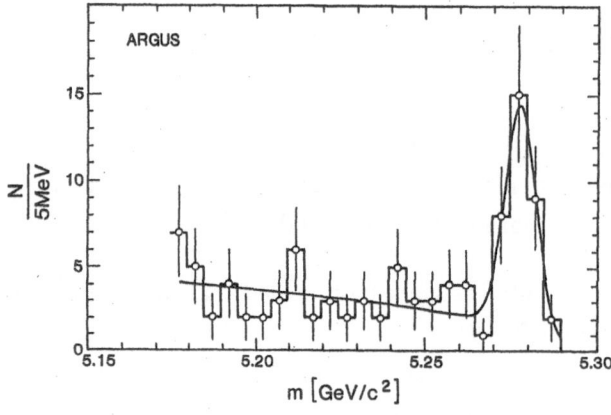

Fig. 45   ARGUS results on exclusive B meson hadronic decay. Channels included are $\bar{B}^0 \rightarrow D^{*+}\pi^-$, $\bar{B}^0 \rightarrow D^{*+}\pi^-\pi^0$, $\bar{B}_0 \rightarrow D^{*+}\pi^- \pi^-\pi^+$, $B^- \rightarrow D^{*+}\pi^-\pi^-$, and $B^- \rightarrow D^{*+}\pi^-\pi^-\pi^0$.

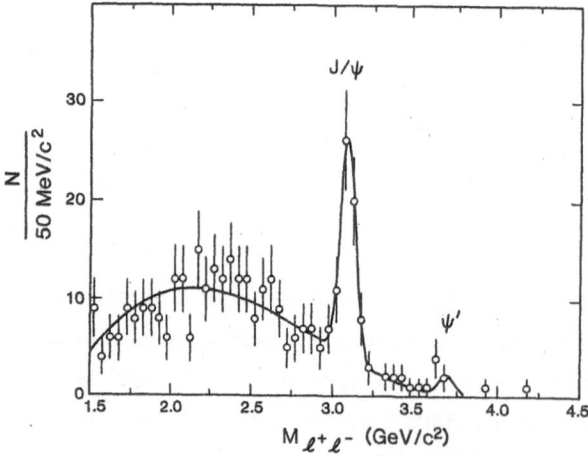

Fig. 46   ARGUS results on J/ψ and ψ' production from B meson decay at the $\Upsilon(4S)$.

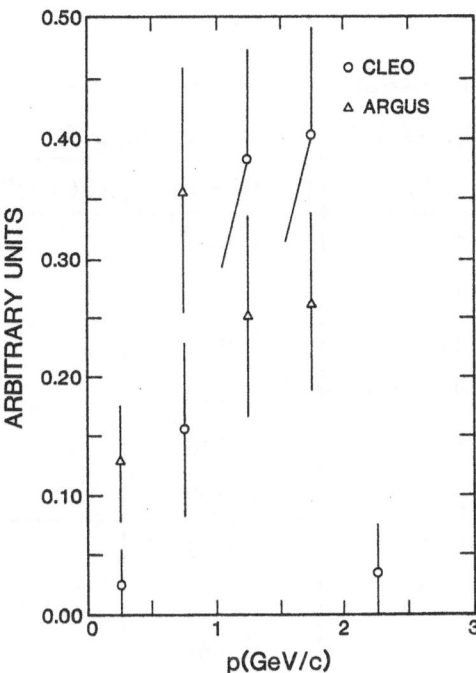

Fig. 47 Momentum spectra of J/ψ produced in B meson decay at the $\Upsilon(4S)$.

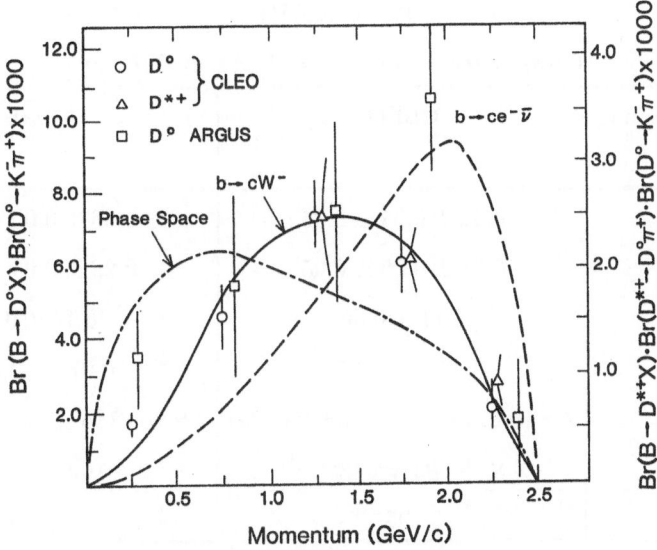

Fig. 48 Momentum spectra of D mesons produced in B meson decay at the $\Upsilon(4S)$.

Table XXV

Limits at 90% Confidence Level on

B Meson Hadronic Branching Ratios

Decay Mode	CLEO[138] $B(\%)$	ARGUS[140] $B(\%)$
$\bar{B}^0 \rightarrow \pi^{\pm}\pi^{\mp}$	< 0.03	< 0.04
$\rho^{\pm}\pi^{\mp}$	< 0.62	
$\rho^0\rho^0$	< 0.05	
$\pi^{\pm}a_1(1260)^{\mp}$	< 0.12	
$\pi^{\pm}a_2(1320)^{\mp}$	< 0.16	
$p\bar{p}$	< 0.02	
$B^- \rightarrow \pi^-\pi^0$	< 0.22	
$\pi^-\pi^+\pi^-$		< 0.25
$\rho^0\pi^-$	< 0.02	< 0.07
$\rho^0 a_1(1270)^-$	< 0.32	
$\rho^0 a_2(1320)^-$	< 0.23	

Table XXVI

Charm Production in B Meson Decay

Charm-Containing Particle	CLEO	ARGUS
D^0	$0.52 \pm 0.04 \pm 0.04$	$0.66 \pm 0.09 \pm 0.10$
D^+	$0.21 \pm 0.05 \pm 0.05$	$0.29 \pm 0.09 \pm 0.07$
D_s^+	0.11 ± 0.06	0.14 ± 0.03
Λ_c^+	0.074 ± 0.029	0.074 ± 0.029[†]
J/ψ	$2 \times (0.0118 \pm 0.0023 \pm 0.0025)$	$2 \times (0.0115 \pm 0.0022 \pm 0.0020)$
ψ'	$2 \times (0.005 \pm 0.0026)$[††]	$2 \times (0.005 \pm 0.0026)$
	.95	1.20

[†] CLEO value
[††] ARGUS value

Both CLEO and ARGUS have isolated inclusive charm signals on the $\Upsilon(4S)$.[143] As there is significant charm production in the continuum, it is necessary to measure this contribution separately and to subtract it in order to isolate the fraction attributable to B meson decay. This is possible because charmed particles from B decay are limited to the region $x = P/E_{\text{beam}} \leq 0.5$. A cut is also made on event shape to emphasize $\Upsilon(4S)$ decays. The D momentum spectra are shown in Fig. 48, compared to the spectra expected from contributing primitive processes. The measured product branching fractions are shown in Table XXVI.

D_s^+ decays are isolated using the $\phi\pi^+$ mode, whose branching ratio, as we have seen, is not directly measured, but depends on several assumptions. The CLEO D_s^+ momentum spectrum is shown in Fig. 49. It appears that D_s^+ production in B decay is dominated by two body decays ($B \rightarrow D_s^+ D$, $D_s^+ D^*$, $D_s^+ D^*$).

Neutral Meson State Mixing

The phenomenon of mixing observed three decades ago in the $K^0 - \bar{K}^0$ system can also occur in meson systems comprised of heavy quarks. This mixing is generated by second-order weak processes: to the extent that such diagrams are calculable, they provide another possible window into the regime of higher masses.

States containing heavy quarks in flavor isosinglet combinations, i.e., $(K^0, \bar{K}^0)$, $(D^0, \bar{D})$, $(B_d^0, \bar{B}_d^0)$, $(B_s^0, \bar{B}_s^0)$ or generically $(M^0, \bar{M}^0)$, which are eigenstates of the (flavor-conserving) strong and electromagnetic interactions, will be mixed by second order weak interactions. M^0 and $\bar{M}^0$ have opposite CP; we will use the convention $|\bar{M}^0\rangle = CP|\bar{M}^0\rangle$. The behavior of these states is easily described in the Wigner-Weisskopf formalism. The effective Hamiltonian H is a 2×2 matrix, with a hermitean part, and an anti-hermitean part which represents decay out of the $(M^0, \bar{M}^0)$ system:

$$H = M - \frac{i}{2}\Gamma.$$

CPT invariance implies $M_{11} = M_{12}$; $M_{21} = M_{12}^*$; $\Gamma_{11} = \Gamma_{22}$ and $\Gamma_{21} = \Gamma_{12}^*$. The physical eigenstates are obtained by diagonalizing H:

$$|M_1\rangle = \frac{1}{\sqrt{2(1+|\epsilon|^2)}}[(1+\epsilon)|M^0\rangle + (1-\epsilon)|\bar{M}^0\rangle]$$

$$|M_2\rangle = \frac{1}{\sqrt{2(1+|\epsilon|^2)}}[(1+\epsilon)|M^0\rangle - (1-\epsilon)|\bar{M}^0\rangle],$$

where

$$\left(\frac{1-\epsilon}{1+\epsilon}\right)^2 = \frac{M_{12}^* - \frac{i}{2}\Gamma_{12}^*}{M_{12} - \frac{i}{2}\Gamma_{12}}.$$

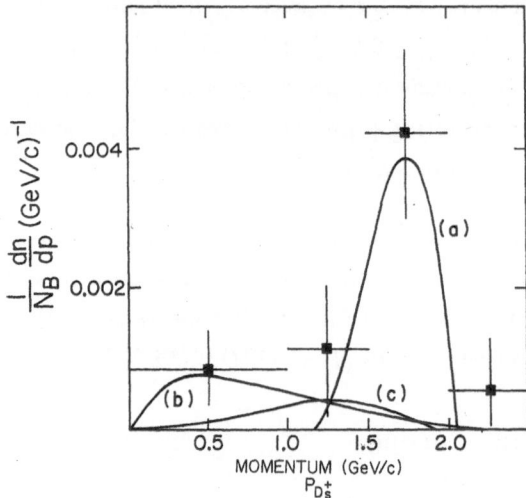

Fig. 49 Momentum spectra of D_s^+ mesons produced in B meson decay, with predicted curves for a) spectator hadronic decay; b) semileptonic decay and c) the W-exchange process (CLEO).

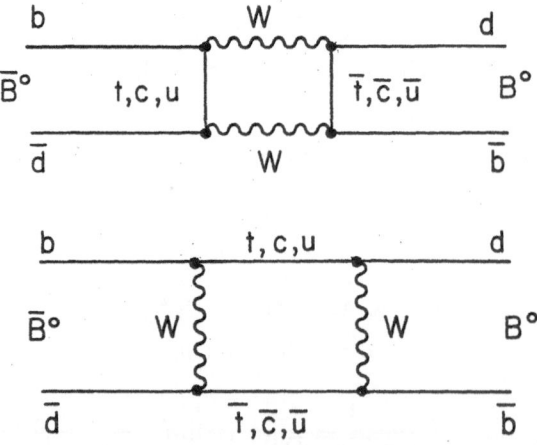

Fig. 50 Second order weak diagrams contributing to $B^0 - \bar{B}^0$ mixing.

The masses m_i and widths γ_i of the physical states M_i are given by

$$m_i - \frac{i}{2}\gamma_i = M_{11} - \frac{i}{2}\Gamma_{11} \pm \left[\left(M_{12}^* - \frac{i}{2}\Gamma_{12}^*\right)\left(M_{12} - \frac{i}{2}\Gamma_{12}\right)\right]^{\frac{1}{2}}.$$

If CP is conserved ($\epsilon = 0$), then M_1 and M_2 are simultaneous eigenstates of CP, M and Γ, and the expressions simplify to

$$m_i = M_{11} \pm M_{12}$$

$$\gamma_i = \Gamma_{11} \pm \Gamma_{12}.$$

If CP is not conserved, then M_1 and M_2 are no longer orthogonal, by an amount

$$\langle M_1 | M_2 \rangle = \frac{2Re\epsilon}{1 + |\epsilon|^2}.$$

As the observation of CP violating effects in the neutral D and B systems is not likely in the near future, let us assume that CP is conserved in these systems.

If we prepare a state $|M^0\rangle$ at $t = 0$ by strong or electromagnetic production, it will evolve as:

$$|M^0(t)\rangle = \frac{1}{2}\left(e^{-im_1 t - \frac{\gamma_1}{2}t} + e^{-im_2 t - \frac{\gamma_2}{2}t}\right)|M^0\rangle$$
$$+ \frac{1}{2}\left(e^{-im_1 t - \frac{\gamma_1}{2}t} - e^{-im_2 t - \frac{\gamma_2}{2}t}\right)|\bar{M}^0\rangle.$$

Thus, in general, we expect $M^0, \bar{M}^0$ mixing to occur. The amount of mixing is seen to be governed by

$$\delta m \equiv m_2 - m_1$$

and

$$\Gamma_\pm \equiv \frac{\gamma_1 \pm \gamma_2}{2}.$$

In the K^0 system, $\gamma_2 \gg \gamma_1$, due to the large phase space difference for the dominant processes $K_1^0 \rightarrow 2\pi$ and $K_2^0 \rightarrow 3\pi$. In heavier systems, such differences do not amount to very much, so that $\gamma_1 \simeq \gamma_2$. Thus $\Gamma_- \simeq 0$ and there will be a single lifetime $\Gamma \cong \Gamma_+$. In general, the amount of mixing is governed by

$$\frac{\delta m}{\Gamma} \equiv x,$$

and

$$\frac{\Delta\Gamma}{\Gamma} \equiv y (\simeq 0 \text{ for } D^0, \bar{D} \text{ and } B^0 \bar{B}^0).$$

We can usefully define a mixing parameter, r, for a state prepared as $|M^0\rangle$ at $t = 0$, by:

$$r = \frac{\int_0^\infty I(\bar{M}^0)dt}{\int_0^\infty (M^0)dt} = \frac{x^2 + y^2}{2 + x^2 - y^2} \, .$$

Evidently, $r = 0$ signifies no mixing, while $r = 1$ implies complete mixing.

In the $K^0, \bar{K}^0$ system, the fast component rapidly decays away, leaving only the slow component, the familiar $K^0, \bar{K}^0$ mixture. In heavier neutral systems, both components have similar lifetimes, so that significant mixing can occur through the $M^0, \bar{M}^0$ transition. We can get a good estimate of Γ and $\Delta\Gamma$ through our familiar spectator picture. In order to estimate δm, we must estimate the various possible contributions:

1) Charged current box diagrams of the type shown in Figure 50 may be calculated in the short distance limit. In the case of $D^0, \bar{D}$ mixing, each diagram is at least doubly Cabibbo-suppressed. The dominant contributions to Δm are:

$$\Delta m(D^0 \bar{D}) = \frac{G_F^2 B_D f_D^2 m_D}{4\pi^2}(m_s^2 - m_d^2)V_{cs}^2 V_{ud}^{*2}$$

$$\Delta m(B^0 \bar{B}^0) = \frac{G_F^2 B_B f_B^2 m_B}{4\pi^2} \left[\eta(m_i^2 + \frac{1}{3}m_b^2 + \frac{3}{4}m_b^2) \ln \frac{m_i^2}{m_b^2} \right]$$
$$\times V_{tb}^2 V_{td}^{*2} \, .$$

The factors B_D and B_B are the so-called "Bag Constants", which enter through the definition

$$\langle D^0 | [(c\bar{s})_L]^2 | \bar{D} \rangle = \frac{4}{3} B_D f_D^2 m_D$$

$$\langle B_d^0 | [(\bar{b}d)_L]^2 | \bar{B}_d^0 \rangle = \frac{4}{3} B_{B_d} f_B^2 m_B$$

$$\langle B_s^0 | [(\bar{b}s)_L]^2 | \bar{B}_s^0 \rangle = \frac{4}{3} B_{B_s} f_B^2 m_B \, .$$

In the vacuum insertion approximation, $B = 1$. There have been numerous estimates of B, particularly in view of the role analogous calculations play in estimates of CP violation in the K^0 system. Values of B obtained in the static quark model, harmonic oscillator model and MIT bag model, sum rules, chiral perturbation theory and lattice calculations range from -0.4 to 2.9, with the majority clustering around 0.4 or so. The dominant opinion seems to be that B is of order 1 multiplied by uncertain scale-dependent logarithmic corrections, which may be substantial.

The Wolfenstein parametrization of the K-M matrix provides a simple method to estimate the relative size of Δm in the neutral D and B systems. The leading terms in

Δm are

$$D^0\bar{D} : \lambda^2 m_s^2 + \lambda^6 m_s m_b + \lambda^{10} m_b^2$$

$$B_d^0\bar{B}_d^0 : \lambda^6 m_t^2 + \lambda^6 m_c m_t + \lambda^6 m_c^2$$

$$B_s^0\bar{B}_s^0 : \lambda^4 m_t^2 + \lambda^4 m_t^2 + \lambda^4 m_c m_t + \lambda^4 m_c^2,$$

leading to the prediction

$$\Delta m_{B_s^0} : \Delta m_{B_d^0} : \Delta m_{D^0} = \lambda^2 \frac{m_t^2}{m_s^2} : \lambda^4 \frac{m_t^2}{m_s^2} : 1$$

$$\simeq 300 : 15 : 1 .$$

With reasonable estimates for other parameters, the box diagrams lead to values of r in the following ranges:

$$r_{D^0} \sim 10^{-7}$$

$$r_{B_d^0} \sim 5 \times 10^{-3} - 2 \times 10^{-2}$$

$$r_{B_s^0} \sim 3 \times 10^{-1} - 9 \times 10^{-1} .$$

Thus, the $B_s^0\bar{B}_s^0$ system appears to be the best place to look for mixing.

2) Long distance contributions may, in the case of the K^0 and D^0 systems, be as large as or larger than both diagrams, although they are almost certainly unimportant in the B^0 sector. Such diagrams, shown schematically in Figure 51, lead to terms in the case of D^0's, of the form

$$\Sigma_I \frac{\langle D^0|H_w|I\rangle\langle I|H_w|\bar{D}\rangle}{m_D^2 - m_I^2 + i\epsilon} .$$

In the SU(3) limit, the GIM cancellation works as well as it does in the short distance approximation, but SU(3) breaking lifts this cancellation in the dispersive sector, leading to estimates of effects for D^0's which are one or two orders of magnitude larger than the box diagrams. Donoghue et al[144] have parametrized the contribution to the D^0 sector as:

$$\Delta m_D^{\text{disp}} \simeq \frac{1}{2\pi} \ln \frac{m_d^2}{\mu^2} \left[\Gamma(D^0 \to K^+K^-) + \Gamma(D^0 \to \pi^+\pi^-) \right.$$

$$\left. - 2 \left(\Gamma(D^0 \to K^-\pi^+)\Gamma(D^0 \to K^+\pi^-) \right)^{\frac{1}{2}} \right]$$

$$\simeq 2 \times 10^{-15} \text{ GeV} \left[1 - 0.65(f_{K^+\pi^-})^{\frac{1}{2}} \right] ,$$

which, for $\mu = 1$ GeV and $f_{K^+\pi^-} = 1$, is more than an order of magnitude larger than the box diagram contribution. Other intermediate states, involving 3 or more pseudoscalars, surely also must contribute.

3) Mechanisms not contained in the Standard Model can also generate mixing, either through direct first order flavor-changing neutral currents (Fig. 52) or, as in the case of right-left symmetric models, through additional box diagrams which can lead back to $r_D \sim O(.1) - O(1)$ with $g_L, g_R \sim 10^{-3}$.

Thus, both through the dependence of mixing on the t quark mass and couplings, and through the sensitivity to nonstandard mechanisms for the generation of flavor-changing neutral currents, the search for evidence of mixing in neutral heavy quark systems is well motivated. Should such evidence appear, however, particularly in the $D^0 \bar{D}$ system, a more complete understanding of the long distance effects will be necessary before definite conclusions can be drawn.

Experimental Signatures for Mixing

Searches for mixing fall into three general classes: those which use 1) a lepton tag, 2) a daughter quark tag, through identification of a kaon, D or D_s meson, and 3) those which use full reconstruction of the complete final state. The first two methods depend on the GIM mechanism to ensure a unique signature.

The **lepton tag** method makes use of the fact that the sign of a lepton produced in D or B quark decay tags the sign of the $W^{\pm}$ and, therefore, the quark or antiquark current to which the W couples uniquely (Fig. 53). Thus, an l^+ arises from a c or $\bar{b}$ which an l^- signifies a $\bar{c}$ or b. Three *caveats* are important:

1) It is necessary to know whether the parent meson was a D or B. This is typically ensured by taking advantage of the quite differing regions of p_T populated by leptons from D and B decay.

2) It is necessary to know that the source of the lepton is the primary vertex. For example, in the following decay sequence:

$$B^0 \to D^- l^+ \nu_l$$
$$\hookrightarrow \bar{K}^0 l^- \nu_l,$$

the l^+ tags the B parent, while the l^- tags the c parent.

3) It is necessary to have high quality lepton identification in order that misidentified hadrons not overwhelm the time lepton signal.

The **daughter quark tag** method makes use of the correlation of kaon charge with D flavor and that of D charge with B flavor. We need to distinguish the D and B cases:

1) $D^0 \bar{D}$

Fig. 51 Long distance contributions to $D^0 - \bar{D}^0$ mixing.

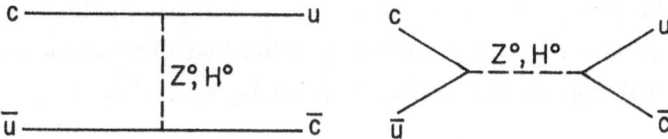

Fig. 52 Flavor-changing neutral current contributions to $D^0 - \bar{D}^0$ mixing.

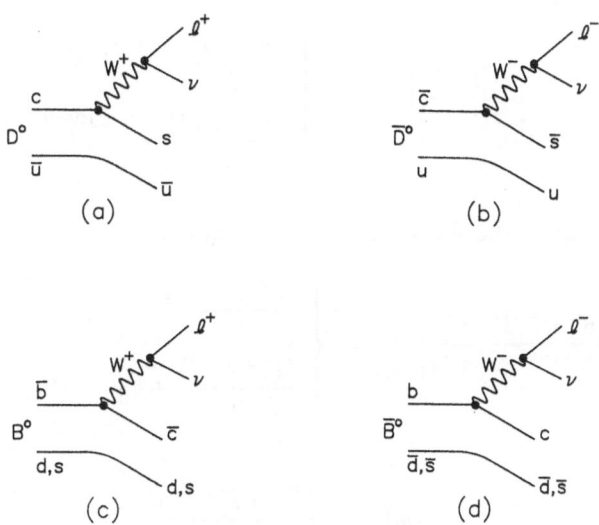

Fig. 53 Lepton tag signatures for: a) and b) $D^0 - D^0$ mixing; c) and d) $B^0 - \bar{B}^0$ mixing.

a) The GIM mechanism provides the primary correlation of, e.g., a K^- daughter with a D^0 parent (Fig. 54a) in Cabibbo-allowed decay.

b) Cabibbo-suppressed decays can, however, produce wrong sign daughters (Fig. 54b). Thus, inclusive K's are not an unambiguous tag of the parent D producing a background at the level of $\sim \tan^2 \theta_c$. This problem can be evaded by reconstruction of exclusive D decay modes (see below).

c) There is, in addition, an irreducible physics background due to double-Cabibbo-suppressed decays (Fig. 54c), which produces a signal indistinguishable from true flavor oscillation at the level of $\sim \tan^4 \theta_c$ i.e., at $10^{-3} - 10^{-2}$.

Thus, while the exclusive production of $D\bar{D}$ pairs at the $\psi(3770)$ resonance in e^+e^- annihilation would seem to provide an ideal place to search for mixing through complete reconstruction of both D's, one must bear in mind that observation of, for example, two $K^-\pi^+$ pairs at m_{D^0} signifies **either** $D^0\bar{D}$ mixing **or** $\tan^4 \theta_c$ decay.

There are two possible ways around this limitation. The first is to reconstruct exclusive final states in which, e.g., the D^0 decays hadronically while the $\bar{D}$ decays semileptonically, since the lepton in D decay unambiguously identifies the parent quark flavor. The second makes use of the detailed study of quantum statistics made by Bigi and Sanda.[145] We can distinguish two cases:

a) Both D^0 and $\bar{D}$ decay semileptonically. Here, the size of the effect depends on the relative angular momentum of the pair:

$$\left| \frac{N(l^\pm l^\pm)}{N(l^+l^-)} \right|_{D^0\bar{D}} = \begin{cases} \dfrac{x^2 + y^2}{2} & l \text{ odd} \\[3mm] \dfrac{3(x^2 + y^2)}{2} & l \text{ even}. \end{cases}$$

b) Both D's decay to the $K\pi$ final state, for example:

$$\left| \frac{N(K^-\pi^+, K^-\pi^+)}{N(K^-\pi^+, K^+\pi^-)} \right|_{D^0\bar{D}} = \begin{cases} \dfrac{x^2 + y^2}{2} & l \text{ odd} \\[3mm] \dfrac{3(x^2 + y^2)}{2} + 4|\rho|^2 + 8y\rho & l \text{ even}, \end{cases}$$

where

$$|\rho|^2 = \left| \frac{A(\bar{D} \to K^-\pi^+)}{A(\bar{D} \to K^-\pi^+)} \right|^2 .$$

In the useful case of $e^+e^- \to D^0\bar{D}^0$, l must be odd. Therefore, only mixing and not $\tan^4 \theta_c$ processes can produce $s = \pm 2$ final states, where the two final states are identical, e.g., $K\pi$, $K\rho$, $K^*\pi$.

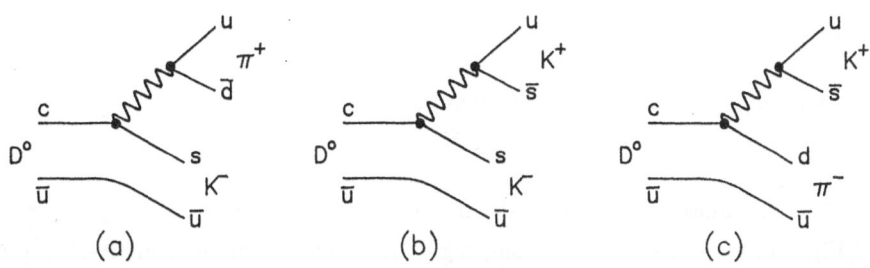

Fig. 54 Daughter quark tag signatures for $D^0 - \bar{D}^0$ mixing: a) normal Cabibbo-allowed amplitude; b) wrong-sign kaon originating from Cabibbo-suppressed amplitude. This background requires that the D^0 be reconstructed in an exclusive mode; c) wrong-sign kaon originating from doubly-Cabibbo-suppressed amplitude. This produces a signal identical to that from $\bar{D}^0$ decay, producing a background which can be isolated by study of the time dependence of the reconstructed events.

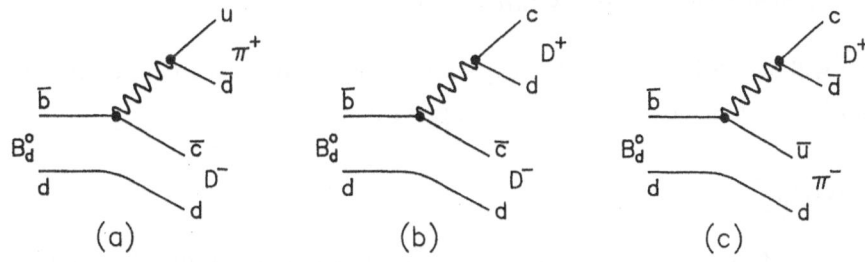

Fig. 55 Daughter quark tag signatures for $B_d^0 - \bar{B}_d^0$ mixing: a) normal K-M-favored amplitude; b) wrong-sign D^+ originating from K-M-suppressed amplitude; c) final state identical to mixing signature arising from doubly-K-M-suppressed process.

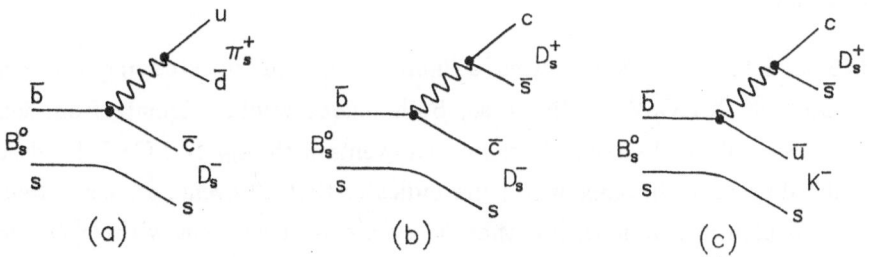

Fig. 56 Daughter quark tag signatures for $B_s^0 - \bar{B}_s^0$ mixing. a) normal K-M-favored amplitude; b) wrong-sign D_s originating from another K-M favored (albeit phase-space-hindered) process; c) wrong-sign D_s originating from singly-K-M-suppressed decay. Note that final state is distinct and can be separated by complete reconstruction of exclusive final state.

2) $B^0\bar{B}^0$

The case of $B^0\bar{B}^0$ mixing is more complicated. We have further to distinguish the two types of neutral B mesons, B_d^0 and B_s^0:

a) B_d^0. The normal K-M-favored decay of the B_d^0 yields a ("right sign") D^- (Fig. 55a). However, a "wrong sign" D^+ can originate from K-M-suppressed decay (Fig. 55b), which is also phase-space suppressed. Double K-M-suppressed decay can yield a final state identical to the naïve mixing signature (Fig. 55c).

b) B_s^0. The normal K-M-favored decay here yields a D_s^- (Fig. 56a). However, in this case, there is a K-M-allowed (albeit phase-space suppressed) mode which yields a wrong sign D_s^+ (Fig. 56b). There is also a singly K-M-suppressed mode which produces a wrong sign D_s^+ (Fig. 56c), but the complete final state is distinguishable from mixing, in that it also contains a K^- rather than a π^-.

Experimental Searches for Mixing

We will here briefly review the experimental situation in searches for mixing, as an illustration of some of the different techniques just described.

$D^0\bar{D}$ Mixing

DELCO[146] has used its good particle identification capability to search its $\sqrt{s} = 29$ GeV e^+e^- sample for wrong sign $K\pi$, using tagged D^*'s. In the decay $D_s^+ \rightarrow D^0\pi^+$, the sign of the π tags the charm of the D. A subsequent $K^-\pi^+$ decay is then "normal", while a $K^+\pi^-$ signifies either mixing or doubly Cabibbo-suppressed decay. They observe 97 normal combinations and 15 of the wrong sign. The latter are consistent with combinatorial background and particle misidentification, so there is no evidence for mixing. A maximum likelihood fit yields $r_D < 8.3 \times 10^{-2}$ at 90% confidence level.

E615 at FNAL[147] has produced a limit for mixing by studying the reaction $\pi N \rightarrow \mu\mu X$ at 225 GeV. In this case, both muons would originate from semileptonic D decays. They observe 3973 same sign events with $m_{\mu\mu} > 2$ GeV. Study of the $|\cos\theta|$ distribution of the same sign pairs indicates that all events are consistent with uncorrelated pion interactions; less than 63 events are consistent with $D^0\bar{D}^0$ mixing. Assuming $\sigma(D^0\bar{D}^0) = (3.8 \pm 0.5)\mu b/\text{nucleon}$, this leads to a limit $r_D < 5.6 \times 10^{-3}$ at 90% confidence level.

Better limits have recently been set by ARGUS[148] ($r_D < 0.023$) and HRS[149] ($r_D < 0.040$), using the sign of the pion in, e.g., $D^{*+} \rightarrow \pi^+D^0$ events to tag the neutral D species.

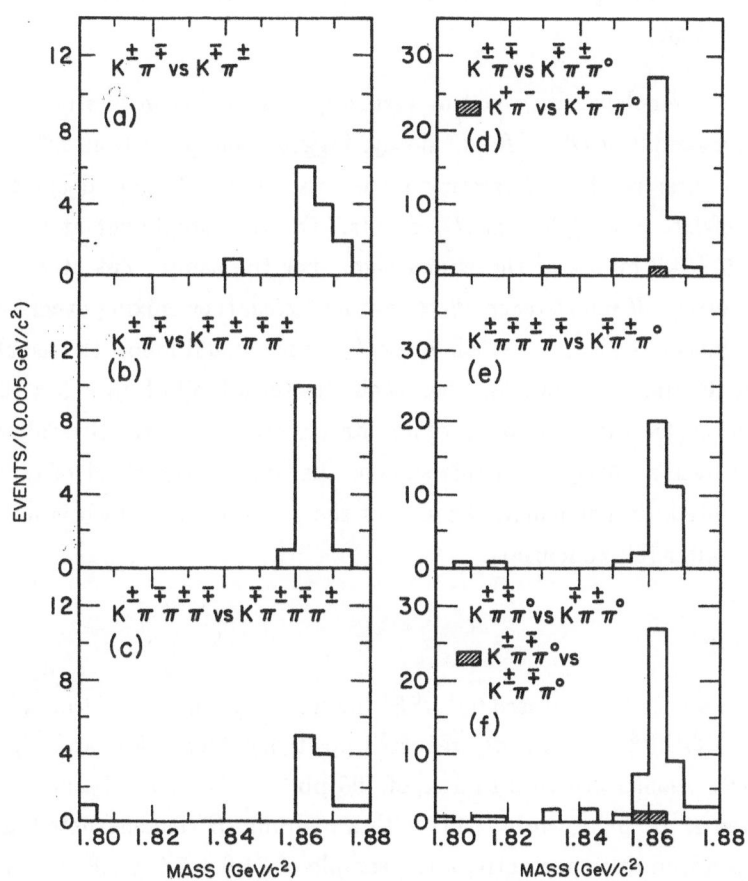

Fig. 57   Mark III double-tag exclusively reconstructed D^0 hadronic events at the $\psi(3770)$. The shaded regions show a single $K^+\pi^-$ vs. $K^+\pi^-\pi^0$ event and two $K^\pm\pi^\mp\pi^0$ vs. $K^\pm\pi^\mp\pi^0$ events which could be evidence either for $D^0\bar{D}^0$ mixing or large doubly-Cabibbo-suppressed decays.

Mark III has provided the only example[150] of an attempt to fully reconstruct $D^0 \bar{D}^0$ final states. Using the 9 pb^{-1} sample of $D\bar{D}$ events produced at the $\psi(3770)$, 162 $D^0\bar{D}^0$ events having total strangeness 0 have been fully reconstructed with very stringent particle identification requirements. These are shown in Figure 57, which also shows three events which have total strangeness ± 2. Events in which both a K and π are misidentified should yield a background of 0.4 ± 0.1 events. Less than one event is expected from $\tan^4 \theta_c$ processes. *If* the events in excess of these background estimates are attributed to mixing, then $r_D \simeq 10^{-2}$, but the statistics are too meager to reach a definite conclusion.

E691[151] has employed its excellent vertex spatial resolution to study the time dependence of a sample of $D^0 \rightarrow K^- \pi^+$ decays tagged using pions from D^*'s. Figures 58 and 59 show these results. Correcting for events missed at short decay times, there are a total of 597 ± 38 right sign D^* events. For the sample cut at $t > 0.22$ psec, there are 9.6 ± 4.9 events of the wrong sign. For the sample cut at $t > 0.88$ psec, there is one event. *If interference effects are neglected*, then mixing events would have a time distribution $\sim t^2 exp(-\Gamma t)$ while $tan^4 \theta_c$ events would follow the usual $exp(-\Gamma t)$ distribution. Fitting to all events, E691 is able to set a limit of $r_D < 5 \times 10^{-3}$ at 90% confidence level. On the assumption of no interference, the Mark III result would have implied ~ 7 events. Analysis of this sample allowing for the effect of interference is clearly warranted. In particular, the sign of the ρ parameter, which is unknown, can dramatically alter this conclusion.

$B^0 \bar{B}^0$ Mixing

The primary technique used in $B^0 \bar{B}^0$ mixing experiments is that of same sign dileptons. ARGUS[152] has recently reported positive evidence for same sign dileptons on the $\Upsilon(4S)$ resonance with a sample of 106 pb^{-1}. They also observe a single fully reconstructed $B^0 B^0$ pair, a few events with a hadronic B^0 reconstructed and a wrong sign lepton and, most importantly, a net sample of $24.8 \pm 7.6 \pm 3.8$ events with same sign dileptons with $270.3 \pm 19.4 \pm 5.0$ unlike sign events. This leads to an overall value of

$$r_{B_d} = 0.21 \pm 0.08 \,.$$

Mark II[153] has searched for $B^0 \bar{B}^0$ mixing using their 220 pb^{-1} data sample taken at $\sqrt{s} = 29$ GeV at PEP. Here, both B_d^0 and B_s^0 are produced, as are B baryons. A sample of 64 same sign dimuons is isolated containing leptons from both b and c quark parents, which are separated by p_T. The interpretation in terms of a mixing parameter depends on b quark fragmentation functions, which are parametrized for B_d^0 and B_s^0 as

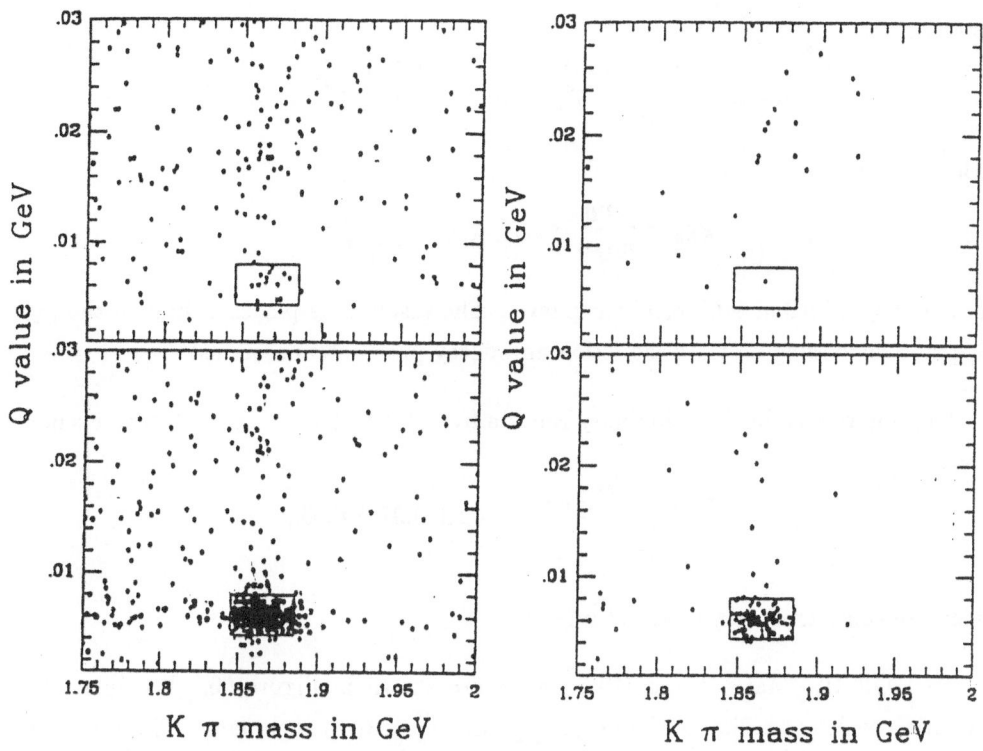

Fig. 58 Scatter plots of $K\pi$ mass vs. $Q[= M_{K\pi\pi}-M_{K\pi}-M_\pi]$ for E691 $K\pi\pi$ events with $t > 0.22$ psec. The upper plot shows $K^-\pi^+\pi^-$ events; the lower plot shows $K^-\pi^+\pi^+$ events. Charge conjugate modes are also shown. The box contains approximately 90% of $D^* \to \pi D$, $D \to K\pi$ events.

Fig. 59 Scatter plots of $K\pi$ mass vs. Q for E691 $K\pi\pi$ events with $t > 0.88$ psec.

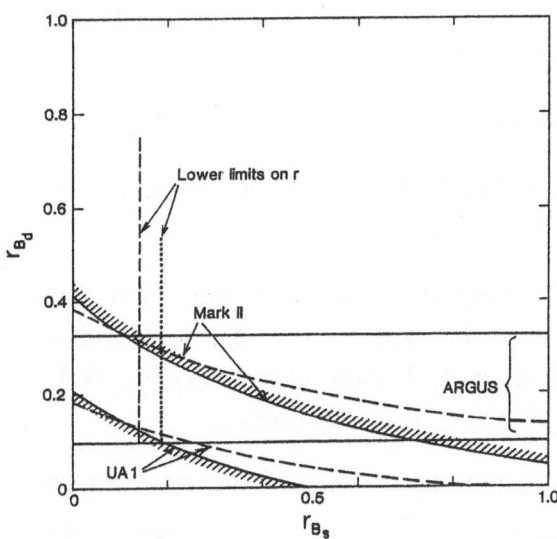

Fig. 60 Regions of r_{B-d} vs. r_{B_s} parameter space delimited by the ARGUS observation of $B^0\bar{B}^0$ mixing candidates, together with limits placed by the Mark II and UA1 results.

follows:

$$\chi_{\text{meas}} = \frac{B(B_d^0 \to l)}{\langle B(B \to l)\rangle} p_d \chi_{B_d} + \frac{B(B_s^0 \to l)}{\langle B(B \to l)\rangle} p_s \chi_{B_s} \, ,$$

with

$$\chi_{B_d} = \frac{r_{B_d}}{r_{B_d} + 1} \quad , \quad \chi_{B_s} = \frac{r_{B_s}}{r_{B_s} + 1} \, .$$

They find $\chi < 0.12$ at 90% confidence level. The result thus places a limit in the r_{B_d}, r_{B_s} plane depending on the production and decay parameters assumed.

There have also been preliminary reports from UA1 of same sign dilepton events:

$$\frac{N_{\mu^+\mu^-} + N_{\mu^-\mu^-}}{N_{\mu^+\mu^-}} = 0.42 \pm 0.07 \pm 0.03 \, ,$$

which are consistent with large B_s^0 mixing.

The ARGUS, Mark II and UA1 results are shown in Figure 60. Together, they delimit a consistent region in the r_{B_d}, r_{B_s} plane. The surprisingly large B_d^0 mixing seen by ARGUS, when naïvely inserted in the box diagram estimate, implies a large top quark mass, perhaps $m_t > 50$ GeV. The results are consistent with the possibility of complete $B_s \bar{B}_s$ mixing.

REFERENCES

1. H. J. Schnitzer, *Phys. Lett.* **134B**, 253 (1984); M. Frank and P. J. O'Donnell, *Phys. Lett.* **157B**, 174 (1985).

2. M. Aguilar-Benitez *et al.* (Particle Data Group) *Phys. Lett.* **170B**, 1 (1986).

3. G. T. Blalock *et al.* (Mark III Collaboration), SLAC-PUB-4158 (1987), submitted to *Physical Review Letters.*

4. A. DeRujula, H. Georgi and G. Glashow, *Phys. Rev.* **D12**, 147 (1975); D. B. Lichtenberg, *Nuovo Cim.* **28A**, 563 (1985); J. M. Richard, *Ann. Phus.* **150**, 267 (1983); E. Eich *et al., Z. Phys.* **C28**, 225 (1985); D. Izatt, C. Detar and M. Stephenson, *Nucl. Phys.* **B199**, 269 (1982).

5. M. Kobayashi and T. Maskawa, *Prog. Theor. Phys.* **49**, 652 (1973).

6. L. Maiani, in *Proceedings of the 1977 International Symposium on Lepton and Photon Interactions at High Energy*, DESY, Hamburg, 1977, p. 867; L.-L. Chau, *Phys. Rep.* **95**, 1 (1983).

7. S. K. Bose and E. A. Paschos, *Nucl. Phys.* **B169**, 384 (1980); V. Barger, K. Whimsat and R. Phillips, *Phys. Rev.* **D23**, 2773 (1981); M. Gronau and J. Schechter, *Phys. Rev.* **D31**, 1668 (1985); F. J. Botella and L.-L. Chau, *Phys. Lett.* **168B**, 97 (1986).

8. K. Kleinknecht and B. Renk, Mainz Preprint MZ-ETAP/86-4 (1986); see also F. J. Gilman and K. Kleinknecht, in *Review of Particle Properties* (Particle Data Group), *Phys. Lett.* **170B**, 74 (1986).

9. D. M. Coffman, Ph.D. Thesis, California Institute of Technology (1986).

10. L. Wolfenstein, *Phys. Rev. Lett.* **51**, 1945 (1984).

11. For a recent compilation, see S. Godfrey and N. Isgur, *Phys. Rev.* **D32**, 189 (1985).

12. R. H. Schindler *et al.* (Mark III Collaboration) *Proceedings of the 23rd International Conference on High Energy Physics*, Berkeley, 1986.

13. V. Lüth, SLAC-PUB-4222 (1987). To appear in *Proceedings of the APS Division of Particles and Fields Meeting*, Salt Lake City, 1987.

14. N. Ushida *et al., Phys. Rev. Lett.* **56**, 1767 (1986); N. Ushida *et al., Phys. Rev. Lett.* **56**, 1771 (1986).

15. M. Adamovich *et al.* CERN/EP86-77, submitted to *Phys. Lett. B*; V. Castillo-Gimenez, Thesis Doctoral, Universidad de Valencia (1986).

16. K. Abe *et al., Phys. Rev.* **D33**, 1 (1985).

17. M. Aguilar-Benitez *et al., Phys. Lett.* **122B**, 312 (1983).

18. A. Badertscher *et al., Phys. Lett.* **123B**, 471 (1982).

19. M. Aguilar-Benitez *et al.*, CERN-EP/86-167 (1986), submitted to *Z. Phys. C*.

20. E. Albini *et al., Phys. Lett.* **110B**, 339 (1982); S. R. Amendolia *et al.*, submitted to the *23rd International Conference on High Energy Physics*, Berkeley, 1986.

21. R. Bailey *et al., Z. Phys.* **C28**, 357 (1985); E. Belau *et al., Proceedings of the International Conference on Hadron Spectroscopy*, College Park, 1985.

22. H. Palka *et al.*, submitted to the *23rd International Conference on High Energy Physics*, Berkeley, 1986.

23. H. Becker *et al.*, submitted to the *23rd International Conference on High Energy*

Physics, Berkeley, 1986; H. Becker *et al.*, *Phys. Lett.* **184B**, 277 (1987).

24. J. C. Anjos *et al.*, *Phys. Rev. Lett.* **58**, 311 (1987); J. C. Anjos *et al.*, FERMILAB-PUB-87/29E (1987).

25. H. Yamamoto *et al.*, *Phys. Rev.* **D22**, 2901 (1985).

26. L. Gladney *et al.*, *Phys. Rev.* **D34**, 2601 (1986).

27. S. Abachi *et al.*, ANL-HEP-CP-86-73;S. Abachi *et al.*, ANL-HEP-CP-86-62;S. Abachi *et al.*, ANL-HEP-CP-86-60; S. Abachi *et al.*, *Phys. Rev. Lett.* **56**, 1775 (1986).

28. M. Althoff *et al.*, *Z. Phys.* **C32**, 343 (1986); W. Braunschweig *et al.*, RAL 86-109 (1986).

29. S. E. Csorna *et al.*, CLNS 86/747 (1986).

30. P. Padley, presented at the *APS Division of Particles and Fields Meeting*, Salt Lake City, 1987.

31. S. Barlay *et al.*, *Phys. Lett.* **184B**, 283 (1987).

32. J. C. Anjos *et al.*, UCSB-HEP-87-7 (1987).

33. P. Coteus *et al.*, COLO-HEP-140 (1986).

34. S. F. Biagi *et al.*, *Phys. Lett.* **122B**, 455 (1983); S. F. Biagi *et al.*, *Phys. Lett.* **150B**, 230 (1985).

35. J. Jaros, *Proceedings of Physics in Collision IV*, Santa Cruz, 1984.

36. W. Ash *et al.*, *Phys. Rev. Lett.* **58**, 640 (1987).

37. D. E. Klem *et al.*, SLAC-PUB-4025 (1986), submitted to *Phys. Rev. D*.

38. D. Blockus *et al.*, ANL-HEP-PR-86-144 (1986).

39. W. Bartel *et al.*, *Z. Phys.* **C31**, 349 (1986).

40. M. Althoff *et al.*, *Phys. Lett.* **149B**, 524 (1984).

41. For a detailed exposition, see R. Rückl, Habilitationsschrift, University of Munich (1983).

42. M. K. Gaillard and B. W. Lee, *Phys. Rev. Lett.* **33**, 108 (1974); G. Altarelli and L. Maiani, *Phys. Lett.* **52B**, 351 (1974).

43. F. J. Gilman and M. G. Wise, *Phys. Rev.* **D20**, 2392 (1979).

44. G. Altarelli, G. Curci, G. Martinelli and R. Petrarca, *Phys. Lett.* **99B**, 141 (1981) and *Nucl. Phys.* **B187**, 461 (1981).

45. R. M. Baltrusaitis *et al.* (Mark III Collaboration) *Phys. Rev. Lett.* **54**, 1976 (1985).

46. R. H. Schindler *et al.*, *Phys. Rev.* **D21**, 2716 (1980).

47. J. M. Feller, Ph.D. Thesis, University of California, Berkeley, LBL-9017 (1979).

48. Y. Zhu, private communication.

49. H. J. Lipkin, *Phys. Lett.* **179B**, 278 (1986).

50. I. Peruzzi *et al.*, *Phys. Rev. Lett.* **39**, 1301 (1977); D. Scharre *et al.*, *Phys. Rev. Lett.* **40**, 74 (1978).

51. R. H. Schindler *et al.*, *Phys. Rev.* **D24**, 78 (1981).

52. A. J. Hauser, Ph.D. Thesis, California Institute of Technology (1984); R. M. Baltrusaitis *et al.*, *Phys. Rev. Lett.* **55**, 150 (1985).

53. M. Aguilar-Benitez *et al.*, *Phys. Lett.* **146B**, 266 (1984).

54. R. Bailey *et al.*, *Phys. Lett.* **132B**, 237 (1983).

55. H. Albrecht *et al.*, *Phys. Lett.* **153B**, 343 (1985).

56. P. Avery *et al.*, *Phys. Rev. Lett.* **51**, 1139 (1983).

57. M. Aguilar-Benitez *et al.*, *Phys. Lett.* **135B**, 237 (1984).

58. M. Aguilar-Benitez *et al.*, *Phys. Lett.* **168B**, 170 (1986).

59. B. R. Kumar, presented at the Lake Louise Winter Institute, Alberta, Canada, 1987.

60. R. A. Partridge, Ph.D. Thesis, California Institute of Technology (1984).

61. R. M. Baltrusaitis *et al.*, *Phys. Rev. Lett.* **55**, 150 (1985).

62. P. Rapidis *et al.*, *Phys. Rev. Lett.* **39**, 526 (1977).

63. H. Sadrozinski, in **High Energy Physics - 1980**, ed. L. Durand and L. Pondrom, AIP Conference Proceedings No. 68 (AIP, New York, 1981), p. 681.

64. G. T. Blalock, Ph.D. Thesis, University of Illinois (1986).

65. R. M. Baltrusaitis *et al.*, *Phys. Rev. Lett.* **56** 2140 (1986).

66. J. J. Becker *et al.*, SLAC-PUB-4291 (1987), submitted to *Physical Review Letters.*

67. H. Albrecht *et al.*, *Phys. Lett.* **158B**, 525 (1985).

68. C. Bebek *et al.*, *Phys. Rev. Lett.* **56**, 1983 (1986).

69. R. M. Baltrusaitis *et al.*, *Phys. Rev. Lett.* **56**, 2136 (1986).

70. J. J. Becker *et al.*, SLAC-PUB-0000 (1987), submitted to *Physics Letters.*

71. R. H. Schindler *et al.*, *Phys. Rev.* **D24**, 78 (1981).

72. W. Bacino *et al.*, *Phys. Rev. Lett.* **45**, 329 (1980).

73. C. Sliwa, presented at the *APS Division of Particles and Fields Meeting*, Salt Lake City, 1987.

74. B. Grinstein, M. B. Wise and N. Isgur, *Phys. Rev. Lett.* **56**, 298 (1986).

75. W. Buchmüller and D. Wyler, *Phys. Lett.* **177B**, 377 (1986).

76. H. Palka *et al.*, CERN-EP/87-10 (1987), submitted to *Physics Letters*; K. Riles *et al.*, *Phys. Rev.* **D35**, 2914 (1987).

77. R. Brandelik *et al.*, *Phys. Lett.* **80B**, 412 (1979).

78. R. Ammar *et al.*, *Phys. Lett.* **94B**, 118 (1980).

79. N. Ushida *et al.*, *Phys. Rev. Lett.* **45**, 1053 (1980).

80. D. Aston *et al.*, *Phys. Lett.* **100B**, 91 (1981).

81. C. Chen *et al.*, *Phys. Rev. Lett.* **51**, 634 (1983).

82. W. Braunschweig *et al.*, RAL-86-109 (1986).

83. H. Albrecht *et al.*, *Phys. Lett.* **153B**, 343 (1985).

84. M. Derrick *et al.*, *Phys. Rev. Lett.* **54**, 2568 (1985).

85. S. Wasserbaech *et al.*, SLAC-PUB-4289 (1987).

86. H. Aihara *et al.*, *Phys. Rev. Lett.* **53**, 2465 (1984).

87. R. Bailey *et al.*, *Phys. Lett.* **139B**, 320 (1984).

88. H. Albrecht *et al.*, *Phys. Lett.* **179B**, 398 (1986).

89. M. Bauer, B. Stech and M. Wirbel, *Z. Phys.* **C34**, 103 (1987).

90. L.-L. Chau and H.-Y. Cheng, UCD-86-32-R (1987), submitted to *Physical Review*.

91. A. N. Kamal, *Phys. Rev.* **D33**, 1344 (1986).

92. R. D. Peccei and R. Rückl, *Proceedings of the Ahrenhoop Symposium* (1981).

93. M. B. Voloshin and M. A. Shifman, *Yad. Fiz.* **41**, 187 (1985).

94. A. J. Buras, J. M. Gérard and R. Rückl, *Nucl. Phys.* **B268**, 16 (1986).

95. I. I. Bigi and M. Fukugita, *Phys. Lett.* **91B**, 121 (1980).

96. J. Donoghue, *Phys. Rev.* **D33**, 1516 (1986).

97. F. Hussain and A. N. Kamal, ALBERTA-THY-1-86 (1986).

98. U. Baur, A. J. Buras and J. M. Gérard, *Phys. Lett.* **175B**, 377 (1986).

99. M. A. Shifman *et al.*, *JEPT Lett.* **22**, 55 (1975); M. A. Shifman *et al.*, *Nucl. Phys.* **B120**, 316 (1977); J. Finjord, *Nucl. Phys.* **B181**, 74 (1980); B. Guberina, D. Tadic and J. Trampetic, *Nucl. Phys.* **B202**, 317 (1982).

100. G. S. Abrams *et al.*, *Phys. Rev. Lett.* **44**, 10 (1980).

101. M. Piccolo *et al.*, *Phys. Rev. Lett.* **39**, 1503 (1977).

102. P. Padley, *op. cit.*

103. J. M. Weiss, *Proceedings of Baryon 80*, Toronto (1980).

104. M. Basile *et al.*, *Nuovo Cim.* **62A**, 14 (1981).

105. A. N. Aleev *et al.*, *Z. Phys.* **C23**, 233 (1984).

106. C. Baltay *et al.*, *Phys. Rev. Lett.* **42**, 1721 (1979).

107. T. Kitagaki *et al.*, *Phys. Rev. Lett.* **45**, 955 (1980).

108. T. Bowcock *et al.*, *Phys. Rev. Lett.* **55**, 923 (1985).

109. S. F. Biagi *et al.*, *Phys. Lett.* **122B**, 455 (1983).

110. S. F. Biagi *et al.*, *Z. Phys.* **C28**, 175 (1985).

111. E. G. Cazzoli *et al.*, *Phys. Rev. Lett.* **34**, 1125 (1975).

112. C. Baltay *et al.*, *Phys. Rev. Lett.* **42**, 1721 (1979).

113. H. Grassler *et al.*, *Phys. Lett.* **99B**, 159 (1981).

114. J. G. Körner, G. Kramer and J. Willrodt, *Phys. Lett.* **78B**, 492 (1978) (Erratum *ibid.* **81B**, 419 (1979)); *Z. Phys.* **C2**, 117 (1979).

115. V. Barger, J. P. Leveille and P. M. Stevenson, *Phys. Rev. Lett.* **44**, 226 (1980).

116. D. Ebert and W. Kallies, *Phys. Lett.* **131B**, 183 (1983) (Erratum, *ibid.* **148B**, 502 (1984)); *Z. Phys.* **C29**, 643 (1985).

117. H.-Y. Cheng, *Z. Phys.* **C29**, 453 (1985).

118. J. G. Körner, MZ-TH/86-11 (1986).

119. E. N. Vella *et al.*, *Phys. Rev. Lett.* **48**, 1515 (1982).

120. J. P. Albanese *et al.*, *Phys. Lett.* **158B**, 186 (1985).

121. C. Chen *et al.*, *Phys. Rev. Lett.* **52**, 1084 (1984).

122. C. Klopfenstein *et al.*, *Phys. Lett.* **130B**, 444 (1983).

123. G. Levman *et al.*, *Phys. Lett.* **141B**, 271 (1984).

124. S. Weseler, Ph.D. Thesis, University of Heidelberg, IHEP-HD/86-02.

125. M. Althoff *et al.*, *Z. Phys.* **C22**, 219 (1984).

126. H. J. Behrend *et al.*, *Z. Phys.* **C19**, 291 (1983).

127. W. Bartel *et al.*, *Z. Phys.* **C33**, 339 (1987).

128. B. Adeva *et al.*, *Phys. Rev. Lett.* **51**, 443 (1983).

129. M. E. Nelson *et al.*, *Phys. Rev. Lett.* **50**, 1542 (1983).

130. E. Fernandez *et al.*, *Phys. Rev. Lett.* **50**, 2054 (1983).

131. D. Koop *et al.*, *Phys. Rev. Lett.* **52**, 970 (1984).

132. H. Aihara *et al.*, *Z. Phys.* **C27**, 39 (1985); H. Aihara *et al.*, *Phys. Rev.* **D31**, 2719 (1985).

133. D. L. Kreinick, *Proceedings of the International Symposium on Production and Decay of Heavy Hadrons*, Heidelberg, 1986.

134. G. Altarelli *et al.*, *Nucl. Phys.* **B208**, 365 (1982).

135. K. Schubert, presented at the *23rd International Conference on High Energy Physics*, Berkeley, 1986.

136. S. Behrends *et al.*, submitted to the *23rd International Conference on High Energy Physics*, Berkeley, 1986.

137. S. Behrends *et al.*, *Phys. Rev. Lett.* **50**, 881 (1983).

138. C. Bebek *et al.*, CLNS 86/142, submitted to *Physical Review*.

140. S. Weseler, *Proceedings of the International Symposium on Production and Decay of Heavy Quarks*, Heidelberg, 1986.

141. H. Albrecht *et al.*, IHEP-HD/86-3.

142. M. S. Alam *et al.*, CLNS 86/739.

143. M. G. D. Gilchreise, presented at the *23rd International Conference on High Energy Physics*, Berkeley, 1986.

144. J. F. Donoghue, E. Golowich and B. R. Holstein, *Phys. Rev.* **D33**, 179 (1986).

145. I. I. Bigi and A. I. Sanda, *Phys. Lett.* **B171**, 320 (1986).

146. H. Yamamoto *et al.*, *Phys. Rev. Lett.* **54**, 522 (1985).

147. C. Biino *et al.*, *Phys. Rev. Lett.* **56**, 1027 (1986).

148. N. Kwak, *Proceedings of the 23rd International Conference on High Energy Physics*, Berkeley, 1986.

149. S. Abachi *et al.*, *Phys. Lett.* **182B**, 101 (1986).

150. G. Gladding (Mark III Collaboration), presented at *Physics in Collision V*, Antun, 1985.

151. J. C. Anjos *et al.*, UCSB-HEP-87-7.

152. H. Albrecht *et al.*, DESY 87-029.

153. T. Schadd *et al.*, *Phys. Lett.* **160B**, 188 (1985).

154. C. Albajar *et al.*, *Phys. Lett.* **186B**, 247 (1987).

HERA: PHYSICS, MACHINE AND EXPERIMENTS

Günter Wolf
Deutsches Elektronen-Synchrotron, DESY
Hamburg, Germany

ABSTRACT

With HERA ep collisions at a c.m. energy of 314 GeV and $Q^2_{max} = 10^5$ GeV^2 will become possible. The physics opportunities, the design and status of the machine and the planned detectors are discussed.

INTRODUCTION

DESY is presently constructing a new storage ring HERA that will provide collisions between 30 GeV electrons and 820 GeV protons[1]. Experimentation is scheduled to begin in 1990.

HERA offers exciting and unique physics opportunities. Electrons and quarks can be probed for substructure down to distances of a few 10^{-18} cm. A search for new mediators of the neutral and charged current is possible up to W' and Z' masses of 800 GeV. The large center of mass energy permits the detection of leptoquarks up to 180 GeV, of families of excited quarks and leptons up to 250 GeV, and squarks and sleptons can be produced with masses up to 180 GeV - provided they exist.

The following lectures present an introduction to the physics expected at HERA and describe briefly the machine and the planned experiments.

2. HERA PHYSICS: GENERAL CONDITIONS

The large momentum transfers possible between electron and proton, $Q^2_{max} = 10^5$ GeV^2, make HERA first and foremost an electron quark collider. The relevant diagram is shown in Fig. 1. The incoming electron emits a

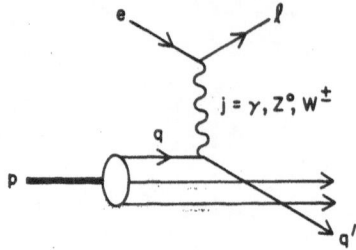

Fig. 1 Lepton proton scattering.

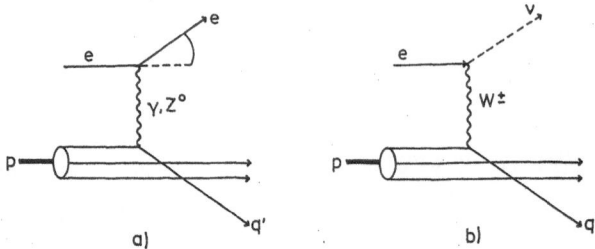

Fig. 2 Diagrams for neutral (a) and charged current (b) scattering.

lepton ℓ and exchanges a current j with one of the quarks of the incoming proton. This leads to the emission of a quark q' called the current quark. Depending on whether a neutral current (γ, Z^0) or a charged current (W^-) is exchanged the final state lepton is either an electron or a neutrino (Fig. 2).

The scattering process is not confined to the known currents, quarks and leptons. New currents may contribute and new quarks and leptons with masses up to the kinematic limit of 314 GeV may be produced. Any particle with electromagnetic and/or weak charge which is within the kinematic limits can be produced.

Next to electron quark scattering, current-gluon fusion (see Fig.3) will play an important role at HERA. Photon-gluon fusion depicted in Fig.4

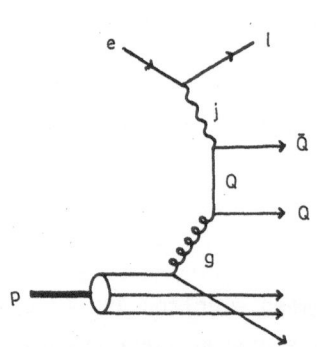

Fig. 3 Quark pair production by current gluon fusion.

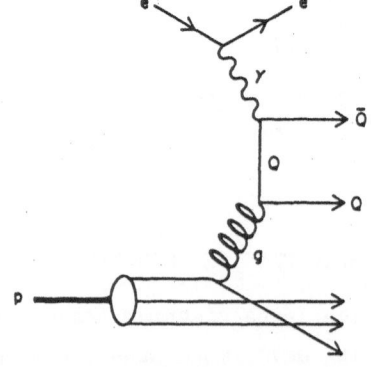

Fig. 4 Quark pair production by photon gluon fusion.

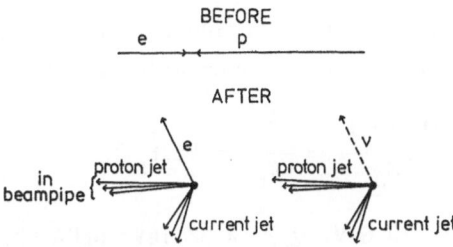

BEFORE

AFTER

in beampipe

proton jet

e

current jet

proton jet

ν

current jet

Fig. 5 Topology of deep inelastic ep scattering events.

will presumably be the dominant process for the production of heavy quarks
$Q = c,b,t...$.

    We shall discuss a few reactions that are archetypical for the dif-
ferent types of processes. More detailed studies can be found in Refs. 2,3.

3.   ELECTRON SCATTERING AND THE PROTON STRUCTURE FUNCTIONS

3.1   Kinematics

    The diagram that describes ep scattering in lowest order is shown
in Fig. 1 (taking $j = \gamma$). The event topology is illustrated in Fig. 5. The
final state partons (quarks, gluons) and the "spectator" remnants from the
incident proton are assumed to develop into jets of hadrons. The lepton
and the current jet emerge on opposite sides of the beam axis, balancing
each other in transverse momentum. The debris of the proton is emitted in
a very narrow cone (of order 10 mrad) around the proton beam direction.

    Apart from the total center-of-mass energy squared,

$$s = (p_p + p_e)^2 = m_e^2 + m_p^2 + 2(E_e E_p + p_e p_p) \approx 4 E_e E_p$$

E_e, E_p energies of incoming electron and proton

there are two kinematic variables that describe the inclusive scattering
process:

$$q^2 = (p_e - p_e')^2 = -Q^2 \qquad \text{square of the four momentum transfer}$$

$$W^2 = (q + p_p)^2 \qquad \text{square of the total mass of the}$$
final hadronic system.

or equivalently

$$x = \frac{Q^2}{2(q \cdot p_p)} = \frac{Q^2}{2m_p \nu} \qquad \text{the Bjorken scaling variable.}$$

$$y = \frac{(q \cdot p_p)}{(q \cdot p_e)} = \frac{\nu}{\nu_{max}}, \qquad \text{note that } Q^2 \approx sxy.$$

In the rest system of the incoming proton ν measures the energy transferred by the current. The maximum value which ν can reach is

$$\nu_{max} = \frac{s-(m_e+m_p)^2}{2m_p} \approx \frac{s}{2m_p} = 2E_e E_p/m_p$$

For $E_e = 30$ GeV, $E_p = 820$ GeV: $\nu_{max} = 52$ TeV. HERA therefore is equivalent to a fixed target experiment with an incident lepton beam of 52 TeV.

The variables Q^2,x,y can be determined either from the energy E_e' and scattering angle Θ_e of the outgoing electron, or from the energy E_j and production angle Θ_j of the current jet.

From the electron:

$$Q^2 = 2E_e E_e' (1+\cos\Theta_e)$$

$$x = \frac{E_e'\cos^2 \Theta_e/2}{E_p (1-(E_e'/E_e) \sin^2\Theta_e/2)}$$

$$y = 1-E_e'/E_e \sin^2\Theta_e/2$$

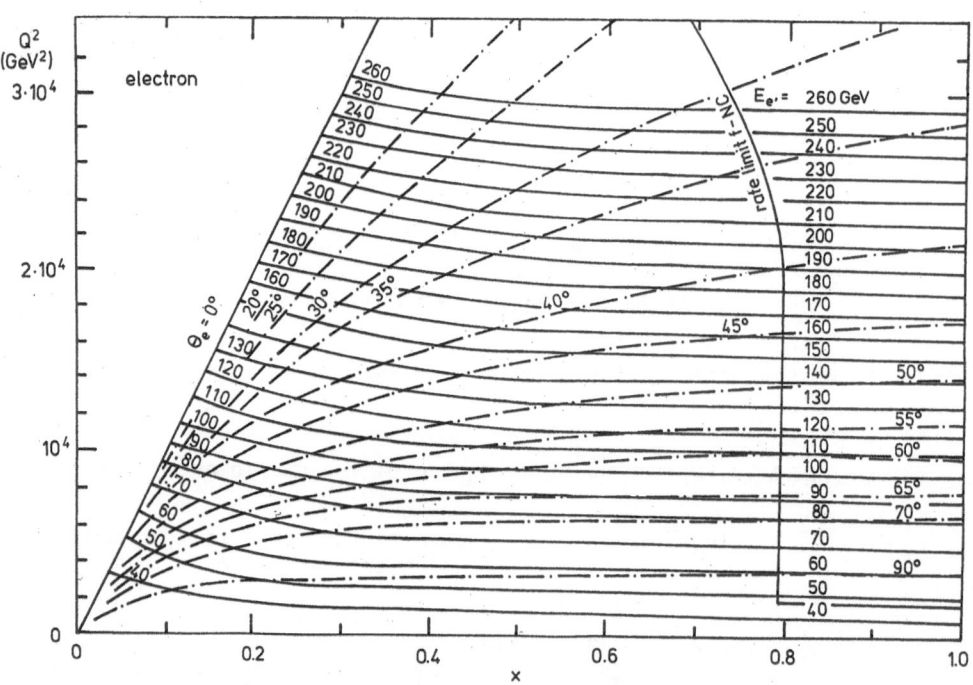

Fig. 6   30 GeV e + 820 GeV p: The x-Q^2 plane with lines of fixed electron scattering angle and energy.

From the current jet:

$$Q^2 = \frac{E_J^2 \sin^2\Theta_J}{1 - \frac{E_J}{2E_e}(1-\cos\Theta_J)}$$

$$y = \frac{E_J}{2E_e}(1-\cos\Theta_J) = \frac{E_J}{E_e}\sin^2\Theta_J/2$$

$$x = \frac{Q^2}{sy} = \frac{E_J \cos^2\Theta_J/2}{E_p\left(1 - \frac{E_J}{E_e}\sin^2\Theta_J/2\right)}$$

The angles are always measured with respect to the proton beam. In deriving the current jet results the mass of the jet has been neglected. As an illustration of the kinematics for 30 GeV electron colliding on 820 GeV protons, Figs. 6, 7 show in the x, Q^2 plane curves for fixed energy and angle of electron and current jet.

3.2 Qualitative Behaviour of the Scattering Cross Sections

It is instructive to make the following crude guess for the cross sections of neutral current (NC) and charged current (CC) scat-

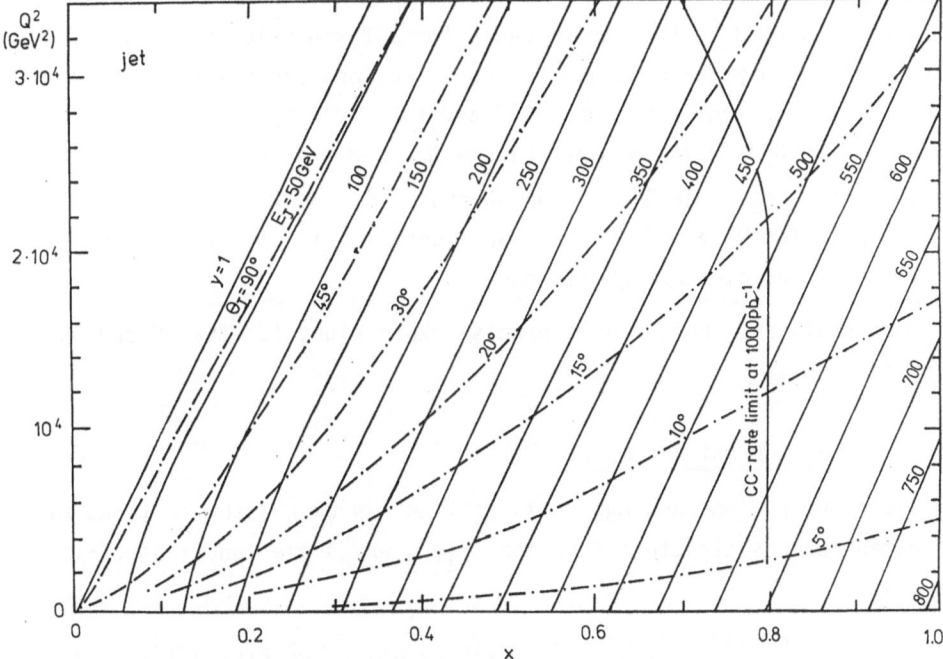

Fig. 7 30 GeV e + 820 GeV p: The x-Q^2 plane with lines of fixed current jet angle and energy.

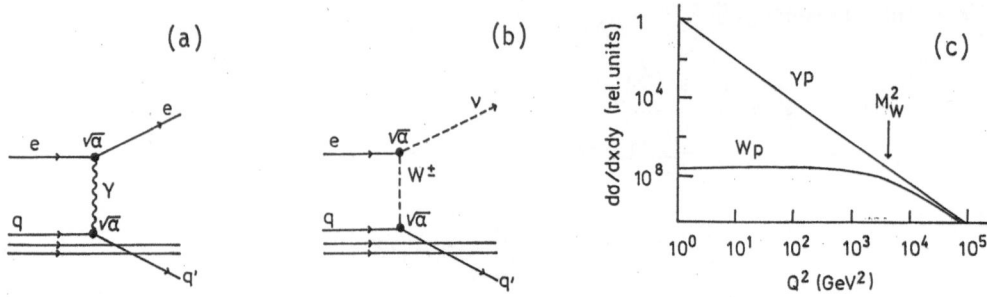

Fig. 8a,b NC and CC scattering by γ and W exchange, respectively.
        c Qualitative behaviour of NC scattering by γ exchange
           and CC scattering by W exchange.

tering. Considering only γ and W exchange, respectively (Fig. 8a,b),
they can be approximated by

$$\frac{d\sigma(\gamma p)}{dxdy} \approx \alpha^2 s \left(\frac{1}{Q^2}\right)^2 F(x,y) \qquad\qquad NC$$

$$\frac{d\sigma(Wp)}{dxdy} \approx \alpha^2 s \frac{1}{(Q^2+M_W^2)^2} F(x,y) \qquad\qquad CC$$

The W has been assumed to couple to leptons and quarks with the same
coupling strength e as the γ and hence the appearence of the factor
α^2 in both cases. The structure function F(x,y) measures the momentum
fraction x carried by the struck quark for a fixed value of y. The
major difference between NC and CC cross sections stems from the different propagator contributions, Q^{-4} and $(Q^2 + M_W^2)^{-2}$, respectively,
where M_W = 82 GeV is the W mass. Figure 8c compares the Q^2 dependence of
the two processes. Near Q^2 = 0 the NC cross section is ~10^8 times
larger while for $Q^2 \gtrsim 10^4$ GeV^2 the weak contribution is of the same
magnitude as the electromagnetic one.

        We shall now present more precise expressions for the NC and CC
cross sections.

3.3   Neutral Current Processes

        The one photon exchange cross section can be written in terms of
two dimensionless structure functions F_1, F_2 which are functions of ν,
Q^2 or x,y:

$$\frac{d^2\sigma(\gamma)}{dxdy} = \frac{4\pi\,\alpha^2}{s\,x^2y^2} \left[(1 - y)F_2(x,Q^2) + xy^2\,F_1(x,Q^2)\right]$$

In the approximation that scattering on spin 1/2 partons dominates F_1

can be expressed in terms of F_2 using the Callan - Gross relation: $2xF_1 = F_2$.

This leads to

$$\frac{d^2\sigma(\gamma)}{dxdy} \approx \frac{4\pi \, \alpha^2}{s \, x^2 y^2} \; (1 - y + \frac{y^2}{2}) \; F_2(x,Q^2)$$

In the quark picture, ep scattering is described as incoherent eq scattering with x measuring the fractional momentum of the struck quark. The structure functions can then be expressed in terms of the quark distribution functions u(x), d(x),...[*] where u,d,... give the number of u,d,... quarks with fractional momentum between x and x + dx:

$$F_2(x) = 2xF_1(x) = x[q(x) + \bar{q}(x)]$$

where for the electromagnetic current

$$q(x) = \frac{4}{9}u(x) + \frac{1}{9}d(x) + \frac{1}{9}s(x) + \frac{4}{9}c(x) \; .$$

At Q^2 values above 10^4 GeV^2 the contribution from Z^0 exchange becomes comparable to that from photon exchange. Formally, the cross section can be written as before,

$$\frac{d^2\sigma(\gamma + Z^0)}{dxdy} = \frac{4\pi \, \alpha^2}{s \, x^2 y^2} \left[(1 - y)F_2(x,Q^2) + y^2 x F_1(x,Q^2) \right]$$

where now the structure functions F_1 and F_2 receive contributions from γ exchange, from Z^0 exchange and from the interference between the two processes. In terms of the quark distribution functions the cross section reads for left (right) handed electrons, allowing for left handed (L) and right handed (R) incoming electrons:[4]

$$\frac{d^2\sigma}{dxdy} = \frac{\pi\alpha^2}{sx^2y^2} \sum_q \{xq(x)(A_q + (1-y)^2 B_q) + x\bar{q}(x)(B_q + (1-y)^2 A_q)\}$$

with

$$A_q = \left(-Q_q + g_{Lq}g_{Le} \frac{Q^2}{Q^2 + M_Z^2}\right)^2 + \left(-Q_q + g_{Rq}g_{Re} \frac{Q^2}{Q^2 + M_Z^2}\right)^2$$

$$B_q = \left(-Q_q + g_{Rq}g_{Le} \frac{Q^2}{Q^2 + M_Z^2}\right)^2 + \left(-Q_q + g_{Lq}g_{Re} \frac{Q^2}{Q^2 + M_Z^2}\right)^2$$

---

[*] The quark distribution functions as well as the structure functions depend on x and Q^2. For ease of writing the Q^2 dependence is not shown.

and

$$g_{Li} = \frac{I_L^3 - Q_i \sin^2\Theta_W}{\sin\Theta_W \cos\Theta_W} \qquad\qquad g_{Ri} = \frac{-Q_i \sin^2\Theta_W}{\sin\Theta_W \cos\Theta_W}$$

where the Q_i, $i = q,e$ measure the charges of the incoming quark and lepton, I_L^3 is the weak isospin of the lepton ($I_{e-}^3 = -1/2$, $I_{e+}^3 = 1/2$), and Θ_W is the weak mixing angle, $\sin^2\Theta_W = 0.225 \pm 0.005 \pm 0.006$. [5]

The final expression of the cross section for left (L) or right (R) handed electron proton scattering reads[6]

$$\frac{d^2\sigma}{dxdy}\left(e^-\begin{Bmatrix} L \\ R \end{Bmatrix} + p \to e^-\begin{Bmatrix} L \\ R \end{Bmatrix} X\right) =$$

$$\frac{4\pi\alpha^2}{s(xy)^2}\left[Q_u^2(u(x) + \bar{u}(x)) + Q_d^2(d(x) + \bar{d}(x))\right] x\,(1 - y + y^2/2)$$

$$+ \frac{\alpha}{xy}\frac{\sqrt{2}\,G_F\,M_Z^2}{sxy + M_Z^2}\begin{Bmatrix} 1 - 2\sin^2\vartheta_W \\ -2\sin^2\vartheta_W \end{Bmatrix} \cdot$$

$$\cdot\,\{\,[Q_u(1 - 4\,Q_u\sin^2\vartheta_W)(u(x) + \bar{u}(x))$$

$$- Q_d(1 + 4\,Q_d\sin^2\vartheta_W)(d(x) + \bar{d}(x))]\cdot(1 - y + y^2/2)\,\} +$$

$$+ \frac{\alpha}{xy}\frac{\sqrt{2}\,G_F\,M_Z^2}{sxy + M_Z^2}\begin{Bmatrix} 1 - 2\sin^2\vartheta_W \\ 2\sin^2\vartheta_W \end{Bmatrix} \cdot$$

$$\cdot\,\{Q_u(u(x) - \bar{u}(x)) - Q_d(d(x) - \bar{d}(x))]\,xy\,(1 - y/2) +$$

$$+ \frac{s}{4\pi}\frac{G_F^2\,M_Z^4}{(sxy + M_Z^2)^2}\begin{Bmatrix} (1 - 2\sin^2\vartheta_W)^2 \\ 4\sin^4\vartheta_W \end{Bmatrix} \cdot$$

$$\cdot\,\{\,[(1 - 4\,Q_u\sin^2\vartheta_W + 8\,Q_u^2\sin^4\vartheta_W)(u(x) + \bar{u}(x)) +$$

$$+ (1 + 4\,Q_d\sin^2\vartheta_W + 8\,Q_d^2\sin^4\vartheta_W)(d(x) + \bar{d}(x))]\cdot(1 - y + y^2/2)\,\} +$$

$$+ \frac{s}{4\pi}\frac{G_F^2\,M_Z^4}{(sxy + M_Z^2)^2}\begin{Bmatrix} (1 - 2\sin^2\vartheta_W)^2 \\ -4\sin^4\vartheta_W \end{Bmatrix} \cdot$$

$$\cdot\,\{\,[(1 - 4\,Q_u\sin^2\vartheta_W)(u(x) - \bar{u}(x)) +$$

$$+ (1 + 4\,Q_d\sin^2\vartheta_W)(d(x) - \bar{d}(x))]\,xy\,(1 - y/2)\,\}$$

with $Q_u = 2/3$ and $Q_d = -1/3$.

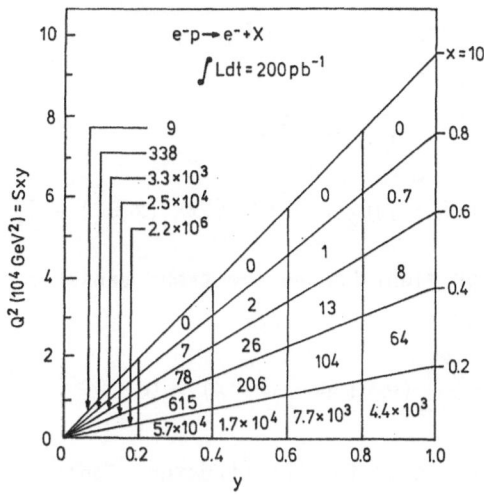

Fig. 9   Event rates for $e^-p \rightarrow e^-X$ with $x > 0.01$
and $y > 0.01$.

Here use has been made of the prediction by the standard theory,

$$M_Z = \frac{\pi\alpha}{G_F/\sqrt{2}} \frac{1}{\sin\Theta_W \cos\Theta_W}$$

where G_F is the fermi coupling constant, $G_F = \dfrac{1.02 \cdot 10^{-5}}{M_p^2}$. To simplify the writing, only the $u,\bar{u}$ and $d,\bar{d}$ contributions are shown.

The expected event rates have been calculated in Ref. 7 using the structure functions of Ref. 8 for an integrated luminosity of 200 pb^{-1} expected after 2 years of data taking. The results are shown in Fig. 9. Note that events with $x < 0.01$ and $y < 0.01$ have been suppressed. There are $3 \cdot 10^6$ events with Q^2 values between 3 and 10^4 GeV^2. Roughly a thousand events have Q^2 values above 10^4 GeV^2 where Z^0 effects dominate. If we demand a minimum of a hundred events we see that neutral currents can be studied with sufficient statistics up to Q^2 values of $3 \cdot 10^4$ GeV^2. The small dark corner with $Q^2 < 200$ GeV^2, $y < 0.004$ marks the kinematical region that has so far been explored experimentally. HERA will extend the kinematic region by two orders of magnitude in either variable.

3.4   Charged Current Processes

Charged current reactions, $e^-p \rightarrow \nu X$, proceed through W^- exchange. The cross section for left handed electrons can be written in terms of three structure functions F_1, F_2 and F_3:

$$\frac{d^2\sigma}{dxdy}(e_L^- p \to \nu X) = \frac{G_F^2 s}{\pi} \frac{1}{(1 + Q^2/M_W^2)^2}$$

$$[(1 - y)F_2(x,Q^2) + y^2 xF_1(x,Q^2) + (y - \frac{y^2}{2}) xF_3(x,Q^2)]$$

The cross section for right handed electrons vanishes since the neutrino is left handed.

$$\frac{d^2\sigma}{dxdy}(e_R^- + p \to \nu + X) = 0$$

The quark-parton expression for the structure functions is:

$$F_2(x) = 2xF_1(x) = x[q(x) + \bar{q}(x)]$$

$$xF_3(x) = x[q(x) - \bar{q}(x)]$$

where

$$q(x) = u(x) + c(x) + \dots$$

$$\bar{q}(x) = \bar{d}(x) + \bar{s}(x) + \dots$$

This leads to

$$\frac{d^2\sigma}{dxdy}(e_L^- p \to \nu + X) = \frac{G_F^2 s}{\pi} \frac{1}{(1 + Q^2/M_W^2)^2} x\{ u(x) + (1 - y)^2 \bar{d}(x)\}$$

where the higher mass quarks have been omitted. The expected event rates are shown in Fig. 10. They have been calculated[7] for the same luminosity as above. Roughly 10 000 events are produced at $Q^2 < 10^4$ GeV^2 and 1000 events at $Q^2 > 10^4$ GeV^2. Figure 10 shows that charged currents can be studied with sufficient statistics up to $\sim 4 \cdot 10^4$ GeV^2.

A precise determination of the structure functions will open the way to address several fundamental questions.

3.5.1 Test of QCD

The large Q^2 range accessible at HERA will allow a stringent test of QCD which predicts logarithmically falling structure functions. Mass corrections and higher twist contributions which affect present experiments should be negligible at HERA. Gluon radiation as illustrated in Fig. 11 leads to scale breaking of the form

$$F(x) \to \frac{F(x)}{1 + c\ell n(Q^2/\Lambda^2)}$$

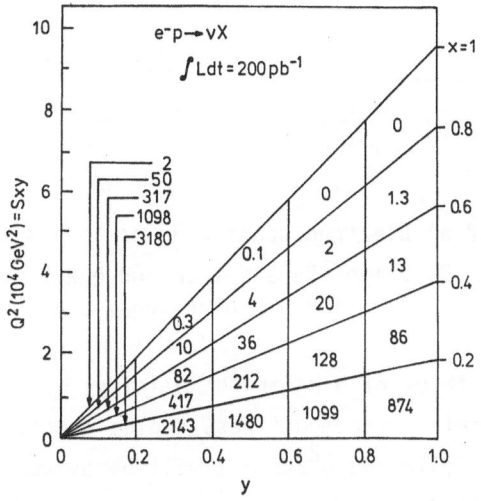

Fig. 10 Event rates for $e^- p \rightarrow \nu X$ with $x > 0.01$
and $y > 0.01$.

The available structure function data span the Q^2 range 0-300 GeV^2; at HERA it will increase to 40 000 GeV^2. This will enhance greatly the sensitivity to power terms (in Q^2) whose presence would be at variance with QCD.

The accuracy with which one can hope to measure the QCD scale parameter is around ±40 MeV for Λ_{QCD} = 200 MeV.

3.5.2 Structure of quarks and electrons

If quarks and/or electrons are extended objects the structure functions will show power law type deviations from their QCD predicted values. The sensitivity to possible structure of quarks and electrons has been estimated with the help of a model[9] which assumes quarks and electrons to be composites and to have common constituents (Fig. 12). The interchange of the constituents leads to a contact term of the form

$$\mathcal{L} \sim \frac{g^2}{\Lambda^2} \quad \bar{e}_\alpha \gamma^\mu e_\alpha \bar{q}_\beta \gamma_\mu q_\beta$$

Fig. 11 eq scattering with accompanying gluon bremsstrahlung.

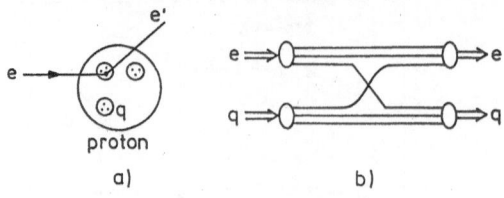

Fig. 12 a) Electron scattering on composite quarks.

b) Composite electron and quark scattering by
   constituent interchange.

where α, β denote states of a definite helicity, $\alpha = L, R$; $\beta = L, R$. The mass parameter Λ sets the compositness scale; g measures the coupling strength. For the following $g^2/4\pi = 1$ will be assumed. Figure 13 shows the ratio of the structure functions F_2 obtained with the contributions from $\gamma + Z^0$ exchange and the contact interaction, and for γ exchange alone:

$$F_2(\gamma + Z^0 + C.I.)/F_2(\gamma)$$

The ratio is shown as a function of Q^2 and x for different Λ values. At Q^2 values above 10^4 GeV^2 a contact interaction with $\Lambda = 1$ TeV leads to large deviations - factors of 5 to 10 - from the standard results. Given two years of data taking HERA will be sensitive up to Λ values of ~7 TeV corresponding to distances of $3 \cdot 10^{-18}$ cm. The present lower limits on Λ are 2-4 TeV on e,μ from $e^+e^- \to \mu^+\mu^-$ and 300 GeV for quarks from the

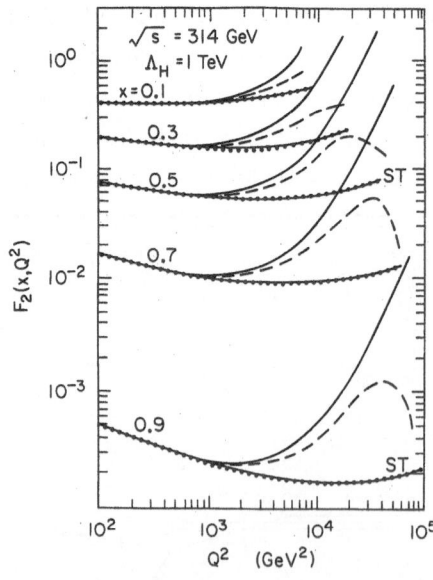

Fig. 13 The ratio $F_2(\gamma + Z^0 + C.I)/F_2(\gamma)$ for

composite electron quark scattering.

total cross section for e^+e^- annihilation into hadrons.[10] A theoretical analysis of some composite models has indicated that the energy of HERA would still be insufficient to see structure in the quarks and electrons.[11] It suggests a lower as well as an upper limit on Λ:

$$\lambda(20 - 30)\,\text{TeV} < \Lambda < 250\,\text{TeV}$$

The lower limit is deduced from the limit on the decay $K^+ \rightarrow \pi^+ \mu e$; λ is a number of order unity. The upper limit stems from cosmological considerations.

3.5.3 New currents

Altarelli et al.[3] have estimated the effect of a second W_2 on the rate of charged current events. The amplitude for the exchange of the standard $W(W_1)$ can be written as

$$A(Q^2) \sim \frac{g^2}{8(Q^2 + m_W^2)} \sim \frac{G_F/\sqrt{2}}{1 + Q^2/m_W^2}\;.$$

Under the assumption that the second W couples in the same way to leptons and quarks the amplitude can be written as

$$A_{1+2}(Q^2) \sim \frac{g_1^2}{8(Q^2 + m_1^2)} + \frac{g_2^2}{8(Q^2 + m_2^2)}\;.$$

The coupling constants g_1, g_2 have to be chosen such that the low Q^2 region remains unchanged, i.e.

$$A(0) \equiv \frac{G_F}{\sqrt{2}} = A_{1+2}(0) = \frac{g_1^2}{8m_1^2} + \frac{g_2^2}{8m_2^2} \equiv \frac{g_2^2}{8m_W^2}\;.$$

Taking $m_1 = m_W$ and defining $r = g_2^2/g^2$ the ratio of the cross sections for two W's over one W is given by

$$\frac{\sigma(W_1 + W_2)}{\sigma(W_1)} = \left[1 - r\,\frac{m_1^2}{m_2^2} + r\,\frac{m_1^2}{m_2^2}\,\frac{1 + Q^2/m_1^2}{1 + Q^2/m_2^2} \right]$$

The ratio is shown in Fig. 14 for different W_2 masses assuming $r = 1$. For m_2 around 200 GeV large deviations from the standard theory result will be observed. The deviations become smaller as m_2 increases. With two years of data taking one will be sensitive to W_2 masses up to ~800 GeV.

A similar result holds for neutral currents and additional Z^0's.

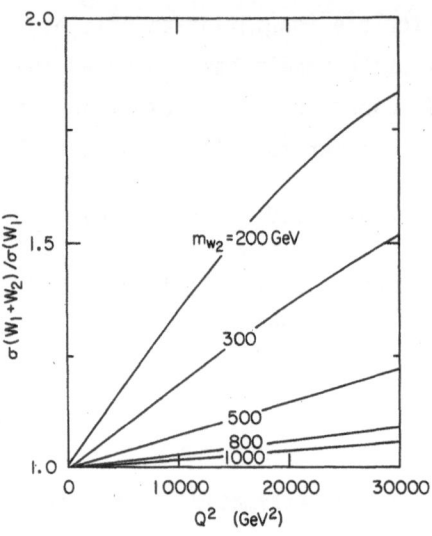

Fig. 14   The ratio of the cross sections for two and one W.

3.5.4 Right handed currents

One of the great mysteries of weak interactions is the asymmetry between left and right handedness; there are left handed neutrinos and left handed currents and no right handed counter parts. It has long been speculated that left-right symmetry is restored at some higher energy; namely there exist heavy right handed neutrinos N_R that couple to a right handed W_R.[12] The mass of N_R has been related to the masses of the corresponding charged lepton, m_e, and the left handed neutrino m_ν:[13] $m_{NR} \approx m_e^2/m_\nu$. For electrons this leads to the following prediction for m_{N_R}:

m_ν (eV)	0.1	1	10
m_{N_R} (GeV)	2611	261	26

Several experimental and theoretical limits have been put on the mass of W_R:

<u>β decay of μ's</u>
if ν is a Dirac spinor: $M_R > 380$ GeV   (Ref.14)
if ν is a Majorana spinor: no limit

<u>Nonleptonic decays</u> $M_R > 200$-300 GeV (Ref.15)

<u>K_S^0 - K_L^0 mass difference</u>
educated guess: $10 < M_R < 100$ TeV (Ref.16)

The lower limit on a right handed Z^0 is $M_{Z_R} \geq 150$ GeV provided $\sin^2\theta_W < 0.25$ (Ref.15). However, a simple Higgs structure leads to

$M(Z_R)/M(W_R) \approx 1.6$. This suggests also $M(Z_R)$ to be in the region of tens of TeV (Ref. 17).

3.5.5 Search for right-handed currents with longitudinally polarized electrons

Longitudinally polarized electrons make HERA ideally suited to search for right handed currents. Any nonzero cross section contribution to $e_R^- p \to \nu X$ or $e_L^+ p \to \bar{\nu} X$ scattering signals the presence of a right handed charged current (see Sect. 3.4). In the absence of a W_R contribution:

$$\frac{d^2\sigma}{dxdy}\begin{pmatrix} e_L^- p \\ e_R^+ p \\ e_R^- p \\ e_L^+ p \end{pmatrix} = \frac{G_F^2 s}{\pi} \frac{x}{(1 + Q^2/M_W^2)^2} \begin{pmatrix} (u+c)+(1-y)^2(\bar{d}+\bar{s}) \\ (d+s)+(1-y)^2(\bar{u}+\bar{c}) \\ 0 \\ 0 \end{pmatrix}$$

For neutral current reactions the situation is different. The photon is blind to the helicity of the electron but the Z^0 gives different contribution to e_L and e_R scattering as shown in sect. 3.3. Figure 15 shows the difference between the cross sections for $e_{L,R}^-$ and $e_{L,R}^+$ scattering. They amount to ~60 % at Q^2 values of 10^4 GeV^2.

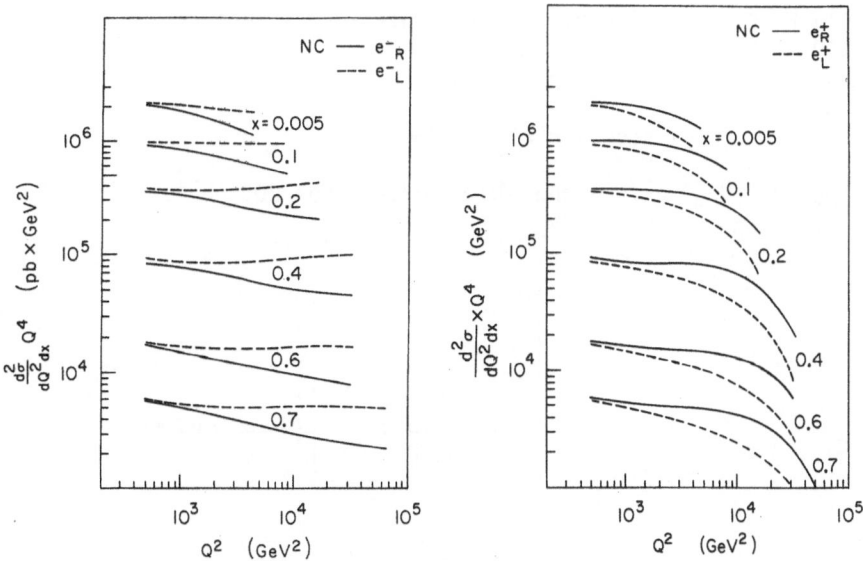

Fig. 15 The neutral current cross section for left and right handed $e^{\pm}$.

4. PRODUCTION OF NEW PARTICLES

4.1 Pair Production of Heavy Quarks

Photon-gluon fusion (Fig. 4) will be a strong source of heavy quarks. A discussion of the process can be found in Ref. 18. The total cross section depends on the quark mass approximately as

$$\sigma(eq \rightarrow Q\bar{Q}X) \sim M_Q^{-4}$$

Given a t quark mass of 40 GeV and a luminosity of 200 pb^{-1} the expected number of $t\bar{t}$ events is a few hundred. Top-like quarks can be searched for up to masses of 100 - 120 GeV.

The heavy quarks will essentially be photo produced ($Q^2 \approx 0$) and will be emitted in the direction of the incoming proton. Given the high mass these heavy quarks will decay into many particles (~20 - 25 for m_t = 40 GeV) which are isotropically distributed in the plane perpendicular to the beams. Thus, the production of new heavy quarks should be discernible from other processes. Figure 16 shows for illustration the production of a $t\bar{t}$ pair assuming M_t = 40 GeV.

Another possible source of Q production might be an intrinsic $Q\bar{Q}$ component in the nucleon. It has been argued for instance[19] that the $c\bar{c}$ component could be as large as 1 % and could become an important contributor once Q^2 exceeds ~4 M_c^2.

4.2 Vector Boson Production

Z^0 and $W^\pm$ production is expected to proceed predominantly through the diagrams shown in Fig. 17a. The total cross sections are predicted to be[20]

$$\sigma(ep \rightarrow Z^0 X) \approx 0.1 \text{ pb}$$

$$\sigma(ep \rightarrow W^\pm X) \approx 0.05 \text{ pb}$$

leading to ~20 Z^0 events and ~10 $W^\pm$ events for 200 pb^{-1}.

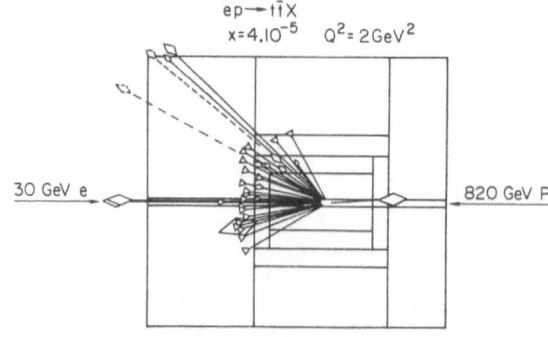

Fig. 16 Example of a $t\bar{t}$ production event.

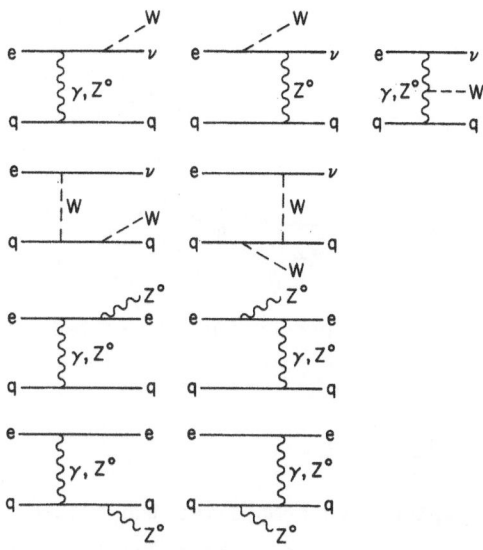

Fig. 17a Diagrams for Z^0 and $W^\pm$ production.

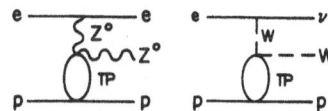

Fig. 17b Diagrams for Z^0, $W^\pm$ production by pomeron exchange.

In the (unlikely) case that Z^0 and $W^\pm$ have also a strongly inter-acting component the time honoured pomeron exchange (Figs. 17b) may add to their production rate.

4.3 Higgs Production

Possible diagrams for production of Higgs particles are shown in Fig. 18. For a total of 200 pb^{-1} the expected yield of Higgs produced is not overwhelming[21]:

$$M_H = 10 \qquad 20 \qquad 50 \qquad 100 \text{ GeV}$$

number of events = 10   10   10   3

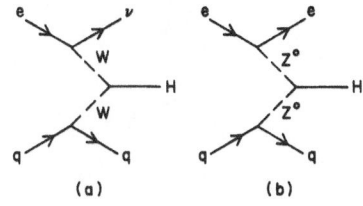

Fig. 18 Diagrams for Higgs production.

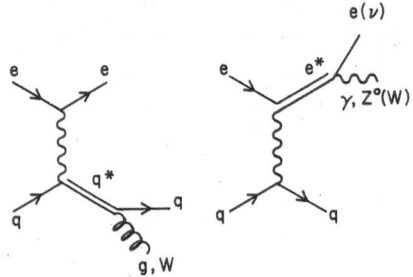

Fig. 19 Diagrams for the production of excited quarks and leptons.

4.4 Excited Quarks and Leptons

If quarks and leptons are composites the existence of excited quarks and leptons q* ,e* appears to be natural (remember p,Δ). Their production would follow the diagrams of Fig. 19. The production rates are sufficient to search for q* ,e* up to masses of 250 GeV.[22]

4.5 New Currents, Quarks, Leptons

Suppose there exist new currents X^o, $X^{\pm}$ which link the known quarks and leptons to new quarks $Q = U,D,S,C,B,T$ and leptons L, with coupling strengths equal to that of the standard weak interactions. The cross section for the dominant associated production of D and L^o as depicted in Fig. 20 is given by

$$\frac{d\sigma}{dxdy} = \frac{G_F s}{2} \left(\frac{m_W}{m_X}\right)^4 \frac{1}{(1 + \frac{sxy}{M_X^2})}$$

$$\left\{ xq(x) \left(1 - \frac{m_L^2 + m_D^2}{sx}\right) + x\bar{q}(x) \left(1 - y - \frac{m_L^2}{sx}\right) \left(1 - y - \frac{m_D^2}{sx}\right)\right\}$$

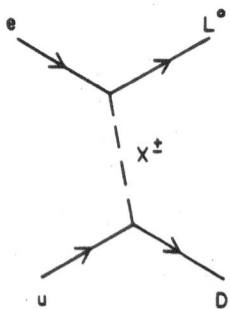

Fig. 20 Diagram for new quark and lepton production.

where

$$q(x) = u(x) + c(x) + \ldots$$

$$\bar{q}(x) = \bar{d}(x) + \bar{s}(x) + \ldots$$

The kinematically allowed regions for x and y are

$$\frac{(m_L + m_D)^2}{s} < x < 1$$

$$y_- < y < y_+$$

$$y_\pm = \frac{1}{2sx} \left[sx - m_D^2 - m_L^2 \pm \sqrt{(sx - m_D^2 - m_L^2)^2 - 4m_D^2 m_L^2} \right]$$

The process will lead to the event rates shown in Fig. 21 as a function of m_D for different lepton masses m_L assuming $m_x = 83$ GeV and 10 days of running at $L = 10^{31}$ cm^{-2}s^{-1}; 200 pb^{-1} would probe up to $m_L + m_D \approx 220$ GeV.

4.6 Leptoquarks

Technicolour offers an alternative though not undisputed explanation for the Higgs mechanism. In this theory the Higgs is a pion-like composite: a state where a new quark and antiquark are bound together by the super strong technicolour force. Besides new quarks U,D,... new leptons E,M,... are also predicted which together lead to a new universe of particles. What is most exciting for HERA is the occurrence of lepto-

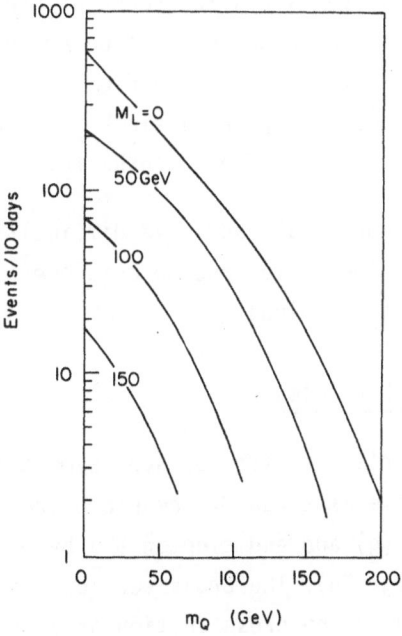

Fig. 21   Production rates for $e^- p \to DL^0 X$.

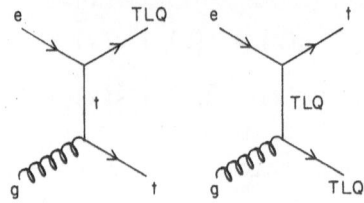

Fig. 22 Diagrams for the production of leptoquarks.

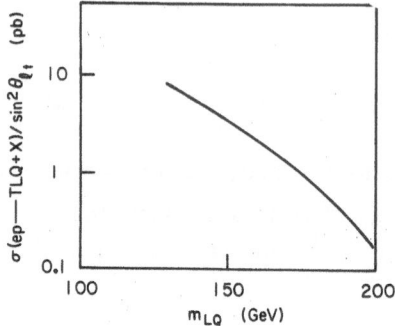

Fig. 23 Cross section for the production of leptoquarks.

quarks TLQ = ūE,... The expected mass hierarchy is as follows:

		mass
- technipions (UD̄,...)		O(10 GeV)
- leptoquarks (UĒ,...)		O(160 GeV)
- technicolour Octets (UD̄,...)		O(240 GeV)

Leptoquarks would be within the reach of HERA experiments. The diagrams expected to dominate the production are given in Fig. 22. The cross section for ep → TLQX is shown in Fig. 23 as calculated by[23]. Here Θ_{et} is the mixing angle. Given a $\sin^2\Theta_{et}$ of ~0.05 leptoquarks with masses up to 180 GeV can be studied at HERA (Fig. 23).

Technicolour theories already have difficulties explaining the small $K_L^0 \to \mu^+\mu^-$ branching ratio, and no evidence has been found at PETRA or PEP for low-lying technipions.

4.7. Supersymmetric Particles

HERA is a hunter's paradise for supersymmetric particles - if they exist in the accessible mass range. The prototype reaction is associated production of squark (q̃) and and slepton (ℓ̃) by photino (γ̃), zino (Z̃) or wino (W̃) exchange (Fig. 24). The cross section has been calculated in Ref. 24. Figure 25 shows the cross section as a function of the squark mass for different slepton mass values. If we demand at least 10 events per 200 pb^{-1} then squark plus slepton production can be studied up to mass values of $m_{\tilde{q}} + m_{\tilde{\ell}} = 160$ GeV.

The production of scalar leptons has been studied in Ref. 20. The dependence on the $e^{\pm}$ polarization has been analyzed in Ref. 25.

5. DETECTION OF NEW PROCESSES

The mass range that is accessible to HERA for new particles was estimated in the previous sections by requiring the production of at least a few tens of these events a year. Since the yield of standard neutral and charged current events is many orders of magnitude larger one might wonder whether their detection will be as difficult as e.g. that of the top quark at the S$\bar{p}$pS collider.

The events should be visible if the dominating background is ordinary neutral current scattering (eq → eq). This is a consequence of

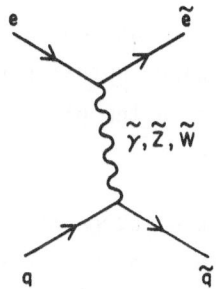

Fig. 24 Diagram for eq → $\tilde{e}\tilde{q}$.

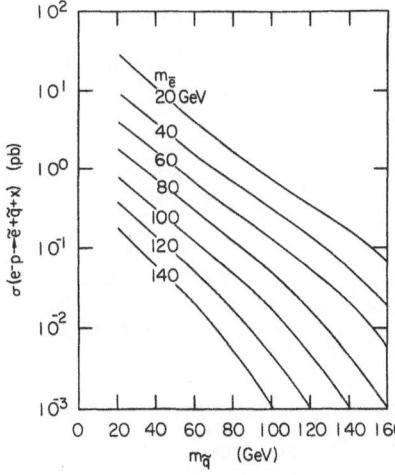

Fig. 25 Cross section for the production of supersymmetric quarks and leptons.

two facts: the kinematics in eq → eq is well defined, and, the decay of a new heavy object will spread the emerging quark (lepton) over a large volume in phase space. For definiteness consider the supersymmetric reaction discussed before,

$$eq \rightarrow \tilde{e}\tilde{q}$$

The event rate relative to the standard neutral current reaction, eq → eq, is

$$\text{for } M_{\tilde{q}} + M_{\tilde{e}} = \left\{ \begin{array}{l} 80 \text{ GeV} \\ 180 \text{ GeV} \end{array} \right. \quad \frac{\sigma(\tilde{e}\tilde{q})}{(eq)} \approx \left\{ \begin{array}{l} 10^{-3} \\ 10^{-5} \end{array} \right.$$

The expected decay chains for $\tilde{q}$ and $\tilde{e}$ are

$$\tilde{q} \rightarrow q + \tilde{\gamma}$$

and

$$\tilde{q} \rightarrow q + \tilde{g} \rightarrow qg\tilde{G}$$

and

$$\tilde{e} \rightarrow e + \tilde{\gamma} \text{ or } \tilde{e} \rightarrow e + \tilde{G} .$$

Under the assumption that $\tilde{\gamma}$ and $\tilde{G}$ are weakly interacting and leave the experiment undetected the final state consists of an electron and a quark jet as in the case of an ordinary neutral current event. The difference lies in the kinematics as seen in the plane perpendicular to the beams. This is illustrated in Fig. 26.

In the standard case the electron and the quark jet are anti-collinear and balance each other in transverse momentum. For the super-symetric process e and q are neither anticollinear nor do they balance each other in transverse momentum. These two facts can be used to suppress the background. How efficiently that suppression can be done is demonstrated in Fig. 27. Here the acollinearity angle Δφ is plotted versus Δy which measures the difference in y values computed from the

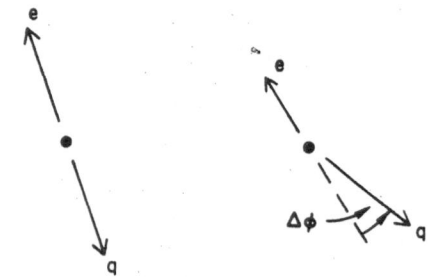

Fig. 26 Comparison of the kinematics
for eq → eq
and for
eq → $\tilde{e}\tilde{q}$ → eq + missing particles

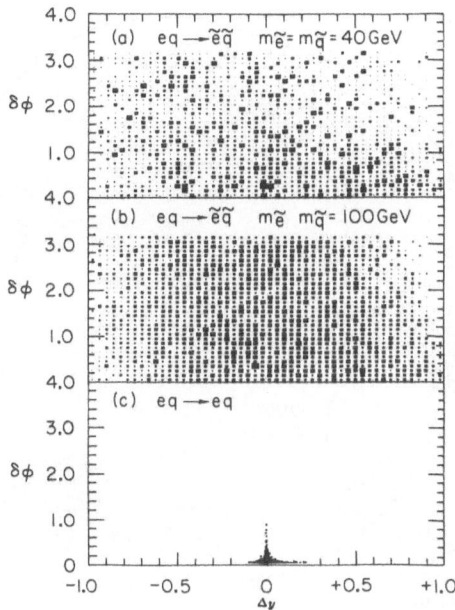

Fig. 27 Distribution of $\Delta\varphi$ versus Δy for the supersymmetric process $eq \to \tilde{e}\tilde{q}$ with $m_{\tilde{e}} = m_{\tilde{q}} = 40$ GeV (a) and $m_{\tilde{e}} = m_{\tilde{q}} = 100$ GeV (b) and for the standard process $eq \to eq$ (c).

current jet and the scattered electron:

$$\Delta_y = y_q - y_e$$

$$y_q = \frac{\Sigma E_i - P_{\parallel i}}{2E_e}$$

$$y_e = \frac{E'_e - P'_{\parallel e}}{2E_e}$$

$E_i, P_{\parallel i}$ are the energies and longitudinal momentum components of the current jet particles; $E'_e, p'_{\parallel}$ are the energy and longitudinal momentum component (measured w.r.t. the incoming proton direction) of the scattered electron.

The distributions shown in Fig. 27 have been computed by a Monte Carlo technique for a reasonably realistic detector assuming a finite hadronic energy resolution ($\Delta E/E = 60$ %/$\sqrt{E}$) and a blind angular region around the beam pipe. Fig. 27(c) shows the result for $\sim 10^4$ standard neutral current events. As expected, the standard NC events cluster near $\Delta\varphi = \Delta y = 0$. In contrast, for the supersymmetric process (shown in Fig. 27(a) for $m_{\tilde{q}} = m_{\tilde{e}} = 40$ GeV and in Fig. 27(b) for $m_{\tilde{q}} = m_{\tilde{e}} = 100$ GeV). The

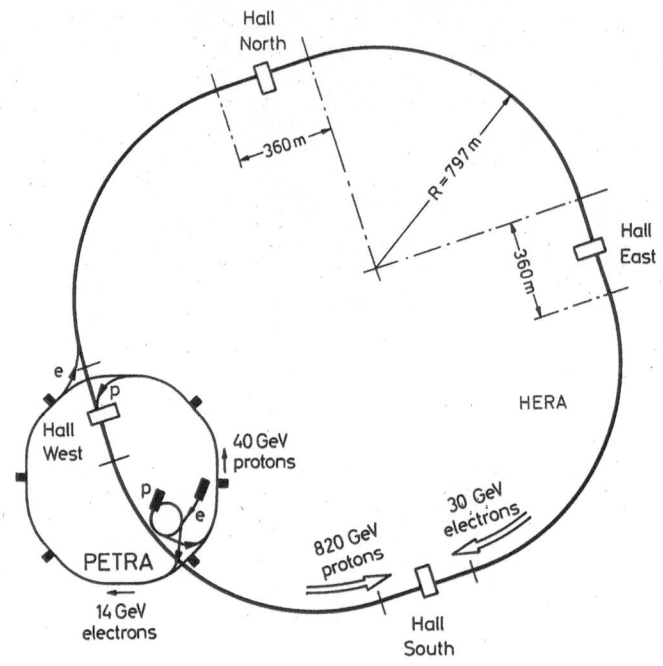

Fig. 28 Layout of HERA.

Table 1: Parameters of HERA

	p-ring		e-ring	units
Nominal energy	820		30	GeV
c.m. energy		314		GeV
Q^2_{max}		98.400		GeV^2
Luminosity		1.5·10^{31}		cm^{2}s^{-1}
Polarization time			25	min
Number of interaction points		4		
Crossing angle		0		
Free space for experiments		±5.5		m
Circumference		6336		m
Bending radius	588		608	m
Magnetic field	4.65		0.165	T
Energy range	300-820		10-33	GeV
Injection energy	40		14	GeV
Circulating current	163		58	mA
Total number of particles	2.1·10^{13}		0.8·10^{13}	
Number of bunches		210		
Number of bunch buckets		220		
Time between crossings		96		ns
Beta function (β^*_x/β^*_y)	10/1.0		2/0.7	m
Beam size at crossing σ_x	0.29		0.26	mm
Beam size at crossing σ_y	0.07		0.070	mm
Beam size at crossing σ_z	11		0.8	cm
Energy loss / turn	1.4·10^{-10}		127	MeV
Max. circumf. voltage	0.2/2.4		260	MV
Total RF power	1		13.2	MW
RF frequency	52.033/208.13		499.667	MHz
Filling time	20		15	min
Tune shift Q_x	0.001		0.016	
Tune shift Q_y	0.00042		0.025	

events are spread all over the $\Delta\varphi, \Delta y$ map. Thus, supersymmetric events can be well isolated from the standard ones without significant loss in efficiency.

6. THE HERA COLLIDER

The layout of HERA is shown in Fig. 28 (Ref. 26). Two separate magnet systems guide the e⁻ and p beams around the 6.3 km long ring. The beams cross each other in four interaction points. The salient machine parameters are listed in Table 1.

6.1 Civil Engineering

The machine houses in a tunnel that is between 15 and 20 m deep underground (Fig. 29). The tunnel is in a plane which is tilted by ~1 % w.r.t. to the vertical. About one half of the tunnel is below the water table. Most of the tunnel runs underneath public land; the stretch where HERA intersects PETRA is on DESY property. As a consequence, most of the installations for cryogenics, electrical and cooling power etc. are located on the DESY site and feed the rings through Hall West while the surface activities near the other halls - which are on public land - are restricted so as to keep the disturbance of the environment there at a minimum.

A cross section of the tunnel is shown in Fig. 30. It has an inner diameter of 5.2 m. The proton ring is located above the electron ring. Fig. 30 also indicates the supply lines for liquid and gaseous Helium and for cooling water.

6.2 Injection System

The injection system uses the rebuilt synchrotron DESY and the storage ring PETRA (Fig. 31). H⁻ions are accelerated to 50 MeV in a newly constructed LINAC, to 7.5 GeV in the new synchrotron, DESYIII, to 40 GeV in PETRA (which requires bypasses around the RF cavities for the e± beams) and injected into HERA. Electrons pass from a LINAC (50 - 400 MeV) to the newly constructed synchrotron DESY II (9 GeV), to PETRA (14 GeV) and finally to HERA. The filling takes for each beam between 15 and 20 min.

6.3 Electron Ring

At 30 GeV a bending field of 0.165 T is required. This can be achieved with conventional magnets. In Fig. 32 a half cell of the magnet system is shown. Dipole, quadrupole and sextupole are premounted on one common support, the total length of a half cell being 11.8 m. Figure 32 also shows

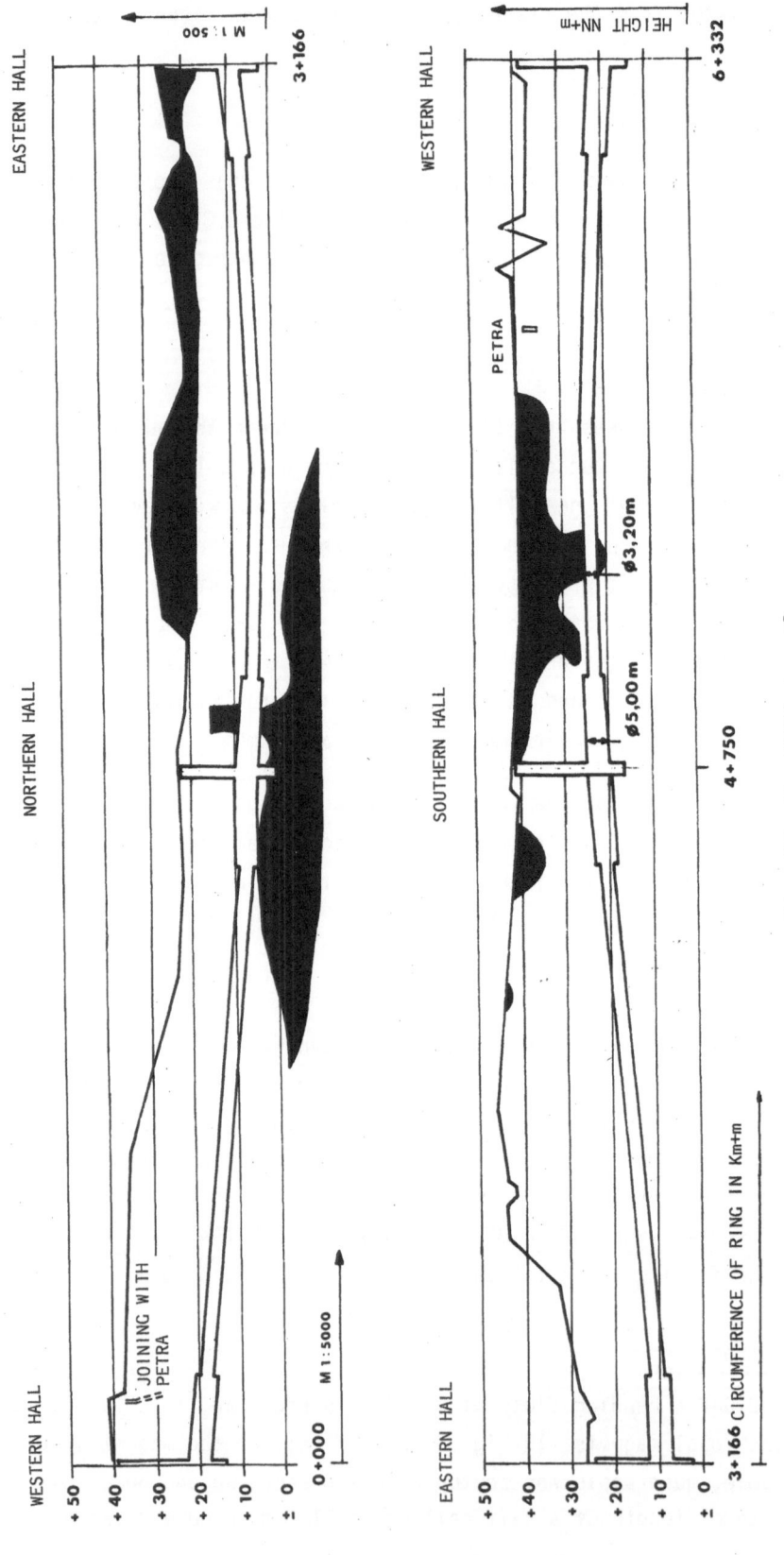

Fig. 29 The HERA tunnel.

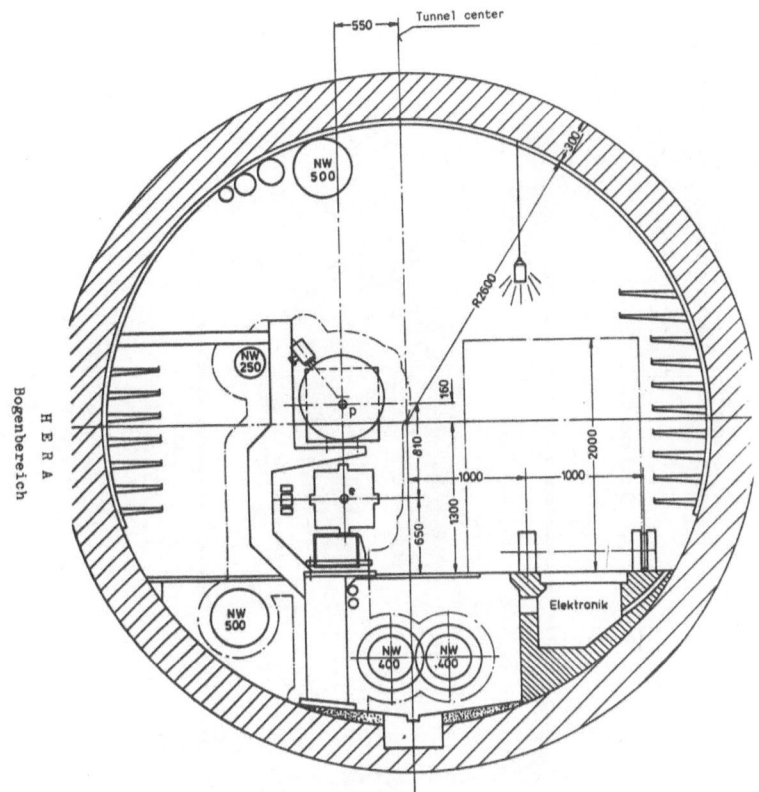

Fig. 30   Cross section of the HERA tunnel.

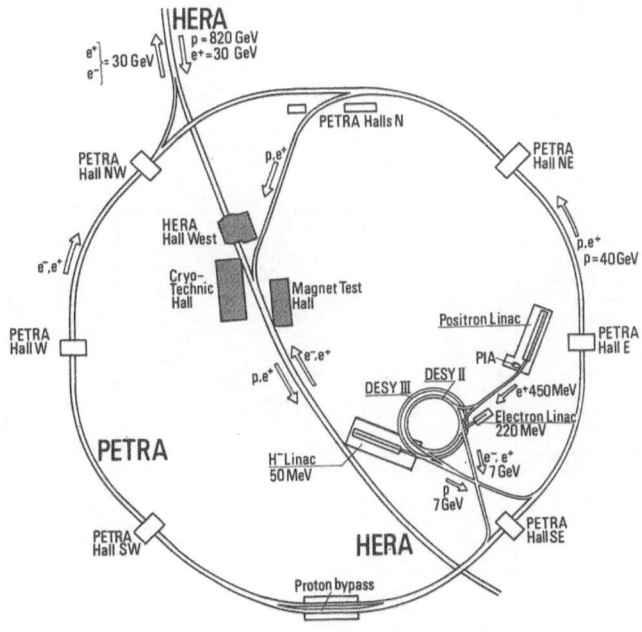

Fig. 31   The injection system for HERA.

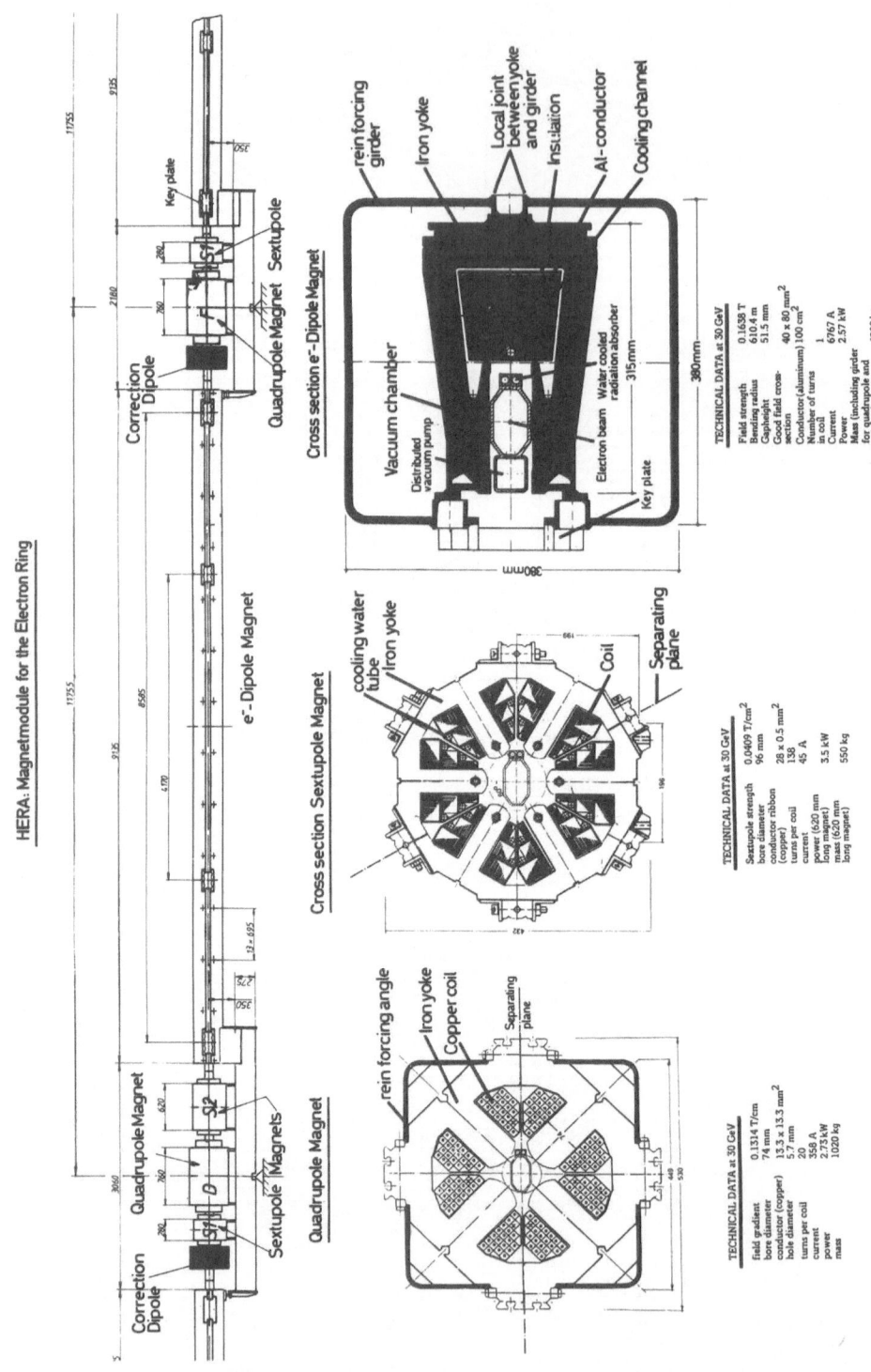

Fig. 32 A half cell of the electron ring.

cross sections for the three different types of magnets. The dipole magnet is excited by a single conductor which carries at 30 GeV a current of 6767 A.

6.4   Proton Ring

The technological challenge of HERA lies in the superconducting magnet system for the proton beam. A bending field of 4.65 T is required to keep an 820 GeV proton beam on orbit. At this field strength iron cannot be used for field shaping and homogeneity has to be achieved by proper shaping of the superconducting coils. Figure 33a,b indicates schematically the manner in which the windings are wrapped around the beam pipe. In order to obtain a homogeneous field the current density has to follow a $\cos\varphi$ distribution, $I = I_0 \cos\varphi$.

The mechanical accuracy required is quite demanding. Consider the case where the height of the left coil halves is larger by 2δ and that of the right coil halves is shorter by 2δ (Fig. 33c). If $\delta = 20$ μm, the skew quadrupole component is $a_2 = 1.7 \cdot 10^{-4}$ which is at the tolerable limit. Hence, a mechanical precision of typically 20 μm has to be kept.

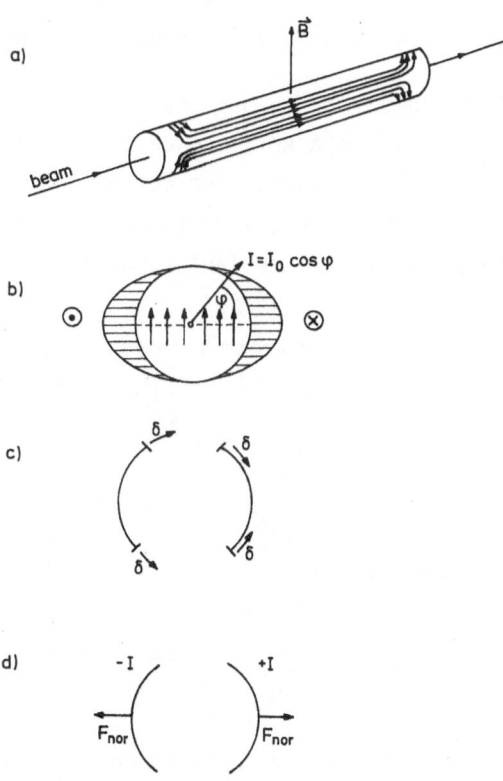

Fig. 33   The mechanics of a superconducting dipole.

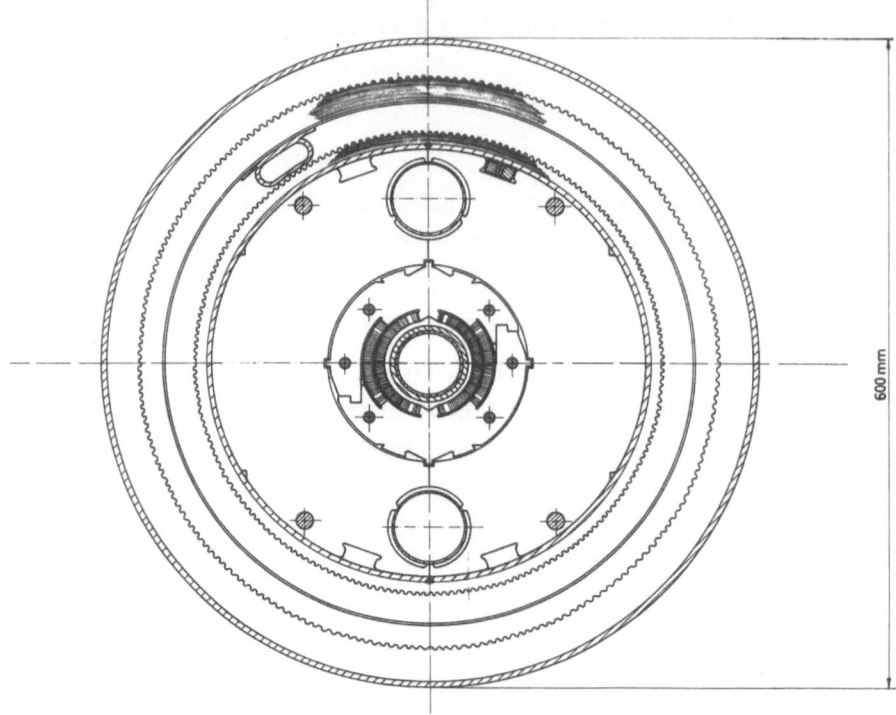

Fig. 34a Cross section of the superconducting dipole.

It has to be kept in the presence of a magnetic pressure which tries to push the two coil halves apart with a force of ~100 t per meter (Fig. 33d).

The design of the HERA magnets has started from the FNAL-TEVATRON and BNL magnets and has lead to a novel concept shown in Fig. 34a. The coils are clamped by aluminum collars. The collaring is reinforced by the cold iron yoke which surrounds the collar tightly. Vacuum and superinsulation shield the iron yoke from the ambient temperature. The dipole magnet has an overall length of 9 m.

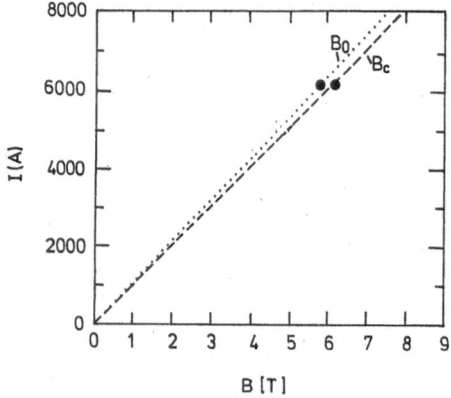

Fig. 34b Load line for dipole Nr. 3. B_o field on axis, B_c maximum field at conductor, at 4.6K (S. Wolff, priv. comm.).

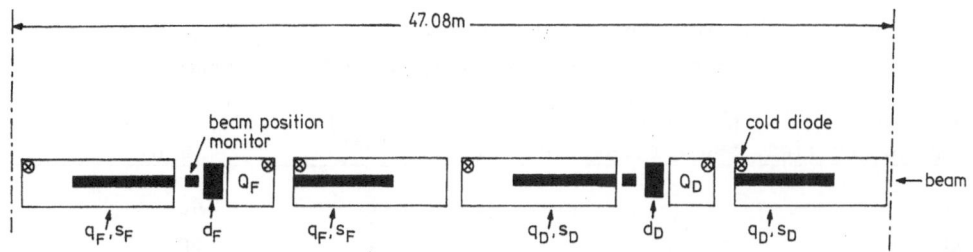

Fig. 35   Sketch of a cell of the proton ring with four dipoles plus quadrupole (q) and sextupole (s) correction coils, and two quadrupoles (Q).

In Fig. 34b the load line measured for dipole Nr. 3 is shown. At
4.6 K the quench occurred at 5.8 T which is about 25 % higher than the
field required for 820 GeV.

The advantages over a warm iron design are a 12 % gain in field, a
smaller static heat load at 4 K and better quench protection. The disadvantage is a 5 times larger cold mass; however, the cooldown-warmup
cycle takes the same amount of time in both cases, about 30 - 40 h.

A schematic drawing of a proton ring cell with the dipole, quadrupole and correction magnets is shown in Fig. 35. The length of one cell
is 47.08 m.

6.5   RF System

6.5.1 Electron ring

The circulating electrons of energy E loose energy by synchrotron
radiation:

$$\Delta E/turn = \frac{0.088\ E^4}{\rho}$$

where ρ is the bending radius in m, ρ = 608 m. For E = 30 GeV,
ΔE = 112 MeV/turn. For an electron current of 58 mA the power lost by
synchrotron radiation is 6.5 MW. The power loss is compensated by RF
which is fed into cavities that are traversed by the electron beam.

The frequency chosen for the RF of HERA is 500 MHz, which is the
same as for DESY, DORIS and PETRA (see Table 1). For operation below
30 GeV HERA will be equipped with cavities formerly used at PETRA. There
is, however, a strong physics desire to run the electron beam at
somewhat higher energy: at E = 27 GeV the polarization time (see sect. 6.6)
is 43 min compared to 12 min at E = 35 GeV.

Table 2:   RF for the electron ring.
A total available RF power of 7 MW is assumed.

number of normal c.cavities	number of super c.cavities	E (GeV)	max I (mA)	polarization time (min)
88	-	27	46	42
88	-	29	11	30
88	8	30	27	25
-	32	35	30	12

In order to reach 35 GeV it is planned to add superconducting cavities (see Table 2). Superconducting 500 MHz cavities are under development at DESY (Fig. 36). With a 2 x 4 cell cavity in a single cryostat an accelerating field of 12 MV or a gradient of 5 MV/m is expected. About 240 kW/cryostat are fed to the beam. The expected heat loss is 85 W/cryostat at 4K and 5 MV/m.

In a prototype test a 1 GHz 9 cell superconducting cavity with 2.7 MV/m has been installed in PETRA. It has been run for 5 months in normal operation with very encouraging results.[27]

6.5.2 Proton ring

The loss of synchrotron radiation by the proton beam is negligible (remember $\Delta E \sim mass^{-4}$). For the proton beam new cavities and clystrons have to be developed. The frequency of the RF will vary from 52 MHz at injection (40 GeV) to 208 MHz at 820 GeV.

6.6   Electron Polarization

After coasting for some time the electrons will become polarized transverse to the ring plane (Fig. 37a). The build-up time for polarization follows from the relation

$$P(t) = P_0 \ (1 - e^{-t/\tau_p},$$

with

$$P_0 = 92 \ \%$$

and

$$\tau_p = \frac{98 \ \rho^2 \ R}{E^5}$$

where τ_p in s, ρ bending radius in m, R average radius in m and E in GeV. For E = 30 GeV, τ_p = 25 min.

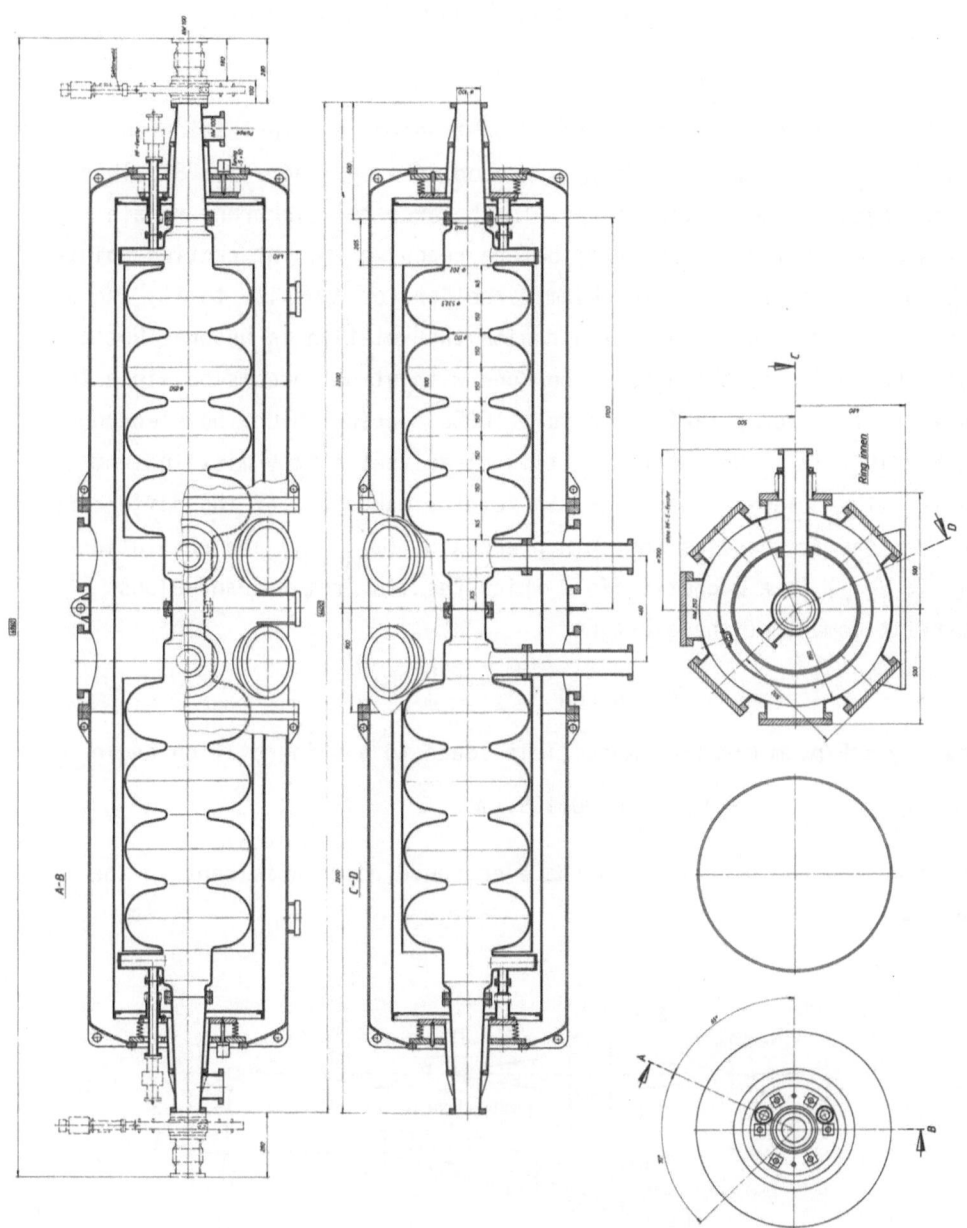

Fig. 36 Layout of the 500 MHz superconducting cavity.

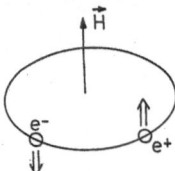

Fig. 37a Transversely polarized $e^{\pm}$ beams.

For particle physics, instead of transverse polarization an electron beam of definite helicity is needed (see eg. sect. 3.5.5). This can
be achieved in the following way. In the arcs the electrons remain
transversely polarized. Shortly before reaching the interaction point
the spin is rotated into the beam direction (or opposite to it) and as
the beam leaves the interaction region the rotation is undone. Notice,
that the two rotations have to be done with utmost precision since the
build up time for polarization is O(1000s) during which the electrons
pass $2 \cdot 10^{8}$ times the rotators. It is clear that a tiny misalignment can
destroy the polarization. The spin rotator makes use of the difference
between the frequencies for spin precession (ω_{Larmor}) and revolution
($\omega_{cyclotron}$) in a magnetic field which results from the anomalous
magnetic moment of the electron:

$$\omega_{Larmor} - \omega_{cyclotron} = \gamma \frac{g-2}{2} \frac{e}{m} B$$

where $\gamma = E/m$, m electron mass. This leads to a spin rotation angle of

$$\Theta_{rot} = 0.001159 \; \gamma\varphi$$

where φ is the angle by which the electron direction is bent in the
magnet.

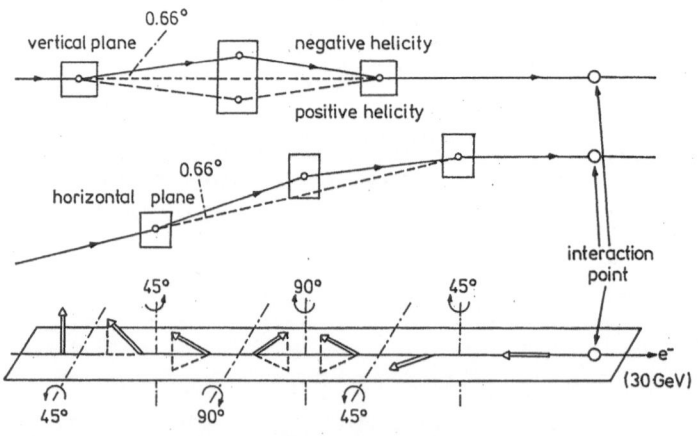

Fig. 37b Spin rotator.

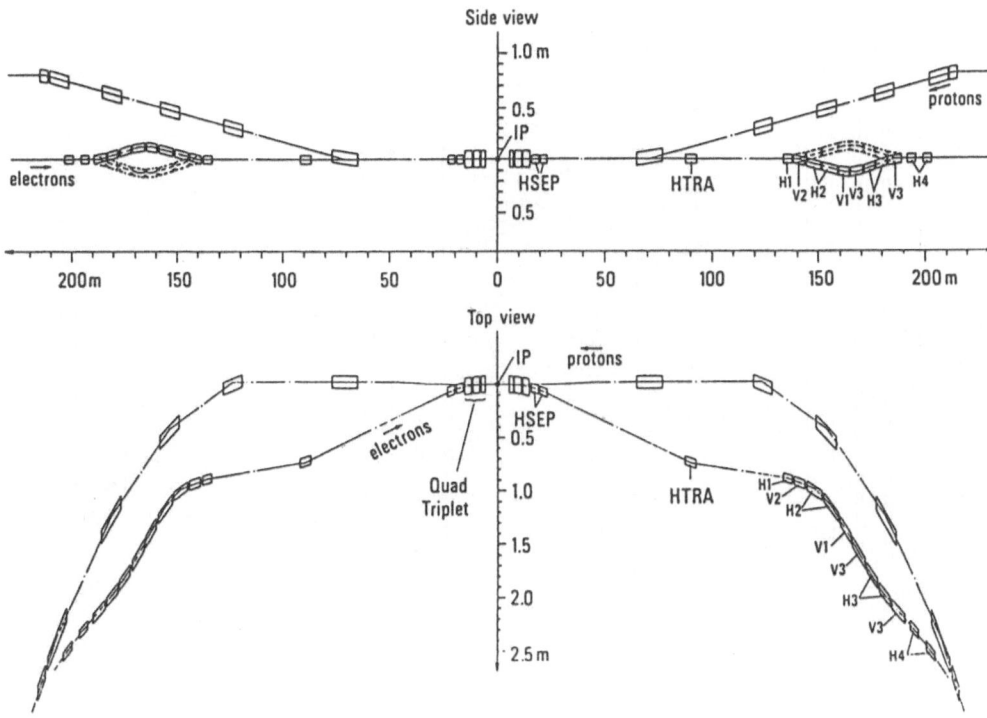

Fig. 38a   Layout near the interaction region with spin rotators and zero crossing angle geometry.

Numerous design studies have been made to find the optimum spin rotator (Ref. 28). A spin rotator of the siberian snake type is shown in Fig. 37b; V and H define vertically and horizontally bending magnets. The total length of the rotator is about 70 m. The maximum degree of longitudinal polarization is expected to be $P_L \approx 80 \%$.

6.7   Interaction Region

The interaction region was originally planned with a finite crossing angle (~20 mrad). When simulations showed that this will lead to beam blow-up via betatron oscillations (Ref. 29) a geometry with zero crossing angle (head-on collisions) was chosen. This necessitated the introduction of additional bending magnets close to the interaction point in order to bend the electron beam away from the proton beam ahead of the first proton ring magnet (see Fig. 38a). The necessary bending power is provided by three quadrupoles which also act as a low beta insertion. The bending of the electron beam produces a strong beam of synchrotron radiation (total power at 35 GeV: 9 KW with critical energy 80 keV, $8 \cdot 10^{17}$ photons/s with k > 20 keV) which passes through the interaction region (see Fig. 38b). For a safe operation the γ flux in a typical drift chamber at a distance of 15 cm from the inter-action point should not exceed ~10^8 γ/s, k > 20 keV. This can be achieved

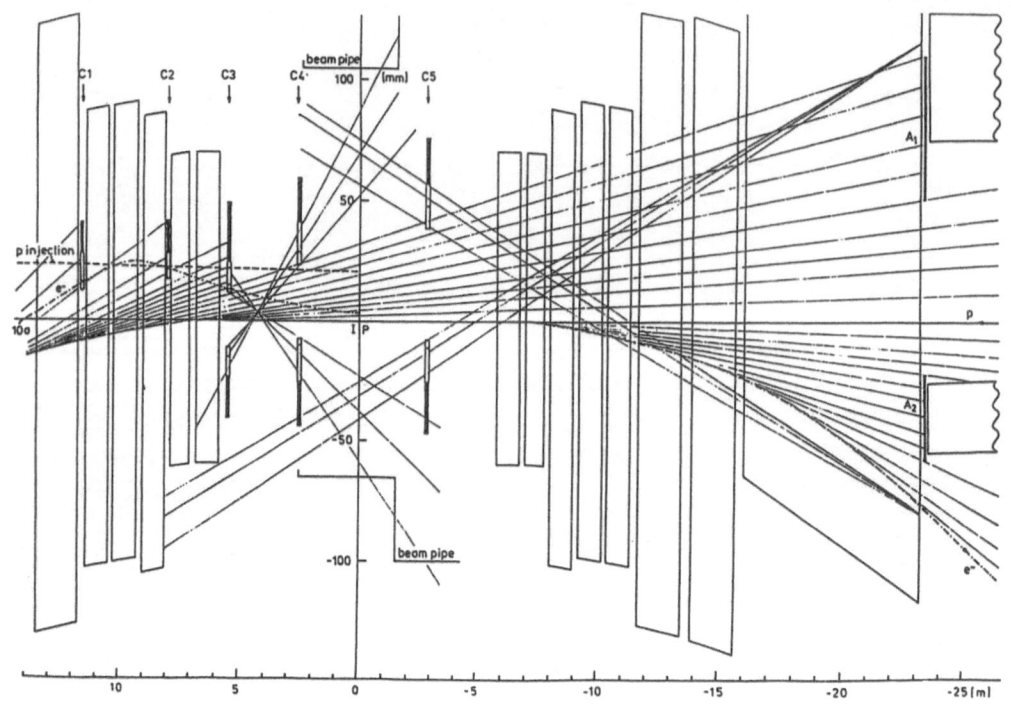

Fig. 38b   Arrangement of magnets and collimators (C_1, C_2,..) and synchrotron radiation absorbers (A_1, A_2) close to the interaction region.

Table 3:   The HERA schedule

Authorization	4/1984
Start civil engineering	4/1984
DESY II assembled	1985
Delivery of first e^- series magnets	8/1986
Start installing machine components in tunnel	10/1986
DESY II operational	end 1986
Electron injection into first quadrant	4/1987
Commissioning of proton LINAC	4/1987
Delivery of last e^- magnet	9/1987
Civil engineering completed	12/1987
Commissioning of DESY III	12/1987
Refrigeration plant operational	3/1988
Proton injection into first quadrant	4/1988
Delivery of last superconducting magnet	10/1988
Commissioning of proton ring	6/1989
Electron-proton physics	1990

by placing the beam pipe asymmetrically with respect to the beam axis so that the synchrotron radiation fan can pass unhindered and by installing suitable masks and collimators (Fig. 38b), see e.g. Ref. 7.

6.8   Construction Schedule and Status of HERA

The schedule foreseen for the construction of HERA is given in Table 3. The status of the project as of June 1986 is as follows:

Buildings.

tunnel   : one half of the tunnel is done; the quadrant South Hall to West Hall is ready.

halls   : South Hall and West Hall are nearly ready; the other two halls are under construction.

DESY II   : is being tested.

e^- ring   : series production of dipoles, quadrupoles, sextupoles has started.

H^- LINAC : all components have been ordered. The ion source works within specification.

p ring   : three 9 m long dipoles have been built and tested; the quality of the field is good. They reach fields 25 % above their nominal value. The quality of the field is well within specification. Saclay has constructed two prototype quadrupoles. They reach gradients well above the nominal value with acceptable field quality. Production of the preseries of the correction magnets has started in the Netherlands.

cryogenic plant: The components are being installed. The first of the three plants will start in October 1986.

6.9   Cost and Financing

The capital costs for HERA in 1981 prices are

General: site, buildings, power	260 MDM
Electron ring	112 MDM
Proton ring	282 MDM
Sum	654 MDM

The Federal Government of Germany and the city of Hamburg share the financial burden in the ratio 9 : 1. However, 15 - 20 % of the expenditures are paid by foreign contributions. These are from

Canada	: beam transport, proton RF
China	: staff
France	: development of the proton quadrupole
	and construction of a part of the quadrupoles
Israel	: currents leads for proton magnets
Italy	: half of all proton dipoles
Netherlands	: correction magnets for the proton ring
Poland	: staff and components
Switzerland	: prototype of the helium transfer line
UK	: technical help
USA	: test of the superconducting cables and
	cryogenic components.

7. EXPERIMENTATION AT HERA

Construction of a good detector for HERA is a challenge. Table 4 lists the requirements for several classes of physics. Particles and jets with energies up to 820 GeV have to be measured with precision. The large momentum imbalance between incident protons and electrons and the nature of space like processes send most particles into a narrow cone around the proton direction (see Fig. 39). Thus the calorimeter has to achieve simultaneously hermeticity, energy resolution and segmentation in a way not yet realized in any existing experiment. The detector has also to cope with high particle densities and with a time distance between bunch crossings of only 96 ns.

7.1 Calorimetry

Good calorimetry will be vital for HERA experiments. For this reason we shall discuss the properties of calorimeters in some detail. For reasons

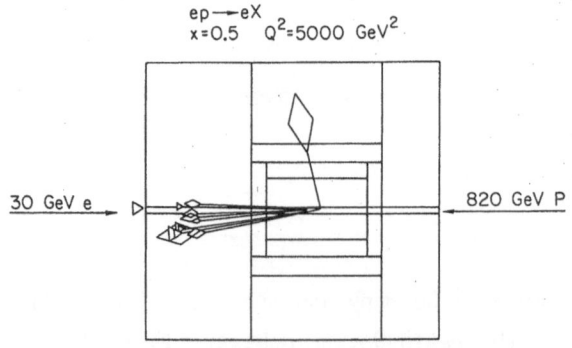

Fig. 39 Example of a neutral current event.

Table 4: HERA physics and detector requirements

topic	comments	requirements
• structure of proton, quark, e	• safe prediction of standard theory; look for deviations • reactions occur at parton level large $Q^2 \lesssim 40\,000$ GeV^2 accessible • polarized $e^{\pm}_{L,R}$ provide excellent handle on preon structure and on new currents	• inclusive measurement of e for NC, current jet for NC,CC → hermetic electromagnetic and hadronic calorimeters with uniform response to photons and hadrons, best possible hadron energy resolution → precise measurement of E_{jet}, Θ_{jet}
• new particles t,LQ,e^*,q^*,N_R, SUSY	• mass range: up to 120-200 GeV • very clean signatures • particles produced by CC alone are unique to HERA	• measurement of jets, jet topology and missing p_T → calorimetry as above • measurement of isolated e, μ and on jets → tracking detectors over full solid angle; excellent e,μ identification
• photoproduction	• c.m. energies up to 250 GeV attainable	→ forward electron tagging

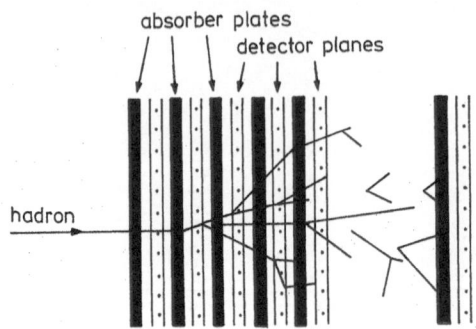

Fig. 40   Sketch of a sampling calorimeter.

of economy the calorimeter has to be of the sampling type pictured in Fig.40: it consists of absorber plates interleaved with detection layers to measure the electromagnetic and hadronic showers.

The energy resolution of a calorimeter results from a combination of different effects. Event-to-event fluctuations which determine the intrinsic resolution and sampling fluctuations usually dominate.

For <u>electromagnetic</u> showers the ultimate resolution is determined by the sampling fluctuations of the electromagnetic cascade, which are well understood and can be described by simulation programs such as EGS (Ref. 30). For a calorimeter made of high Z material (e.g. Pb, U), the energy resolution is

$$\sigma/E \approx 0.05 \; \frac{\sqrt{\Delta E(1 + 1/N_{pe})}}{\sqrt{E}}$$

where E is the incident electron energy in GeV, ΔE is the energy loss per layer in MeV and N_{pe} the number of electrons seen by the detector per minimum ionizing particle (mip) and layer. For instance, for 3.0 mm uranium plates, 2.5 mm scintillator and N_{pe} = 4 one finds

$$\sigma/E \approx \frac{14 \ \%}{\sqrt{E}}$$

The performance of a calorimeter for <u>hadrons</u> depends strongly on its relative responses to electrons and hadrons. This can be understood by considering a simple example. Assume a standard calorimeter with a response ratio of e/h = 1.4, and hadronic showers which consist only of $\pi^{\pm}$ and π^{0}. On the average the energy carried by $\pi^{\pm}$'s, $E_{\pi}\pm$, will be twice that carried by π^{0}'s, $E_{\pi}o$. In individual events the $E_{\pi}\pm$: $E_{\pi}o$ ratio can fluctuate widely. This will lead to a large spread of the energy measured by the

414

calorimeter and consequently to a poor energy resolution:

$$E_{\pi^\pm} : E_{\pi^0} = 3 : 0 \quad \text{rel. energy measured}: \quad 1.0$$
$$3 : 1 \qquad\qquad\qquad 1.13$$
$$1 : 2 \qquad\qquad\qquad 1.27$$
$$0 : 3 \qquad\qquad\qquad 1.40$$

Obviously, the energy resolution can be strongly improved if the calorimeter has equal response to electrons and hadrons, $e/h = 1$ (compensated calorimeter). It has been shown (Ref. 31) that in this case a hadron energy resolution as good as $\sigma/E = (0.30 - 0.35)/\sqrt{E}$ can be achieved while noncompensating calorimeters yield typically $(0.5 - 0.6)/\sqrt{E}$.

In hadronic showers, the energy is deposited through electromagnetic cascades (E_{em}), ionization by charged hadrons (E_{had}) and nuclear breakup and excitation energy (E_{nuc}), (Ref. 32). The detection layers of the calorimeter respond differently to these different components and compensation requires a delicate balance between them. Up to now equal response to electrons and hadrons has been achieved only in calorimeters made from depleted uranium (DU) and scintillator (Ref. 31).

7.2 Compensation in a DU-Scintillator Calorimeter

The following effects determine the response of a DU-scintillator calorimeter to hadronic showers (see Fig. 41):

1. Ionization by charged particles produces energy loss in the detector (dE/dx).

2. Electromagnetic showers are the result of the decay of π^0's. Electromagnetic energy (mainly low energy γ's) is predominantly lost in the high Z material. The difference in response to electromagnetic showers

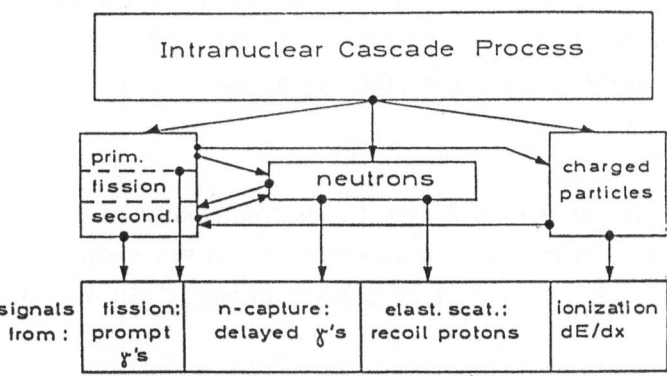

Fig. 41 Interactions contributing to compensation. Not shown is the electromagnetic component of hadronic showers.

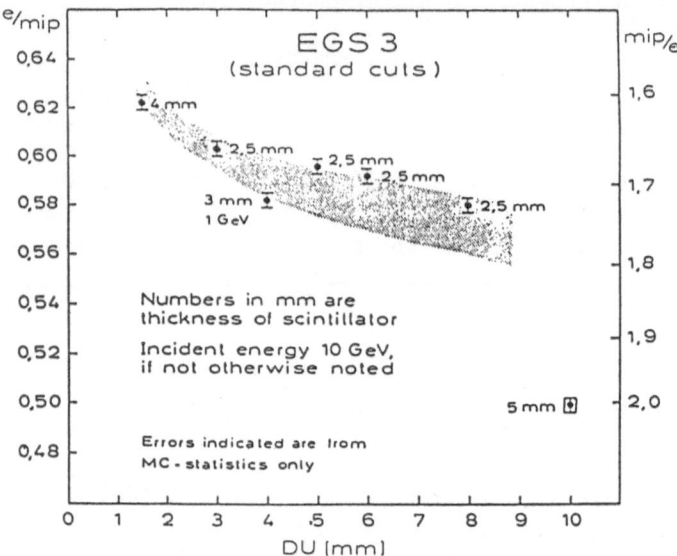

Fig. 42   The e/mip ratio for several DU-scintillator structures,
calculated with EGS (Ref. 7).*

and minimum ionizing particles (e/mip ratio) (Ref. 33) is due to the
different Z dependence of the cross sections for ionization ($\sim Z^1$), pair
production ($\sim Z^2$), bremsstrahlung ($\sim Z^2$), photoelectric effect ($\sim Z^5$) and
Compton effect ($\sim Z^1$). In a DU-scintillator sampling structure the e/mip
ratio is around 0.6, weakly dependent on the plate thickness (see Fig.42).
Crudely speaking, almost all of the energy of low energy photons is
absorbed via the photoelectric effect in the nonactive high Z absorber
material.

3. The hadron nucleus scattering in hadronic showers leads to nuclear breakup and fission; e.g. in massive uranium 0.6 GeV/c protons produce 9.8
   fissions per GeV, 300 GeV/c pions produce 4.1 fissions per GeV (Ref.34).
   The missing binding energy E_{bind} is taken from the energy E_{inc} of the
   incident particle. In uranium for E_{inc} = 10 GeV the energy lost due to
   binding is typically E_{bin} = (15-20%) E_{inc} (Ref. 31). Some energy E_ν is
   also lost to neutrino emission. The sum of both shall be called E_{nuc},

$$E_{nuc} = E_{bind} + E_\nu$$

4. In material with very high Z like DU, hadronic showers produce a significant number of neutrons with energies in the MeV range (Fig. 43a).
   For instance at E_{inc} = 10 GeV an average of about 450 neutrons is pro-

---

* Recent calculations with EGS4, using an improved stepping for electrons,
  indicate a larger e/mip value, e.g. e/mip = 0.67 instead of e/mip = 0.58.

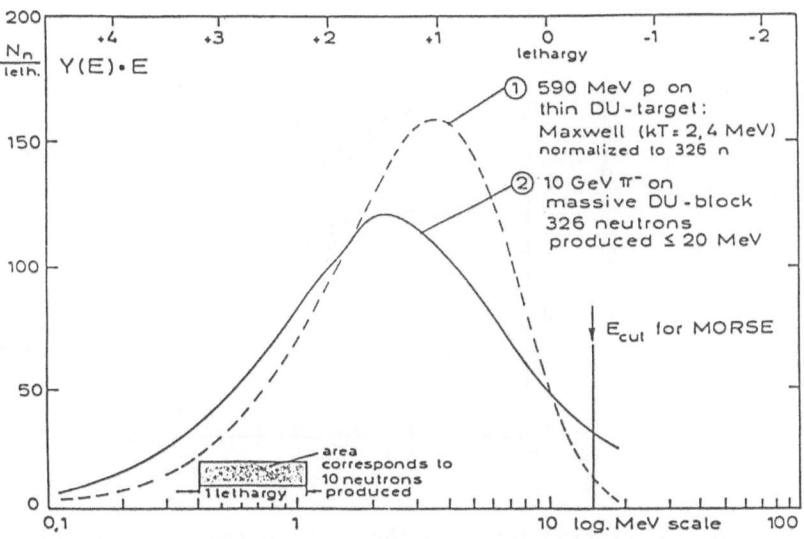

Fig. 43a The neutron yield spectrum for a massive DU block (Ref. 7).

duced with a mean energy around 3 MeV. The hydrogen content of an
organic scintillator converts most of the kinematic energy of the
neutrons into measurable proton-recoil energy. Thus the hadron response
is boosted considerably. In contrast, if liquid argon were used as detector material, the maximum energy transfer to the argon nucleus in
an elastic n-Ar collision is 1/40 ($\equiv$ 1/A) of that of an np collision
and consequently the boost of the hadron signal would be negligible.

5. Not all of the proton energy is detected in the scintillator, however,
   due to saturation of the scintillator. The average neutron energy is
   only 3 MeV, that of the recoil protons is even less. The velocity of
   a 3 MeV proton is β = 0.08. Since the ionization loss dE/dx ~ $1/\beta^2$ the
   resulting ionization density is very high. The light output of the
   scintillator saturates according to Birk's law

$$\frac{dL}{dx} = \frac{S\ dE/dx}{1 + KB\ dE/dx}$$

   where the factor KB has to be determined empirically. For NE102A
   KB = $1.0 \cdot 10^{-2}$, for SCSN38 KB = $0.85 \cdot 10^{-2}$ (Ref. 7) in units of
   $gcm^{-2}\ MeV^{-1}$. Fig. 43b shows the ionization loss and the light output
   for protons as a function of their kinetic energy.

6. The neutrons in the MeV range are moderated in energy mainly by
   elastic collisions with hydrogen. At <u>very</u> low neutron energies the

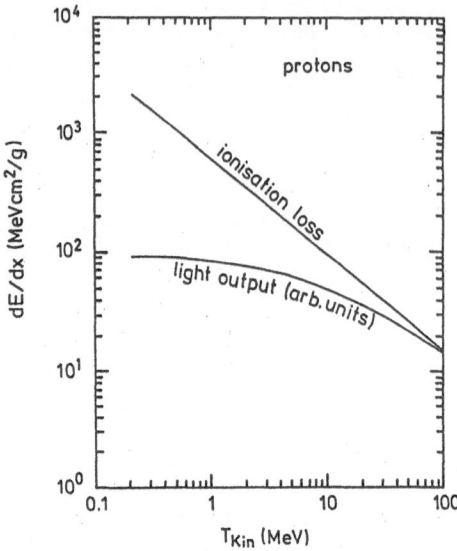

Fig. 43b The ionization loss and the light output (in arb. units) for protons for a KB value of 0.01 $gcm^{-2}MeV^{-1}$.

neutron capture process dominates which yields delayed gamma radiation and hence a delayed boost of the signal.

7. Every spallation and every fission process is accompanied by prompt nuclear gamma deexcitation. Although the γ efficiency of scintillator is very low, the effect still contributes to the compensation. Liquid argon would detect a larger fraction of this and the delayed γ components due to its shorter radiation length (14 cm vs. 42 cm).

The preceding discussion shows that the calorimeter response to electrons and hence also to,photons is suppressed by about 40 % relative to a mip. The response to hadrons, in general, is suppressed even more. About 20 % of the energy is lost in the nuclear breakup. However, the breakup of uranium either by primaries or by secondaries (e.g. low energy neutrons) produce extra neutrons which release their energy mainly in the scintillator by elastic np scattering. What fraction of the neutron's energy finally is seen by the photomultiplier depends on the saturation properties of the scintillator. As shown in the following, the uranium and scintillator layers can be tuned to yield equal response to electrons and hadrons.

We shall express the e/h ratio in terms of the pulse height contributions for the electron and the hadron:

Electrons of energy E yield a pulse height $G_e(E)$;

Hadrons of energy E yield:

- π^0's with a total energy of E_1 yielding a pulse height of $G_e(E_1)$,

- charged hadrons with a total energy E_2 yielding via dE/dx a pulse height $G_x(E_2)$,

- neutrons with total energy E_3 yielding via proton recoil a pulse height $G_{rec}(E_3)$,

- energy E_4 lost to binding and neutrinos.

Note, the small contributions from prompt and delayed γ's have been neglected; in the case of scintillator their energy can be included in E_4.

Now make the following assumptions:

1. The pulse height $G_i(E)$ (i = e,x,...) is proportional to E, i.e.
$$G_i(E) = g_i E.$$

2. The energy carried by neutrons is proportional to the energy lost to the binding energy (plus neutrinos),
$$E_3 = aE_4$$

and
$$E = E_1 + E_2 + E_3 + E_4$$

This yields

$$\frac{h}{e} = \frac{G_h(E)}{G_e(E)} = \frac{g_e E_1 + g_x E_2 + g_{rec} E_3}{g_e E}$$

$$= 1 - \left[1 - \frac{g_x}{g_e}\right] \frac{E_2}{E} - \left[(1 + \frac{1}{a}) - \frac{g_{rec}}{g_e}\right] \frac{E_3}{E}$$

We shall now make a further and not yet well founded assumption:

3. The pulse heights produced by electrons and produced via dE/dx by hadrons are equal,
$$g_x = g_e,$$

and obtain
$$\frac{h}{e} = 1 - \left[(1 + \frac{1}{a}) - \frac{g_{rec}}{g_e}\right] \frac{E_3}{E}$$

Compensation can now be achieved for any energy partition E_3/E and therefore separately for each event if

$$\frac{g_{rec}}{g_e} = (1 + \frac{1}{a})$$

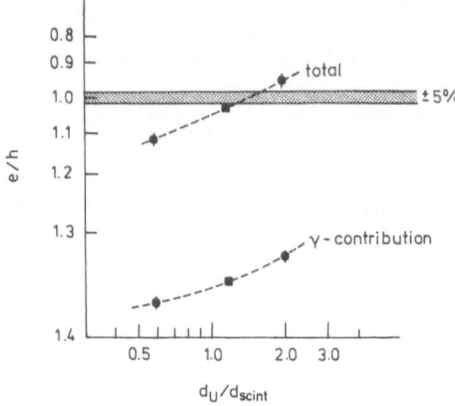

Fig. 44   The e/h ratio as a function of the ratio of the DU to
          scintillator thickness (Ref. 7).

With an e/h ratio equal to unity for each individual event we expect to
obtain the best energy resolution possible.

Figure 44 illustrates how the ratio g_{rec}/g_e might be tuned to
achieve compensation. It is the result of a model calculation for a
DU-scintillator calorimeter with d_u = 3 mm thick DU plates and variable
scintillator thickness d_s (Refs. 7, 35). The air gaps between scintillator and DU plates are assumed to be small (< 0.5 mm). The scintillator
is assumed to have a kB factor of $0.85 \ 10^{-2}$ g/MeV cm^2 (as is the case
e.g. for SCSN38). The gate length is assumed to be 100 ns. Compensation
is predicted for d_u/d_s = 1.4 or a scintillator thickness of 2.1 mm. This
is close to the 2.5 mm that has been used by Ref. 31 for their DU-
scintillator calorimeter. Figure 44 shows that for a fixed DU plate thickness a thicker scintillator will increase the e/h ratio, a thinner one will
decrease it.

The energy resolution in a compensated DU-scintillator calorimeter
can be written as (Ref. 36):

$$\frac{\sigma(E)}{E} = \frac{22 \ \%}{\sqrt{E}} \oplus \frac{0.09 \ \sqrt{\Delta E(1 + 1/N_{pe})}}{\sqrt{E}}$$

where ΔE measures in MeV the energy loss of a mip per absorber plate.
The $\oplus$ sign stands for quadratic addition. The second term under the
square root represents the contribution from photon statistics: N_{pe} is the
number of photoelectrons seen by the phototube per mip and layer. For
instance, for 3 mm DU, 2.5 mm scintillator, N_{pe} = 4, the expected energy resolution is

$$\frac{\sigma(E)}{E} = \frac{22 \ \%}{\sqrt{E}} \oplus \frac{25 \ \%}{\sqrt{E}} = \frac{33 \ \%}{\sqrt{E}}$$

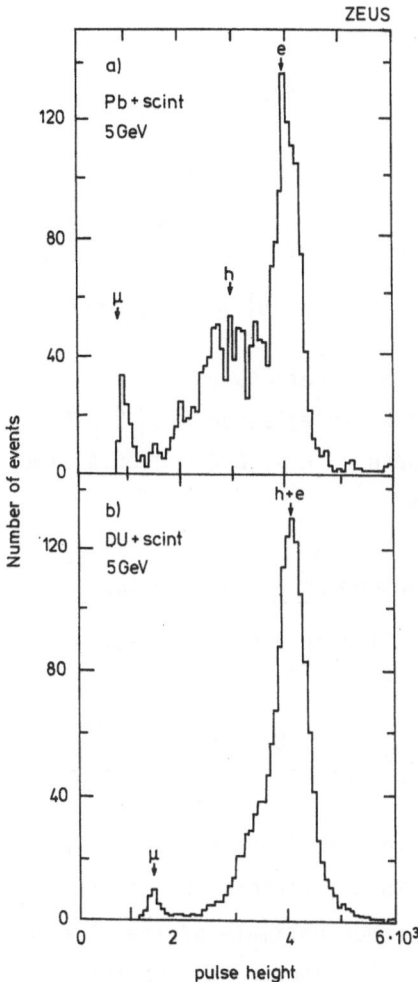

Fig. 45   Pulse height distributions for 5 GeV electrons, muons and
hadrons observed with a lead - scintillator (5 mm Pb, 5 mm scint)
calorimeter (a), and with a uranium - scintillator (3.2 mm DU,
5 mm scint) calorimeter (b). From ZEUS collaboration.

The potential of a uranium scintillator calorimeter is illustrated in
Figs. 45a,b. On top is shown the pulse height distribution measured by
ZEUS (Ref. 7) with a lead-scintillator calorimeter (5 mm Pb plates, 5 mm
scintillator) for a 5 GeV beam composed of electrons, muons and hadrons.
The electron signal is considerably larger than that for the hadrons,
$e/h \approx 1.3$. The bottom distribution shows the result obtained for the same
beam with the same calorimeter after replacing the lead plates with 3.2 mm
thick uranium plates. The electron and hadron signals are almost equal,
$e/h \approx 1.07$. (Note, that the choice of scintillator thickness was not
optimum). It is also evident that the hadron signal has become narrower.

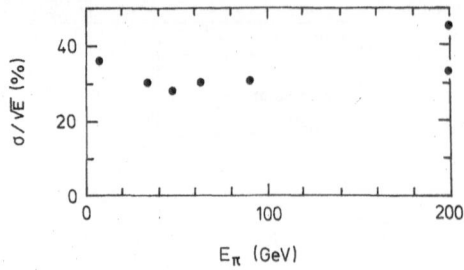

Fig. 45c The energy resolution of a DU-scintillator calorimeter (Ref. 7).

Figure 45c shows measurements (Ref. 37) from the NA34 experiment at CERN using 3 mm DU plates interleaved with 2.5 mm scintillator plates. The resolution varies between $\sigma/\sqrt{E}$ = 28 % and 36 %, the average being 33 % for pions from 8 to 200 GeV.

7.3   Weighting in a Noncompensating Calorimeter

The energy resolution in a noncompensating calorimeter such as Fe-scintillator is given by

$$\frac{\sigma(E)}{E} \approx \frac{40\ \%}{\sqrt{E}} \oplus \frac{0.09\ \sqrt{\Delta E(1 + 1/N_{pe})}}{\sqrt{E}} + 5.5\ \%$$

The last term arises from the event-to-event fluctuations in the energy deposited by π^0's and the difference in response to e and h. One can get rid of this term by a crude estimate of the π^0 content of the shower. The idea is the following: The shower of a charged hadron is typically spread over 3 - 5 absorption lengths, (e.g. λ(Pb) = 17.0 cm) while the energy

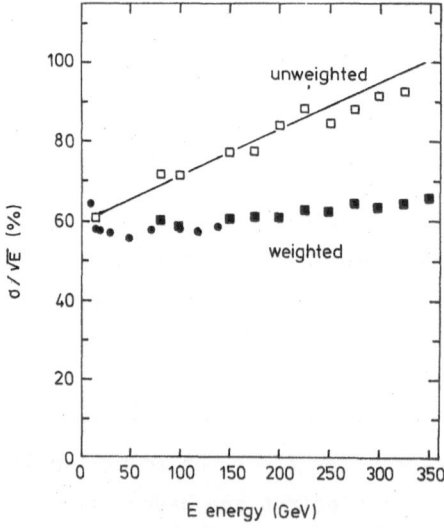

Fig. 46   The energy resolution of a Fe-scintillator calorimeter before and after weighting (Ref. 38).

of a π^0 is absorbed within 5 to 10 radiation lengths, X_0 (e.g. $X_0(Pb) = 0.58$ cm). The π^0 energy deposits are therefore very localized in a hadronic shower; they form hot spots. By measuring the energy content of hot spots the energy deposited by π^0's can be estimated.

In practice it is sufficient to measure the first and second moment of the energy release along the shower. Figure 46 shows the result from an iron-scintillator calorimeter (Ref. 38) using 2.5 cm thick Fe plates ($\lambda(Fe) = 16.8$ cm). Every 5 layers were readout by a separate phototube i yielding the measured energy E^i_{meas}. The corrected energy E^i_{corr} is found by (Refs. 38, 39)

$$E^i_{meas} = E^i_{corr} (1 - \alpha E^i_{corr})$$

$$E_{meas} = \sum_i E^i_{meas} = \sum_i E^i_{corr} - c \frac{\sum_i (E^i_{corr})^2}{\sqrt{\sum_i E^i_{corr}}}$$

The constant c depends on the e/h ratio.

As shown by Fig. 46 the energy resolution in the unweighted case increases from $\sigma(E)/\sqrt{E} = 60$ % near E = 10 GeV to ~100 % at 200 GeV incident pion energy. After weighting the energy resolution remains 60 % over the full energy range.

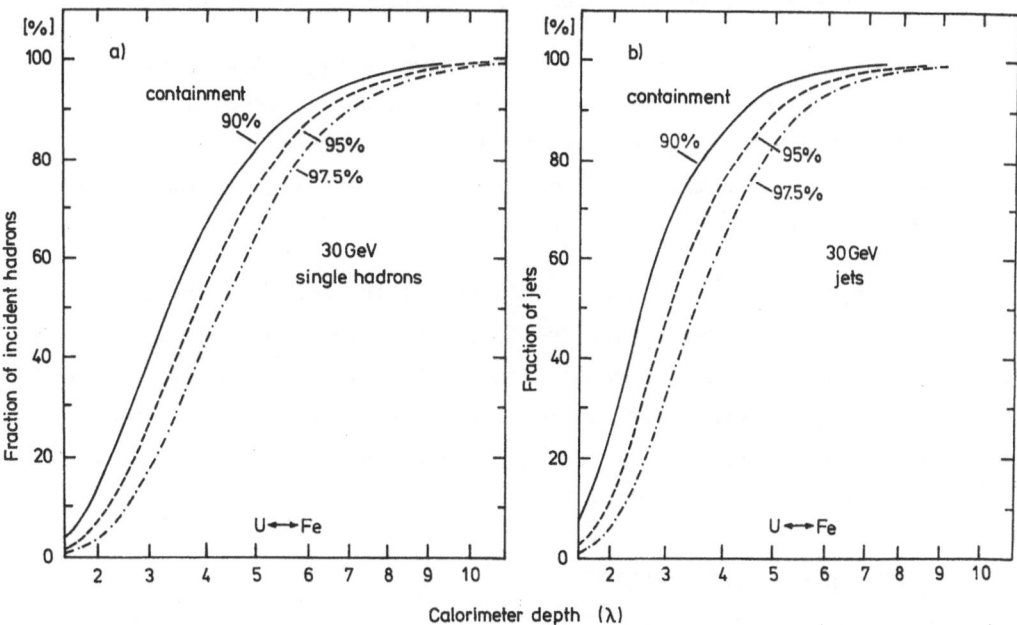

Fig. 47   Fraction of 30 GeV single hadrons (a) and of 30 GeV jets with 90%, 95% and 97.5% energy containment versus calorimeter depth (J. Krüger, priv. comm.).

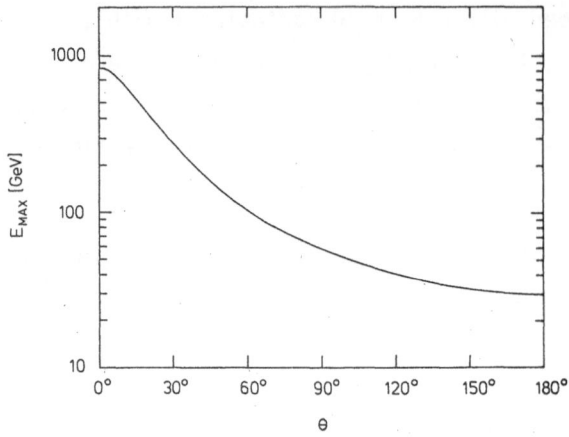

Fig. 48 Maximum jet energy as a function of polar angle.

7.4 Depth of the Calorimeter

In order to preserve the optimum energy resolution at all energies, the calorimeter has to be sufficiently deep. The depth profile of hadronic showers was measured experimentally in the WA78 experiment at CERN using a 5.2 λ deep DU-scintillator calorimeter followed by a 7.5 Fe-scintillator calorimeter. Fig. 47 shows as an example the fraction of 30 GeV/c single hadron showers and of jet showers with 90, 95 and 97 % energy containment as a function of the calorimeter depth. In order to have 95 % containment for 90 % of 30 GeV/c hadrons a depth of 6 λ is required. For jets the necessary depth is about 1 λ smaller.

The maximum jet energy for HERA is shown in Fig. 48 as a function of the production angle Θ. It reaches 800 GeV near the forward direction, falling to 300 GeV at 30°, 100 GeV at 60° and less than 50 GeV beyond 90°. Requiring 99 % containment leads to a depth of ~11 λ in the forward direction, 7 λ at 90° and 6 λ at 180°.

In general, cost considerations prohibit the construction of a high resolution calorimeter over the full depth. Instead, the calorimeter is divided into a high resolution front part and a backing calorimeter with moderate energy resolution. The backing calorimeter recognizes late showers and can be used to correct for energy leakage or to select a high resolution sample.

8. PROPOSED HERA EXPERIMENTS

Two collaborations, ZEUS and H1, have proposed to construct detectors for HERA. In 1985 they have submitted letters of intent (Ref. 40, 41) which have been followed in 1986 by technical proposals (Refs. 42, 43).

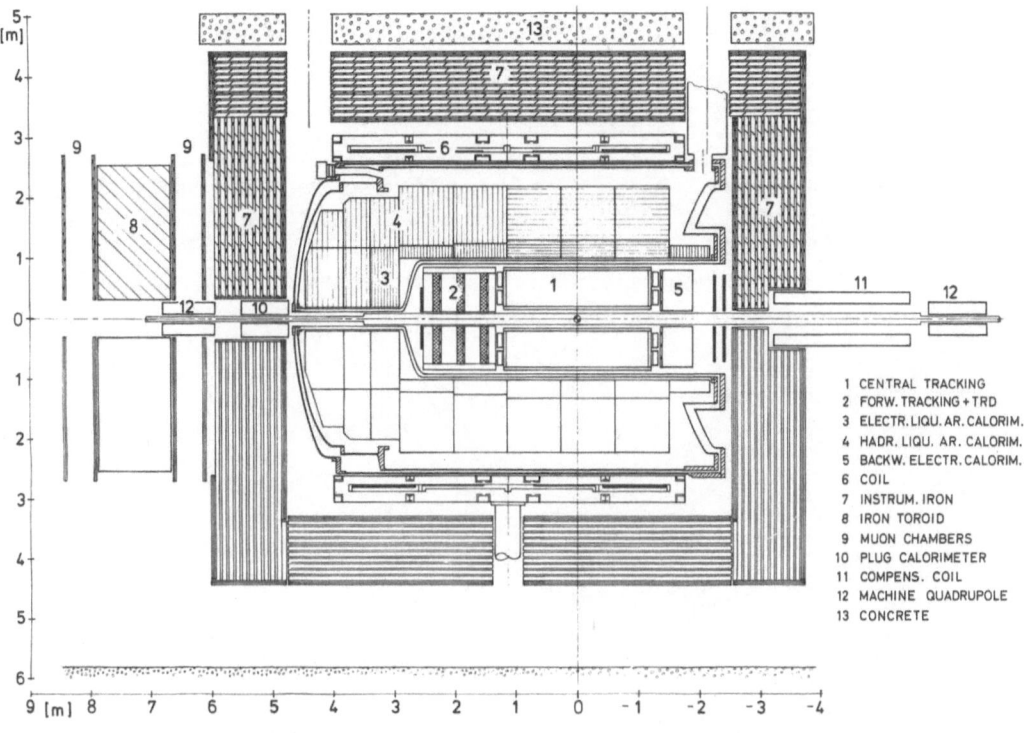

Fig. 49 Section of the H1 detector along the beam.

These proposals are presently under review by the DESY Physics Research
Committee and approval is expected for summer of this year.

Both detectors require large efforts in terms of personnel and
investment. The ZEUS collaboration consists of 268 physicists and estimates the material budget required to build the detector at 135 MDM. The
corresponding numbers for H1 are 165 physicists and 92 MDM. Both groups
envisage completion of the detector by 1990.

8.1   The H1 Detector

The layout of the detector is shown in Figs. 49, 50. Starting
from the interaction point,charged particle tracking in the central and
forward region is provided by a central jet-chamber interleaved with two
z-chambers, and MWPC's (1) (inner, outer radii: 15, 90 cm) and a forward
tracker (2) consisting of a series of radial and planar drift chambers
interleaved with three layers of MWPC's and transition radiators. In combination with a magnetic field of 1.2 T the momentum resolution expected for
isolated charged particles between 7° and 150° is $\sigma(P)/P^2 \leq 0.003$.

The EM calorimeter utilizes Pb plates with liquid argon as readout
in the barrel and forward region (3) and a lead scintillator sandwich in

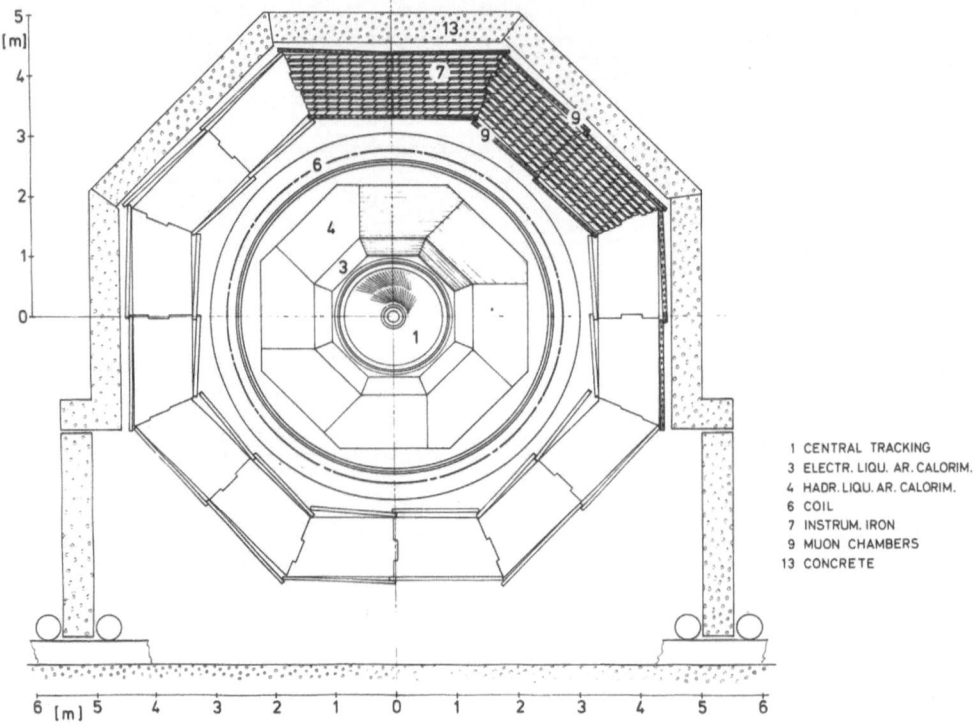

Fig. 50   Section of the H1 detector perpendicular to the beam.

the backward region (5). The liquid argon part of the EM calorimeter is highly segmented. It consists of 3×3 cm^2 towers and 4 longitudinal readouts in the forward direction and of 8×8 cm^2 towers and 3 readouts in the backward barrel region.

The hadron calorimeter uses stainless steel absorber plates with liquid argon readout. The EM and hadron calorimeters provide an energy resolution of $\sigma(E)/E \leq 10 \%/\sqrt{E}$ for electrons and, using a weighting procedure, $\sigma(E)/E = 55 \%/\sqrt{E} \oplus 2 \%$ for hadrons.

A superconducting coil (inner, outer radii 260, 304 cm, length 575 cm) encloses the calorimeter and provides a longitudinal field of 1.2 T.

The solenoid and calorimeters are surrounded by a set of iron plates (7) to contain the return flux. The iron is instrumented with plastic streamer tubes and acts as a backing calorimeter and muon filter.

Muon detection is provided by three layers of muon chambers (9) in the barrel and forward region complemented by a forward muon spectrometer consisting of a magnetized iron toroid (8) and four layers of drift chambers (9).

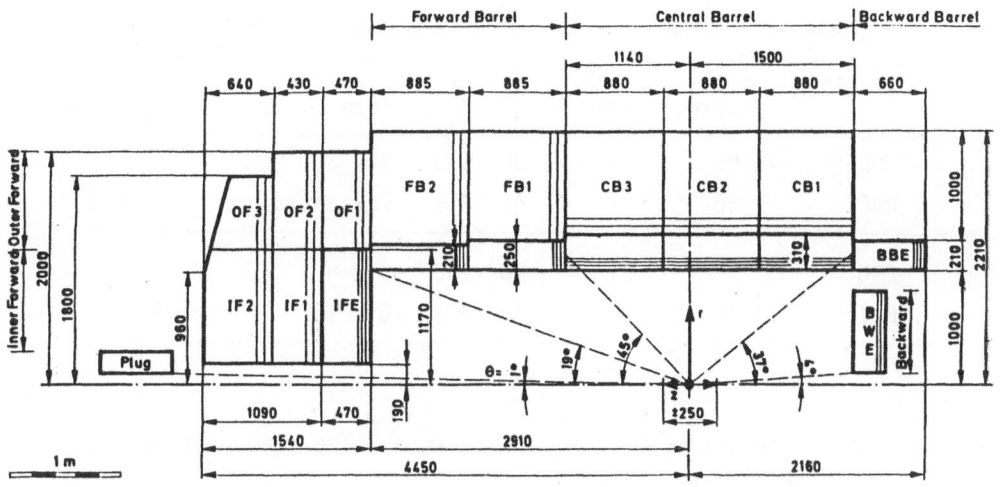

Fig. 51 Layout of the H1 liquid argon calorimeter.

A plug calorimeter (10) detects hadronic energy at small angles down to 0.7°. It consists of a copper silicon sandwich.

The field produced at the beam by the large coil is compensated by an additional solenoid (11).

We shall describe the calorimeter and the tracking system in more detail.

8.1.1 The liquid argon calorimeter

The arrangement of the absorber plates and the thickness in absorption length of the H1 LA calorimeter is indicated in Figs. 51, 52. The electromagnetic part (EMC) uses 2.4 mm lead plates and 3.0 (or 2 x 1.5) mm argon gaps. The hadronic part (HAC) is made of 12 mm stain-

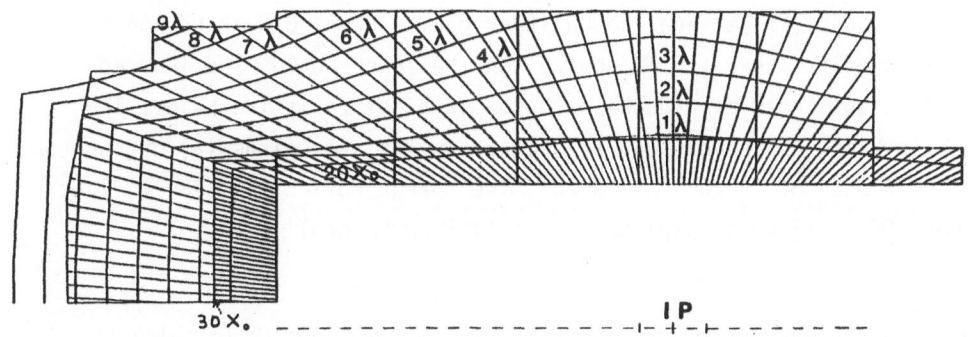

Fig. 52 Amount of absorber material in the EM sections of the H1 calorimeter.

Table 5: Properties of the LA calorimeter for H1

Θ	section	n.of layers	depth (cm)	X_o	λ	tower size (cm^2)	number of longt.segm.
near 0°	EMC	65	47	30	1.5	3 x 3	4
	HAC	58	107	43	4.5	7x7→15x13	5 or 6
90°	EMC	43	31	20	1.0	8 x 9	3
	HAC	48	90	34	3.8	15 x 13	4

less steel plates separated by 2 x 1.5 mm argon.[*] The number of layers and depths are summarized in Table 5.

Liquid argon (LA) offers several advantages as a readout medium. The calibration can be kept constant over several years at the one-percent level provided the argon does not get contaminated. The response is uniform across the readout area. LA readout does not suffer from radiation damage. Furthermore it allows in an easy way fine transverse and longitudinal segmentation. Its major drawback is that up to now it has not been possible to design a compensating LA calorimeter. Another drawback is the long charge collection time.

The weight of the calorimeter is 57 t lead, 327 t Fe and 80 t LA.

The calorimeter is highly segmented. The transverse segmentation will be helpful for the recognition of electrons within jets. The longitudinal segmentation is chosen for the application of the weighting technique since the calorimeter will be noncompensating. In total there are 42 100 readout channels for the EMC and 18 238 for the HAC.

The readout gaps are sketched in Fig. 53: with two alternative solutions for the EMC (a,b) and one for the HAC (c). For the EMC the high voltage is applied between the Cu layer of the upper G10 board and a resistive paint layer (10 MΩ/square) on the lower board. The induced signal is collected by pads beneath the layer of resistive paint. The leads to bring the signal out are either in the same plane (a) or run on the lower side of the G10 board (b). The charge collection time will be either 600 ns (3 mm LA gap) or 300 ns (1.5 mm LA gap).

---

[*] In the meantime the H1 collaboration has decided for 18 mm thick Fe plates and 2 * 2.5 mm LAr gaps.

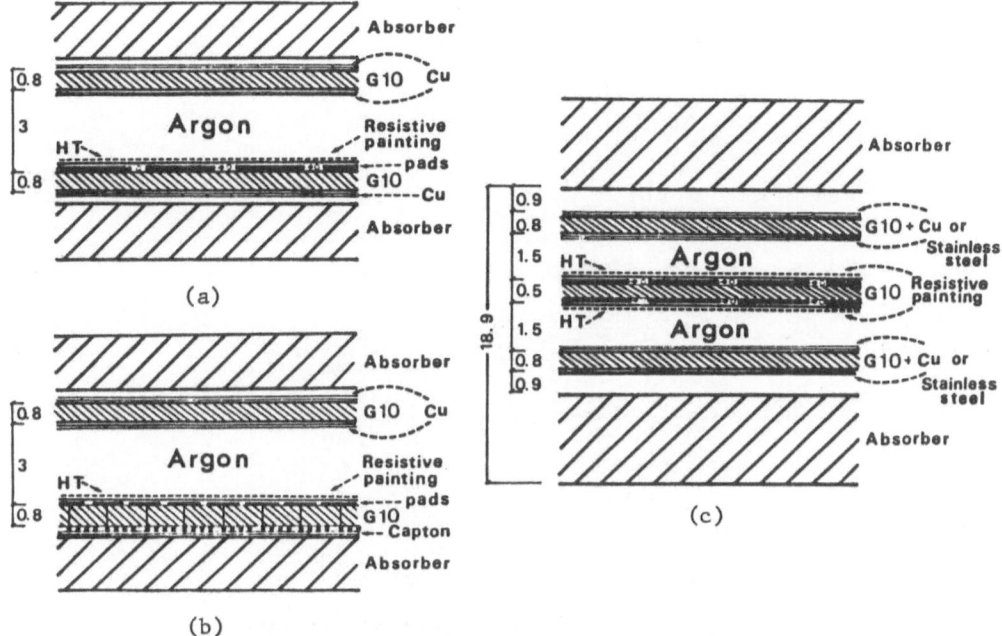

Fig. 53   Read out gaps for the H1 liquid
argon calorimeter. a) and b) show
two different solutions for the
EM gap, c) shows a gap in the
HAC section.

The design of the electronics is faced with the problem of noise
pick-up. Two solutions are under study. In the warm solution the preamplifiers are located on the outside of the cryostat. In the cold solution
preamplifier and multiplexor are next to the stack inside of the liquid
argon tank. The latter solution should have less problems with noise but
requires a certain optimism w.r.t. the lifetime of the electronic components.

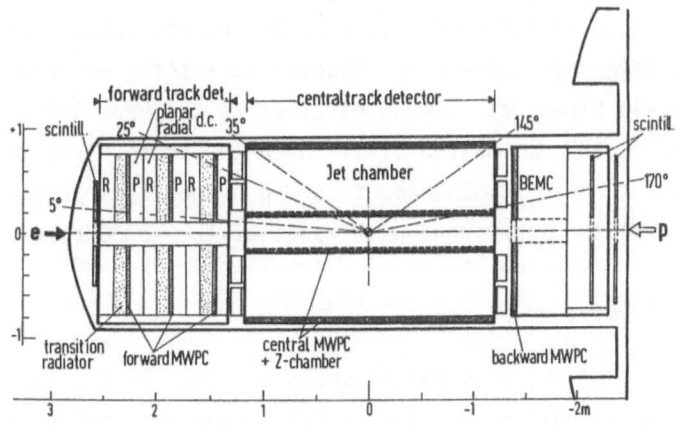

Fig. 54   Layout of the tracking system for H1.

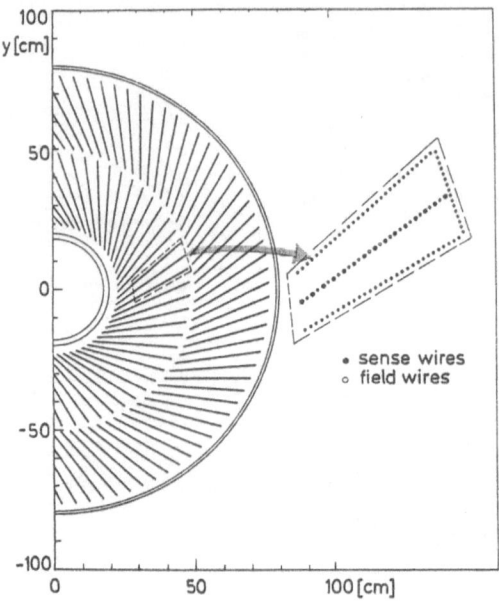

Fig. 55   Cross section of the H1 central tracking chamber.

8.1.2 The tracking system

An overview of the inner tracking detectors is shown in Fig. 54. The central region is surrounded by a jet type drift chamber. Plane drift chambers measure the paths of forward and backward going particles. Transition radiators followed by multiwire proportional chambers help to identify electrons.

Figure 55 indicates the wire arrangement in the jet chamber. The drift cells do not point to the interaction point but are tilted to account for the Lorentz angle. The maximum drift path is 5.1 cm corresponding to 1.5 µs drift time or 15 beam crossings. The chamber has an outer radius of 80 cm, a length of 264 cm and is operated at atmospheric pressure. There are 64 layers of sense wires leading to a total of 2560 sense wires and 10 000 field wires, all strung parallel to the beam axis. The expected $r\varphi$ resolution is 100 µm. Charge division on the sense wires yields some z-information. Precise z measurements are provided by separate drift chambers attached at the inner and outer radii of the jet chamber (see Fig.54).

8.2   The ZEUS Detector

The layout of the detector is shown in Figs. 56-58. The essential elements are a central track detector (CTD) plus transition radiation (TRD) and planar chambers (FTD, RTD) within a thin magnetic solenoid

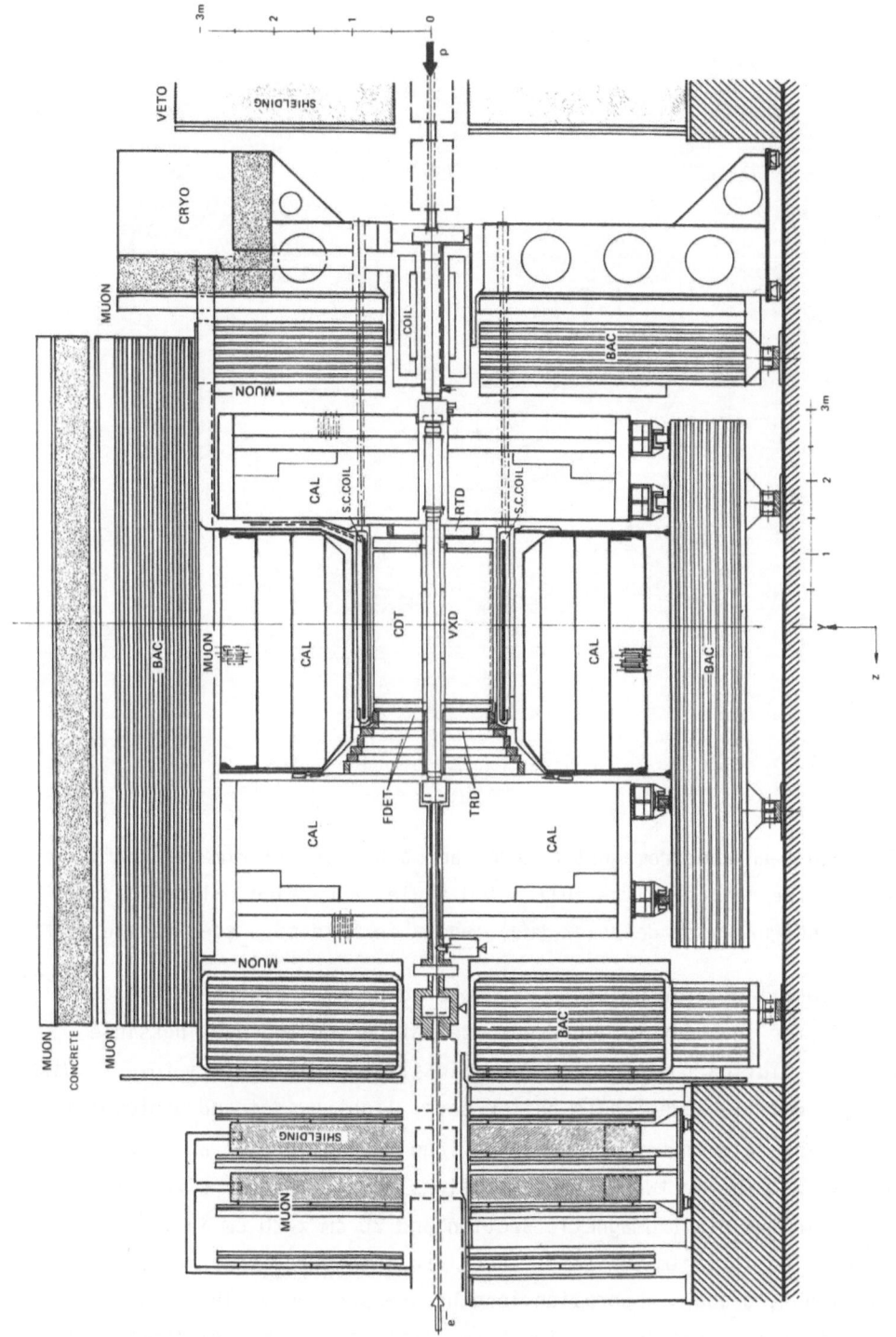

Fig. 56 Section of the ZEUS detector along the beam.

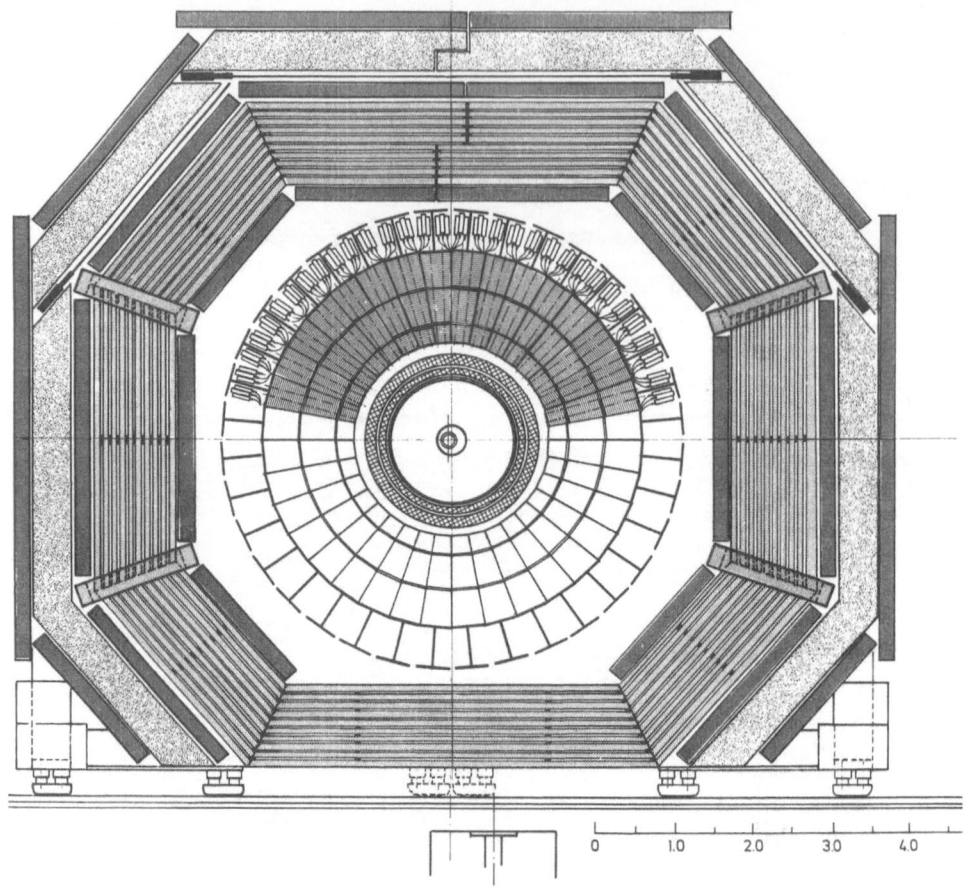

Fig. 57 Section of the ZEUS detector perpendicular to the beam.

(SOLENOID), an electromagnetic (EMC) and a hadron calorimeter (HAC) surrounding the coil over the full solid angle, a backing calorimeter (BAC),
barrel and rear muon detector (MU), and a forward muon spectrometer (FMU).

The calorimeter consists of depleted uranium plates with plastic
scintillator in order to achieve compensation and the best possible
energy resolution for hadrons. The scintillator plates form towers which
are read out via wave length shifer bars, light guides and photomultipliers. The calorimeter is segmented longitudinally into an electromagnetic and one or two hadronic sections. Typical tower sizes are 5 cm
x 20 cm in the electromagnetic section and 20 cm x 20 cm in the hadronic
section. The calorimeter is divided into a forward, a barrel and a rear
part with 7, 5 and 4 absorption lengths, respectively. The active area
in the forward direction (proton beam direction) starts at about 60 mrad.
The solid angle coverage corresponds to 99.8 % in the forward hemisphere
and 98 % in the backward hemisphere. The expected energy resolutions are

for electrons $\sigma(E)/ = 0.15/\sqrt{E} \oplus 2\%$ and for hadrons $\sigma(E)/E = 0.35/\sqrt{E} \oplus 2\%$.

In order to identify electrons within dense jets a silicon pad detector is under study which could be inserted in the calorimeter at a depth of 3-5 radiation lengths.

The high resolution calorimeter is surrounded by a backing calorimeter of moderate energy resolution. Its purpose is to measure the energy of late showering particles. The backing calorimeter uses as absorber the iron plates which form the magnet yoke. Limited streamer tube chambers are used for read out. The expected energy resolution for hadrons is $\sigma(E)/E = 1.0/\sqrt{E}$.

The central track detector consists of a cylindrical jet type drift chamber with an outer radius of 85 cm and an overall length of 240 cm. Track position and dE/dx loss are measured in 9 superlayers each with 8 layers of sense wires. Four of the superlayers have stereo wires. A

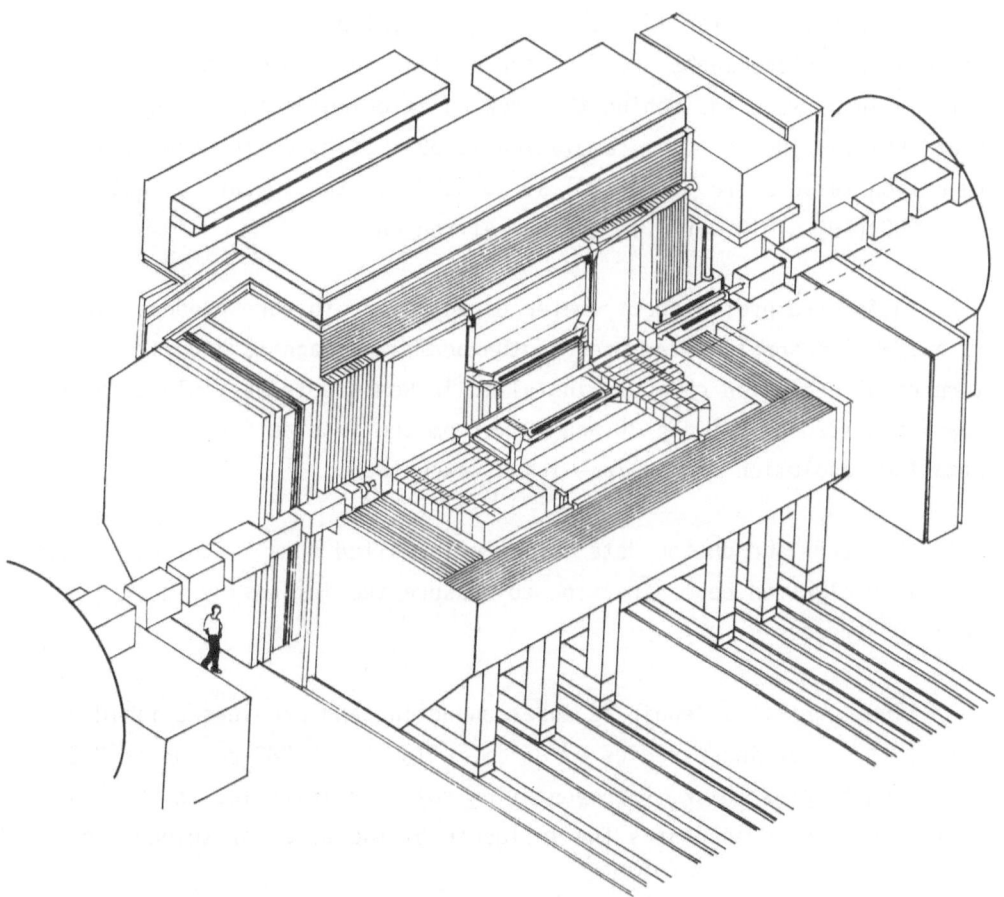

Fig. 58 Isometric view of the ZEUS detector.

resolution of 100 μm is expected, leading to a momentum resolution of $\sigma(p)/p = 0.0018 \cdot p \oplus 0.0027$ (p in GeV/c) for a magnetic field of 1.8 T. Particle tracking at small forward and backwards angles to the beam is aided by four planar drift chambers providing a momentum measurement with $\sigma(p)/p = 0.01 \cdot p$ at a foward angle of 120 mrad.

For the detection of the decays of short lived particles two possible designs for a vertex detector are under study.

Electron identification is performed with dE/dx information from the tracking detectors and with the calorimeter. In the forward direction a transition radiation detector, consisting of four modules, yields an additional hadron rejection factor of more than 100 for momenta below 30 GeV/c. The combined hadron rejection is well above 10^3.

Muons are detected in the forward direction in a spectrometer using drift and limited streamer tube chambers plus scintillator counters interspersed between the magnetized iron yoke and magnetized iron toroids. The momentum resolution for 100 GeV/c muons is $\sigma(p)/p = 30$ %. In the barrel and rear detectors muons are detected by limited streamer tube chambers before, in between and behind the backing calorimeter and behind the concrete shield. The momentum resolution is 35 % at 20 GeV/c. The pion (kaon) rejection factors are 1000 (100) at 40 GeV/c in the forward direction, and 5000 (50) at 10 GeV/c in the barrel region.

A forward proton spectrometer detects very forward produced protons $x_L > 0.3$. The spectrometer uses proton beam line magnets and five miniature high resolution chambers installed in Roman pots very close to the beam at distances between 20 and 90 m from the interaction point. A momentum resolution of $\sigma(p)/p \sim 1$ % is expected.

Electron and photon detectors are installed some 30 to 100 m downstream in electron beam direction to measure the luminosity and tag small Q^2 processes.

The magnetic solenoid is superconducting and provides a field of 1.8 T. It has an inner radius of 86 cm, a length of 280 cm and is 0.8 radiation lengths thick. A compensating solenoid installed in the rear of the detector compensates the influence of the detector solenoid on the beam.

We shall take a closer look at some of the detector components.

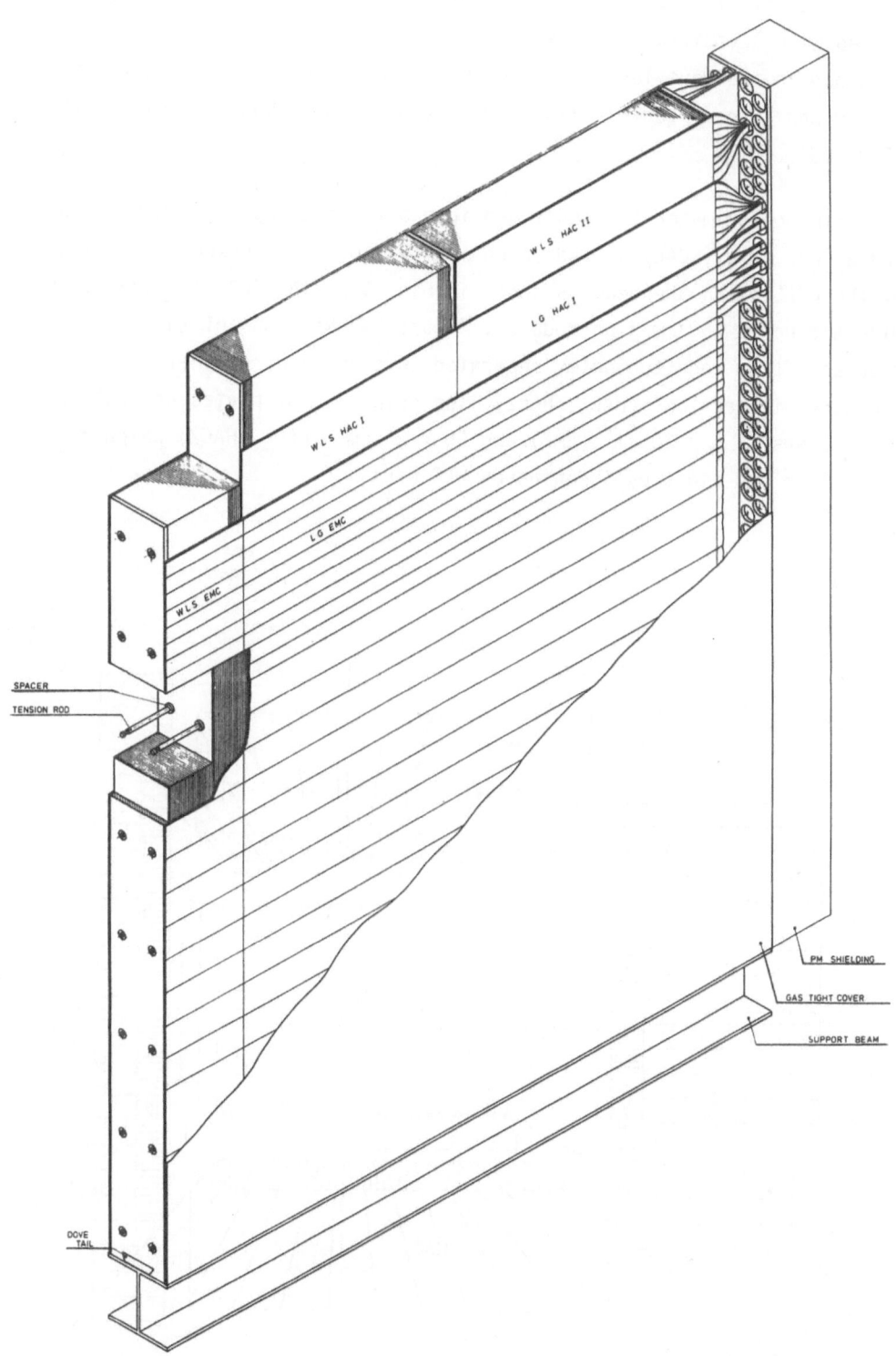

Fig. 59 View of an FCAL module; particles enter from the left.

8.2.1 The uranium calorimeter

The calorimeter consists of depleted uranium (DU) sheets interleaved with scintillator plates. Light generated in the scintillator is
absorbed in the wave length shifter bar (WLS), reemitted at lower frequency and transported via light guide (LG) to the photomultiplier (PM)
(Fig. 59).

The calorimeter is subdivided into a forward (FCAL), a barrel (BCAL)
and a rear part (RCAL) as indicated in Fig. 60. A cross sectional view
of the FCAL along the beam is shown in Fig. 61a. The FCAL is made of 20 cm
wide and up to 260 cm high modules. Figure 59 shows one of the FCAL
modules. It is longitudinally separated into an electromagnetic (EMC) and
two hadronic sections (HAC1, HAC2). The scintillator plates are cut to
form towers of 5 x 20 cm^2 (EMC) and 20 x 20 cm^2 (HAC1, HAC2) which are
readout on two sides by independent WLS bars.

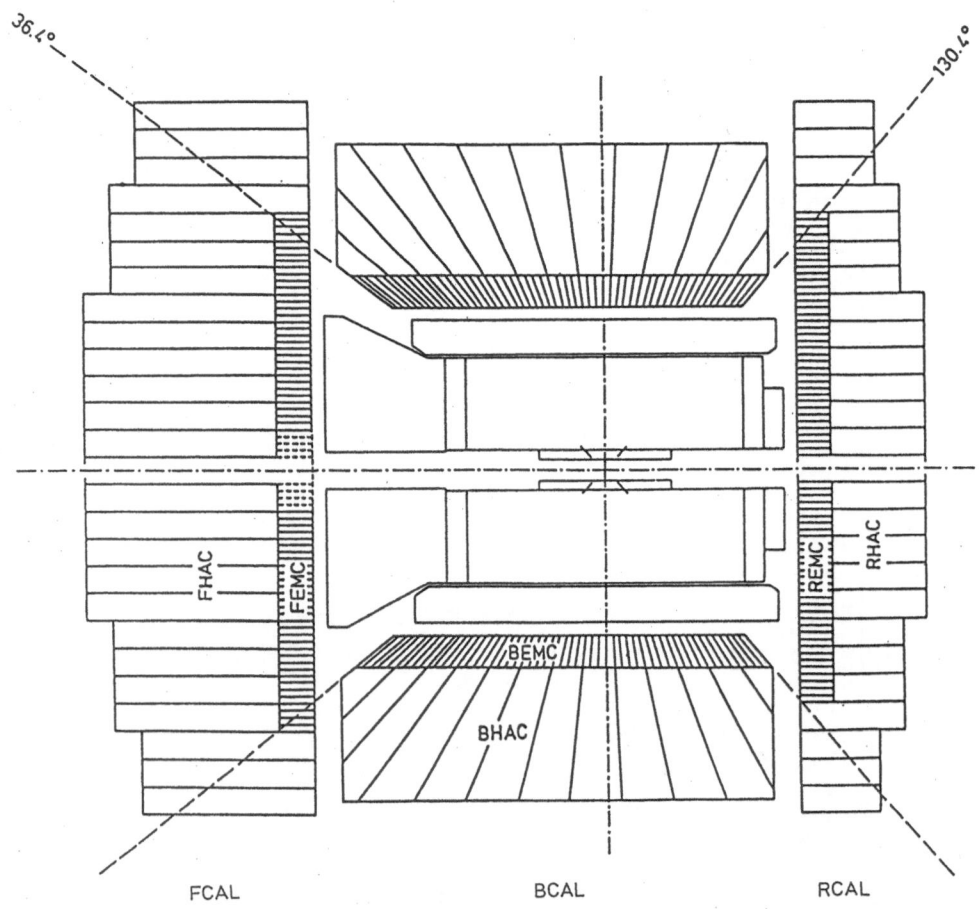

Fig. 60 Layout of the DU-scintillator calorimeter of ZEUS.

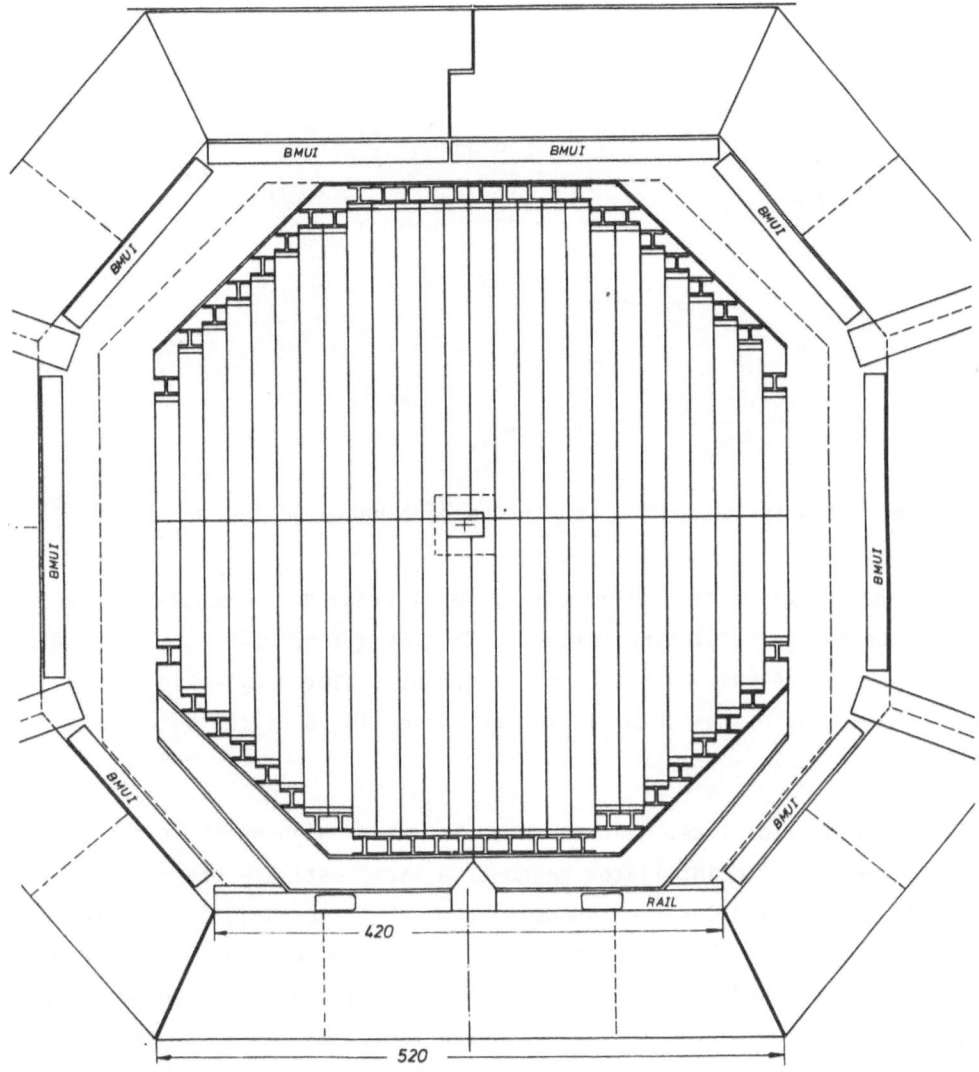

Fig. 61a   The ZEUS forward calorimeter FCAL as seen along the proton beam.

The barrel calorimeter is subdivided into 36 identical modules, each subtending 10° in azimuth (Fig. 61b). The scintillator readout is arranged in towers which are typically 5 x 20 cm^2 (EMC) and 20 x 20 cm^2 (HAC) measured at the entrance face of the calorimeter.

Throughout the calorimeter the layer structure is kept the same, namely 3.2 mm DU sheets and about 3.0 mm scintillator plates in order to ensure full compensation and maintain optimum energy resolution for hadrons. Table 6 summarizes the number of layers and depths for the different calorimeter sections. The calorimeter has in total 7768 EMC and 5118 HAC readout channels.

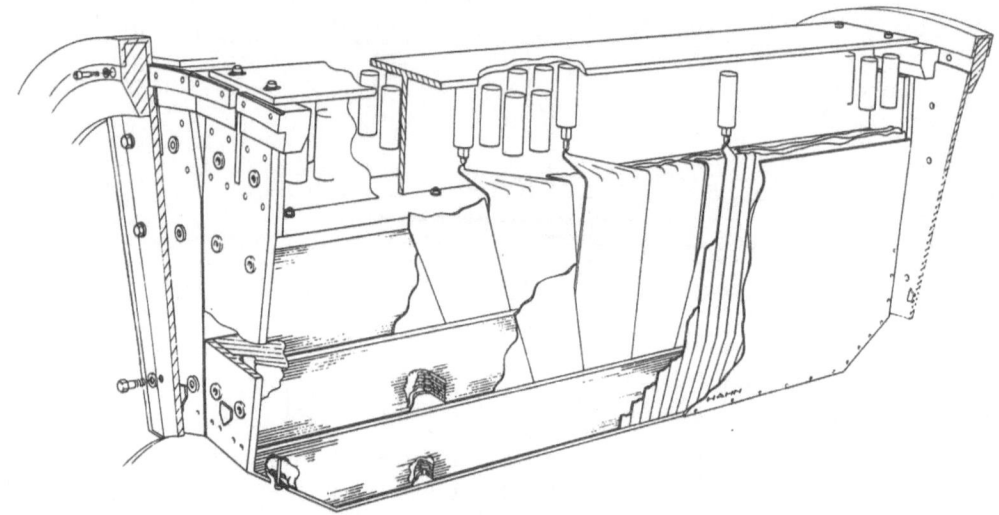

Fig. 61b Cut away isometric view of the ZEUS barrel calorimeter (BCAL).

The design of the calorimeter aims at keeping the constant term (c) in the energy resolution, $\sigma(E)/E = 35\ \%/\sqrt{E} \oplus c$, below c = 2 %. This puts a premium on uniform response. It can be achieved by taking special care in wrapping of the scintillator and inserting a filter between scintillator and WLS.

The radiation level near the beam is rather high which is a potential problem for a scintillator readout. A first estimate based on experience at the $Sp\bar{p}S$ and Tevatron shows that most of the radiation is produced during injection of the beams and machine steering. For this reason FCAL and RCAL are divided vertically and each half will be retracted by 40 cm from the beam line during these beam operations. This is done without opening the iron yoke. With this measure the yearly dosage at the worst place is about $3 \cdot 10^4$ rad.

Table 6:   Properties of the DU calorimeter for ZEUS

calorimeter	Θ	section	n.of layers	depth (cm)	X_0	λ	tower size (cm^2)	n.of longt. segm.
FCAL	$2.3^0 - 39^0$	EMC	26	25.5	26	1.0	5x20	1
		HAC	154	141.6	162	6.0	20x20	2
BCAL	$39^0 - 130^0$	EMC	25	23.8	25	1.1	5x20	1
		HAC	96	92.4	103	4.0	20x20	2
RCAL	$130^0 - 172^0$	EMC	26	24.5	26	1.0	5x20	1
		HAC	77	70.8	81	3.0	20x20	1

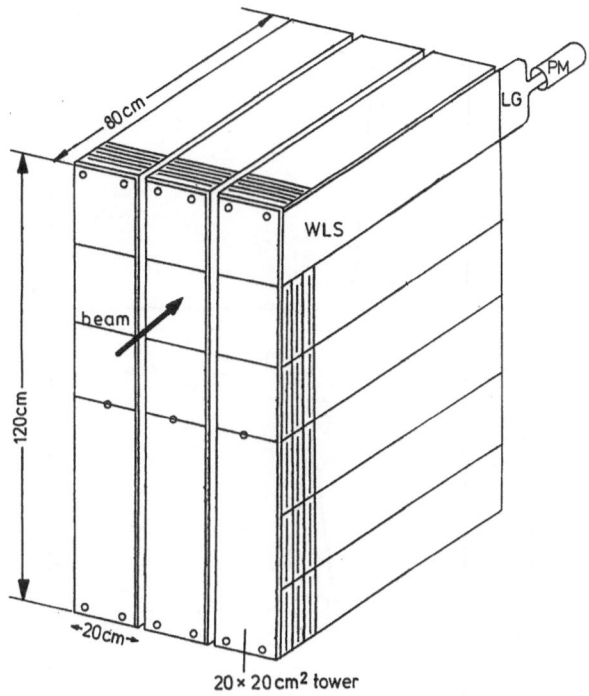

Fig. 62a Setup of a DU-scintillator test calorimeter for ZEUS made of
material from the AFS experiment at CERN.

The dosage from the DU depends on the cladding of the DU plates. A
yearly level of 10^2 rad is expected.

The radiation resistance of scintillator depends strongly on the
base material. Most hadron calorimeters for collider experiments have
employed acryllic based scintillator (AFS, UA1, UA2, CDF) whose performance deteriorates in a normal atmosphere above 10^5 rad. Aromatic scintillator such as NE110, SCSN38 to be used in ZEUS can sustain at least
an order of magnitude more. Thus, the lifetime of the ZEUS calorimeter is
expected to be well above 10 years.

For calibration the calorimeter modules have to be placed in a test
beam. When installed in the detector calibration will be maintained using
the radioactivity of the DU plates.

Test measurements have been performed with a DU calorimeter shown
in Fig. 62a.   Pulse height distributions for electrons and hadrons
(mostly pions) are shown in Figs. 62b,c. An energy resolution of
33 - 34 %/$\sqrt{E}$ has been obtained as shown in Table 7. The e/h ratio is
between 1.09 and 1.06. It has to be corrected for nonuniform response of

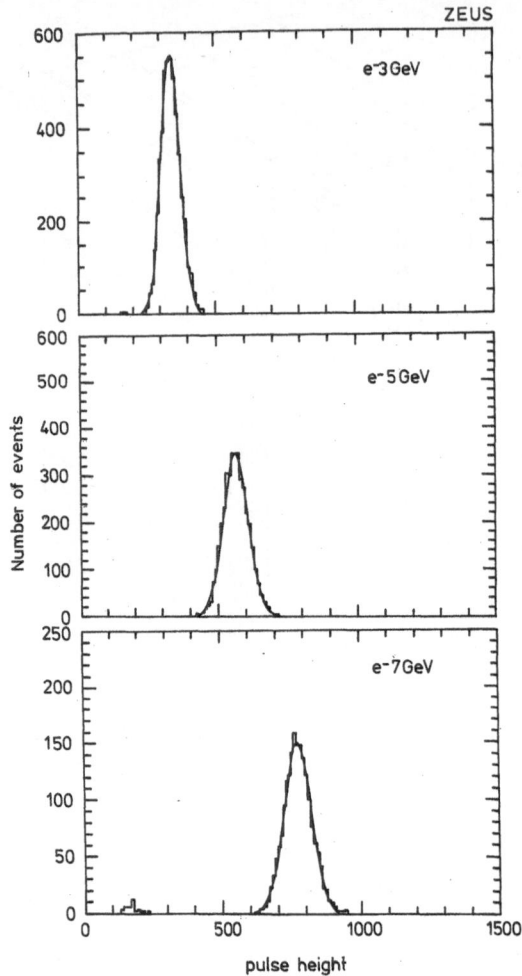

Fig. 62b   Pulse height distributions measured with a DU - scintillator
calorimeter (3 mm DU, 2.5 mm scint) for 3, 5 and 7 GeV electrons.
(From ZEUS collaboration).

the WLS and for side and rear leakage estimated to be around 5 - 8 % for
pions which will bring e/h close to unity.

8.2.2 The tracking system

     The arrangement of the inner tracking chambers is depicted in
Fig.63. In the central region charged particles are being tracked by a
drift chamber (CTD) which has an outer radius of 85 cm and a length of
240 cm. A sector of the CTD is shown in Fig. 64. There are 9 superlayers
each with 8 layers of sense wires. The drift cells are tilted to account
for the Lorentz angle. The maximum drift length is 3 mm corresponding to
500 ns drift time or 5 beam crossings. Five of the superlayers have wires
parallel to the beam axis; in four they are tilted by a small angle for

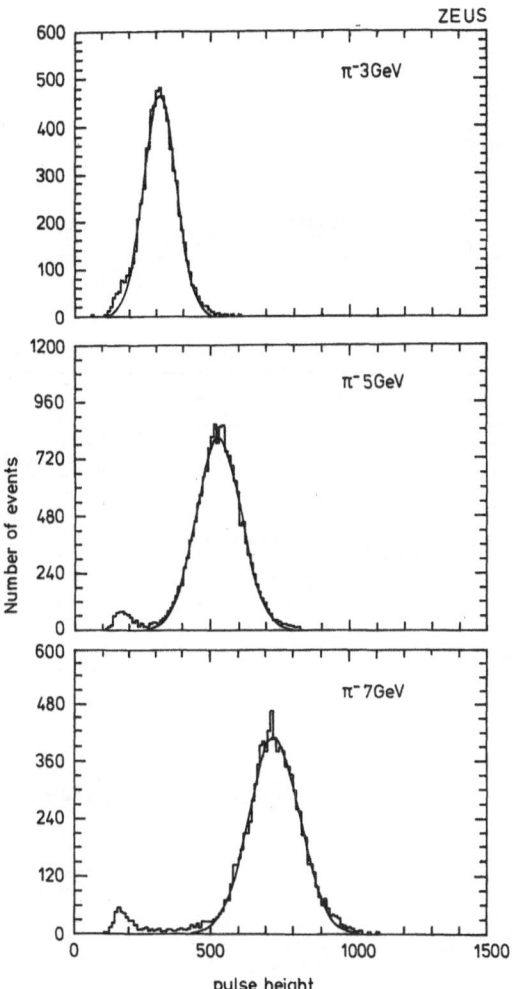

Fig. 62c Pulse height distributions measured with a DU - scintillator
calorimeter (3mm DU, 2.5 mm scint) for 3, 5 and 7 GeV pions.
(From ZEUS collaboration).

Table 7:   Performance of a preprototype calorimeter for ZEUS

	3 GeV	5 GeV	7 GeV
electrons, central tower	$\sigma/\sqrt{E}$ = 15.5 %	16.5 %	15.7 %
pions, all towers	$\sigma/\sqrt{E}$ = 33.3 %	34.1 %	33.0 %
e/h	1.09	1.07	1.06

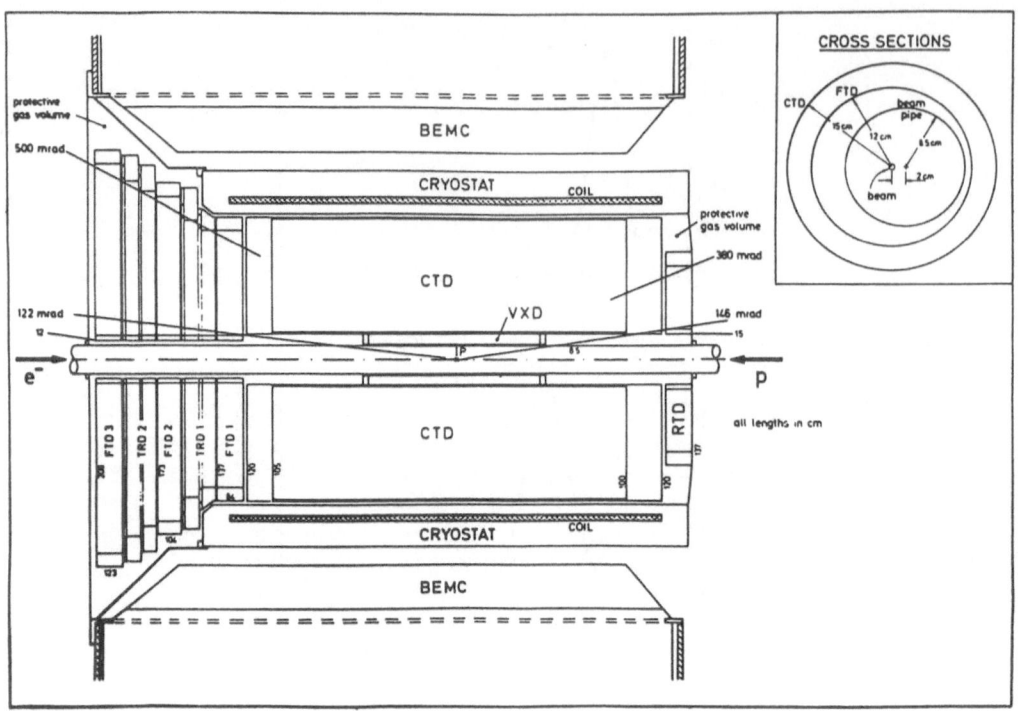

Fig. 63 Layout of the inner tracking system for ZEUS.

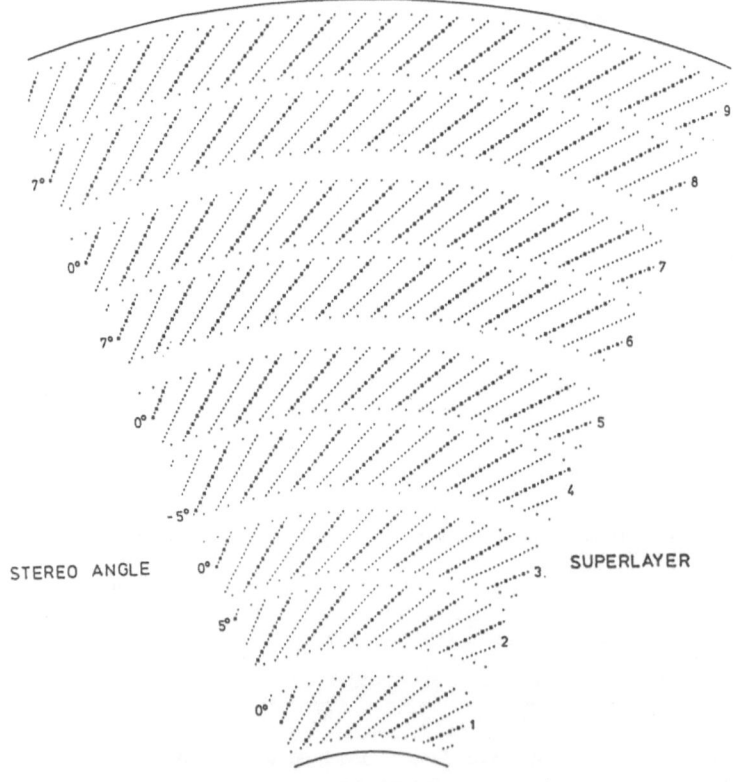

Fig. 64 Cell arrangement in the ZEUS central tracking chamber.

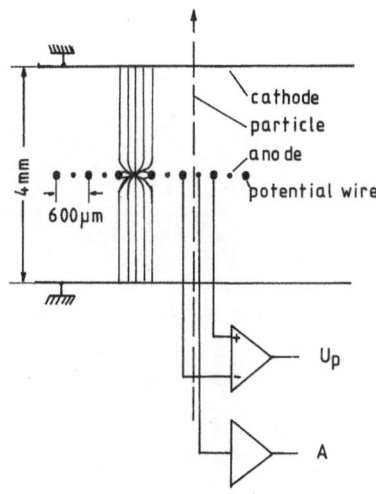

Fig. 65 Wire arrangement for the ZEUS vertex detector
in the induction chamber design.

z measurement. The chamber has a total of 4608 sense and 19584 field wires.
The chamber is operated at atmospheric pressure yielding a precision of
< 100 μm.

The ZEUS collaboration is studying two alternative designs for a
vertex detector. The first is of the time expansion chamber type, the
second one uses a novel concept, the induction chamber (IC) (Ref. 44).
Figure 65 shows the wire configuration for one layer. Anode and field wires
are placed between two cathode planes. The distance between sense and field
wires is only 300 μm. From the anode signal alone a position accuracy of

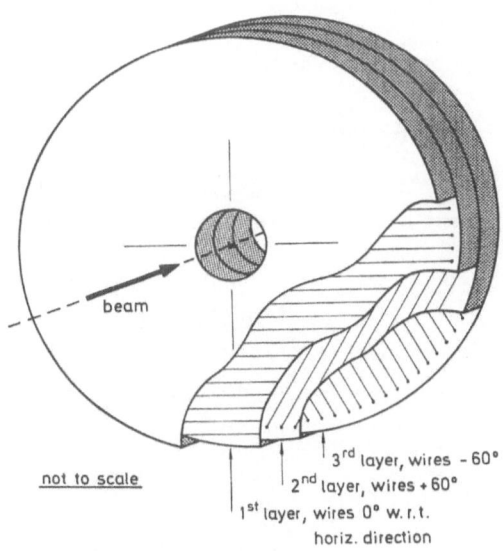

Fig. 66a. Layout of a plane drift chamber for the ZEUS forward detector.

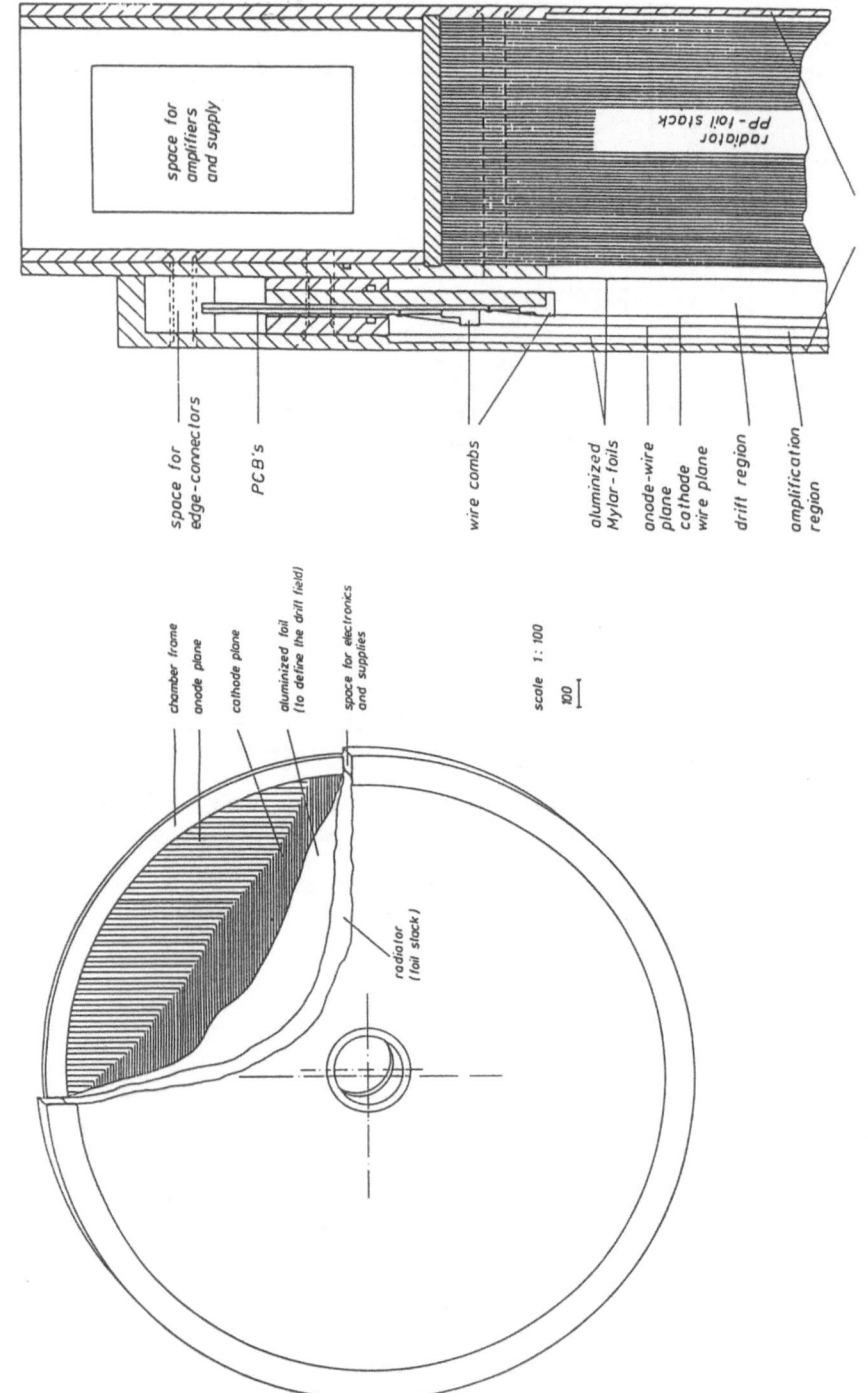

space for
amplifiers
and supply

radiator
PP-foil stack

C-fibre (Cu-coating)

space for
edge-connectors

PCB's

wire combs

aluminized
Mylar-foils

anode-wire
plane
cathode
wire plane

drift region

amplification
region

chamber frame

anode plane

cathode plane

aluminized foil
(to define the drift field)

space for electronics
and supplies

radiator
(foil stack)

scale 1:100

100

Fig. 66b The transition radiation detector for ZEUS. Longitudinal and transverse
view of a chamber module.

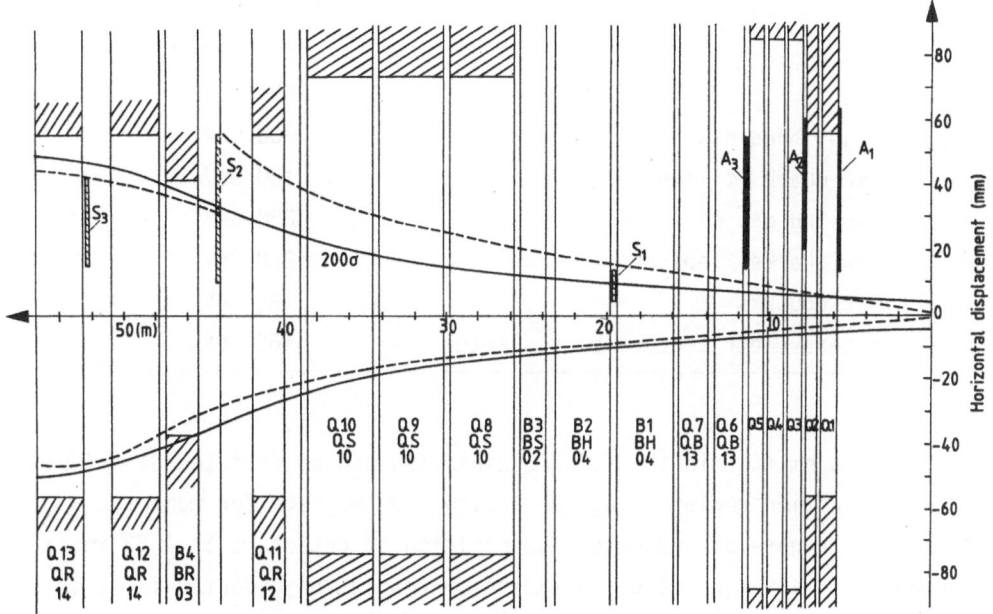

Fig. 67. Horizontal projection of the beam transport system and the detectors S1,S2,S3 of the ZEUS forward proton spectrometer.

600 μm/√12 = 176 μm is obtained. By comparing the signals produced in the left and the right field wires a resolution of ~20 μm is expected.

The forward and backward regions are equipped with planar drift chambers (see Fig. 63). Each chamber consists of three planes with 6 layers of signal wires each (Fig. 66a). For electron detection, in between the drift chambers of the forward region four layers of transition radiation detectors (TRD) with polypropylen fibres as radiator and drift chambers filled with X_e gas for TR detection are installed (Fig.66b). A hadron rejection of > 100 is expected for momenta below 30 GeV/c.

8.2.3 Forward proton spectrometer

In most ep events the debris of the proton is emitted at very small forward angles and escapes down the beam pipe. It is possible to gain access to this region for a class of events which contains a leading proton. In Lund simulation roughly 25 % of all NC and CC events are expected to have a leading proton. Measuring the leading proton together with the scattered electron and the current jet will give a class of fully contained NC events. For CC scattering this will allow to measure the mass of the undetected neutral particle (mostly the neutrino).

Table 8: Expected sensitivity in mass

Table 8: Expected sensitivity in mass

new W	800 GeV
new Z^o	800 GeV
right handed W	500 GeV
new quark (t like)	120 GeV
excited q*, e*	250 GeV
new quarks, leptons	220 GeV
leptoquarks	180 GeV
supersymmetric quark plus slepton	160 GeV

Protons emitted in the forward direction can be detected downstream of the interaction region using the proton ring magnets for momentum measurement. Figure 67 indicates the position of detectors S1 - S3 installed in Roman pots along the ring. The momentum resolution expected is ~1 %.

9.   SUMMARY

HERA will give access to a new territory of physics. Structure functions and charged current reactions can be studied for Q^2 values up to 30 000 - 40 000 GeV^2. The Q^2-ν space will be extended by two orders of magnitude in either variable over what presently can be reached. Precise structure function values will provide a stringent test of QCD, will probe for substructure of quarks and leptons down to distances of $3\cdot10^{-18}$ cm, and could detect new neutral or charged pieces of the current. In the search for right handed weak currents longitudinally polarized electrons will be extremely useful.

HERA is potentially a rich source of new particles. Basically any state that has electric and/or weak charge can be produced up to masses of 200 - 250 GeV. The sensitivity in mass is shown in Table 8.

In many aspects HERA will be complementary to the e^+e^- colliders SLC and LEP which are also under contruction. It probes primarily the space like region and gives access to charged currents over a unique kinematic range. For a number of the newly proposed particles the mass range open to HERA experiments exceeds that of LEP II.

The design of an optimal detector for HERA poses a veritable challenge. The large momentum imbalance between incident electron and proton and the nature of space like processes throw most final state particles into a narrow cone around the proton direction. On the other hand, the standard neutral curent reaction in HERA, ep → eX, offers an important advantage over p̄p/pp interactions. By tagging the scattered electron the kinematics of the relevant subsystem is determined. It is this handle which provides HERA experiments with a special sensitivity to new exotic particles.

ACKNOWLEDGEMENTS

The interaction with the students at the Institute has been very inspiring and stimulating, thanks to the gracious organizing of its director, T. Ferbel. A sizable part of the material on calorimetry presented here has resulted from the work of the ZEUS collaboration. I am indebted to P. Schmüser and B.H. Wiik for discussions on the HERA machine. Mrs. Hell has carried the burden of getting the manuscript ready.

REFERENCES

1. Wiik, B.H., "Electron-Proton Colliding Beams, the Physics Programme and the machine, proc. 10th SLAC Summer Institute, ed.A.Mosher, 1982, p.233.

   Proc. 1984 ICFA Seminar on Future Perspectives in High Energy Physics, May 14-20, 1984, KEK, Japan, eds. S.Ozaki, S.Kurokawa, Y.Unno, p.23.

   Voss, G.-A., Status of the HERA Project, 12th Int.Conf.High Energy Acc., FNAL, 1983 and DESY-HERA Report 83-25 (1983).

2. Llewellyn-Smith, C.H. and Wiik, B.H., DESY Report 77/36 (1977).

3. Proc. Study of an ep facility for Europe, ed. U.Amaldi, DESY Report 79/48 (1979);

   HERA, proposal report DESY-HERA 81/10 (1981);

   Résumé Discussion Meeting "Physics with ep Colliders in view of HERA", Wuppertal, report DESY-HERA 81/18 (1981);

   Proc. Workshop "Experimentation at HERA", NIKHEF, Amsterdam, June 9-11, 1983, report DESY-HERA 83/20 (1983);

   Proc. Discussion Meeting on "HERA Experiments", Genova, Oct.1-3, 1984, report DESY-HERA 85/01 (1985);

   Altarelli, G., Mele, B. and Rückl, R., CERN Report TH3932 (1984);

   Bagger, J.A., Peskin, M.E., SLAC-PUB-3447 (1984);

   Cashmore, R.J et al., "Exotic Phenomena in High Energy ep Collisions", Physics Reports (1985).

   Schmüser, P., "Tief Inelastische Lepton-Nukleon-Streuung", U.Hamburg, 1984, Lecture notes.

4. Bagger, J.A. and Peskin, M., Phys.Rev. D31, 2211 (1985);

   Altarelli, G., Mele, B. and Rückl, R., CERN Report TH3932 (1984);

   Proc. Large Hadron Collider in the LEP tunnel, Vol.2, 549 (1984);

   Rückl, R., Phys.Lett. 129B, 363 (1983); Nucl.Phys. B234, 91 (1984).

5. see Sciulli, F., Proc. 1985 Int.Symp.Lepton Photon Int. High Energies, ed. M.Konuma, K.Takahashi, p.8.

6. see e.g. Lohrmann, E., Mess, K.-H., DESY-HERA 83/08 (1983).

7. The ZEUS detector, Technical Proposal of the ZEUS Collaboration, March 1986.

8. Duke, D. and Owens, J.F., Phys.Rev. D30, 49 (1984)

9. Eichten, E.J., Lane, K.D., Peskin, M.E.,Phys.Rev.Lett. 50,811 (1983);
Also see Rückl, R., Phys.Lett. 129B, 363 (1983).

10. MAC Collaboration, Fernandez, E. et al.,Phys.Rev.Lett. 50,123 (1983);
TASSO Collaboration, Althoff, M. et al., Z.Phys. 22, 13 (1984);
Phys.Lett.138B, 441 (1984); MARK J Collaboration, Adeva, B. et al.,
Phys.Rep. 109, 131 (1984); HRS Collaboration, Bender, D. et al.,
ANL-HEP-PR-03; 71; PLUTO Collaboration, Berger, Ch. et al., Z.Phys.
27, 341 (1985).

11. Bars, I., Bowick, M.J., Freese, K., Phys.Lett. 138B, 159 (1984).

12. c.f. Pati, J.C., Salam, A., Phys.Rev. D10, 275 (1974);
Mohapatra, R.N., Pati, J.C., ibid, 566, 2559 (1975);Mohapatra, R.N.,
Marshak, R.E., Phys.Rev.Lett. 44, 1316 (1980).

13. Mohapatra, R.N., Senjanovic, G., Phys.Rev.Lett. 44, 912 (1980).

14. Carr, J. et al., Phys.Rev.Lett. 51, 627, 1222(E) (1983).

15. Senjanovic, G., Proc. SSC Fixed Target Workshop, The Woodlands,Tx,
(1984);
Stoker, D.P. et al., LBL-18935 (1985) and Phys.Rev.Lett.

16. Beall, G., Bander, M. and Soni, A., Phys.Rev.Lett. 48, 848 (1982);
Harari, H. and Leurer, M., Nucl.Phys. B233, 221 (1984);
Ecker, G. and Grimus, W., Z.Phys. C30, 293 (1986).

17. Harari, H., private communication.

18. Leveille, J.P., Weiler, T., Nucl.Phys. B147, 147 (1979).

19. Brodsky, S.J., Gunion, J.F., SLAC-PUB-3527 (1984).

20. Altarelli, G., Martinelli, G., Mele, B., Rückl, R., CERN TH 4094
(1985).

21. Ellis, J., Gaillard, M.K., Nanopoulos, D.V., Nucl.Phys. B106, 292
(1976).

22. See e.g. Kühn, J.H., Tholl, H.D., Zerwas, P.M., CERN TH 4131 (1985).

23. Rudaz, S., Vermaseren, J.A.M., CERN TH 2961 (1981).

24. Jones, S.K., Llewellyn-Smith, C.H., Nucl.Phys. B217, 145 (1983).

25. Harrison, D.P., Nucl.Phys. B249, 704 (1985).
Marlean, L., SLAC-PUB-3533 (1984).

26. For a recent review see e.g. Wiik, B.H., Progress with HERA, DESY-
HERA 85-16 (1985).

27. Dwersteg, B. et al., DESY Report 85-08 (1985) and Proc. 1985 Part. Acc. Conf. Vancouver, IEEE Trans. Nucl. Sci. Vol. NS-32, 3596.

28. See e.g. Buon, J., Steffen, K., DESY report 85-128 (1985).

29. Piwinski, A., Proc. 1985 Part.Acc.Conf., Vancouver, 1985.

30. Ford, R.L. and Nelson, W.P., SLAC report 210 (1978).

31. Fabjan, C. and Willis, W., Phys.Rev.Lett. 60B, 105 (1975);
    Fabjan, C. et al., Nucl.Inst.Meth. 141, 61 (1977);
    Akesson, T. et al., Nucl.Inst.Meth. A241, 17 (1985).

32. See e.g. Gabriel, T.A. et al.,IEEE Trans.Nucl.Sci. NS-32, 697 (1985);
    Fesefeldt, H., The simulation of hadronic showers, PITHA 85-02 (1985).

33. Lankford, A.J., Thesis Geneva 1978, CERN Internal Report EP 78-3;
    Kondo, T. et al., 1984 Summer Study on the Design and Utilization of the Superconducting Super Collider, Snowmass, Colorado, June 1984;
    Brau, J., Proc. of Workshop on Compensated Calorimetry, Pasadena 1985;
    Flauger, W., Nucl.Instr.Meth. A241, 72 (1985).

34. See e.g. Leroy, C., Sirois, Y. and Wigmans, R., CERN Report EP/86-66 (1986), submitted to Nucl.Inst.Meth.

35. Brückmann, H. et al., private communication.

36. See e.g. Fabjan, C.W. and Ludlam, T., Ann.Rev.Nuc.Part.Sci. 32,335 (1982).

37. Wigmans, R., Proc. of Workshop on Compensated Calorimetry, Pasadena 1985.

38. Abramowicz, H. et al., Nucl.Inst.Meth. 180, 429 (1981).

39. WA-78 Collaboration, De Vincenzi, M. et al., CERN Report EP/86-12 (1986), submitted to Nucl.Inst.Meth.

40. ZEUS: A Detector for HERA, Letter of Intent, June 1985.

41. Letter of Intent for an Experiment at HERA, H1 Collaboration, June 28, 1985.

42. The ZEUS Detector, Technical Proposal, March 1986.
    Carleton, Manitoba, McGill, Toronto, York, Bonn, DESY, Freiburg, Hamburg I, Hamburg II, Siegen, Weizmann Institute, Bologna, Florence, Frascati, L'Aquila, Lecce, Milan, Padua, Rome, ENEA-Rome, Torin, NIKHEF Amsterdam, Cracow, Warsaw, Madrid, Bristol, London (IC), London (UC), Oxford, Rutherford, Argonne, Columbia, Illinois, Ohio, Pennsylvania State, Virginia, Wisconsin.

43. Technical Proposal for the H1 Detector, March 25, 1986,
    Aachen, Davis, DESY, Dortmund, Ecole Polytechnique, Glasgow, Hamburg, Houston, Lancaster, Liverpool, Manchester, Moscow, MPI München, Northeastern, Orsay, Paris, Rome, RAL, Saclay, Wuppertal, Zeuthen, Zürich.

44. Walenta, A., University of Siegen Report Si 86-5 (1986).

THE PRINCIPLES AND CONSTRUCTION OF LINEAR COLLIDERS*

JOHN REES

Stanford Linear Accelerator Center
Stanford University, Stanford, California 94305

1. INTRODUCTION TO LINEAR COLLIDERS

The linear collider in its simplest form consists of two linear accelerators aimed at one another so that their beams collide in the space between them, the interaction region, as shown schematically in Fig. 1. Their beam energies, or more properly their mean beam energies, since each beam has some energy spread, are the same so that the centers-of-mass of the particle-particle collisions are stationary on the average. One of the linacs (linear accelerators) is equipped with a positron source so that the colliding system is an electron and a positron, a more fruitful system to study than two electrons. In order to develop high enough luminosities for high-energy particle physics, the linacs must be far more sophisticated than linacs of the past, and they must have ancillary damping rings to condense their beams to tiny lateral dimensions. We shall discuss the problems posed to the designers and builders of high-energy linear colliders in the following sections, but first a little history will explain why we are studying these new machines.

Although the linear collider was first suggested in print in 1965,[1] it did not emerge as a candidate to supplant the colliding-beam storage ring until the late nineteen seventies. It was with storage rings that colliding-beam physics was started, developed and exploited, beginning in the mid-fifties and continuing to the present. But the costs of building storage rings rise approximately in proportion to the second power of the energy of the ring,[2] while the costs of building linear colliders rise only as the first power of their energy. As the collision energies required to explore the frontiers of particle physics go up and up, the linear collider eventually becomes

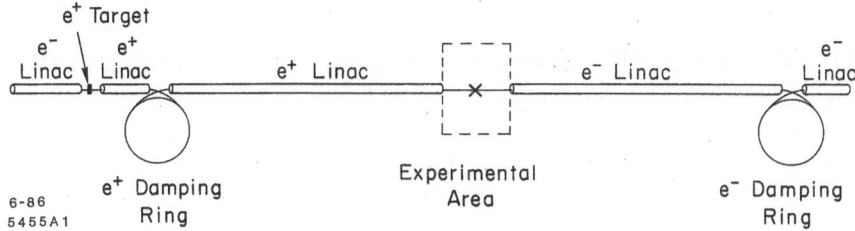

Fig. 1. Schematic design of a linear collider.

6-86
5455A1

*Work supported by the Department of Energy, contract DE-AC03-76SF00515.

the more economical choice, assuming that equal performance (luminosity) can be attained with it. Interest in the linear collider was revived in 1976,[3] and it was declared the machine of preference for collision energies of several hundred GeV and up by 1980.[4]

2. THE SCALING LAWS OF LINEAR COLLIDERS

Scaling laws are the equations which relate experimental conditions at the interaction point (collision energy, luminosity, energy spread) to accelerator physics parameters including economic factors. Given the experimental use the collider is intended for, these laws tell the builder certain accelerator parameters he must produce. In particular, the dimensions of the bunch when it reaches the interaction point are specified.

2.1 RESTRICTION TO ROUND BEAMS AT THE COLLISION POINT

In general, beam bunches of colliders of different design may have a wide variety of distributions in phase space at the collision point. However, it will clarify our introductory studies to choose a simple distribution and stick with it throughout our work. We shall choose *round beams,* where the term is shorthand for a tri–Gaussian spatial distribution which is circularly cylindrical in the transverse dimensions. That is, the bunch has a particle density at the interaction point proportional to

$$\exp\left\{-\tfrac{1}{2}\left(\frac{x^2+y^2}{\sigma_r^2} + \frac{z^2}{\sigma_z^2}\right)\right\} \quad ,$$

where x and y are the transverse coordinates and z is the coordinate in the direction of motion (the longitudinal coordinate), σ_r is the radial standard deviation and σ_z is the longitudinal standard deviation. Generally $\sigma_z >> \sigma_r$. In phase space the distribution is a six-dimensional Gaussian. This assumption is somewhat restrictive, but it is useful for our purposes, since it permits us to concentrate on the basic physical phenomena of colliders in terms of the simplest formulas.[5,6] Such distributions do not, in fact, prevail in linear colliders, but they are close enough to give sensible, realistic results for our purposes.

Flat beams — beams having one lateral dimension much greater than the other — offer the advantage that their peak electric and magnetic fields are lower than those of round beams.[7] Consequently beamstrahlung energy losses are diminished at given luminosity, although attainable luminosity enhancement through the pinch effect is significantly reduced.[8] We shall discuss beamstrahlung loss and the pinch effect later. Suffice it for now to note that the advantage of flat beams is greater at lower beam energies and lower bunch fields than at higher energies and bunch fields.

2.2 LUMINOSITY

The luminosity of a colliding-beam system gives the reaction rate per unit cross section for a given reaction; for bunched beams colliding head-on, its formula is

$$L = \frac{fN^2}{A} \ , \tag{1}$$

where L is the luminosity, f is the frequency with which the bunches collide at the interaction point, N is the number of particles in a bunch (its population), and

A is the effective interaction area. Figure 2 shows the beam envelope at the interaction region and two interpenetrating bunches (idealized as cylinders). Equation (1)

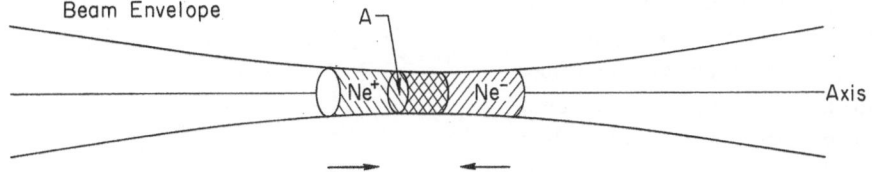

Fig. 2. The interaction region of a collider.

can be thought of as the flux of one beam, fN, multiplying the target density, N/A, of the other, and it assumes that the bunches are alike. In practice, of course, although the bunches may be alike in shape and population, they will have a non-uniform transverse density distribution and A must be obtained by carrying out an integral. For round beams,

$$A = 4\pi\sigma_r^2 \ , \tag{2}$$

where σ_r is the radial standard deviation.

Inspecting Eq. (1), we see that f and N should be big and A should be small to make high luminosity. But we must attend to another formula before going on.

2.3 BEAM POWER

The average power which must be imparted to each of the two beams is

$$P_b = fN\gamma m_e c^2 \ , \tag{3}$$

where $\gamma m_e c^2$ is the beam energy. Our electric power bill for running the collider after we get it built, as well as many of the elements of the construction cost, will be proportional to P_b. We must keep these costs within bounds, which becomes more difficult the higher the energy is. The upshot is that, in practice, there is always great pressure to make the effective interaction area very small — tiny in fact. This is the message of these first two equations.

In the next section, we shall take up the problems of attaining small interaction areas, but first we must consider two phenomena that occur while the bunches interact, which influence the effective area and which impose important limitations on the precision with which particle physics can be done with the machine. They are called beam disruption and beamstrahlung.

2.4 DISRUPTION

The basic process which leads to beam disruption is depicted in Fig. 3. which shows a particle of one beam being deflected by the collective electromagnetic field of the counter-moving bunch. Incident particles at different impact parameters and different incident angles are deflected by different amounts. If the incident particle is close to the axis, though, the fields of the opposing bunch are lens-like (they vary linearly with the impact parameter) with a focal length F. We characterize beam disruption by the disruption parameter D which is just the ratio of the bunch length σ_z to that focal length.

$$D = \sigma_z/F \ . \qquad (4)$$

If D is small compared to one, there is little deflection, and the beams do not alter each other's motions very much. On the other hand, if it is about equal to one, particles entering parallel to, but well separated from, the axis, leave the back of the opposing bunch very near the axis with a relatively large angle. In other words, the bunch has focused those particles to a point at its tail. That constitutes substantial disruption. For a round beam,

$$D = \frac{4\pi r_e \sigma_z N}{A\gamma} \ , \qquad (5)$$

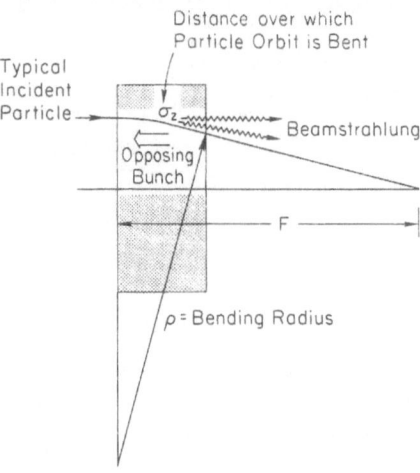

Fig. 3. The motion of a typical particle of one beam passing through the opposing bunch.

where r_e is the classical radius of the electron. Beam disruption is troublesome for experiments, because it can cause background events in the detector if it is not allowed for. The disrupted beam occupies a much larger volume in phase space than the incoming beam does if D is substantial in comparison to one. The detector surrounds the interaction point and has a hole running through it for the beams to pass through. This hole has to be large enough to accommodate the largest beam, and the largest beams are the disrupted beams. Otherwise, the detector would be showered with back- ground. Thus beam disruption determines how close to the interaction point an event can be tracked.

On the other hand, disruption has a beneficial effect. Since each beam has a generally focusing effect on the other, the bunches are *pinched*, and their transverse densities are increased in the interaction region. The effective interaction area is reduced by the pinch and the luminosity is correspondingly increased.[9] (By the way, the reader is warned that the author is using the term "pinch effect" in a way which may not be exactly consistent with the usage of that term in plasma physics; however it is descriptive enough of the phenomenon to warrant its use here.) Figure 4 shows a computer simulation of the time sequence of spatial distributions of two bunches as they pinch each other and then fly apart. Both the increase in density during interaction and the disruption after it are evident in the sequence. In order to describe the enhancement of luminosity by the pinch, we need to introduce the *incoming* area, A_o. This effective area corresponds to that of the topmost distributions in Fig. 4. It is the cross sectional area provided by the accelerators and their focusing systems and would be the interaction area if the beams were sufficiently weak that they did not pinch each other. Corresponding to A_o there is an incoming disruption parameter

$$D_o = \frac{4\pi r_e \sigma_z N}{A_o \gamma} \ . \qquad (6)$$

Now the actual luminosity can be written in terms of the unenhanced luminosity

$$L = \frac{fN^2}{A_o}\frac{A_o}{A} \; , \qquad (7)$$

and we see that the enhancement is just the factor A_o/A. The process has been studied by computer simulation, and curves of the enhancement of the luminosity have been obtained. An example pertinent to the SLAC Linear Collider is shown in Fig. 5.

Fig. 4. Computer simulated collision of intense relativistic beams illustrating the pinch effect. From Hollebeeck.

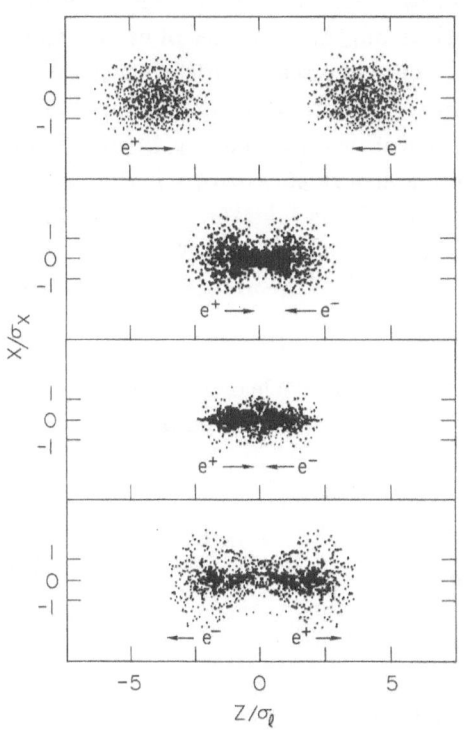

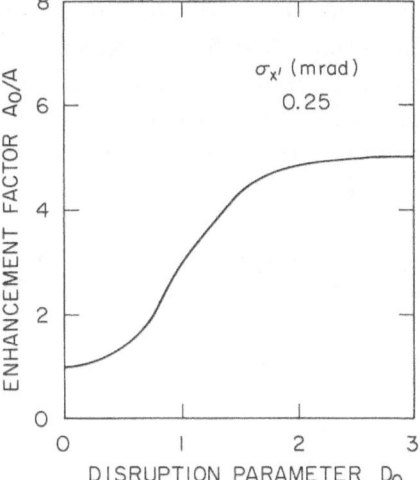

Fig. 5. The enhancement of luminosity, A_o/A, as a function of the incoming disruption parameter, D_o (from Ref. 8).

As an alternative approach to computer simulation, the interpenetrating bunches may be considered as a relativistic neutral plasma whose plasma instabilities can be calculated.[9] From such studies, we conclude that values of D_o up to 10 should be comfortably stable and usable. Perhaps even larger values are permitted, but they do not promise increased luminosity enhancement according to Fig. 5.

2.5 BEAMSTRAHLUNG

Beamstrahlung is the emission of acceleration radiation by the individual particles as a result of the collective electromagnetic fields of the opposing bunches passing them. Since the acceleration is almost perpendicular to the particle velocity, beamstrahlung is often treated as synchrotron radiation in the bunch field. It is distinguished from beam-beam bremsstrahlung which considers only close particle-particle encounters that give rise to emissions of photons having energies comparable to the particle energies. Beamstrahlung may be regarded as considering only comparatively distant encounters where the fields of many particles are superimposed.

Referring to Fig. 3, as a particle moves through the opposing bunch, it is deflected with radius of curvature $\rho(z)$; the radius varies during the passage. At each moment it is radiating electromagnetic power

$$P = \frac{2}{3} \frac{ce^2\gamma^4}{\rho^2(z)} \ .$$

(8)

The result is that particles with different trajectories experience different energy losses to the radiation, and an energy spread is created among the particles of each bunch. The incoming bunches already have some energy spread, created in the accelerators themselves, and the beamstrahlung energy spread adds to it in quadrature. The more the net energy spread, the less precisely the interaction energy is known for any event. Therefore we regard the beamstrahlung energy spread as a performance-limiting parameter. As a measure of the energy spread, we take the mean fractional energy loss, averaged over impact parameters, which we denote by δ. Now, δ has very different formulas for high and low values of γ^3/ρ, because of the nature of synchrotron radiation.

The spectrum of synchrotron radiation, provided the particle energy is sufficiently low, is characterized by a single photon-energy parameter, ϵ_{crit}, given by

$$\epsilon_{crit} = \frac{3}{2} \hbar c \frac{\gamma^3}{\rho} \ .$$

(9)

When the critical energy is very low compared to the particle energy, a classical treatment of the problem gives the correct answer, but when the critical energy [as defined by Eq. (9)] approaches and exceeds the particle energy, it does not. The fact that a particle cannot radiate a real photon whose energy is greater than the particle's energy forces us to use a quantum mechanical treatment. As a consequence, we have two equations for δ.[10]

Referring to the topmost graph in[10] Fig. 6 and in particular to the curve labeled "classical", we see that the classical synchrotron radiation spectrum is a broad spectrum with a maximum near the critical energy (10^0 on the abscissa). When the particle energy greatly exceeds the critical energy — by a lot more than a factor of ten — the classical spectrum agrees quite precisely with the quantum mechanical spectrum.

When the particle energy is only ten times the critical energy, the two spectra begin to differ as shown in the top graph. "QM" labels the exact quantum mechanical result. In these graphs, E is the particle energy and E_c is the critical energy. The bottom two graphs show that the (correct) quantum mechanical spectrum differs radically from the classical spectrum when the particle energy is equal to or less than the critical energy.

The classical equation for the fractional energy spread is [11]

$$\delta_c = 2.71 \, \frac{r_e^3 N^2 \gamma}{A \sigma_z} \, . \qquad (10)$$

This formula applies specifically to our round Guassian beams. Using a somewhat different bunch shape and the approximation of a sharp cut-off on the synchrotron radiation spectrum, Himel and Siegrist obtain the following formula for the quantum mechanical case. [10]

$$\delta_q = 1.63 \left(\frac{\alpha^4 r_e N^2 \sigma_z}{A \gamma} \right)^{1/3} , \qquad (11)$$

where α is the fine structure constant. Although the bunch model used in their treatment is different, the dependencies on the variables are reliable in the region of variables in which the approximation is good. The complete quantum mechanical result (shown in Fig. 6 as the curves labeled "QM") is, of course, valid at all values of the ratio, ϵ_{crit}/E, but it is very complicated.

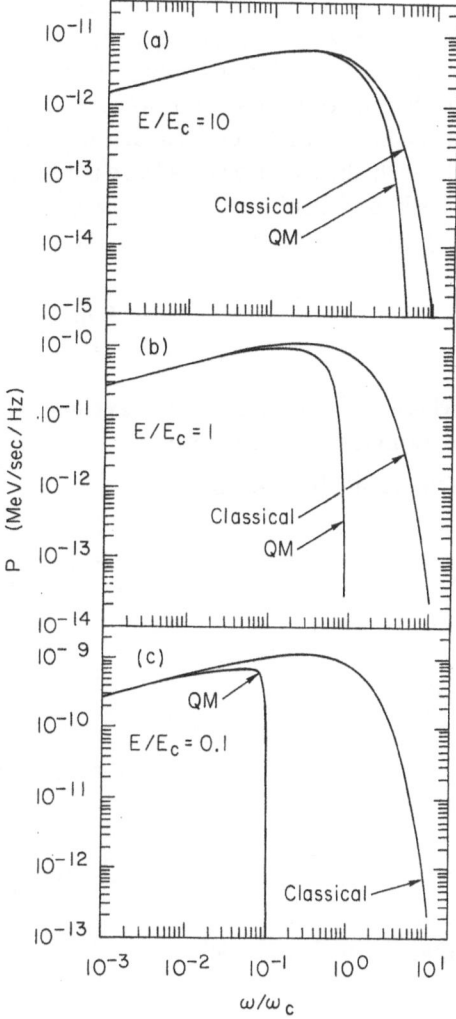

Fig. 6. Synchrotron radiation spectra taken from Ref. 10.

2.6 THE SCALING LAWS COLLECTED

We have now written down all of the scaling laws. For the case of the beam energy being far above the critical energy we have Equations (1), (3), (6) and (10), and for the case of the beam energy being below the critical energy, we have Equations (1), (3), (6) and (11). However they are not in the form in which we would like to have them, because they give the luminosity, beam power, disruption parameter and beamstrahlung energy spread in terms of the beam energy and what we regard as accelerator parameters: f, N, A and σ_z. We wish to specify the energy, the energy spread, the luminosity and perhaps the disruption parameter on the grounds of the physics we want to do with the collider, and we wish to specify the beam power on economic grounds; then we want to derive the accelerator parameters from those desiderata. This we can do by solving the equations for the accelerator parameters. But we must remember that we have *two* sets of equations, and we will not know which set is valid until we know the critical energy of the beamstrahlung — which we will not know until we have the equations solved. We would have only one set of equations and no such dilemma if we had used the complicated quantum mechanical result for δ (which would, in that case, need no subscript). What we can do with the equations we have is to solve both sets. Then for any given case, we must check, by calculating ϵ_{crit}/E, whether either set is valid. For this purpose we combine Eq. (9) with Eq.(4) and a little geometry in Fig. 3 to get

$$\frac{\epsilon_{crit}}{E} = \frac{3}{4\pi^{1/2}} \frac{\lambda_c A^{1/2} D\gamma^2}{\sigma_z^2} \; ,$$

(12)

where λ_c is the Compton wavelength. The two sets of equations for the accelerator parameters are as follows.

Classical $(\epsilon_{crit} \ll E)$

$$N = \frac{1}{2.71(4\pi)} \left(\frac{P_b^2 D\delta_c}{r_e^4 E^2 L^2} \right)$$

(13)

$$A = \frac{1}{2.71(4\pi)} \left(\frac{P_b^3 D\delta_c}{r_e^4 E^3 L^3} \right)$$

(14)

$$f = 2.71(4\pi) \left(\frac{r_e^4 E L^2}{P_b D\delta_c} \right)$$

(15)

$$\sigma_z = \frac{1}{4\pi} \left(\frac{P_b D}{r_e m_e c^2 L} \right)$$

(16)

Quantum $(\epsilon_{crit} > E)$

$$N = \frac{4\pi}{(1.63)^3 \alpha^4} \left(\frac{\delta_q^3}{D} \right)$$

(17)

$$A = \frac{4\pi}{(1.63)^3 \alpha^4} \left(\frac{P_b \delta_q^3}{EDL} \right)$$

(18)

$$f = \frac{(1.63)^3 \alpha^4}{4\pi} \left(\frac{P_b D}{E\delta_q^3} \right)$$

(19)

$$\sigma_z = \frac{1}{4\pi} \left(\frac{P_b D}{r_e m_e c^2 L} \right)$$

(16)

Now that we have put the scaling laws in a convenient form, let us fix these ideas in our minds by working through a couple of examples. First we shall consider the SLAC Linear Collider (SLC), the only extant (or nearly extant) specimen. Its objective specifications are as follows.

$$E = 50 \text{ GeV}, \qquad\qquad L = 6 \times 10^{30} \text{ cm}^{-2} \text{ s}^{-1},$$

$$P_b = 0.072 \text{ MW}, \qquad D = 2.5, \qquad \delta = 0.0019.$$

We try the "Classical" equations first, and the results are

$$f = 181 \text{ Hz}, \qquad\qquad A = 7.4 \times 10^{-8} \text{ cm}^2,$$

$$N = 5 \times 10^{10}, \qquad\qquad \sigma_z = 0.10 \text{ cm}.$$

When these numbers are used in Eq. (12), the classical case proves to be the valid one. These are the accelerator parameters given by the equations, but one of them, A, is not the transverse bunch area that the collider system delivers to the interaction region; it is rather the *pinched* area. Instead of A, the accelerator builder needs to know A_o, and to get it, he must find values of A_o and D_o which correspond to A and D. to do this, he uses the curves of Fig. 5 and the relation $A/A_o = D/D_o$. When this is done, it turns out that D_o is about one and A_o is 2.2×10^{-7} cm^2.

A final remark about the SLC example: the total energy spread in the interacting bunches is some combination of the beamstrahlung energy spread and the *incoming* energy spread — the energy spread created in the acceleration process itself. If we assume that both are Guassian and uncorrelated, they combine as the sum of squares. At worst, they could add. In the SLC, the energy spread due to the accelerator is intended to be ±0.002 to ±0.005, so it is dominant.

Now let's do another example: that of a 1-TeV collider, one which gives a mean center-of-mass energy of 2 TeV. For this machine, we choose the parameters

$$E = 1 \text{ TeV}, \qquad\qquad L = 10^{33} \text{ cm}^{-2} \text{ s}^{-1},$$

$$P_b = 1 \text{ MW}, \qquad D = 0.1, \qquad \delta = 0.3.$$

We have chosen a rather small value of the disruption parameter, because we anticipate that this collider will operate under the conditions for which the "quantum" equations will be valid, and those equations place a premium on small values of D to keep the area large and the repetition frequency low. For the same reasons, we have chosen a rather large energy spread: 0.3 — much larger than that we would expect from the accelerators. Indeed a fractional energy spread of 0.3 in the collision energies would seriously weaken experiments done with the collider. However, the spread in collision energies is not the same as the mean beam-energy spread. The rms center-of-mass energy spread amongst collisions has been treated by Yokoya[13] and by Noble[14]. The fractional spread is indeed less than 0.3. It is about 0.15. This will not permit us to use the interaction energy as a strong constraint in fitting data, but it may be tolerable.

Using these parameters in the "quantum" equations, we obtain

$$f = 22,600 \text{ Hz}, \qquad A = 1.7 \times 10^{-12} \text{ cm}^2,$$

$$N = 2.8 \times 10^8, \qquad\qquad \sigma_z = 3.5 \times 10^{-4} \text{ cm}.$$

In this case we have used so small a disruption parameter that we can consider that $A = A_o$ and $D = D_o$.

These parameters are certainly beyond present practice and technology, and whether they can be achieved is under study in many laboratories. We shall address some of the problems of achieving them in subsequent sections of our study.

3. THE ATTAINMENT OF SMALL INTERACTION AREAS

In order to address the problem of attaining tiny cross-sectional areas of the beams at the interaction point, we must introduce the emittance of the beam and the beta function of the beam transport system. The emittance characterizes the organization of the beam in phase space, and the beta function characterizes the focusing properties of the transport system.

3.1 EMITTANCE

We have already referred to phase space and remarked that it is a six-dimensional space. For the purposes of this section it will prove convenient to think of transverse phase space as the four dimensional space that describes particle motion on the two transverse coordinates, and indeed, even further, to think of the two-dimensional phase space that describes the motion on one transverse axis. (See Fig. 7.) The phase space commonly used to describe particle motion in accelerators and beam-transport systems has the particle's transverse coordinate as abscissa and the angle of the particle's trajectory, projected on the plane of the coordinate axis and the central orbit, as ordinate. These variables are not canonically conjugate, but they prove most useful and convenient.[15] We use z to denote the coordinate along the direction of motion, the coordinate called s in Ref. 15.

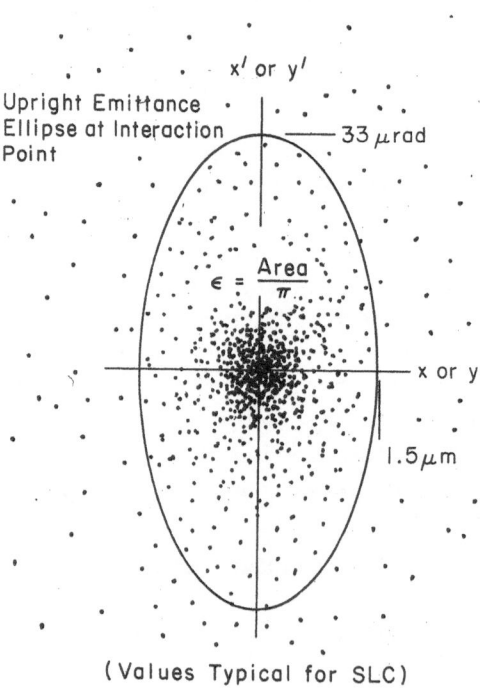

(Values Typical for SLC)

Fig. 7. The phase space commonly used to describe particle motion in accelerators and beam-transport systems. The ellipse surrounding a fraction of the particle points represents the emittance.

The emittance of a beam is a measure of its concentration in phase space; the smaller the emittance, the more concentrated is the beam. Figure 7 shows a swarm of particle points in phase space, a certain fraction of them being surrounded by an ellipse. The area of the ellipse, divided by π, is the emittance. (Warning: some authors do not divide by π.) For different distribution functions, different fractions of the swarm are customarily chosen to define the emittance. For our Gaussian distributions, we choose the fraction to be 47%, which means that, if the emittance ellipse is *upright,* the emittance is $\epsilon_x = \sigma_x \sigma_{x'}$ in the x–coordinate.

The use of an ellipse in defining the emittance is motivated, in part, by a special property the ellipse possesses for the case of a drifting beam — a beam that is not being accelerated and is not radiating — in a beam-transport system which provides linear focusing forces. In that case, the emittance ellipse has the same area everywhere along the drift path, although it changes its eccentricity and orientation as we move along. The emittance is conserved.

Even if the focusing forces are not linear, Liouville's Theorem tells us that particle points, once confined within a closed figure, must remain forever in a closed figure of the same area, although the figure does not necessarily remain an ellipse. This is a property of a system that obeys a conservative Hamiltonian.

If the beam is being accelerated in a linear accelerator, the emittance is not conserved. It shrinks, and it shrinks in a simple way. The emittance in each transverse coordinate varies inversely with the particle energy. In the absence of transverse focusing forces, it is easy to see qualitatively why some damping should occur. The accelerating force is purely longitudinal, so the transverse momentum is constant and the transverse velocity goes down as γ goes up.

Since we are concerned with linear accelerators, we shall find it useful to define the normalized emittance

$$\epsilon_n = \epsilon \gamma \ . \tag{20}$$

The normalized emittance will remain constant throughout acceleration provided accelerating forces and linear focusing forces are the only ones that need be considered.

3.2 THE BETA FUNCTION

The beta function characterizes the transverse focusing provided by the beam transport system. There is a beta function for each transverse coordinate: β_x and β_y. The beta function has been taken over to linear-collider use from storage-ring practice and from general "round" accelerator theory, where it appears as the modulating function of a WKB solution of the harmonic oscillator equation with varying wavelength.[15] In such a solution, the transverse coordinate, say x, is given by

$$x(z) = a \beta_x^{1/2}(z) \cos[\phi_x(z) + b] \ , \tag{21}$$

where a and b are arbitrary constants and

$$\phi_x(z) = \int \frac{dz'}{\beta_x} \ .$$

(I have used a slightly different definition of ϕ_x than that of Ref. 15, but one in common use.) Since a beam consists of many particles, each with its own a and

b, $\beta^{1/2}$ describes the envelope of the beam's transverse motion (provided there is no momentum dispersion — the case in a linear accelerator). Where β is small, the beam is thin; where β is large, the beam is fat. In terms of the emittance,

$$\sigma_x(z) = [\epsilon_x \beta_x(z)]^{1/2} \tag{22}$$

and similarly for y. We design the optics of the final focusing system of a linear collider so as to have no dispersion at the interaction point so that dispersion will not add to the lateral dimensions of the beam. Thus the spot size is just proportional to $\beta^{1/2}$ there, and we must make β as small as possible to get a small interaction area. Assuming that $\epsilon_x = \epsilon_y = \epsilon$, a round beam requires $\beta_x = \beta_y = \beta^*$. The superscript star indicates values at the interaction point. Now we can write the incoming effective interaction area

$$A_o = 4\pi\epsilon\beta^* = \frac{4\pi\epsilon_n\beta^*}{\gamma} \ . \tag{23}$$

As we have seen, A_o must be very small, and therefore one of our chief tasks in building a collider is to make *both* the emittances *and* the interaction-point beta functions as small as they can be made. The measures we take to secure and to maintain small emittances will be discussed later.

There are limits on how small the beta functions can be made. The limits arise from chromatic aberrations. In a particle-optical system, momentum is the analog of frequency in a light-optical system, and the dependence of the focal length of a quadrupole magnet on momentum is the analog of chromatic aberration. Figure 8 shows a parallel beam being focused to a waist by a lens. Particles of the central momentum, p_o, stay within the envelope shown, which has the algebraic form

$$\sigma(z) = \sigma^*\sqrt{1 + \left(\frac{z}{\beta^*}\right)^2} \ . \tag{24}$$

In the case of a typical collider final focus system, $L \gg \beta^*$, so

$$\sigma_o \approx \sigma^*\frac{L}{\beta^*} \ . \tag{25}$$

The reader should be aware that we are using the symbol L in this section as the distance from the interaction point to the nearest lens — not as the luminosity — but the meaning should be clear from the context and no confusion should arise.

Consider a particle that is at the edge of the beam envelope as it passes through the lens and that is brought to the axis at the waist by the action of the lens. That particle is bent by the lens through an angle σ_o/L. If the particle had a momentum $(p_o + \Delta p)$, it would be bent by a smaller angle as shown by the dashed line in Fig. 8, the diminution of the angle being by the fraction, $\Delta p/p_o$. Another way of looking at it is that the waist produced by the lens for a higher momentum is farther to the right. Different momenta yield different spot sizes at the interaction point.

The consequence of the energy spread in the incoming beam, then, is to "fuzz out" the spot at the interaction point so that the smallness of the beta function (for

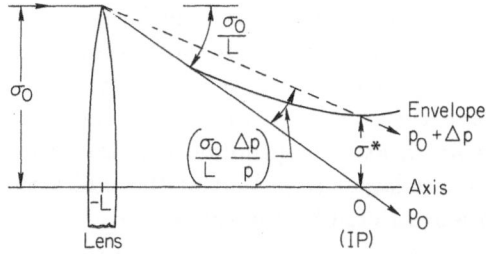

Fig. 8.   Chromatic aberration at a waist in the beam.

the central momentum) is vitiated. In order to keep this chromatic aberration from dominating the spot size

$$L\frac{\sigma_o}{L}\frac{\Delta p}{p_o} = \sigma^* \frac{L}{\beta^*}\frac{\Delta p}{p_o} < \sigma^* \quad .$$

For this simple system, we get the result

$$\frac{\Delta p}{p_o} < \frac{\beta^*}{L} \quad . \tag{26}$$

Since L is the space on either side of the interaction point which is left free for detector components to be snug against the beam pipe, we cannot make it too short without interfering with the ability of the machine to do physics. This means that making β^* small places demands on the accelerator to keep the incoming energy spread correspondingly small. For example, if $L = 3$ m and $\beta^* = 1$ cm, then $\Delta p/p_o < 0.3\%$.

This simple system does not represent the best of present technology. For example, the rather complex final focus system of the SLC does a factor of two or three better. However this simple example reveals the physical origin of the effect and estimates its magnitude reasonably well.

Now that we know about how far we can reduce β^*, we can figure out how small the emittances need to be for our examples, the SLC and the 1-TeV collider.

In the SLC, the beta function at the interaction point is designed to be $\beta^* = 0.5$ cm. We calculated earlier that the required incoming interaction area was $A_o = 2.2 \times 10^{-7}$ cm^2. From Eq. (23), then, we find that $\epsilon = 3.5 \times 10^{-10}$ m $-$ rad, and $\epsilon_n = 3.5 \times 10^{-5}$ m $-$ rad. By the way, we shall use meter-radians as the units of emittance, because those are the common units in the literature of colliders.

As an exercise, the student should work out the normalized emit- tance for the 1-TeV collider considered earlier, assuming the incoming $\Delta p/p_o = 0.5\%$ and choosing a sensible value for L.

To summarize, we have seen that both the emittance and the interaction-point beta function must be made as small as possible in TeV–range colliders, and since the smallness of the beta function is limited by optical aberrations, we are left with the problem of creating very small emittances.

An important design restraint arises from the form of Eq. (24). The waist is only small over a longitudinal distance that is short compared to β^*. If interactions between colliding bunches take place outside this short region, they do so at decreased lateral particle density and therefore at reduced local luminosity. For example, the cross-sectional area of a beam at a longitudinal location removed one β^* from the interaction point is two times larger than that at the interaction point. The upshot is that the bunch length is restricted by the value of β^*.

$$\sigma_z \ll \beta^* \tag{27}$$

This limitation has not proved to be troublesome in linear colliders designed to date.

4. DAMPING RINGS

A damping ring is a storage ring for electrons or positrons which is designed to condense its bunches in phase space and thus to decrease their emittances.

We have discussed normalized emittance, a quantity which is conserved during linear acceleration and drifting. The normalized emittance which reaches the interaction region of a linear collider will be just the normalized emittance that was injected into the linac, so we must inject beams with small enough values into the linacs. What determines the emittances of electron beams and positron beams? Electrons are obtained from electron guns, and such guns do not produce sufficiently low emittances for collider service when emitting the high currents required for collider service.[6] Positrons are usually collected from the electromagnetic shower produced in a heavy-metal target which has been struck by a bunch of high-energy electrons. The resulting distribution in transverse phase space is very broad and the emittances are high — much higher than those from electron guns. Consequently, both electrons and positrons must be "cooled" in damping rings.

4.1 THE DAMPING PROCESS AND DAMPING TIME

In a storage ring, the particles are continually being accelerated transversely to their directions of motion by the bending magnets which cause them to go around a closed orbit. The centripetal acceleration causes synchrotron radiation which results in a loss of energy. The lost energy is continually restored by the radio frequency (rf) accelerating system. Because of the radiation, the particles in a storage ring do not obey a conservative Hamiltonian; the emittance of the stored beam shrinks toward an equilibrium value. Figure 9 shows an idealized diagram of a damping ring in which the particles are guided in a circle of radius ρ and acted on by a radio frequency accelerating cavity. For simplicity, we shall imagine that the accelerating force provided by the cavity is uniformly distributed around the ring. The magnets that bend the orbits also afford lateral focusing forces to keep the beams confined in the damping ring's vacuum chamber, and the resulting transverse motions are quasi-sinusoidal betatron oscillations as

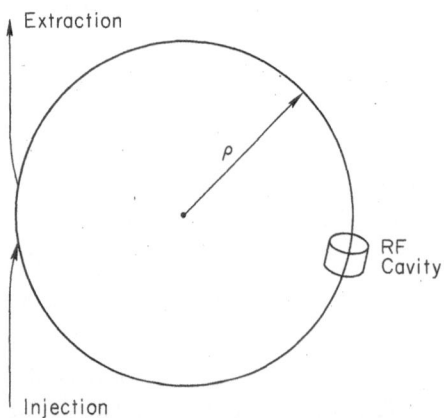

Fig. 9. Simplified diagram of a damping ring.

shown in Fig. 10. The synchrotron radiation is always emitted in the instantaneous direction of motion (within the conical angle $1/\gamma$), so the radiation reaction force is opposite to the direction of motion. But the rf force that restores the lost energy is always in the direction of the central orbit — the z-axis in Fig. 10. The vector diagram in Fig. 10 shows that a transverse force results, which is proportional to the slope of the particle's trajectory, $\dot{x}/c$. That force introduces damping terms into the two transverse equations of motion.

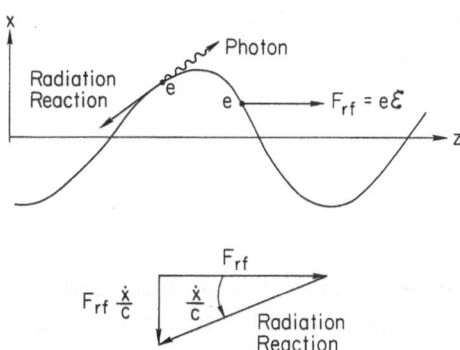

Fig. 10. A particle executing betatron oscillations in a damping ring.

We can estimate the magnitude of the damping rate. Assume that, in the absence of damping, the particles obey a simple harmonic oscillator equation in the lateral degrees of freedom.

$$\ddot{x} = -\left(\frac{c}{\beta}\right)^2 x \ , \tag{28}$$

where β is the average value of the beta function in the ring. Requiring power balance and using Eq. (8), we can figure out the damping term which must be added to Eq. (28).

$$e\mathcal{E}c = P \tag{29}$$

ensures power balance. From Fig. 10 we see

$$F_\perp = -\frac{e\mathcal{E}}{c}\dot{x} \ , \tag{30}$$

and the complete equation of motion becomes

$$\ddot{x} + \frac{e\mathcal{E}}{\gamma m_e c}\dot{x} + \left(\frac{c}{\beta}\right)^2 x = 0 \ . \tag{31}$$

Solving this equation, we can get the exponential damping rate and its reciprocal, the damping time.

$$\tau = \frac{3\rho^2}{cr_e\gamma^3} \ . \tag{32}$$

An exact treatment for a real damping ring is somewhat more complicated,[16] but this formula gives the correct magnitude of the damping times.

4.2 EQUILIBRIUM EMITTANCE

If radiation damping were the only effect at play in the damping ring, the particle motions would all die out, and the transverse emittances would shrink to zero. Of course, this does not happen. We have considered only what might be called the classical or smooth aspect of the radia- tion process and ignored its quantum nature. The radiation is emitted in quanta which cause statistical fluctuations in the motion. The particle motions are stirred up by the fluctuations and damped by the damping, and after many damping times, statistical equilibrium is reached. This equilibrium state of the particle swarm has the lowest emittances attainable in the damping ring. The derivation of the equilibrium emittance is beyond the scope of these lectures, but it may be found in Ref. 16. The operation of a damping ring proceeds as follows. A bunch with transverse emittances that are too large for collider service is injected into the ring and left there for a while to damp. The emittances damp at twice the damping rate of the oscillation amplitudes.

$$\epsilon(t) = \epsilon_{initial} e^{-2t/\tau} + \epsilon_{equilibrium} \left(1 - e^{-2t/\tau}\right) \tag{33}$$

After many damping times, the emittance becomes equal to its equilibrium value, so the equilibrium emittance must be below the desired emittance. If it is, the first term is dominant in determining how long the particles must be allowed to damp before the bunch is extracted from the damping ring. The challenge of designing damping rings for high energy colliders is to achieve rapid damping and the smallest possible equilibrium emittances. These problems are being studied by several workers.[17]

5. THE PRESERVATION OF EMITTANCE DURING ACCELERATION

After a bunch has been damped to the desired emittance in the damping ring, it is launched into the linear accelerator where it will be accelerated. Unfortunately, the linear accelerator is a hostile environment for the compact, well organized bunch and tends to disorganize it in such a way that the emittances are effectively increased and the luminosity is reduced. This process takes place through the agency of the wake field of the bunch — the electromagnetic field excited in the accelerator structure by the bunch current. topsimple terms, the wake field of the head of the bunch acts on the tail of the bunch. The wake field has components which act along the direction of motion of the bunch (longitudinal wakes) and components which cause transverse deflections of the particles (transverse wakes). Longitudinal wakes alter the accelerating field and lead to energy spread in the bunch. Transverse wakes cause particle motions which, in effect, increase the transverse emittance.

Although wake fields exist in all kinds of linear accelerators, they have been studied and dealt with extensively only in conventional disk-loaded microwave linac structures, and we shall confine our discussion to those.

5.1 WAKE FIELDS[6]

Figure 11 shows two particles moving down the bore of a linac structure. Charge q at radius r_q creates a wake field which is experienced by charge e following it and displaced from the axis by r. The wake is expressed in terms of the radii at which the charges are located and the difference ϕ between their azimuthal angles. It also depends on the longitudinal dis- tance by which e follows q. We let

$$c\tau = z_q - z_e \ , \tag{34}$$

and describe the longitudinal dependence of the wake by a function, $W(\tau)$, called the wake potential.

Fig. 11. Two particles moving down the bore of a linac structure. Charge q at radius r_q creates a wake field which is experienced by charge e following it and displaced from the axis by r.

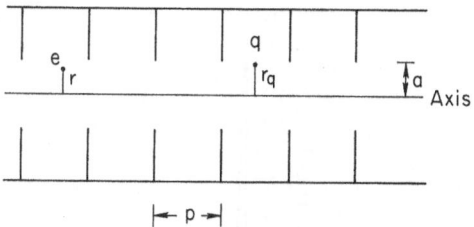

The longitudinal wake does not depend on either the radii or the azimuth. The average field on the trailing particle is given by

$$E_z p = -q W_L(\tau) \quad , \tag{35}$$

where W_L is the longitudinal wake potential of a unit charge and p is the length of a period. The wake potential multiplied by the source charge gives the voltage loss of the test charge in the length of one period of the linac structure. Since the test charge is an electron, that is just its energy loss or gain in electron-volts as it follows charge q at a distance $c\tau$ through one period. Figure 12 shows the longitudinal wake potential for the SLAC linac structure for particles following the source particle by 10 picoseconds or less, and Fig. 13 shows it for longer following times.

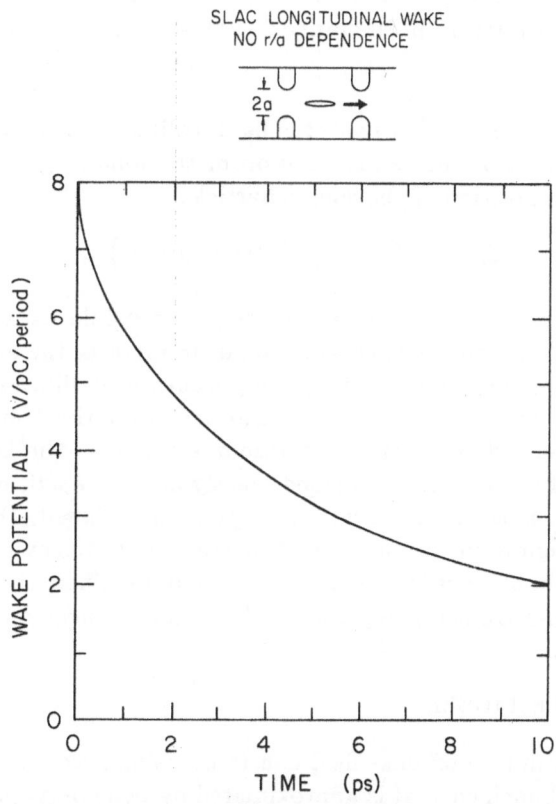

Fig. 12. Longitudinal wake potential per cell for the average cell in the SLAC disk-loaded structure in the range 0-10 ps (from Ref. 6).

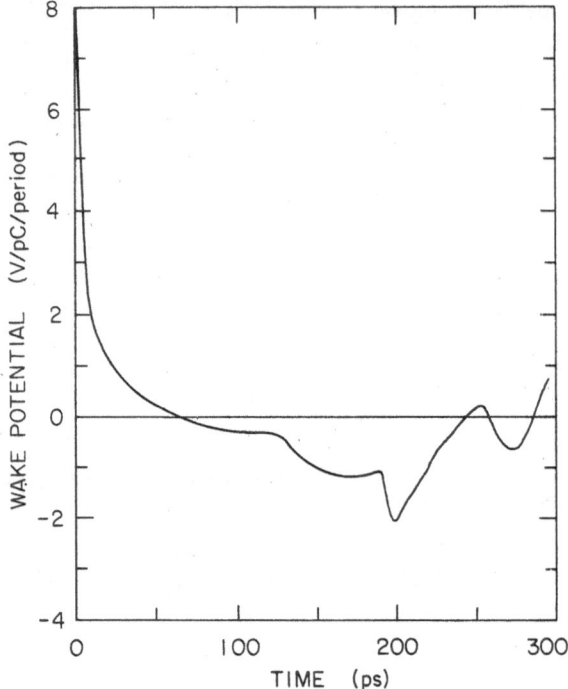

Fig. 13. Longitudinal wake potential per cell for the
SLAC structure in the range 0–300 ps (from Ref. 6).

The transverse wake field that concerns us in collider design is called the dipole
wake. It is independent of the radial position of the following charge but depends
linearly on the radial position of the source particle.

$$\vec{E}_{Tp} = qW_T(\tau)\frac{r_q}{a}\left(\hat{r}cos\phi - \hat{\phi}sin\phi\right) \ , \tag{36}$$

where $\hat{r}$ and $\hat{\phi}$ are unit vectors. Figure 14 shows the dipole wake potential for the
SLAC structure. It is the dipole wake which tends to increase the effective emittance.
It "whips the tail" of the bunch to large amplitudes of oscillation in the focusing
system of the linac as shown in Fig. 15. The fact that the transverse wake field
is proportional to the radial position of the source charge is the key to avoiding its
deleterious effects. If the bunch is launched exactly along the axis of the accelerating
structure and if the axis of the structure is perfectly aligned, the dipole wake is
suppressed. These conditions are never met in practice, but they can be approached
within tolerances. The stronger the focusing system of the linac, the less is the growth
of emittance, all other things being equal. More and stronger quadrupole magnets
along the linac help.

5.2 TWO PARTICLE MODEL

We can make estimates of wake field effects by using a very simple model of the
bunch in which the bunch current is approximated by two point charges, each having
half the charge of the bunch and the second (tail) following the first (head) by $2\sigma_z$.

Turning first to longitudinal effects, we can calculate the wake potential at
the tail due to the head using Eq. (35). Taking the total bunch population to be

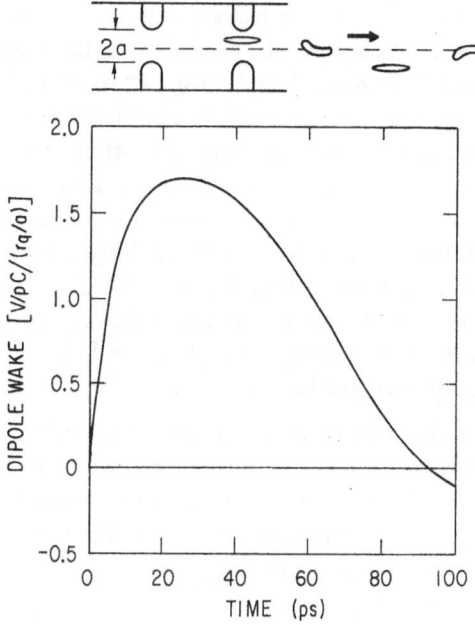

Fig. 14. Dipole wake potential per cell for the SLAC structure in the range 0–100 ps (from Ref. 6).

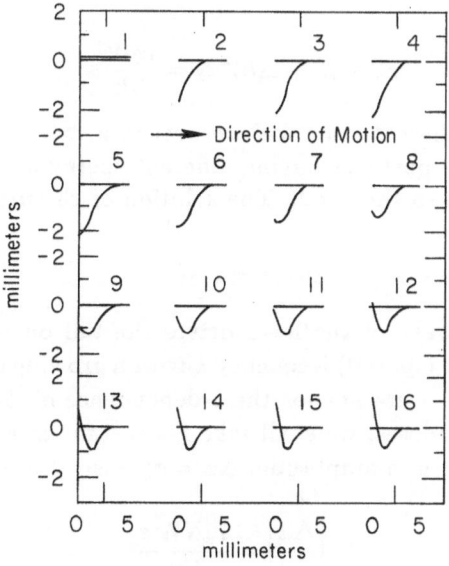

Fig. 15. The shape of a bunch of 5×10^{10} electrons injected with an initial error of 0.1 mm in transverse position. The bunch is viewed once each betatron wavelength (from Ref. 18).

5×10^{10} electrons, $\sigma_z = 1$ cm and the total number of cells of the structure to be 10^5 — numbers appropriate to the SLAC Linear Collider — we find that the total energy loss of the tail particle is about 1.1 GeV. But the head particle loses energy too — it must in order to create the wakes which leave energy behind stored in the structure. The head loses half the energy that would be lost by a test particle travelling an infinitesimal distance behind it.[6] According to Fig. 12, the head loses about 1.6 GeV, so the difference in energy loss, head-to-tail, is about 0.5 GeV. The student should verify these results. Fortunately, there is something that can be done to ameliorate this effect. The bunch can be placed off-crest on the accelerating wave so that the accelerating wave itself compensates the difference. Thus with the two particle model, the effect of the longitudinal wake field can be entirely cancelled at the expense of some total energy lost due to accelerating the bunch off-crest. This is not the case with a real bunch; some energy spread always arises due to the longitudinal wake, and we have seen in Section 3.2 that energy spread in the beam entering the final focus system works against small effective interaction area.

Next, let us turn to the effects of the transverse wake. In order to treat the transverse motion of particles moving down the linac, we must describe the transverse focusing system. In the SLC, the focusing system consists of a series of quadrupole magnets which may be adjusted in strength to provide a variety of beta functions to suit special requirements. For our present purpose we shall consider the case of a constant beta function — one that would be achieved by adjusting the quadrupoles' strengths to increase in proportion to the particle energy as it increases along the accelerator. Letting subscripts 1 and 2 refer to the head and tail respectively, the equations of transverse motion are the following.[6]

$$x_1'' + k^2 x_1 = 0 \tag{37}$$

$$x_2'' + (k + \Delta k)^2 x_2 = \frac{eqW}{2E} x_1 \tag{38}$$

where primes indicate differentiation with respect to z, $k = 1/\beta_x$, Δk is a shift in the focusing force due to the particles having different energy and W is the dipole wake potential at the tail due to the head. The solution of the first equation is simply a free oscillation.

$$x_1(z) = x_{10} e^{ikz} \tag{39}$$

If $\Delta k = 0$, the dipole wake of the head drives the tail on resonance, and the tail executes an oscillation at (spatial) frequency k with a growing amplitude. The growth would be linear in z if it were not for the z-dependence of E, the energy of the tail particle; and for our purposes, we shall take E constant at some appropriate value. In that case, the difference in amplitude, $\Delta x = x_2 - x_1$, grows as follows.

$$\left| \frac{\Delta x}{x_{10}} \right| = \frac{r_e N W z}{4k\gamma} \tag{40}$$

If x_{10} is zero, there is no growth. Growth arises because of errors in the initial conditions — the launching errors — as we saw in the sequence in Fig. 15. Even if there were no launching errors, misalignments of the linac structure itself would cause amplitude growth of this kind.

We may choose to operate the collider with a deliberately created difference in energy between head and tail so that Δk is not zero and the frequencies of the two particles are not exactly the same. Doing so reduces the rate of growth of amplitude. It is sometimes referred to as Landau damping. The SLC makes use of this strategy.

While the results we have obtained using the two bunch model reveal the features of the effects of the dipole wake quite nicely, they are not sufficiently detailed or precise to calculate the tolerances required for a given collider. They do however show the qualitative feature that decreasing the beta function (increasing the focusing strength) eases the tolerances, and that remains true when Landau damping is used.

6. THE SLAC LINEAR COLLIDER

The SLAC Linear Collider is the first linear collider system to be undertaken. It is not strictly linear, because it seeks to use a single linac — the existing two-mile machine — as both of the linacs that would be used in the sort of linear collider described in the Introduction. Figure 16 is a schematic drawing of the SLC that shows its main systems. The electron gun and booster provide short, intense bunches of 50-MeV electrons. The first sector of the linac accelerates the electrons from the booster and also positrons from the positron source to about 1.2 GeV for injection into the damping rings. After being damped in the damping rings, bunches of electrons and positrons are accelerated simultaneously by the rest of the linac. The positron source intercepts one bunch of electrons from which it produces positrons that are transported at 200 MeV back to the first sector of the linac. The other two bunches — one of electrons and one of positrons — are accelerated to the end of the linac where they are separated to the left and right and conducted around curved beam transport paths, the arcs, which aim them at one another. Then the bunches pass through the final focus system which demagnifies them to small dimensions to produce the required effective interaction areas at the interaction point where they collide.

This scheme for using the same linac for both positrons and electrons is quite feasible for the energy of the SLC, 50 GeV. At that energy both the energy loss and the emittance growth due to synchrotron radiation in the arcs are tolerable. (The energy loss is about 1 GeV.) But both the energy loss and the emittance growth rise very rapidly with energy, and the scheme is not suitable for energies much in excess of 50 GeV. Fortunately, the SLAC accelerator was readily capable of being upgraded in energy from 30 GeV to 50 GeV, and a promising experimental program, built around the Z^0, made the SLC an attractive project.

The SLC operating cycle proceeds as follows. Beginning when the bunches have been damped, one positron bunch and two electron bunches are extracted from the damping rings and launched down the linac with the positron bunch leading the procession. The bunches are about twenty meters apart (60,000 picoseconds), a large enough distance to allow the wake fields to die out between bunches and to allow a fast-kicker-magnet pulse to rise between the second and last bunches. When the three bunches reach the two-thirds point of the linac, the fast kicker magnet extracts the trailing bunch of electrons which is transported to a heavy-metal target where the electron bunch makes an electromagnetic shower. Positrons are selected out of the shower and accelerated to 200 MeV by the positron collection and booster systems and are sent back down the positron return line. When the positrons arrive at the first sector, the electron source is fired twice at the right times to establish in the first sector another procession of bunches like the one described above but in

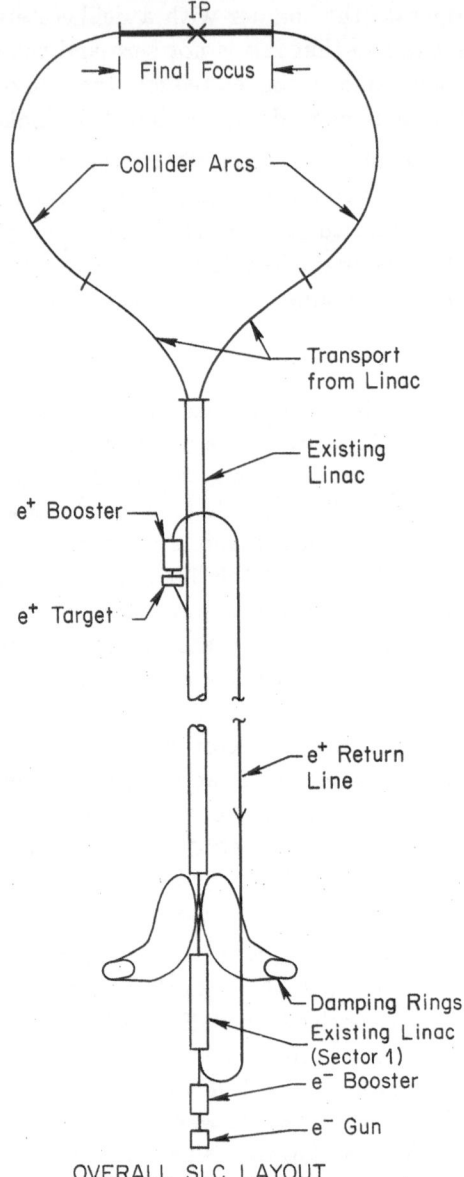

OVERALL SLC LAYOUT

Fig. 16. Schematic layout of the SLAC Linear Collider.

reverse order. These are accelerated to 1.2 GeV and injected into the damping rings, restoring the conditions of the beginning of the cycle. Meanwhile, the positron bunch and the electron bunch that were not extracted at the two-thirds point of the linac have been accelerated to the end of the linac, transported around the arcs and brought into collision at the interaction point.

The SLC damping rings are designed for service at a repetition rate of 180 Hz, which means that electrons are left in the rings to damp for about 5.6 milliseconds. The positrons, which have much larger initial emittances than the electrons, are left in their damping ring for two interpulse periods or 11 milliseconds. The rings have a

472

damping time of 3 milliseconds which is achieved by operating the bending magnets at 2 Tesla for a bending radius of 2 meters. (The observant student will note that the damping time is three times longer than that given by Eq. (32). The reason is that, for a real damping ring, the equation must be corrected by a multiplicative factor of the ratio of the circumference of the ring to the sum of the lengths of the bending magnets. For the SLC damping rings, this factor is three.) The output emittance desired from these rings is 1.3×10^{-8} m-rad which corresponds to a normalized emittance of 3×10^{-5} m-rad, and the equilibrium emittance is somewhat lower than that.

In the linear accelerator itself, a more powerful beam focusing and guidance system has been installed. In order to avoid emittance growth of the kind discussed in the preceding section, the axis of the beam must be maintained within a few tenths of a millimeter of the axis of the accelerating structure. The beam tends to wander from the axis because the accelerator is not straight, because of launching errors and because the rf accelerating fields steer it. The beam guidance system is modular. Each module comprises a quadrupole magnet, a high-precision beam position monitor (located in the bore of the quadrupole), and two steering magnets, one horizontal and one vertical. A quadrupole magnet assembly is shown in Fig. 17. There are about 300 of these distributed along the two-mile length of the accelerator.

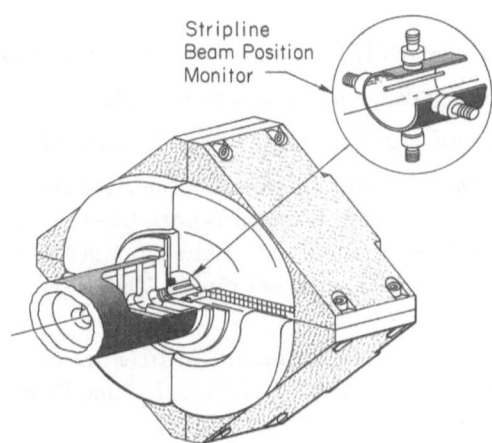

Stripline
Beam Position
Monitor

Fig. 17.   SLC linac quadrupole magnet assembly.

REFERENCES

1. M. Tigner, Nuovo Cim. **37**, p.1228 (1965).

2. B. Richter, Nucl. Inst. Methods, **136**, p. 47 (1976).

3. U. Amaldi, Phys. Lett. **B 61**, p. 313 (1976).

4. J.-E. Augustin *et al., Proc. of the Workshop on the Possibilities and Limitations of Accelerators and Detectors, Fermilab, p.87 (April 1979)*.

5. J. Rees, *"Linear Colliders—Prospects 1985,"* to be published in Particle Accelerators.

6. P. Wilson in *Laser Acceleration of Particles: Proc. of the 2nd Int. Workshop on Laser Acceleration of Particles, Los Angeles, Calif., Jan 7–18, 1985.* Eds. C. Joshi and T. Katsouleas, N.Y., American Inst. Phys.,1985, (AIP Conference Proceedings, 130). Unfortunately, the book contains several errors, and Wilson has prepared a revision of his chapter, SLAC–PUB–3674 (Rev), October 1985. It is available from the author at SLAC.

7. V. Balakin and A. Skrinsky, *Proc. Second ICFA Workshop on Possibilities and Limitations of Accelerators and Detectors, Les Diablerets, p. 31 (October 1979).*

8. R. Hollebeeck and A. Minton, Collider Note–302, SLAC internal memo, June 10, 1985.

9. R. Hollebeeck, Nucl. Inst. and Methods, **184**, p. 333 (1981).

10. T. Himel and J. Siegrist, *Laser Acceleration of Particles: Proc. of the 2nd Int/ Workshop on Laser Acceleration of Particles, Los Angeles, Calif., Jan. 7–18, 1985,. (AIP Conference Proceedings, 130).*

11. M. Bassetti and M. Gygi–Hanney, LEP–Note 221, CERN internal memo (April 1980).

12. A. Sokolov and I. Ternov, *Synchrotron Radiation*, Pergamon Press, N. Y., 1968.

13. K. Yokoya, KEK–Preprint 85–53, October 1985. (Submitted to Nucl. Instr. and Methods.)

14. R. Noble, SLAC–PUB–3871, January 1986. (Submitted to Phys. Rev. Lett.)

15. E. Courant and H. Snyder, Ann. Phys. **3**, p. 1 (1958).

16. M. Sands, *"The Physics of Electron Storage Rings"*, SLAC Report 121 (1970).

17. H. Wiedemann, *11th Int. Conf. on High Energy Accelerators (Birkhäuser Verlag, Basel, 1980)*, p. 693; L. Teng, internal report LS–17, Argonne National Laboratory (March 1985); R. Palmer, internal report AAS Note–15 (Rev.), SLAC (February 1985); U. Amaldi, internal report CERN–EP/85–102, CERN (June 1985).

18. R. Stiening in *Physics of High Energy Particle Accelerators*, Ed. M. Month, N.Y., American Inst. Phys., 1983, (AIP Conference Proceedings, 105), p. 281.

VERTEX DETECTORS – LECTURES PRESENTED AT THE ADVANCED STUDY INSTITUTE ON TECHNIQUES AND CONCEPTS OF HIGH ENERGY PHYSICS, ST CROIX, JUNE 1986

C. J. S. Damerell

Rutherford Appleton Laboratory
Chilton, Didcot
Oxon, England

ABSTRACT

Since the discovery of the J/ψ in November 1974, there has been a strong interest in the physics of particles containing higher-flavour quarks; charm, bottom ...). High precision vertex detectors can be used to identify the decay products of parent particles which have lifetimes of the order 10^{-13} seconds. This paper summarises the performance achieved by a variety of techniques, and proceeds to a detailed discussion of the current status and potential of silicon detectors for high precision tracking.

CONTENTS

1. Physics Motivation
2. Implications for Detectors
3. Non-Silicon Detectors: Performance and Applications
4. Silicon Detectors: General Characteristics
5. Microstrip Detectors and Silicon Drift Chambers
6. Charge Coupled Devices
7. Applications of Silicon Detectors
8. Radiation Damage
9. Future Prospects

1. PHYSICS MOTIVATION

Physicists long ago concluded that the variety of observable particles is so great that one could reasonably expect nature to have provided a simplifying sub-structure. All theoretical progress regarding this sub-structure, of quarks and gluons and beyond, is based on clues provided by the composite particles which we can observe in experiments. One property of the observable particles is their lifetimes, and high precision vertex detectors enable the lifetime range to be pushed below the previously measurable limits. Such measurements may at first sight seem to be unrelated to the interesting problem of hadron structure, but in fact they can be of great importance in precisely this area, as we shall see. But let us start by taking a global look at the physically observable particles and their lifetimes.

The term stable particles is usually taken to include those having lifetimes in excess of about 10^{-8} s. If produced in high energy collisions, such particles have decay lengths of the order 1 metre or more, which means that the electrically charged ones ($\mu^{\pm}$, $\pi^{\pm}$, $K^{\pm}$, $\overset{(-)}{p}$, ...) can be tracked in conventional detectors and often identified by means of Cerenkov counters. The neutral ones (γ, K_L^o, n, ...) are in general observable by calorimetry; however, neutrinos can usually be inferred only by means of missing energy.

We shall use the term long-lived particles to describe those having lifetimes of the order of 10^{-10} s (K_s^o, Λ^o, $\Sigma^{\pm}$, ...) which (depending on the experiment) may be visible directly, or at least will be recognised by having decay products whose tracks clearly do not point back to the production vertex. Such decay products have a projected distance of closest approach to the primary vertex (impact parameter) of typically 1 cm.

Until about 10 years ago, all known particles had lifetimes in one or other of these categories, or else were subject to what we shall call prompt decays, ie lifetimes too short to allow for direct experimental observation by any known technique. In this class we have the π^0 ($\tau \sim 10^{-16}$ s) and all the resonances (η^o, Σ^o, ω, ρ, Δ, ...) with lifetimes 10^{-18} to 10^{-23} s or (more relevant to experiments) mass widths of 1 keV to 100 MeV. Such particles are observable only via their decay products as peaks in effective mass distributions, or in formation experiments via the energy dependence of a measured cross-section, such as $\pi^+p \rightarrow \Delta^{++}$, or $e^+e^- \rightarrow J/\psi$. The observation of resonances in high multiplicity inelastic processes is notoriously difficult due to the problem of combinatorial background; one may for example have so many possible $\pi^+\pi^-$ combinations that the recognition of which pairs (if any) result from the decay $\rho^o \rightarrow \pi^+\pi^-$ becomes impossible. Quite apart from the non-observation of the quark substructure, one may often be unable to disentangle these first generation hadronic states, and be left only with the measured stable particles which are one stage further removed from the fundamental physical processes of interest. This is a limitation of all current high energy experiments, and could be avoided only by operating above the threshold for producing free quarks, if there is one!

During the past decade, a sequence of particles (the τ lepton, and hadrons such as D, F, Λ_c, B, ...) have been discovered which have lifetimes in the region 10^{-13} to 10^{-12} s. Such particles were first observed as effective mass peaks in favourable situations (low combinatorial background) but can be seen much more extensively in experiments where special high precision vertex detectors are used to recognise the finite lifetimes of the parent particles, which we shall hereafter refer to as short-lived. By recognising which of the charged particle tracks emerge from the decay vertex, the parent particle can be reconstructed without the combinatorial background which otherwise could completely obscure the signal. The measurement of the lifetime is a by-product, possibly a very important one.

As we shall see when we start to look quantitatively into the problem, the task of cleanly recognising multiple vertices is technically quite challenging. Unlike general particle tracking, where gaseous drift chambers can be adapted to do the job in virtually all cases, there is not yet a single vertex detecting technique which can be applied to all experimental situations. Details (such as how close one can get to the interaction point) impose very stringent constraints. Some of the detectors we shall discuss are suitable for large area coverage around large diameter beam pipes, but inevitably have no hope of associating tracks to vertices on individual events. Fortunately, nature provides a continuum of interesting physics questions, some of which can be answered with relatively imprecise detectors. For example, a large clean sample of τ leptons allows a precise measurement of the τ lifetime even when the vertex reconstruction is of such limited quality that the apparent lifetime of (almost) half the events is negative (ie the decay seems to emerge from behind the production vertex). We shall briefly review the achievements of such detectors, but spend the majority of the time investigating the achievements and potential of silicon detectors because they are rapidly expanding the range of experimental situations for which a clean topological reconstruction of multi-vertex events can be made.

Of course, even these detectors are of no help in sorting out the large class of promptly decaying particles like ρ mesons. Given that the detectors are quite complex, one might reasonably ask why we go to such trouble to extend our lifetime sensitivity down to cover the region of $\gtrsim 10^{-13}$ s. The reason is simply that the short-lived charm and bottom particles achieve their 10 orders of magnitude lifetime extensions by being ground states of matter containing quarks of higher flavours. As such they are particularly interesting. Not only that, but the predominance of <u>sequential decays,</u>

$$c \to s$$
$$b \to c$$
$$t \to b$$
$$x \to t \text{ (where x is the low-lying member of a possible 4th generation)}$$
$$\cdot$$
$$\cdot$$
$$\cdot$$

ensures that a c or b tag will also enrich the signals for t, x, ... where lifetimes may well be too short for direct measurement.

We shall use the nomenclature c, b, t, ... to signify the heavy quarks and C, B, T, ... to signify hadrons containing these quarks.

Higher energies (SLC, LEP, CERN and Fermilab $\bar{p}p$ colliders) result in generally less clean events, events of high track multiplicity and great complexity, many containing multiple vertices. Already, some ambiguous signals such as the $t \to b$ candidates in UA1 could be transformed into definitive experimental results with the aid of vertex detectors able to see the decays of short-lived particles. Vertex detectors are essential tools for untangling many physics problems in existing experimental conditions, and their value will obviously grow as we move to higher energies in the future.

As a specific physics area where vertex detectors may have an important role, we consider the case of $e^+e^- \to Z^\circ$ in SLC or LEP. The Z° will decay via all kinematically allowed $q\bar{q}$ final states including the higher flavours. The Z° decay involves a large release of energy, the charged particle multiplicities will be high, and decays such as $D \to K\pi$ will be swamped by combinatorial background unless the K and π tracks can be recognised as not coming from the primary vertex. Let us now look at

some areas of physics which can be studied provided we are able to
distinguish the heavy flavour decays.

1.1 Neutral Current Weak Coupling of Quarks

The aim is to measure separately the couplings for the processes
$Z° \rightarrow q\bar{q}$ where q = u, d, s, c, t, b, ... over a good angular range (say
$\theta(q) \gtrsim 20°$, where θ is the polar angle of the produced quark) including the
distinction between quark and anti-quark. Being unable to observe the
decays at the quark level, we cannot really hope to distinguish the
processes involving the light quarks

$$Z° \rightarrow u\bar{u}$$
$$d\bar{d}$$

or $s\bar{s}$ due to the ease with which such $q\bar{q}$ pairs are generated out
of the vacuum. The situation is much more promising for tagging the Z°
decays to massive quark pairs (which are very unlikely to be generated from
the vacuum) viz:

$$Z° \rightarrow c\bar{c}$$
$$b\bar{b}$$
$$t\bar{t}$$
$$\cdot$$
$$\cdot$$

Indeed, we have some techniques which can be extrapolated from our
experience at lower energy. Muons may be used as a signature for D or B
decay, the process $D^* \rightarrow D\pi$ may be used to enrich the $c\bar{c}$ sample, and
kinematic tests based on the mass of the B or T states may be useful. But
the increasing complexity of events with energy, and the presence of
sequential decays, make these procedures at best problematical. What is
needed for a clean signature is (for example) kaon identification in
conjunction with the vertex topology. As shown in Figure 1, the emission
of a positively charged kaon from the final vertex can be used to cleanly
distinguish the $\bar{c}$ or $\bar{b}$ jet and also (in conjunction with kinematic tests on
particles from the primary vertex) the $\bar{t}$ jet. A good vertex detector at
SLC can give flavour tagging efficiencies of 25-50% for all of these quark
states, in contrast with approximately 1% for the branching ratio times
efficiency for c tagging with a D^* tag.

Once one has a good sample of $q\bar{q}$ events, one can deduce the weak
couplings. The situation at SLC is particularly favourable in view of the
possibility of having a longitudinally polarised e^- beam.

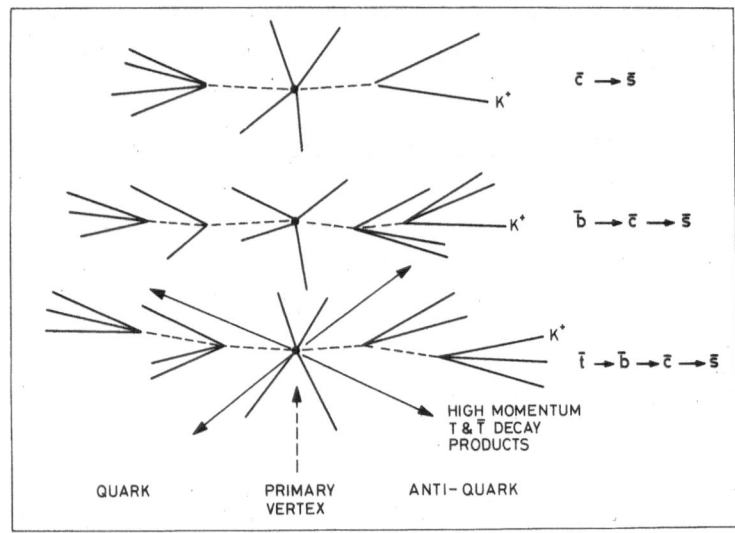

Fig. 1. In events where the vertex topology is observed and a charged kaon emerges from the final decay vertex, the primordial quark content of the jet may be inferred.

1.2 Flavour-Mixing Matrix

The measurement of "the B lifetime" (in reality there may be different lifetimes for the different charge states) and the tightened bounds on $\Gamma(b \rightarrow u)/\Gamma(b \rightarrow c)$ have considerably improved the knowledge of the Kobayashi-Maskawa mixing matrix. Theoretical papers abound, and generally explain such features as the increasing mass separation and decoupling between the higher flavours. In some theories (see for example Stech[1]), the mass of the top quark and the amplitude ratio $b \rightarrow u/b \rightarrow c$ are predicted. A detailed measurement of lifetimes and decay modes of the many undiscovered physical particles (mesons and baryons) containing charm, bottom and top quarks (decay modes only in the case of top) will be of great importance. The signals from these particles in general depend on vertex detection in order that they should be pulled out of the background, and of course the vertex detector is needed for any lifetime measurements.

1.3 Particle-Antiparticle Mixing

The measurement of the off-diagonal elements of the K° mass matrix gave one of the first clues for the existence of a fourth (charmed) quark. In the same way, measurements of $D^\circ \bar{D}^\circ$ mixing, $B_d^\circ \bar{B}_d^\circ$ mixing or $B_s^\circ \bar{B}_s^\circ$ mixing (the first expected to be small and the second to be large)[2] could disprove the 6-quark model even before the direct observation of a higher flavour. The distinction between $B°$ and $\bar{B}°$ or between $D°$ and $\bar{D}°$ is given unambiguously by the identification of the strange particle which emerges in Cabibbo favoured decays from the charm vertex, except in the case where this is a neutral kaon. In all cases, very useful numbers of events should be accessible at SLC, particularly since the vertex system should result in their being observed on negligible background. The observation of a several percent mixing in $D° \bar{D}°$ or $B°_d \bar{B}°_d$ would invalidate current models. Just a few events (on zero background) would be sufficient.

Another very powerful approach to the search for anomalous particle-antiparticle or flavour mixing is based on the study of semi-leptonic decays. Figure 2 from reference 3 indicates the richness of information available, and demonstrates that the simple rules applicable in hadronic collisions where only $c\bar{c}$ production is significant no longer apply. The presence of like sign dileptons (for example) signifies nothing unless one knows from which vertex each lepton emerged. One should further note that decays including τ leptons are particularly characteristic topologically, and will stand out very clearly.

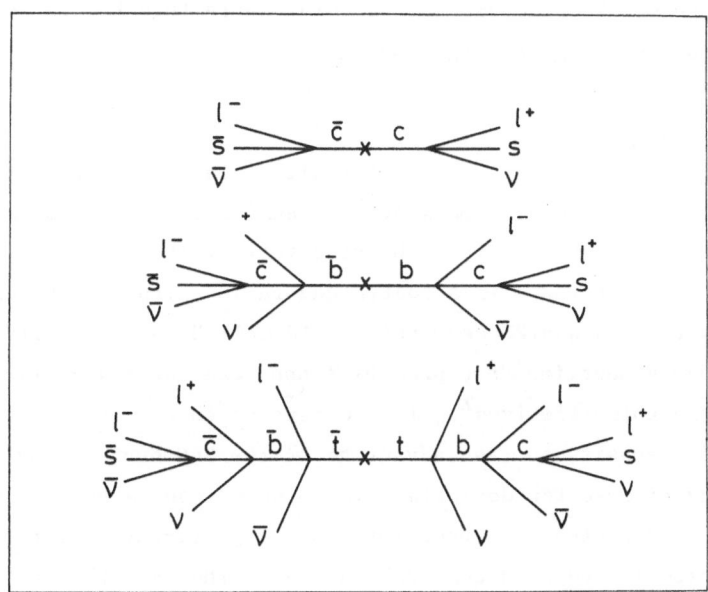

Fig. 2. Semileptonic decay chains. Any lepton pair can be replaced by $q\bar{q}$.

1.4 Higgs Boson Production

This is an interesting example where a vertex detector would be used in "veto mode", to exclude events where the particles of interest emerge from a background of short-lived particle decays.

A promising production mode in searching for Higgs bosons is

$$Z° \rightarrow H°\ell^+\ell^-$$

which would be seen as 2 jets (the Higgs decaying to $b\bar{b}$ or higher flavours if kinematically allowed). Given the tiny sample of H° events expected (less than 100) most estimates suggest major backgrounds from leptons from semileptonic decays of B and C. Using the vertex detector to establish the prompt origin of each lepton is a very powerful element in background rejection. The vertex detector will in addition provide evidence for the expected secondary vertices in the H° decay, where the precise vertex structure of course depends on the H° mass.

The production process $Z° \rightarrow H° \nu\bar{\nu}$ has a factor 3 rate advantage over the first process, but is generally considered to have very large background. Estimates for SLD[4] taking advantage of the almost complete coverage with electromagnetic and hadronic calorimetry, and using the vertex detector to single out $b\bar{b}$ final states, suggest that it may be possible to reduce backgrounds to the point where this becomes a viable process for discovering the Higgs boson.

1.5 Other New States

Particle lifetimes are very difficult to calculate theoretically. We have learned that increasing mass by no means implies decreasing lifetimes (as was commonly believed before the discovery of the J/ψ). It has been pointed out[5] that if there is a fourth generation with its charge $-\frac{1}{3}$ member below the top quark mass, this could well have a long lifetime. Long-lived heavy neutrinos are possible[6] and some supersymmetric particles may have measurable lifetimes[7]. Independent of theory, it is obviously an experimental necessity to take advantage of the SLC environment and set up the most sensitive vertex detection system which can be built, in order to look for the unexpected (and therefore most significant) short-lived particles which may be produced by Z° decays. The fact that all heavy flavours and leptons so far discovered have observable lifetimes in a favourable experimental environment lends confidence that there may be more to be found.

In summary, the first centimetre away from the Z° production point is a rich source of information which will have a vital bearing on many areas of physics, some of which may be completely unexpected on current theory. A vertex detector capable of defining the event topology would provide orthogonal information to that given by all other detectors in the spectrometer. Apart from measuring lifetimes, such information can provide orders of magnitude suppression of combinatorial background which would otherwise obliterate signals from heavy quark states.

In the case of e^+e^- production of the Z°, the experimental situation is particularly favourable due to the democratic coupling of the Z° to many $q\bar{q}$ states. Another promising area for vertex detectors, possibly leading to important new physics could be fixed target photoproduction experiments, especially with the higher energy beams at Fermilab. Current and future hadron colliders may also profit greatly from vertex detectors, provided the beam pipes can be made small enough, and sufficiently clean conditions maintained.

A by-product of high precision is (in general) a very high multi-track capability. Even when the former characteristic is less important, the latter may be an end in itself. For example, the very large track densities expected from high energy heavy ion collisions will require detectors of extremely fine granularity in order even to count the tracks. Pixel-based silicon detectors may (as we shall see) be well suited to this requirement.

2.   IMPLICATIONS FOR DETECTORS

2.1   General Remarks

We would like to specify the required precision, granularity, maximum amount of material, etc, for the ideal vertex detector. Unfortunately there are no simple answers to these questions, for two main reasons.

Firstly, one should specify the experimental application. Thus in a typical high energy fixed target experiment, multiple scattering is generally no problem. Due to the lower momenta of final state particles in a collider such as SLC, one has to be much more careful about the amount of material in the vertex detectors. In LEP (same momenta as SLC, but 10 times larger beam pipe) the effects of multiple scattering are really large, imposing serious limits on the precision of vertex reconstruction.

The second important question is the one of aims. Most _modestly_, one could design a vertex detector which allowed only a short-lifetime tag by measuring the impact parameter of the track for which this happens to be maximal. Such a system can in fact do useful physics. It provides an enriched sample of charm and bottom events, which may be refined by other cuts. Another modest aim is to concentrate attention on processes which can be selected with reasonable cleanliness by other criteria, and to use the vertex detector to measure the lifetime from a large event sample. At modest collider energies, where rich charm and beauty samples can be selected, vertex detectors have made excellent lifetime measurements. But at high energies, events will tend to have multiple vertices, and we really need to do better. Most _ambitiously_, we may design a vertex detector which can efficiently assign the tracks to their respective vertices (primary vertex, B decay vertex, C decay vertex, ...). The efficiency for doing this can never be 100% due to decay tracks which happen to be colinear with the parent track, and very short-lived decay products; remember that the most probable lifetime is always zero. But we shall take this most ambitious aim as our guide to the quality of a given vertex detector, since it corresponds in general to the total physics information which is potentially available from such a detector. Setting a more modest aim can in fact be quite misleading. Thus one may demonstrate that some detector has an 80% efficiency for heavy flavour tagging. This gives the impression that a more precise detector could only pick up an additional 20%, whereas in fact it is also able to sweep aside the ambiguities asociated with all the tracks _not_ used for the flavour tagging. A fairer representation of the two situations is provided by quoting for each detector the percentage of tracks which can be uniquely assigned to their correct parent vertices, for various classes of event ($e^+e^- \rightarrow c\bar{c}$, $b\bar{b}$, etc). Such criteria will be used in the discussions which follow.

In order to relate the discussion to a relevant physics area, we continue with the example of $e^+e^- \rightarrow Z^\circ$ at the SLC. In general, the conclusions regarding impact parameter precision carry over to any processes where cascades of heavy flavour decays are encountered. For example, the impact parameter is invariant between centre-of-mass and fixed target experiments. What mainly varies are the lab momenta of the final state particles and background conditions which influence the practical detector geometries. No very general statements can be made and (fortunately) experimentalists will push in all possible directions. (At the time when the ISR and SPS were being planned, there were endless discussions as to their relative merits, almost all of which were proven

irrelevant in practice. For example, the J/ψ should have been found in the
ISR much more easily than in fixed target hadron collisions at low CM
energy. History proved the opposite.) It should also be remembered that
the really exciting (ie unexpected) physics with vertex detectors may well
come from an environment where they are not technically most comfortable.
The following discussion relating to SLC is not at all intended to imply
that this will necessarily produce the best physics with vertex detectors.

2.2 Implications for SLC

We now turn to the technical requirements for a vertex detector which
could be used in the interesting area of Z° decay physics at the SLC.

The detector requirements are evaluated on the basis of generated
events (using the Lund Monte Carlo) of the type

$e^+e^- \to Z^\circ \to q\bar{q}$ (ie 2-jet events)

For $c\bar{c}$ events we put in the experimental lifetimes for $D^\pm$, D°, Λ_c and
F.

For $b\bar{b}$ events, we take $\tau_B = 9 \times 10^{-13}$ s for all B states (since we
know no better)

For $t\bar{t}$ we assume $M_T = 30$ GeV and prompt T decay to B states.

For decays including τ leptons, we put in the experimental τ
lifetime.

Decay modes and branching ratios are left at the default values which
emerge from the Lund Model (JETSET 5.21 of February 1984 in the CERN
Program Library). For charm decays, these branching ratios typically agree
within a factor 2 with measured values, for those decay modes for which
measurements exist.

In deciding what detector characteristics are needed, we shall use
the criterion of topological efficiency already mentioned. To get a
feeling for the problem, let us begin by looking at some individual events
from the generation program. These events are generally displayed in a
"beam's eye view" with the primordial q jet directed vertically upwards in
azimuth, and the $\bar{q}$ jet downwards, for clarity. The line lengths represent
the track momenta, with the full-scale momentum (primary vertex to edge of
figure) being approximately 5 GeV/c. .Only the long-lived tracks are shown
as lines; the bottom and charm particles (whether charged or neutral) are

not shown explicitly since their tracks are not directly observable in the detectors. Figure 3(a) shows what is in fact a quite typical hadronization according to the Lund model. The tracks emerging from the primary vertex include several soft ones; most of the energy is carried off by the charm quarks and appears in the decay products of the D and $\bar{D}$. The charm particles travel on the order of a few mm before decay. Figure 3(b) shows a b$\bar{b}$ event in which the B happens to decay very close to the primary vertex, a common occurrence. Figure 3(c) shows a t$\bar{t}$ event; the track multiplicities are higher, the track angles are greater, and the track momenta are reduced. In this XY view the event topology is unclear. This again is frequently observed. Figure 3(d) shows the same event rotated by 90°. The topology is now clear, illustrating the desirability of building a vertex detector capable of providing more than one viewing direction for the event. For these topological reconstructions it is helpful if one can rotate the event in space in order to make a clean assignment of tracks to vertices.

After getting a general idea of the events we shall be trying to reconstruct, let us now look at some statistical information based on large numbers of generated events. Figure 4 shows the distribution of impact parameters (distances of closest approach) between the decay tracks for c$\bar{c}$

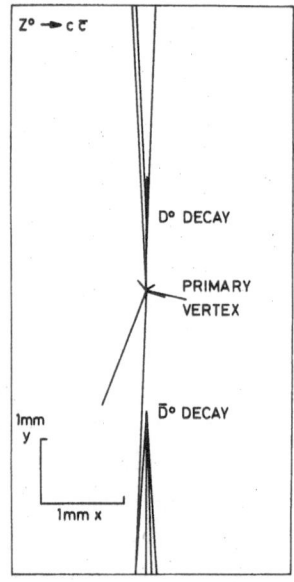

Fig. 3(a). Z° → c$\bar{c}$. Hadronization from the Lund Monte Carlo. Beam's eye view of the event.

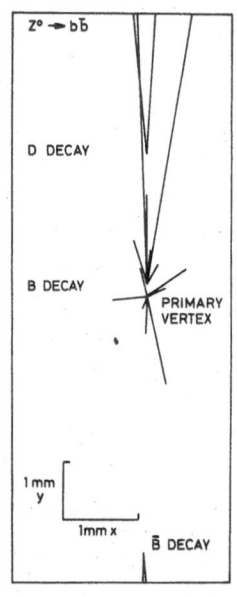

Fig. 3(b). Z° → b$\bar{b}$.

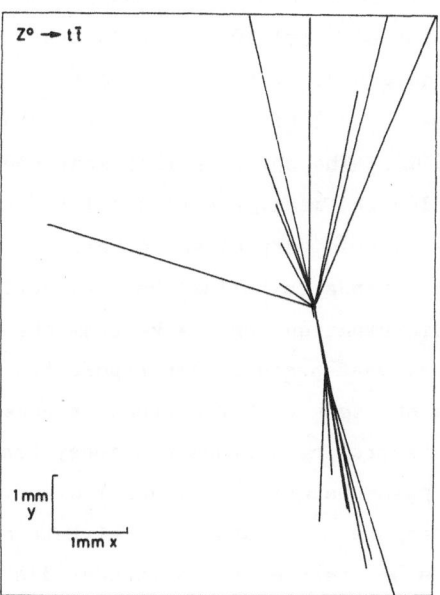

Fig. 3(c). $Z° → t\bar{t}$.   The event topology is rather confused in this XY view.

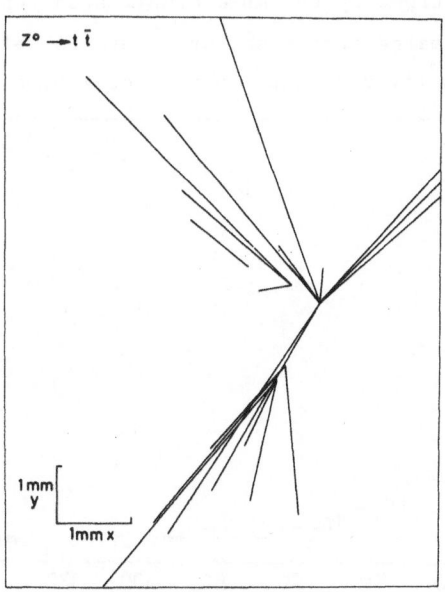

Fig. 3(d). $Z° → t\bar{t}$ YZ projection (the Z axis is the beam direction). The event topology is clearly seen in this view.

events and the primary vertex, as seen in the XY projection. The mean
value is 44 μm, which is the sort of number which has induced in some areas
the idea that a precision of (say) 10 μm represents "overkill". If one is
aiming for efficient assignment of tracks to vertices, however, this is a
very misleading statement. 26% of the tracks have impact parameters below
10 μm, and 16% below 5 μm. The point is that many charm particles decay
with lifetimes well below the mean, and kinematics can lead to situations
in which tracks really do pass very close to vertices other than their own
(eg D* → Dπ, where the π tends to follow the D direction). If one looks at
the impact parameter distribution for tracks from the primary vertex in c$\bar{\text{c}}$
events, then this is somewhat broader than Figure 4. This becomes obvious
from Figure 3(a) where one sees that the prompt tracks are of low momentum
and appear reasonably isotropic, whereas the decay tracks from the high
momentum charm decays point back in the general direction of the primary
vertex. Figure 4 relates to c$\bar{\text{c}}$ events, ie a mixture of Ds, Fs, Λ_cs, etc.
If we look at b$\bar{\text{b}}$ events we might expect a broader distribution due to the
longer lifetime. But if we maintain our aim of assigning tracks to their
correct vertices, we have to consider the impact parameters of the B decay
tracks to the primary vertex and also to the same side charm decay vertex.
If we plot the impact parameters of decay tracks with respect to <u>all</u>
potentially confusing vertices, then we find an impact parameter
distribution which is closely similar to Figure 4. The advantage of the
longer B lifetime is offset by the more complicated topology (see
Figure 3(b)). This remains true also for t$\bar{\text{t}}$ events which are similar to
the b$\bar{\text{b}}$ ones apart from lower momenta in the decay products and higher

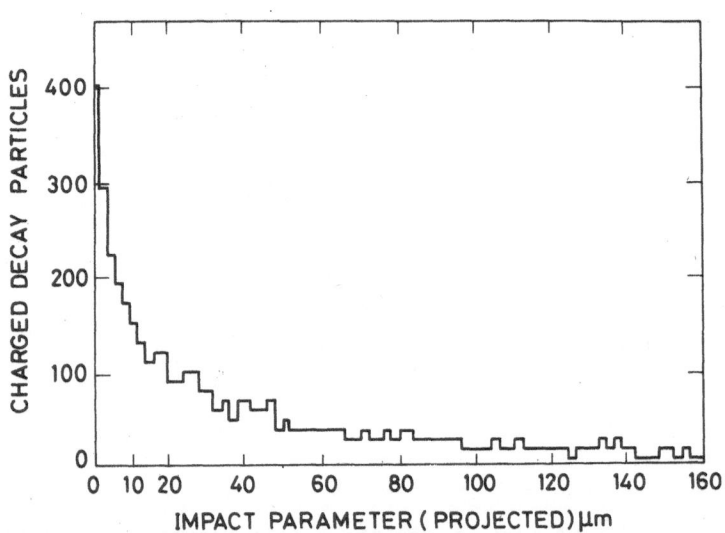

Fig. 4. Impact parameter distributions (in XY projection) of decay
products in Z° → c$\bar{\text{c}}$ events, with respect to the primary
vertex.

momenta in the primary tracks. As is well known, the impact parameter (other things being equal) is momentum independent; higher momentum gives longer decay lengths but smaller decay angles.

Turning now to the momentum distributions of the decay tracks, these are shown in Figures 5(a) to (c). We have the following general trend:

	Mean Decay Track Momentum (GeV/c)
$Z^\circ \rightarrow c\bar{c}$	8.3
$Z^\circ \rightarrow b\bar{b}$	3.3
$Z^\circ \rightarrow t\bar{t}$	1.6

The bottom and charm decay products in the $t\bar{t}$ events will be of even lower momentum if the top mass is greater than 30 GeV, as suggested by UA1. Note also from Figure 5 that the distributions peak well below the mean momentum. Multiple scattering is a very serious consideration for these events, especially for $b\bar{b}$ and $t\bar{t}$ production.

2.3 Conclusions

We may summarise the following implications for vertex detectors which aim to have good topological efficiency for reconstructing Z° decay to heavy flavours:

- Precision of measured impact parameter $\lesssim$ 5 μm
- If possible, 2-view reconstruction so that kinks in space are always visible.
- Minimal material, and the first measuring plane close to the primary vertex, so that the measurement precision is useful over as wide a momentum range as possible. (This may place tight requirements on the 2-track resolution.)

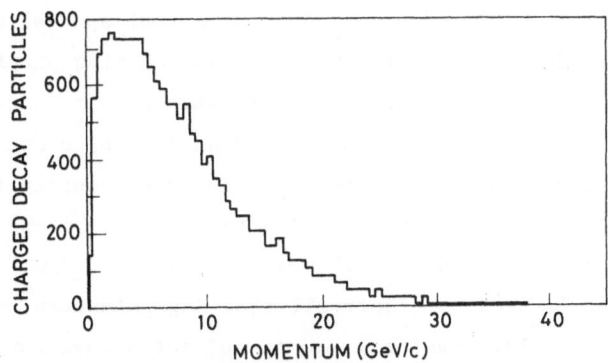

Fig. 5(a). Momenta of charged decay particles from events of the type $Z^\circ \rightarrow c\bar{c}$.

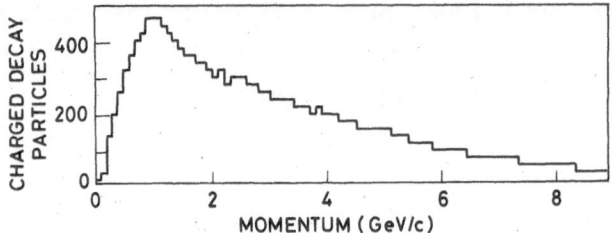

Fig. 5(b). Momenta of charged decay particles from events of the type $Z^\circ \rightarrow b\bar{b}$. (Note the change of momentum scale.)

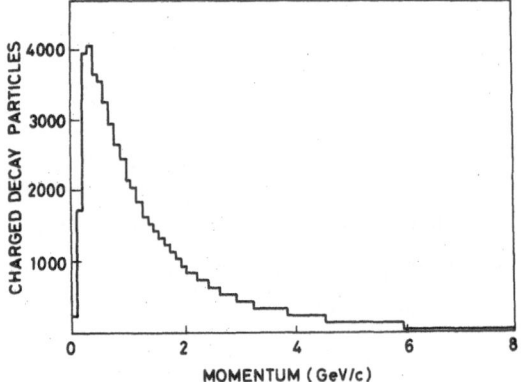

Fig. 5(c). Momenta of charged decay particles from events of the type $Z^\circ \rightarrow t\bar{t}$.

If one considers vertex detectors for high energy fixed target experiments, the first two conditions continue to apply. But the momenta are normally much higher, and multiple scattering is much less serious.

A quantitative evaluation of how well we might be able to do in practice will be given in Section 7, once we have an idea of the range of detectors which are available, and their various capabilities.

3.   NON-SILICON DETECTORS: PERFORMANCE AND APPLICATIONS

The very first tracking detectors for ionizing particles were photographic plates, followed by the development of special nuclear emulsions, having good sensitivity for minimum-ionizing particles. However, the superb spatial precision achievable with this technique was not often required, and its attendant disadvantages by comparison with "electronic detectors" (ie detectors which could be coupled to electronic readout systems) and bubble chambers, led to its decline and near extinction. By 1974, the year of the discovery of the J/ψ, drift chambers were almost uniquely used as electronic tracking detectors, both in fixed target experiments (flat chambers) and in colliders (cylindrical chamber systems). The precision typically achieved (~ 200 μm) was considered adequate for all forseeable applications. The main aim was momentum measurement, and this could always be improved (if necessary) by larger field volumes and longer lever arms.

The November Revolution in 1974 changed all that. The prospect of charm particles with lifetimes $\sim 10^{-13}$ s spurred experimentalists to try to see these particles in the laboratory. There are even hints that charm had already been seen by nuclear emulsions exposed to cosmic rays[8], but the events were not sufficiently fully reconstructed. With hindsight, the observed decays ($\tau \sim 10^{-13}$ s) were probably charm, but these vertex detectors on their own were inadequate; in addition there was the common problem with cosmic ray data of very few events. The essential point is that vertex detectors gain their full power only by being embedded in spectrometers able to measure momenta and identify hadrons.

3.1   Hybrid Emulsion Spectrometers

After the J/ψ discovery, hybrid emulsion experiments (ie an emulsion target followed by a multiparticle spectrometer) began to run at Fermilab and at the CERN SPS. Soon they were achieving their goal of seeing charm decays, with unambiguous reconstruction of specific final states. For example, Figure 6 shows a beautiful event of the type

$$\gamma p \to \bar{D}^\circ \; \Lambda_c^+ \; X$$
$$\hookrightarrow \pi^+ \Lambda$$
$$\hookrightarrow p\pi^-$$
$$\hookrightarrow K^+\pi^-\pi^-\pi^+$$

induced in emulsion by a photon of energy 25 GeV. The event, from an experiment[9] at the SPS is fully reconstructed with the aid of the Omega

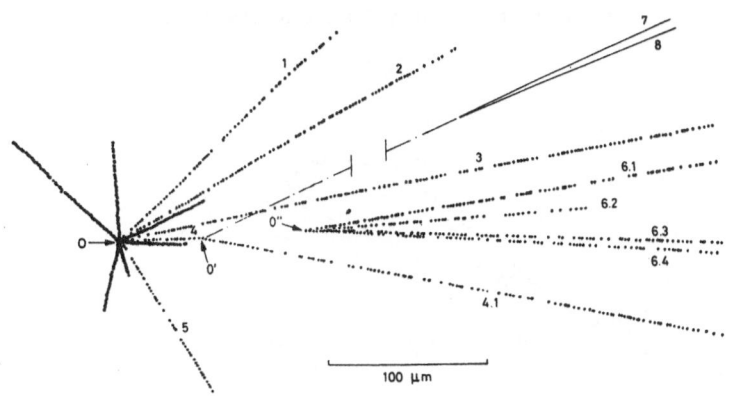

Fig. 6. Associated charm production seen in nuclear emulsion.

spectrometer, where all the small angle tracks from the emulsion, and the
Λ decay products, are seen.

While physicists working with other techniques are always predicting
the demise of nuclear emulsions, these detectors continue to improve.
Certainly they suffer from a very limited incident flux (and are much
easier to use with incident photons or neutrinos than with charged
hadrons). In addition, the linking of tracks into the emulsion, and
scanning the film, has been extremely slow. But great strides have been
made, notably by the Nagoya group, and using fully automatic measuring
techniques the analysis time per event is reduced by a factor 10. What
this means[10] is an analysis time of 27 rather than 280 minutes. From this
we see that nuclear emulsions, in spite of their beautiful precision
(~ 1 μm impact paramters) and clean visualization of events, are likely to
be limited to small charm samples. This statement could be proved
incorrect if some preselection of charm events were possible. One idea is
to back the emulsion up with some electronic vertex detector (CCDs have
been proposed by a group from Florence[11]) which will select charm
candidates on the basis of measured impact parameters, to be looked at in
detail by the emulsion.

Recently[12], nuclear emulsions have had a further success; the first
reconstruction of a $B\bar{B}$ event with measured lifetimes. While all the
charged particles are seen in the emulsion (Figure 7) the following
detector (beam dump) allowed only the momentum analysis of the muons. This
obviously implies some uncertainties on the B momenta, but the conclusion
is that the B^- has a flight time of $(0.8 \pm 0.1) \times 10^{-13}$ s and the $\bar{B}^0$ has a
flight time of $(5^{+2}_{-1}) \times 10^{-13}$ s. These very interesting results show
how important it is to have high statistics experiments in which lifetimes
of specific B charge states can be measured. Nuclear emulsions are likely
to continue to have an important rôle in such fixed target experiments.

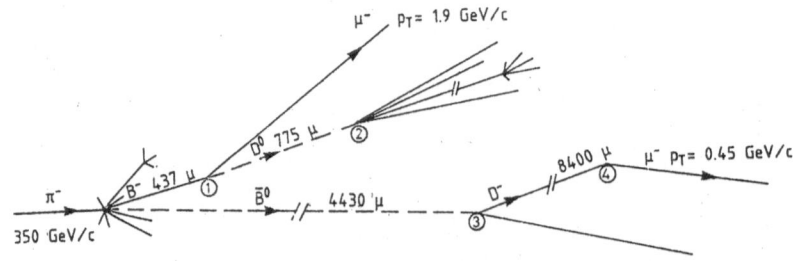

Fig. 7. Sketch of a BB event found in nuclear emulsions.

3.2   High Precision Bubble Chambers

While bubble chambers had early success in finding charm particles
produced in neutrino interactions, this was purely on the basis of an
effective mass reconstruction. The precision of the large bubble chambers
was far too coarse to allow the decays to be seen directly. In spite of
some attempts to improve on this with special holographic chamber optics,
it is generally true of large bubble chambers even today. Nevertheless, a
new breed of rapid cycling small bubble chambers has evolved which do
indeed have the necessary precision. Such chambers with a triggerable
light flash have the further advantage in principle that only events of
interest need be recorded. To date, this feature has been used only to
enhance the selection of interactions in the bubble chamber liquid. The
LEBC chamber in CERN has yielded a very high quality sample of charm
events. With 20 µm diameter bubbles, the precision is excellent. Unlike
the situation with an electronic detector, however, it is out of the
question to digitise every event, finding the charm decays by a
computerised search for tracks whose impact parameters with respect to the
primary vertex differ perceptibly from zero. On the contrary, the charm
candidates have to be found by scanning. This means in practice that at
least one track, seen in the projection of the camera, must have an impact
parameter in excess of 50 µm. Thus, although the tracking precision is
comparable to that of the nuclear emulsion, the ability to recognise
short-lived decays (below about 2×10^{-13} s) is very limited. The LEBC
detector running at 30 Hz, and backed up by the powerful multiparticle
European Hybrid Spectrometer (EHS) has successfully measured about 200 D
decays as well as finding 3 Λ_cs. No Fs have been found. This fact, as
well as the small Λ_c sample, may be due in part to the above-mentioned
scanning loss of short-lived decays. The excellent quality of the data can
be seen in Figure 8, an HPD digitization of an event which had previously
been seen by scanning to contain charm. On the figure are indicated the
primary vertex (A), a 3-prong decay (B) and a one-prong decay (C).

At the present time, no further experiments with high-precision
bubble chambers are planned. Both bubble chambers and nuclear emulsions
are effectively confined to fixed target experiments. The further examples
of vertex detectors to be discussed may all be used both in fixed target
and in collider experiments.

3.3   Drift Chambers

Particle tracking in a gaseous medium suffers from one overriding
disadvantage by comparison with the condensed matter detectors which are

493

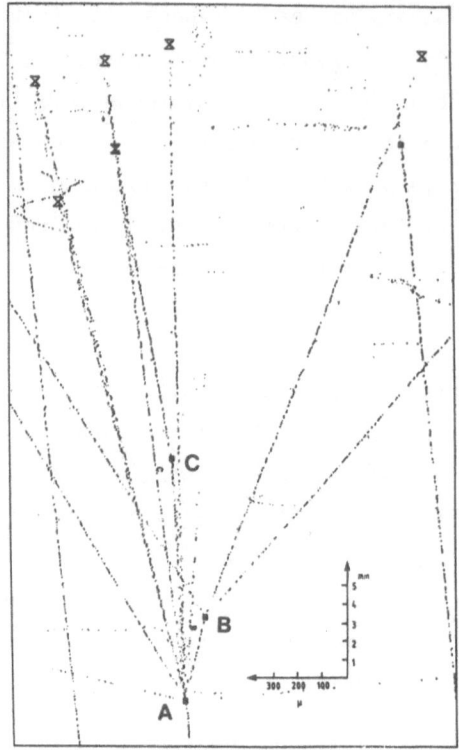

Fig. 8. HPD digitization of a charm production event in the LEBC
bubble chamber, from Reference 13.

used in every other type of vertex detector we shall discuss. That
disadvantage is diffusion of the electrons released by the ionization
process prior to collection on the anode wires. Diffusion limits the
precision, but the effect depends on many parameters (drift length, gas
composition, pressure, electric field etc). High pressure drift chambers
have achieved precision as low as 20 μm per point measured, but this
applies to tracks with normal incidence, and excludes tracks which are
close to sense or field wires. The limit of precision so far achieved in
real experiments is about 100 μm per point, leading (in typical
geometries) to a precision on the impact parameter similar to this. So we
see immediately (recall Figure 4) that drift chambers have little chance of
recognising individual tracks in an event as coming from primary or
secondary vertices. For this reason, they have not been used with much
success as vertex detectors in fixed target experiments. Nevertheless,
none of the other techniques has yet been applicable to colliders, and
small cylindrical drift chambers have been developed to fit between the
beam pipe and the inner wall of the main central tracking drift chamber.

These vertex drift chambers have been optimised for tracking precision
(precise wire alignment, short drift lengths, high operating voltage and in
some cases running above atmospheric pressure). Several vertex chambers
are in use at PEP and PETRA, and one has been tried on UA1. A good example
of such a chamber is the one used for some time in the Mark II detector at
PEP[14], Figure 9. It consists of 7 layers of axial wires, providing a
projection of the events on a plane normal to the beam direction. The
measured precision on the impact parameter projected onto this plane, for
high momentum tracks, is a standard deviation of 98 μm. Such a detector is
obviously not a vertex detector in the sense we have been using for nuclear
emulsions or high precision bubble chambers. It cannot be used to find
secondary vertices by recognising the tracks which emerge from them, on an
event by event basis, except in anomalously long-lived cases.
Nevertheless, such detectors can be very useful for physics if certain
conditions are satisfied, as follows:

(a) Other criteria are available which allow the selection (or at
least the strong enrichment) of a final state containing some
short-lived particles. The best example is $e^+e^- \rightarrow \tau^+\tau^-$ where a
nearly pure sample can be obtained. A marginal example is
$e^+e^- \rightarrow B\bar{B}$. The problems here are that the purity is not so
high, that one has always a mixture of B charge states (and so
lifetimes) and that mixed in are tracks from the further
C decays in the chain B → C → stable hadrons.

(b) A large event sample.

(c) Control of the <u>systematic</u> errors of the track reconstruction to
a level << 1 σ.

(d) A good knowledge of the mean beam centre, and very stable
collider operation.

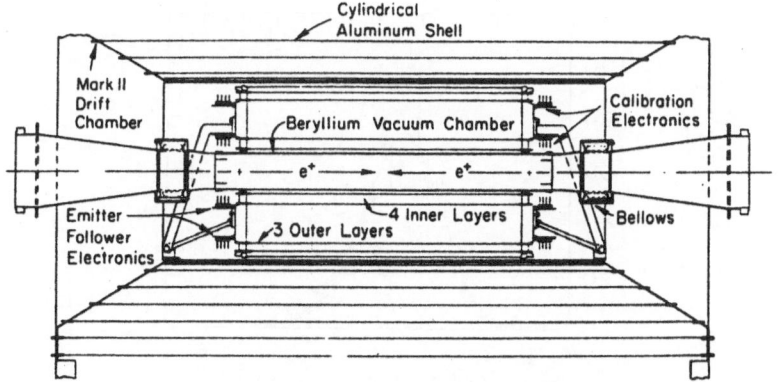

Fig. 9. Cross-section of the vertex drift chamber in the Mark II
detector at PEP.

Then we can reconstruct vertices on an event by event basis and evaluate the apparent flight path of the parent particle. This will frequently be negative due to the measurement errors, but a large sample will reveal the lifetime. This is seen in Figure 10 from Reference 15, which shows the data and a fit to a Gaussian, convoluted with an exponential decay distribution. From the latter component, the τ lifetime is found to be $(2.86 \pm 0.16 \pm 0.25) \times 10^{-13}$ s.

A similar procedure can be followed for charm lifetimes. The case of beauty particles is not so clean; results inevitably depend on models to correct for the decay chains involved.

A novel type of vertex chamber (the "straw chamber") has been developed by the MAC Collaboration[16]. In this chamber the sense wires are each surrounded by thin aluminised mylar cylindrical cathodes. The maximum drift length of 3 mm, operation at 4 atmospheres pressure, and running with high gas gain, give some advantages. In practice the impact parameter precision achieved at PEP is similar to that of the Mark II detector, but the technique may be attractive for hadron colliders. For example, such a chamber has good immunity to electromagnetic pickup from the beam pulse, and can tolerate a high particle flux.

Overall, the vertex chambers in use in e^+e^- colliders are producing lifetime values for charm which are of comparable quality with those available from fixed target experiments, and they are virtually the only source of τ and B lifetimes. Nevertheless, we should remember that this excellent performance is not a justification for being content with these vertex detectors at future colliders. On individual events they are still

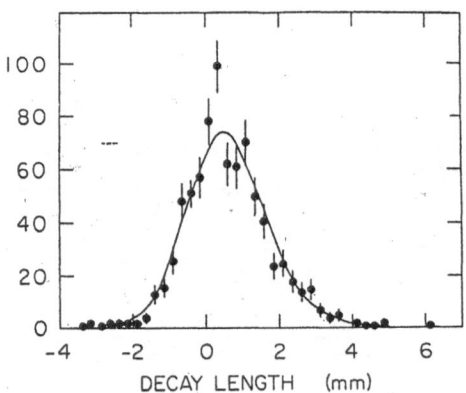

Fig. 10. τ decay length distribution measured by the Mark II detector.

very weak, and there may be a lot of exciting physics to be revealed once one is able to reconstruct individual events in e^+e^- with a clean assignment of tracks to vertices. This will be particularly important at higher energies where events are more complex, and unexpected short-lived particles may be revealed by detectors of adequate precision.

One can express the same hope for hadron colliders; that they become equipped with detectors measuring impact parameters to 5 μm rather that 100 μm. The transfer of the expertise which has been developed for fixed target experiments, to the collider environment, is an extremely important goal. As we shall see, there are non-trivial challenges associated with these aims.

3.4   Scintillating Fibres

The history of scintillation counting is almost as old as the history of ionizing particle detection. Around 1960, there were active efforts to construct tracking detectors based on fibres of ~ 1 mm diameter. These receded under the impact of the successive development of spark chambers, multiwire proportional chambers, and drift chambers, except in a few special areas (eg high rate beam hodoscopes).

The challenge of detecting charm particles led to a revival of this work, now aimed at a much higher level of precision, taking advantage of the technical developments being made in the opto-electronics industries.

Pioneering work with thin fibres has been done by Ruchti and co-workers[17]. One promising application is to fixed target beauty/charm production. In a design for such an experiment[18], the beam passes along the axis of a close-packed cylindrical fibre bundle. The experiment is triggered on interactions in the fibres. The forward-going jet of particles, including the short-lived particles and their decay products, are tracked in the fibres which record a projection of the event onto a plane normal to the beam direction. Thus (as with cylindrical drift chambers) the longitudinal evolution of the event is lost by this procedure. Results reported so far are based on a readout system consisting of proximity-focused photocathode/phosphors feeding a microchannel plate image intensifier, followed by phtographic film. The fibres were square (28 μm × 28 μm cross-section). Signals (amplified light spots seen on the film) are found with a density of 3.5 hits/mm of minimum-ionizing track length. With this setup[19], interactions have been recorded (eg Figure 11) and vertices reconstructed with an impact parameter

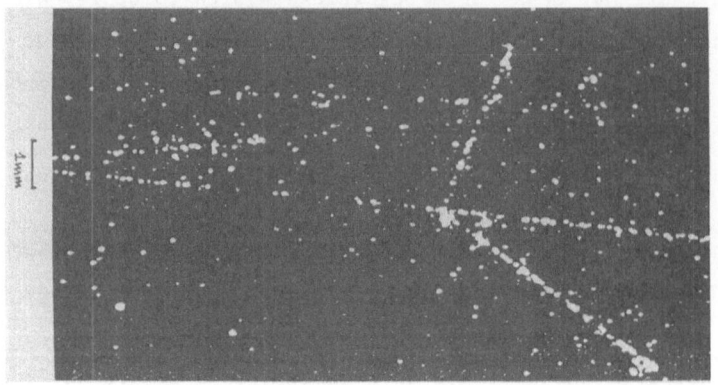

Fig. 11. Interaction seen in a bundle of scintillating fibres.

precision of 10 μm. But note that the beam in this test passed normal to
the fibre direction. In the axial-beam orientation, the precision on
impact parameter is likely to be degraded due to saturation problems. In
order that the technique can be used in an experiment, the photographic
film is being replaced by a fast-readout imaging CCD.

There are ideas of using scintillating fibres of 10-20 μm diameter in
thin shells in collider applications (eg for SSC). This is an attractive
idea since the primary detector has potentially a good radiation hardness
and rate capability. There may however be severe problems. Firstly, the
fibres need to be much longer than any yet tested (of that diameter).
Light attenuation may be serious. Secondly, the shells have to be
optically coupled to readout image intensifiers in a way which preserves
the spatial information. Short readout fibres lead to a large amount of
complex electronics embedded in the heart of the detector, which is
probably unacceptable for several reasons. Long ones imply some delicate
optical coupling from the detector fibres to the thicker readout fibres.
There are thus several tricky problems, but these are deservedly receiving
a great deal of attention from experts developing these detectors.

4.   SILICON DETECTORS: GENERAL CHARACTERISTICS
4.1   Why Silicon? (And Other General Remarks)

Nuclear emulsions and high precision bubble chambers have, as has
been described, certain limitations due to their being non-electronic
detectors, and are not obviously amenable to operation in colliders.
Gaseous detectors have limited precision; they seem destined not to allow
full exploitation of the physics possibilities. There are, however,
several types of <u>electronic condensed matter detectors</u> which have

498

considerable potential as vertex detectors. Scintillating fibres have been mentioned, and in addition there are liquid multiwire proportional chambers and various ideas for germanium detectors. Some of these are under active development and may have a very important role in the future.

What is unique about silicon detectors (and what accounts for their current lead in the field of high precision electronic detectors) is the fact that they are based on the highly developed <u>planar technology</u>. This technology encompasses all aspects of the precise processing (< 1 μm feature sizes) which can be applied to one face of a single crystal of silicon. The planar technology has revolutionised electronics (leading to integrated circuits through MSI, LSI, VLSI and now WSI—wafer scale integration), and is particularly suited to the fabrication of detectors for visible light, X-rays, min-I particles, stopping particles (alphas, etc). Such detectors can have very high precision in spatial position <u>and</u> in the measurement of energy deposited.

Semiconductor detectors have had a long and important history in the field of nuclear physics. The first signals were seen in 1951, from α particles impinging on a reverse-biased point contact germanium diode[20]. This principle - the detection of charge generated within the depletion region of a reverse-biased junction - has been retained in every semiconductor detector since then.

A long-standing advantage of silicon detectors is their intrinsic energy resolution. An ionizing particle in plastic scintillator has to expend 300 eV for every photoelectron generated at the photocathode. A gaseous detector (argon) requires 30 eV per electron liberated. In contrast, the lightly bound valence electrons in silicon are very easily excited into the conduction band; on average, an ionizing particle expends only 3.6 eV for every electron-hole pair liberated. For many nuclear physics applications, the stopping power of silicon is a major advantage. A 10 MeV proton stops in 1 mm of a silicon detector, but has a range of 1 metre in argon gas. This feature will not be directly useful to us, but the high density has the related advantage of yielding a large signal from a very thin detector, and of greatly reducing the range of δ-electrons.

Over 25 years, semiconductor detectors evolved in several forms (intrinsic silicon and germanium, and lithium-drifted varieties) generally in the direction of increasing sensitive volume (up to many cubic centimetres) and improved energy resolution. In some cases detectors were provided with subdivided surface electrodes to achieve modest spatial resolution in 1 dimension ($\gtrsim$ 1 mm). With the November Revolution (the discovery of the J/ψ in November 1974) an enormous interest focused on charm production experiments. By about 1980, it was apparent that high precision vertex detectors would be an enormous asset in such experiments, and some groups started to work on the problem of building tracking detectors with the necessary precision, based on the planar technology which had become increasingly refined since its inception in 1960. These efforts have gone in 3 related directions:

- by incorporating the planar technology into conventional nuclear physics detectors (leading to microstrip detectors)
- by adapting existing photosensitive detectors (leading to particle-detecting CCDs)
- by developing new detector types (leading to the silicon drift chamber).

Before looking at the performance of these detectors, we consider two general topics:

a. What are the fundamental limits to precision in tracking a min-I particle through silicon?

b. What are the principles of operation of silicon detectors? Here I shall spend some time on the basic operating features of MOS devices, of which all our detectors are particular examples.

4.2 Limits to Spatial Precision in Silicon Detectors

4.2.1 Mean Energy Loss. Consider a min-I particle traversing a thin sheet of silicon. It generates electron-hole pairs directly by ionization and indirectly by excitation of atoms and groups of atoms (phonons). These de-excite, emitting X-rays, light, etc. We shall ignore long-range coherent effects such as channelling and Cerenkov radiation since these are at a relatively very low level. We shall discuss these processes with models starting from the very simple (but least physical) and progressing to the more complex (and adequately physical).

We start with the free electron model, in which the electrons of the medium are supposed to be suspended in the material; all atomic binding is ignored. Consider the case of the non-relativistic passage of a charged particle through the material (charge equal in magnitude to the electron charge e, and velocity v = βc), so that the interaction can be handled by classical electrostatics. Take also the case of the particle mass m being much larger than the electron mass m_e, so that the main effect on the incident particle is energy loss rather than scattering through large angles. The maximum force experienced by an electron will be inversely proportional to the square of the impact parameter (see Figure 12)

$$F_{MAX} = \frac{e^2}{b^2} .$$

The components of force in the direction of the particle trajectory average to zero after the passage of the particle, and the overall effect is of a transverse force $\sim F_{MAX}$ for a time duration $t = \frac{2b}{v}$, giving it a momentum

$$p = \frac{2e^2}{bv} \quad \text{and energy } T = \frac{p^2}{2m_e} = \frac{2e^4}{m_e c^2 b^2 \beta^2} \qquad (4.1)$$

$$\therefore \quad 2b \, db = \frac{2e^4}{m_e c^2 \beta^2} \frac{dT}{T^2}$$

Now the probability F(b) of a collision with impact parameter b in thickness dx of material is given by the number of electrons in a cylinder of radius b

$$F(b) \, db \, dx = 2\pi b \, db \, NZ \frac{\rho}{A} dx$$

where N is Avogadro's number

A, Z are the atomic weight and atomic number of the material

ρ is the density

This is equal to the probability $\phi(T)$ of a collision which imparts energy T to an electron where T and b are related by (4.1)

$$\text{Thus } \phi(T)dT \, dx = \frac{2\pi e^4}{m_e c^2 \beta^2} \frac{NZ}{A} \rho \frac{dT}{T^2} dx \qquad (4.2)$$

The term $\frac{1}{m_e}$ expresses the fact that work is most effectively done on the electrons due to their low mass. The energy transfer to nuclei is reduced by an enormous factor $\sim \frac{m_p + m_n}{m_e} \sim 4000$, and can be entirely neglected.

The term $\frac{1}{\beta^2}$ expresses the fact that slow particles have a higher rate of energy loss.

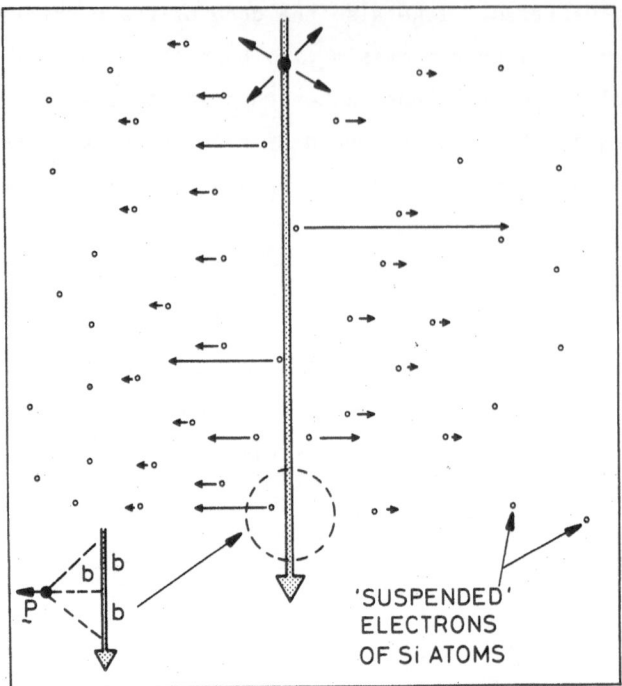

Fig. 12. Passage of a heavy charged particle through matter. Close electrons receive a powerful brief impulse; distant electrons receive a weak impulse which is also much more extended in time.

The term $\frac{1}{T^2}$ expresses the fact that collisions with large impact parameter (low T) take place with a much higher rate than close collisions. But even if we put T = 1 eV (and atomic binding will certainly dominate below this) we find from (4.1) that for $\beta = 1$,

$$b \quad = \quad 3 \times 10^{-10} \text{ cm.}$$

Thus even the most remote ionizing collisions of the most weakly bound electrons will pinpoint the particle trajectory at the level of atomic dimensions. Thus the limit to precision will certainly not be set by the primary ionization process.

Note from (4.2) that in the free electron model, defining $\frac{dE}{dx}$ as the specific energy loss due to all collisions, we have

$$\frac{dE}{dx} \quad = \quad \int T\phi(T) \ dt \ \propto \ - \ \ln(T_{min}) \quad = \quad \infty$$

Due to the long range of the coulomb interaction, this free-electron material would have infinite stopping power.

In moving to a more physically reasonable model, two developments are needed. Firstly, the change to relativistic quantum mechanics for the particle-electron collisions, and secondly a treatment of the atomic binding of the electrons. The calculation proceeds by choosing an energy η (typically 10 to 100 KeV) such that for

$T > \eta$ we have a 'close collision' in which the atomic electrons are effectively free; and

for $T < \eta$ we have a 'distant collision' in which the primary particle is effectively a point, and we handle all transitions leading to excitation or ionization of the atom.

For electron energies $T > \eta$, formula (4.2) holds with little modification. We have[21]

$$\phi(T) \quad = \quad \frac{d^2N}{dTdx} \quad = \quad \frac{2\pi e^4}{m_e c^2 \beta^2} \ \frac{NZ}{A} \ \rho \ \frac{1}{T^2} \ F \qquad\qquad (4.3)$$

where $F \ = \ 1 - \beta^2 \ \dfrac{T}{T_{MAX}}$ for spin-0 projectiles.

T_{MAX} is the maximum kinematically allowed energy transfer, and is given by

$$T_{MAX} \quad = \quad \frac{2 \ m_e \beta^2 \gamma^2 c^2}{1 + 2\gamma \ m_e/m}$$

At low energies, the $\dfrac{1}{T^2}$ form holds down to energies of the order of the mean ionization potential of the atomic electrons, as first defined by Bethe

$$I(Z) \quad \simeq \quad 16(Z)^{0\cdot9} \ eV$$

For still lower energies, the electron energy distribution must flatten and turn over in a way which depends on the details of the atomic structure of the medium. In the case of silicon, the crystalline structure is important. The importance of the low energy electrons can be gauged from the fact that integrating the δ-electron energy distribution (4.3) down to $T = I(Z)$ yields only half of the energy loss given by the Bethe-Block formula[22]

$$\frac{dE}{dx} = \frac{4\pi e^4}{m_e c^2 \beta^2} \frac{NZ}{A} \rho \left[\ln\left(\frac{2\, m_e \gamma^2 \beta^2 c^2}{I(Z)}\right) - \beta^2 \right] \qquad (4.4)$$

In low density gaseous media this formula holds, including the $2 \ln \gamma$ term which implies a _relativistic rise_ in the specific ionization, due partly to the increasing effective range of the electromagnetic interaction with γ, and partly to the increasing kinematic limit on the δ-electron energy. For a condensed material such as silicon, however, the density effect gives rise to an additional term $-D$ in the parentheses, where D rises to $\ln \gamma$ for highly relativistic projectiles; the rise in $\frac{dE}{dx}$ is thus limited to the effect of the increasing kinematic limit for δ-electron production.

4.2.2 Fluctuations in Energy Loss.

So far we have not dealt specifically with the statistical fluctuations in the energy loss, which give rise to the limits to precision in silicon detectors.

For thin detectors, the $\frac{1}{T^2}$ form for the δ-electron energy distribution leads to the situation that collisions over much of the kinematically allowed energy transfer range occur with probability $<< 1$. Most frequently, traversal of the detector will be characterised by a large number of low energy collisions, with an occasional high energy transfer being seen. This gives rise to the familiar asymmetric Landau distribution[23] in energy loss. According to the Landau theory, energy loss fluctuations result only from the collisions described by the Rutherford spectrum (4.3), ie by energy transfers much greater than $I(Z)$. As a result, it _underestimates_ the widths of the energy loss distribution for thin samples, for which even energy transfers around $I(Z)$ occur sufficiently infrequently that statistical fluctuations cannot be ignored.

Blunck and Leisegang[24] introduced corrections to the Landau theory to take account of the atomic binding of the electrons. However, this theory considerably _overestimates_ the widths of the energy loss distribution for thin samples. Chechin and Ermilova[25] showed that both theories were inapplicable in such cases, and Ermilova et al[26], using experimental photoabsorption coefficients, were able to obtain excellent agreement with experiment from samples as thin as 1.5 cm of argon at atmosphere pressure.

The essential point is that the 'distant collisions' previously referred to ($T < \eta$) involve the exchange of soft, nearly on-shell, virtual photons, whose interaction with the material is well described by the measured photoabsorption coefficients. The model is described in detail by Allison and Cobb[27] (applied to gaseous detectors) and has been successfully applied to silicon detectors by G Hall[28]. When used in a Monte Carlo approach (as in Reference 26) it is possible to determine the precise shapes of the energy loss distributions. When used with numerical integration (as in Reference 28) the broadening of the Landau distribution is described by a convoluting normal distribution of which the width is explicitly calculated.

The thinnest samples in which min-I signals have been detected are the depletion layers of CCDs which amount to only about 20 µm of silicon. Figure 13 shows the experimental $\frac{dE}{dx}$ distribution[29] and indicates the good fit achieved by the calculation of Hall. Several earlier calculations predicted much broader distributions, to the point that it would have been impossible to achieve high efficiency with such thin silicon detectors. Fortunately, both experiment and the most recent calculations are in agreement, and achieving high efficiency is not a problem.

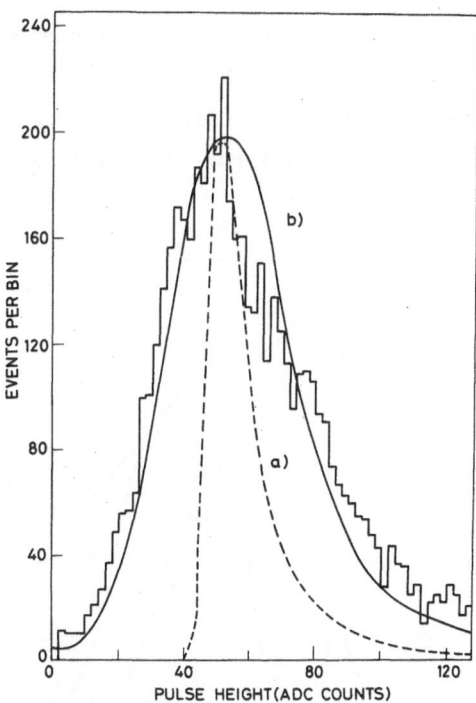

Fig. 13. Experimental energy loss distribution from min-I pions traversing a CCD detector (20 µm of silicon equivalent) together with
(a)   the Landau distribution and
(b)   the distribution calculated by Hall.

Before leaving this topic, we should look briefly at the experimental photoabsorption data on which the calculations are based (Figure 14). Apart from getting a feeling for the range of virtual photon energies whichare important, the figure is instructive in showing the usefulness of silicon for real photon detection.

Almost any conceivable silicon detector has a suspect or dead region due to surface layers etc of depth ~ 1 μm, and most have a total thickness of << 1 mm. If one converts the absorption cross-sections to absorption/μm or absorption/mm, one sees that for much of the photon energy range silicon is either too opaque or too transparent to be useful. The solid curves (a) on the figure show the efficiency (linear scale on the right) for a typical CCD detector (10 μm depletion) and the broken curves (b) show the effect of increasing the depletion depth to 200 μm (achievable on high resistivity silicon wafers). From these one can see that silicon detectors are of no use in the infra-red, where the material is entirely transparent, but are useful for visible light and over a region in the X-ray spectrum from 1 to 10 or 20 KeV. While of no direct relevance to the detection of min-I particles, this information can be very important when thinking about test procedures and backgrounds (eg synchrotron radiation) as well as providing industry with its main impetus for the development of silicon detectors.

4.2.3 <u>Limits to Precision.</u> The basic situation, which follows from the previous section, is summarised in Figure 15. This shows firstly the

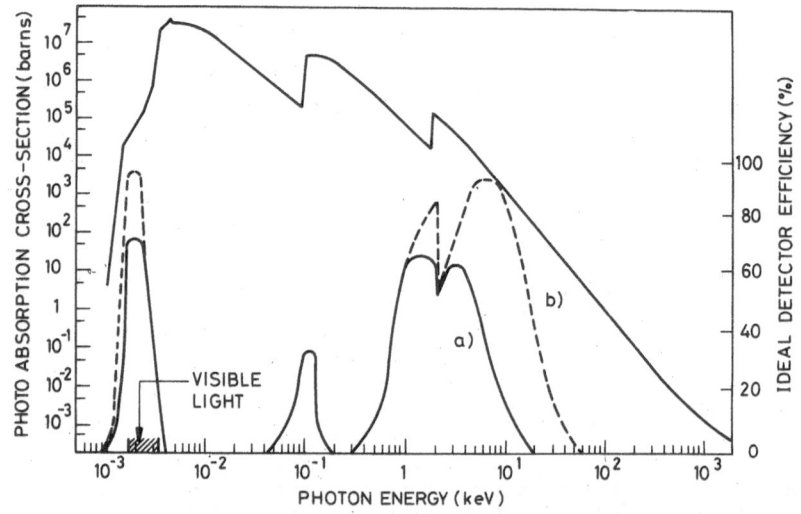

Fig. 14. Experimental photoabsorption cross-section for silicon, indicating the energy ranges of use for the detection of real photons.

probability per micron of ejecting an electron of energy greater than T, and contains the information needed to evaluate the high energy fluctuations in energy loss. As will be seen, these fluctuations in themselves cause problems for position measurement, but these problems are exacerbated once the ranges of the ejected electrons become significant. The electron ranges are included on the figure. For each decade in electron energy up to the kinematic limit, the flux of ejected electrons falls as $\frac{1}{T^2}$, but each ejected electron releases an increasing number of secondaries (one electron-hole pair per 3.6 eV of primary electron energy) with the result that the mean charge Q released in the silicon (averaged over many samples) increases linearly with ln(T), eventually tailing off in the last decade before the kinematic limit T_{MAX}.

Due to the asymmetric nature of the Landau distribution, the <u>mean</u> energy loss in a sample (which is weighted by the rare cases with very large energy loss) is significantly higher than the <u>most probable</u> energy loss. This difference diminishes for thick samples; the mean energy loss

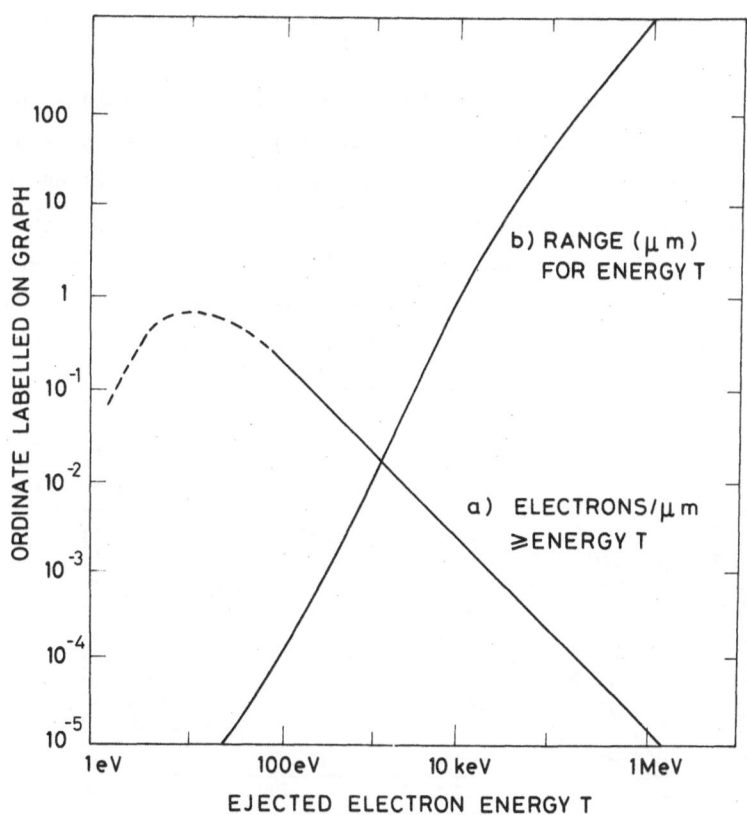

Fig. 15. (a) The number of electrons ejected by a min-I particle per μm of path with energy > T.
(b) The range vs T.

scales with thickness but the most probable loss grows faster. For 5 GeV/c
pions incident on 100 μm of silicon, the mean energy loss is 400 eV/μm,
while the most probable is 240 eV/μm (110 and 67 electron-hole pairs per
micron, respectively). From Figure 15(b) we see that one such measurement
has a probability of 10% of including a δ-electron of energy > 20 keV, ie
of range > 5 μm. The 5500 secondary electrons from this δ-electron will
pull the centroid of the charge distribution off its true position by
typically

$$\frac{5500 \times 2.5}{6700 + 5500} \quad = \quad 1 \ \mu m$$

Figure 16 shows the probability that the centroid be displaced by
some specified values (1 μm and 5 μm are taken as examples) as a function
of detector thickness. A detector with a thin active depth of silicon is
in principle to be preferred to a thicker one, since in the thin detector
there is a much lower probability of generating a δ-electron of dangerously
long range. Although the amount of charge in the main column of ionization
increases with thickness, the effect of δ-electrons becomes more serious
since their range increases approximately as the square of their energy.

Finally, it should be noted that the Landau fluctuations in energy
loss are particularly serious in determining the positions of angled
tracks. Here, one can do no better than assign the centroid of the charge
distribution to the mid-plane of the detector. Even ignoring the effect of
δ-electron range, fluctuations in energy loss along the track immediately
cause errors in the assigned position, as illustrated in Figure 17 for
tracks at 45°. In a thin detector there is a 10% probability of producing

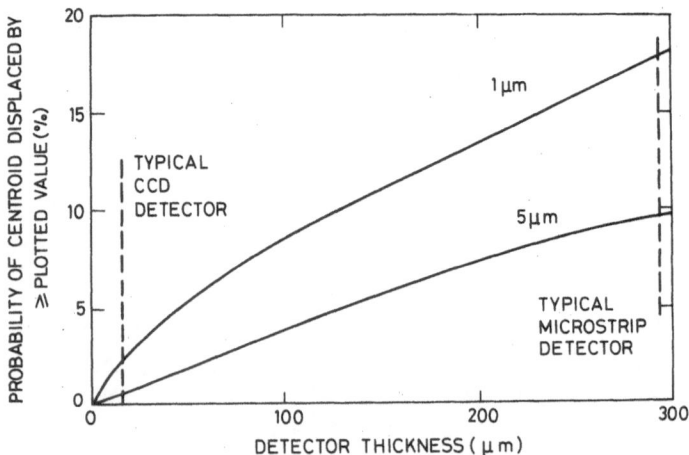

Fig. 16. Detector precision limitations from δ-electrons for tracks
at normal incidence as a function of detector thickness.

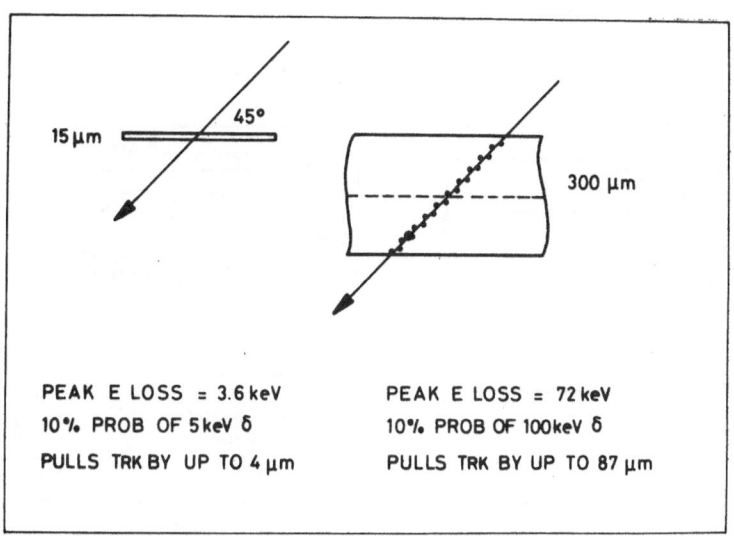

Fig. 17. Effect on detector precision for angled tracks due to energy loss fluctuations.

a δ-electron which, if it occurs near one end of the track, pulls the co-ordinate across by 4 μm. In the thick detector, there is the same probability of producing a δ-electron which can pull the co-ordinate by 87 μm.

Of course, the advantages of thin detectors must be weighed against the relative difficulty of extracting the information from them due to such effects as a poorer signal to noise ratio. These points will be discussed in the context of specific types of detectors.

In summary, the fine trail of electron-hole pairs in a thin silicon detector allows in principle an unprecedented precision of tracking (<< 1 μm) compared even with nuclear emulsion, in which a δ-electron of ~ 400 eV is needed to blacken a grain. Our task is to find the best methodof taking advantage of this information; how can the positions of these minute quantities of charge be measured? There is in fact no completely satisfactory answer at present; the field is wide open for new ideas. But in order to explore the problem further, we shall consider those principles of construction and operation of MOS devices which are relevant to the silicon detectors which have been built so far.

4.3 Principles of Operation of Silicon Detectors

Gaseous silicon has a typical structure of atomic energy levels (see Figure 18). It has an ionization potential of 8.1 eV, ie it requires this

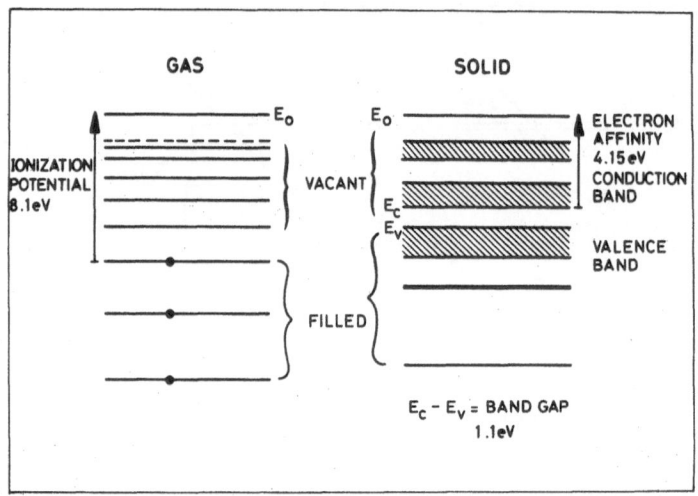

Fig. 18. Diagrammatic sketch of allowed energy levels in gaseous
silicon which become energy bands in the solid material.

much energy to release a valence electron, compared with 15.7 eV for argon.
As silicon condenses to the crystalline form, the discrete energy levels
of the individual atoms merge into a series of energy bands in which the
individual states are so closely spaced as to be essentially continuous.
The levels previously occupied by the valence electrons develop into the
valence band, and those previously unoccupied become the conduction band.
Due to the original energy level structure in gaseous silicon, it turns out
that there is a gap between these two bands. In conductors, there is no
such gap, in semiconductors there is a small gap (1.1 eV in silicon, 0.7 eV
in germanium) and in insulators there is a large band gap. In particular,
the band gap in silicon dioxide is 9 eV. This makes it an excellent
insulator and, coupled with the ease with which the surface of silicon can
be oxidised in a controlled manner, accounts partly for the pre-eminence of
silicon in producing electronic devices.

We shall denote as E_v and E_c the energy levels of the top of the
valence band and the bottom of the conduction band (relative to whatever
zero we like to define). The energy needed to raise an electron from E_c to
the vacuum E_o is called the electron affinity. For crystalline silicon
this is 4.15 eV.

4.3.1 Conduction in Pure and Doped Silicon. To understand the conduction
properties of pure silicon, the liquid analogy is helpful. This is
illustrated in Figure 19. (a) shows the energy levels in silicon under no
applied voltage with the material at absolute zero temperature. All

electrons are in the valence band and under an applied voltage (b) there is no change in the population of occupied states, and so no flow of current; the material acts like an insulator. At a high temperature (c) a small fraction of the electrons are excited into the conduction band, leaving the same number of vacant states in the valence band. Under an applied voltage (d) the electrons in the conduction band can flow to the right and there is a re-population of states in the valence band which can be visualized as the left-ward movement of a bubble (hole) in response to the applied voltage.

Now kT at room temperature is approximately 0.03 eV. This is small compared with the band gap of 1.1 eV, so the conductivity of pure silicon at room temperature is very low. To make a quantitative evaluation, we need to introduce the Fermi-Dirac distribution function $f_D(E)$ which expresses the probability that a state of energy E is filled by an electron. Figure 20(a) shows the form of this function

$$f_D(E) \quad = \quad \frac{1}{1 + e^{(E - E_f)/kT}} \tag{4.5}$$

E_f, the Fermi level, is the energy level for which the occupation probability is 50%. Figure 20(b) shows the density of states g(E) in silicon. The concentration of electrons in the conduction band is given by the product f.g, and the density of holes in the valence band by (1 - f).g, as shown in Figure 20(c). In pure silicon, the Fermi level is approximately at the mid-band gap, and the concentrations of electrons and holes are of course equal. These concentrations, due to the form of f_D, are much higher in a narrow band gap semiconductor (d) than in a wide gap material (e).

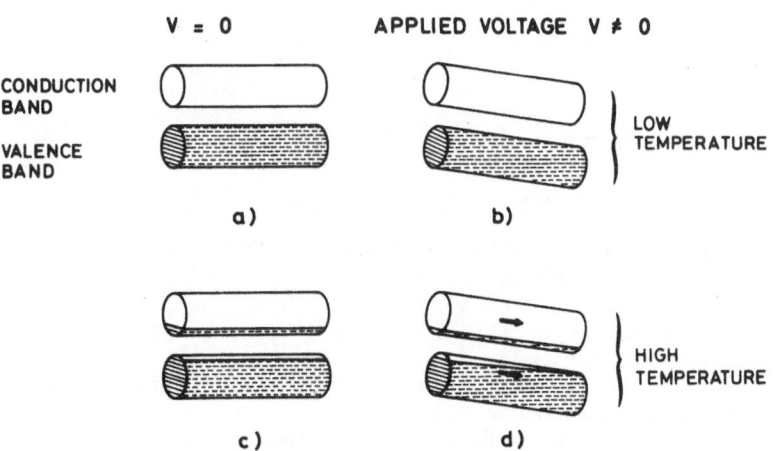

Fig.19. Liquid analogy for a semiconductor.

So far we have been discussing pure (so-called <u>intrinsic</u>) semi-conductors. Next we have to consider the <u>doped</u> or extrinsic semi-conductors. These allow us to achieve high concentrations of free electrons (n-type, Figure 20(f)), or of holes (p-type, Figure 20(g)), by moving the Fermi level very close to the conduction or valence band edge. The procedure for doing this is to replace a tiny proportion of the silicon atoms in the crystal lattice by dopant atoms with a different number of valence electrons.

For example, <u>phosphorus</u> has 5 valence electrons in contrast with 4 for silicon. At absolute zero it holds all 5 and phosphorus-doped silicon is still an insulator. But at a very low temperature the extra electron is shaken free, and at room temperature most of the extra phosphorus electrons are available for conduction.

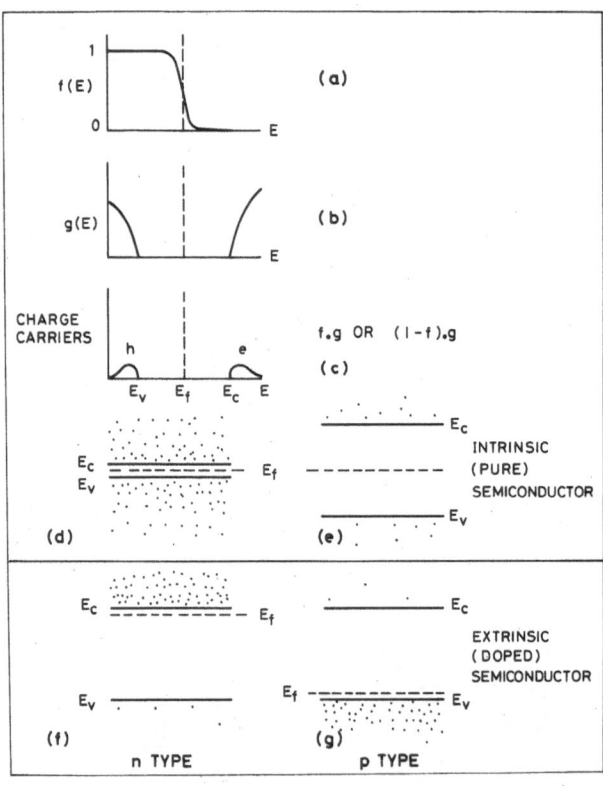

Fig. 20(a). Fermi-Dirac distribution function. The slope increases as the temperature is reduced.
   (b). The density of states below and above the forbidden band-gap.
   (c). Concentration of electrons and holes (charge carriers) available for conduction.
   (d). and (e) Charge carrier distributions in narrow and wide band-gap semiconductors.
   (f). and (g) Charge carrier distributions in n- and p-type semiconductors.

Conversely, <u>boron</u> has 3 valence electrons, leaving one vacant bond, easily filled by the movement of electrons, and so an available path for conduction. This is best visualised as the contrary motion of the vacant bond (hole).

Figure 21 shows the levels associated with various <u>donor</u> atoms (measured relative to E_c) and <u>acceptor</u> atoms (measured relative to E_v). Figure 22 shows the concentration of electrons in n-type silicon (1.15 x 10^{16} dopant atoms, arsenic, per cm^3) as a function of temperature. Below about 100° K one sees the phenomenon of <u>carrier freeze-out</u>, loss of conductivity due to the binding of the donor electrons. This is followed by a wide temperature range over which the electron concentration is constant, followed above 600° K by a further rise as the thermal energy becomes sufficient to add a substantial number of intrinsic electrons to those already provided by the arsenic atoms. This general behaviour is typical of all doped semiconductors.

The <u>resistivity</u> ρ of the material depends on the concentration of free holes and electrons and on their <u>mobilities</u>. These are a function of temperature and of impurity concentration. At room temperature, in lightly doped silicon, we have

electron mobility μ_n = 1350 cm^2 $(V.s)^{-1}$
hole mobility μ_p = 480 cm^2 $(V.s)^{-1}$

and the resistivity is given by

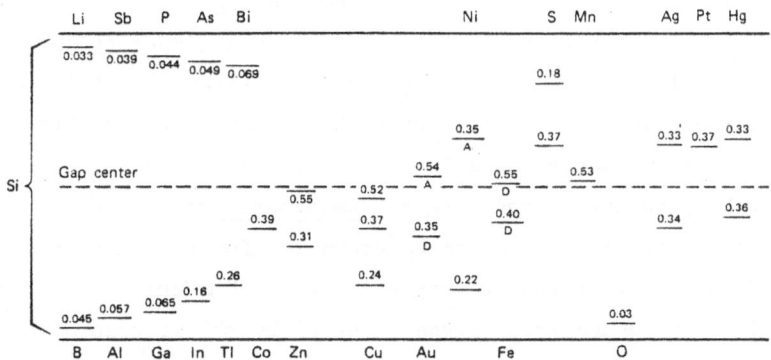

Fig. 21. From Reference 30. Energy levels in electron volts of various impurity elements in silicon. Levels of acceptor atoms are measured from the top of the valence band, and levels of donor atoms are measured from the bottom of the conduction band.

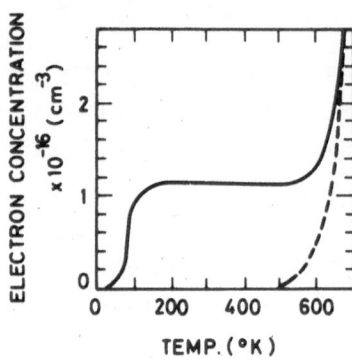

Fig. 22. Electron concentration vs temperature for n-type (arsenic-doped) silicon. The dashed curve shows the concentration for intrinsic material.

$$\rho \quad = \quad \frac{1}{e(\mu_n \cdot n + \mu_p \cdot p)} \qquad\qquad (4.6)$$

$\Big[$e is the charge on the electron and n and p are

the electron and hole concentrations$\Big]$

For pure silicon at room temperature $n_i = p_i = 1.45 \times 10^{10}$ cm^{-3} which gives $\rho_i = 235$ KΩ cm.

The resistivity as a function of impurity concentration is shown in Figure 23. For reasons which will become clear, we are often concerned in silicon detectors with unusually high resistivity material, beyond the range of these graphs. For example, 20 KΩ cm p-type silicon requires a dopant concentration of 5×10^{11} cm^{-3}. Remembering that the crystalline silicon has 5×10^{22} atoms per cm^3, this implies an impurity level of 1 in 10^{11} which even in the highly developed art of silicon crystal growing is a major challenge. The resistivity noted above in connection with pure silicon is entirely unattainable in practice. Very high resistivity n-type silicon can be produced in the form of <u>compensated</u> materials. The most uniformly doped material which can be grown is (for technical reasons) p-type, and this (with a resistivity of about 10 K Ω cm) is used to start with. it is then turned into n-type material by the procedure known as neutron doping. The crystal is irradiated with slow neutrons and by means of the reaction

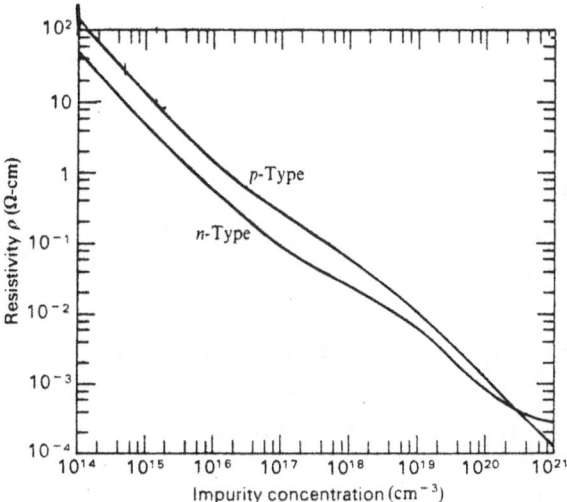

Fig. 23. From Reference 31. Resistivity of silicon at room temperature as a function of acceptor or donor impurity concentration.

$$Si^{30} + n \rightarrow Si^{31}$$
$$\phantom{Si^{30} + n \rightarrow} \downarrow$$
$$\phantom{Si^{30} + n \rightarrow} \rightarrow P^{31} + \beta^- + \bar{\nu}$$

is turned into n-type material. The resistivity is monitored and the irradiation ceases when this, having passed through a maximum, falls to the required value. In this way, material of resistivity as high as 100 K Ω cm can be made. But this is a very difficult art.

There is a useful approximation which relates the carrier concentration to the Fermi level. For n-type material, the carrier concentration is dominated by

N_c the effective density of states at the conduction band edge

and $E_c - E_f$ the energy separation between the Fermi level and those available states.

Thus $n = f \cdot g$

becomes $n = N_c \exp\left[-\left[E_c - E_f\right]/kT\right]$

In fact, the extent to which this is an approximation is illustrated by the temperature dependence of the effective density of states, which can be shown to be proportional to $T^{3/2}$. But it is still true that the overall temperature dependence is dominated by the exponential term. Similarly we can define N_v as the effective density of states at the valence band edge, and

$$p = N_v \exp \left[-\left[E_f - E_v \right]/kT \right]$$

Now for intrinsic material we have

$$n = p = n_i \quad \text{and} \quad E_f = E_i \; (\sim \text{mid band gap})$$

Thus in general (intrinsic or doped silicon) we have

$$
\begin{aligned}
n &= n_i \exp \left[(E_f - E_i)/kT \right] \\
p &= n_i \exp \left[(E_i - E_f)/kT \right]
\end{aligned}
\tag{4.7}
$$

This shows that the deviation of a doped semiconductor from the intrinsic material can be simply represented by the energy separation of the Fermi level from the intrinsic Fermi level.

Notice that

$$np = n_i^2 = N_c N_v \exp \left[- E_g/kT \right] \tag{4.8}$$

where $E_g = E_c - E_v$

This is a particular example of the very important law of mass action which applies as much in semiconductor theory as it does in chemistry, and which states that in a sample in thermal equilibrium the increase in electrons (eg by donor doping) has as a result a decrease in holes (by recombination) such that the np product is constant.

Notice also from (4.8) that the main term in the temperature dependence of the intrinsic carrier concentration is $\exp \left[- E_g/2kT \right]$ which implies a factor 2 increase in n_i for each 8° K increase in temperature around room temperature.

It is generally valid to think of n-type material as containing only electrons and p-type material as containing only holes. These are referred to as the majority carriers in each case. An important point to be aware of in designing semiconductor detectors is the possibility of growing thin

(10 to 50 μm) <u>epitaxial layers</u> of high resistivity material on a low
resistivity substrate, with ~ 1 μm transition region between (Figure 24).
It is even possible to grow epitaxial layers of alternating conductivity
type (n and p), which opens up some very interesting possibilities for
particle detection.

4.3.2 <u>The np Junction.</u> We now need to introduce a most important fact
relating to conducting materials which are electrically in contact with one
another and in thermal equilibrium; <u>they all must establish the same Fermi
energy.</u> This applies to

metal/semiconductor systems

n-type/p-type systems etc.

Charges flow from the high to low energy region until this condition
is established. For example, at an np junction there is exposed a fixed
space charge of ionized donors and acceptors, creating a field which
opposes further drift of electrons and holes.

The <u>depletion approximation</u> says that the semiconductor changes
abruptly from being neutral to being fully depleted. This is far from
obvious and in fact there is a finite length (the <u>Debye length</u>, typically

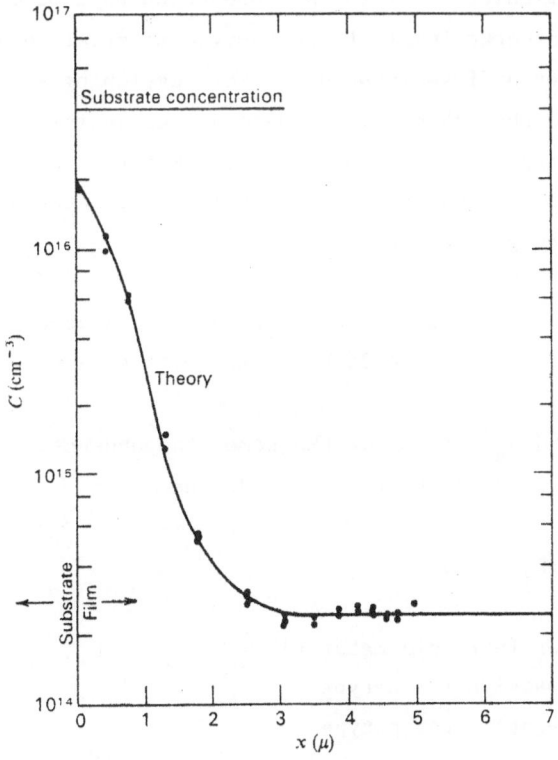

Fig. 24. From Reference 31. Impurity distribution after epitaxial growth.

≈ 0.1 μm) over which the transition takes place. But the depletion approximation will be adequate for all the examples we need to consider.

Let us look in some detail at the important case of the np junction. Before contact (Figure 25(a)) the surface energy E_o is equal in both samples; the p-type Fermi level is close to E_v and the sample is densely populated by holes; the n-type Fermi level is close to E_c and the sample is densely populated by electrons.

On contact, the electrons diffuse into the electron-free material to the left, and the holes diffuse to the right. In so doing the electrons leave exposed donor ions (positively charged) in the n-type material, and the holes leave exposed acceptor ions (negatively charged) in the p-type material. This builds up an electric field which eventually just balances the tendency for current flow by diffusion. Once this condition is reached (Figure 25(b)) the Fermi levels in the materials have become equal. The electrical potentials in the two samples (the potential energy at the surface E_o, or at the conduction band edge, E_c) are now unequal.

Intuitively, this can be understood as follows. Initially, the electrons at a particular level in the conduction band of the n-type material see equal-energy levels in the p-type material which are unpopulated, so they diffuse into them. The developing space charge bends the energy bands so that these levels become inaccessible. Eventually, only very high energy electrons in the n-type material see anything other than the empty states of the band gap of the p-type material and conversely for the holes in the p-type material.

Let us develop this quantitatively, adopting a co-ordinate system in which the np junction of Figure 25(b) is at position x = 0.

E_o, E_c, E_i and E_v all follow the same x dependence. The zero of the electric potential ϕ is arbitrary, so we define

$$\phi = -\frac{1}{e} (E_i - E_f)$$

Thus $\phi = 0$ for intrinsic material
positive for n-type
negative for p-type

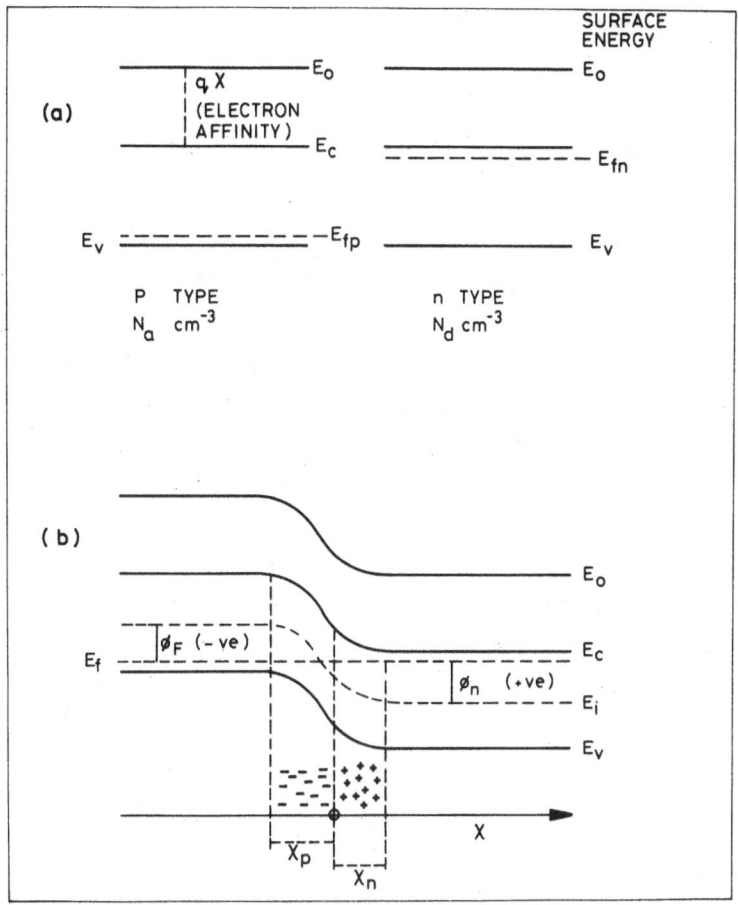

Fig. 25(a). Energy levels in two silicon samples (of p and n-type) when electrically isolated from one another.

　　(b). When brought into contact, the Fermi level is constant throughout the material. The band edges bend in accordance with the space charge generated.

From (4.7)

$$\phi_n = \frac{kT}{e} \ln \frac{N_d}{n_i}$$

$$\phi_p = -\frac{kT}{e} \ln \frac{N_a}{n_i}$$

where N_d and N_a are the concentration of donor and acceptor atoms in the n- and p-type material respectively.

The potential barrier $\phi_i = \phi_n - \phi_p = \frac{kT}{e} \ln \left(\frac{N_d N_a}{n_i^2} \right)$ (4.9)

Notice that the potential barrier falls linearly with temperature since it is sustained by the thermal energy in the system.

We may deduce the electric field strengths near the junction by using Poisson's equation

$$\frac{d^2\phi}{dx^2} = -\frac{e}{\varepsilon_s}\rho(x)$$

ε_s is the permittivity of silicon $= \varepsilon_r\varepsilon_o$
ε_o is the permittivity of space $= 8.85 \times 10^{-14}$ F cm^{-1}
$\hspace{5.5cm} = 55.4$ electron charge/V.μm
ε_r is the dielectric constant or relative permittivity of
$\hspace{4cm}$ silicon $= 11.7$

$$= -\frac{dE}{dx}$$

where E is the electric field

For $x > 0$

$$\frac{dE}{dx} = +\frac{eN_d}{\varepsilon_s} \quad \therefore E(x) = -\frac{eN_d}{\varepsilon_s}(x_n - x)$$

For $x < 0$

$$\frac{dE}{dx} = -\frac{eN_a}{\varepsilon_s} \quad \therefore E(x) = -\frac{eN_a}{\varepsilon_s}(x + x_p)$$

$$\left.\begin{array}{c} \\ \\ \\ \\ \end{array}\right\} \qquad (4.10)$$

The underlined undepleted silicon on either side of the junction is field-free. The depleted silicon close to the junction contains an electric field whose strength is maximum at the junction and is directed to the left, ie opposing the flow of holes to the right and opposing the flow of electrons to the left.

Requiring continuity of the field strength at $x = 0$ implies

$$N_a x_p = N_d x_n \qquad (4.11)$$

Thus, if one wants to make a deep depletion region on one side of the junction (important, as we shall see, for many detectors) we need to have a very low dopant concentration, ie very high resistivity material.

The electric field strength varies linearly with x; the electric potential, by integration of (4.10), varies quadratically.

For $x_n > x > 0$ $\quad \phi(x) = \phi_n - \dfrac{eN_d}{2\varepsilon_s}(x_n - x)^2$ $\left.\rule{0pt}{55pt}\right\}$ (4.12)

For $x_p < x < 0$ $\quad \phi(x) = \phi_p + \dfrac{eN_a}{2\varepsilon_s}(x + x_p)^2$

Requiring continuity of the potential at $x = 0$ implies

$$x_n + x_p = \left[\frac{2\varepsilon_s}{e}\phi_i\left(\frac{1}{N_a} + \frac{1}{N_d}\right)\right]^{\frac{1}{2}} \qquad (4.13)$$

From (4.9) ϕ_i depends only weakly on N_a and N_d.

If for example $N_a \gg N_d$ we have $x_p \approx 0$ and (4.13) gives $x_n \propto \dfrac{1}{N_d^{\frac{1}{2}}}$.

So a factor 2 increase in resistivity leads to a factor $\sqrt{2}$ increase in depletion depth.

Figure 26 summarises these results on the characteristics of an unbiased np junction, with the inclusion of some typical numerical values based on $N_a = 10^{14}$ cm^{-3} and $N_d = 2 \times 10^{14}$ cm^{-3}. The peak field in this case is about 3 kV/cm. By high doping concentrations and large bias voltages it can easily happen that one approaches the limiting field of about 300 kV/cm at which internal breakdown in the silicon sets in.

We now consider the effect of applying a voltage across the junction. Under equilibrium conditions, electron-hole pairs are continually generated by thermal excitation throughout the semiconductor. In the case of zero bias (Figure 27(a)) the electrons and holes generated within the bulk of the semiconductor recombine. Those generated in the depletion region are swept into the undepleted silicon, holes to the left, electrons to the right. This effect would act to reduce the potential barrier and so is compensated by a leakage of majority carriers which diffuse across the barrier in the opposite directions at just the rate needed to cancel the charge generation in the depleted material. The overall effect is of no current flow.

By applying a forward bias (Figure 27(b)) we separate the previously equal Fermi levels by an amount equal to the bias voltage; the system is no longer in thermal equilibrium or this condition could not be maintained. Although there is still an electric field in the depletion region which is

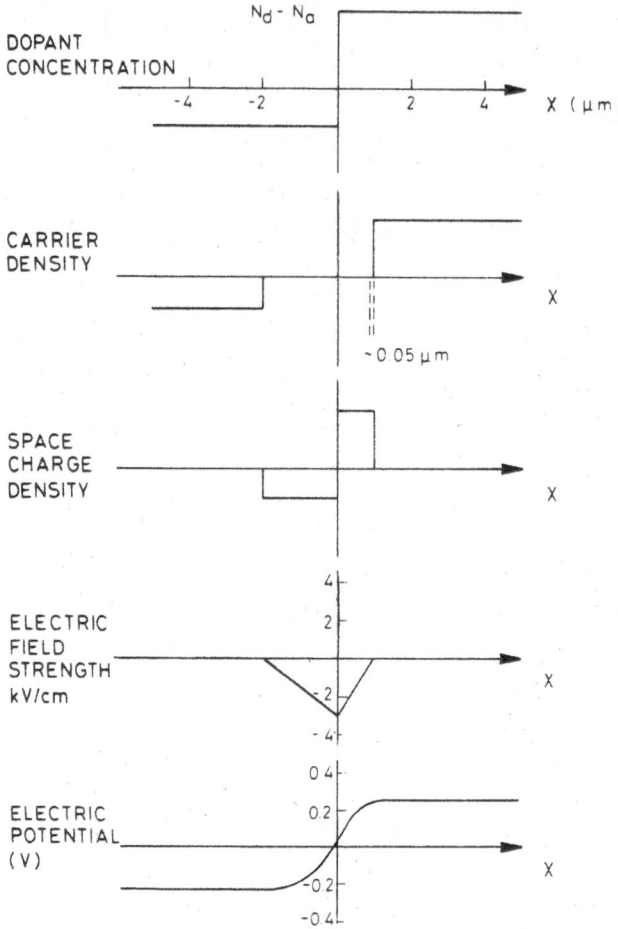

Fig. 26. Summary of various quantities across an unbiased np junction.

directed against the current flow, the depletion region is narrowed and the potential barrier is now inadequate to prevent majority carriers from flooding across it, holes from the left and electrons from the right. Many of these will recombine within the depletion region giving rise to the <u>recombination current</u>. Those which survive are absorbed within one or two diffusion lengths by recombination with the majority carriers on that side of the junction, giving rise to the <u>diffusion current</u>. Beyond these regions there is just a steady flow of majority carriers supplied from the voltage source to keep the current flowing. Notice that in a forward biased junction the current flow is due entirely to electron-hole <u>recombination</u>.

With a reverse bias, we have the situation shown in Figure 27(c). The depletion region is now much wider and electron-hole pairs generated within it are efficiently swept into the undepleted silicon, electrons to the right and holes to the left giving rise to the <u>generation current</u>.

Unlike the case of the unbiased junction, there is now no supply of
majority carriers able to overcome the increased potential barrier across
the junction. On the contrary, the thermal generation of <u>minority carriers</u>
within one or two diffusion lengths of the depletion region leads to some
holes generated in the n region reaching this depletion region and then
being briskly transported across it, and conversely for electrons generated
in the p region. This leads to the so-called <u>diffusion current</u>. In the
case of the reverse-biased junction, the current flow is thus due entirely
to electron-hole <u>generation</u>. The current flow across reverse-biased
junctions is of great importance in determining the noise limits in silicon
detectors. An immediate observation is that, since the current arises from
thermal generation of electron-hole pairs, the operating temperature will
be an important parameter.

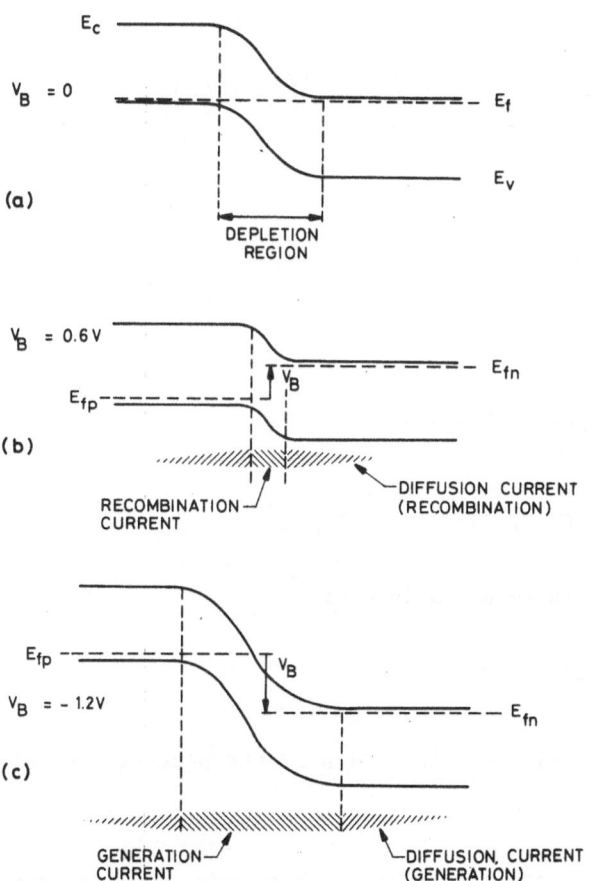

Fig. 27. Effect of an applied voltage on the semiconductor in the
region of the junction.

Before continuing to discuss this point, it is worth noting that we have finally collected up enough information to calculate the characteristics of a typical particle detector, and it is instructive to do so. Referring to Figure 28, we have a silicon detector made of good quality, high resistivity p-type silicon (ρ = 10 KΩ cm). On the front surface we make a shallow implant of donor atoms and on the back surface we make a highly doped p-type implant to provide a good low-resistivity ground electrode. The terms n^+ and p^+ are conventionally used to represent high doping levels. Now we apply a positive voltage V to the n-type surface with the aim of completely depleting the detector. In this way we shall ensure complete collection of the electrons and holes generated by the passage of a charged particle; with incomplete depletion we would lose signal by recombination. Equation (4.13) applies, with the difference that we replace ϕ_i by $V + \phi_i$ since the junction is biased in the direction which assists the previously existing depletion voltage.

We have

$$x_n + x_p \approx x_p = \left[\frac{2\varepsilon_s}{e} \left(V + \phi_i \right) \left(\frac{1}{N_a} + \frac{1}{N_d} \right) \right]^{\frac{1}{2}}$$

$$\approx \left[\frac{2\varepsilon_s}{e} \times \frac{V}{N_a} \right]^{\frac{1}{2}}$$

From Figure 23, we see that $N_a \approx 10^{14} \times \frac{150}{\rho}$ and we require $x_p = \ell$

$$\therefore V = \frac{e}{2\varepsilon_s} \times \frac{1.5 \times 10^{16}}{\rho} \times \ell^2$$

$$= \frac{10^{-4}}{2 \times 55.4 \times 11.7} \times \frac{1.5 \times 10^{16}}{\rho} \ell^2 \times 10^{-8}$$

where ℓ is in μm and ρ in Ω cm

$$\therefore V = \frac{11.6 \; \ell^2}{\rho}$$

For the above example, V = 105 volts is the potential needed to fully deplete the detector.

Returning to the properties of the reverse biased junction, Figure 29 shows the current/voltage characteristics of a typical silicon junction over a wide temperature range.

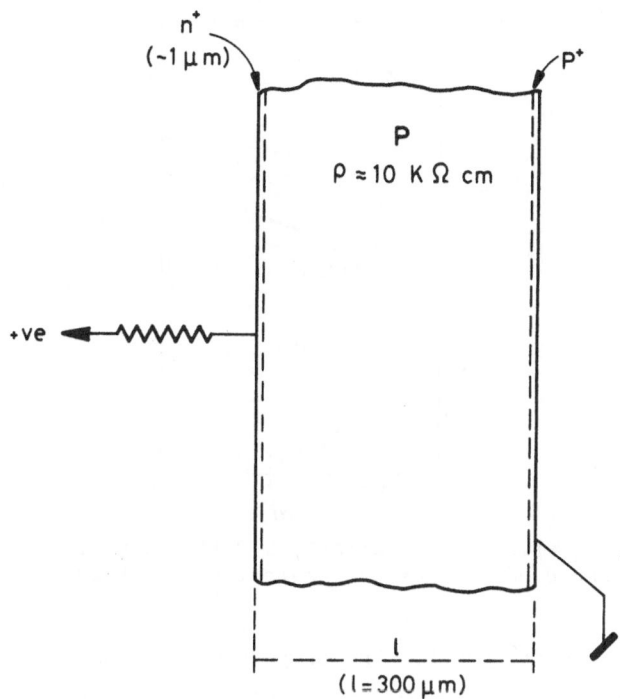

Fig. 28. Typical structure used for particle detection. Essentially it is no more than a diode, reverse biased so as to fully deplete the thick p-type layer.

At high temperatures the leakage current is dominated by thermal electron-hole generation within approximately one diffusion length of the depletion edge. The diffusion length for minority carriers is

$$L_D = \sqrt{D\tau} \qquad\qquad (4.14)$$

where D is the diffusion constant and is related to the mobility μ by $D = \frac{kT}{e}\mu$

For electrons $D_n = 34.6$ cm^2 s-1 $\Big\}$ at room temperature

For holes $\quad D_p = 12.3$ cm^2 s^{-1}

τ is the minority carrier lifetime, and it can vary from ~ 100 ns to > 1 ms depending on the care taken in the silicon processing. This point will be discussed further.

This high temperature leakage current (termed the diffusion current, as previously noted) is almost independent of the reverse bias voltage, but

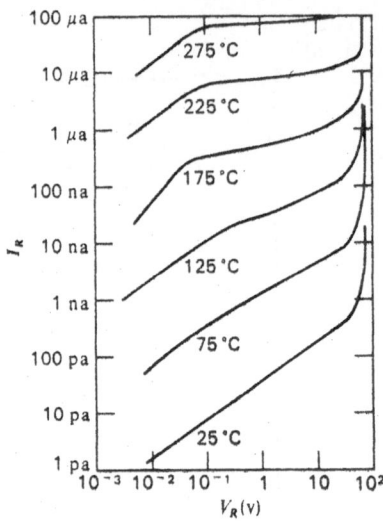

Fig. 29. From Reference 31. Current vs voltage in a reverse biased silicon diode at various operating temperatures.

is highly temperature dependent. The temperature dependence stems of course from the thermal generation of the minority carriers.

At lower temperatures ($\lesssim 100°$ C) the diffusion current becomes negligible and the generation current dominates. This continues to show a similarly fast temperature dependence, but is now also quite voltage dependent, as seen in Figure 29, since the depletion width is proportional to $V^{\frac{1}{2}}$.

The diffusion and generation currents depend on the rate of electron-hole generation, and the diffusion current depends also on the minority carrier lifetime. These quantities are in fact closely related. Direct thermal generation of an electron-hole pair is quite rare in silicon for reasons which depend on the details of the crystal structure. Most generation occurs by means of intermediate generation-recombination centres (impurities and lattice defects) near the band-gap centre. Thus an electron-hole pair may be thermally created in a process where the hole is free in the valence band and the electron is captured by the trapping centre, to be subsequently emitted into the conduction band. These <u>bulk trapping states</u> vary enormously in their density and can be held down to a low level by suitable processing. It is precisely these states which determine the minority carrier lifetime already mentioned. Reducing the density of bulk trapping states does two things. It cuts down the thermal generation of charge carrier pairs in the material, so reducing the concentration of minority carriers available for the generation of current across a reverse-biased junction. It also increases the minority carrier lifetime and so the diffusion length (but only at $\tau^{\frac{1}{2}}$). The first effect vastly outweighs the second, so that a low density of bulk trapping states is highly advantageous in ensuring low leakage current. As we shall see later, even originally high grade silicon can deteriorate due to the production of bulk trapping states by radiation damage.

Mid band-gap impurities such as gold (see Figure 21) are a particularly serious source of bulk trapping centres. As shown in Figure 30, even in low concentrations, gold atoms strongly reduce the carrier lifetimes, and lead to greatly increased dark current.

These effects are obviously not serious in cases where one is collecting large signals promptly. But in cases of small signals and/or long storage times (such as in a silicon drift chamber, or CCD), particular care is needed. One important design criterion is to keep the stored charges well away from the surface of the silicon, since the surface region is more likely to have picked up some undesirable impurities as well as having a high level of lattice defects.

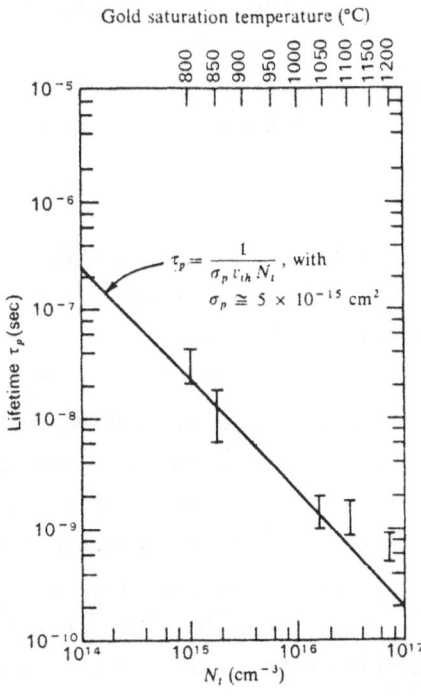

Fig. 30. From Reference 31. Effect of gold impurities of various concentrations on the minority carrier (hole) lifetime in n-type silicon.

4.3.3 Electron Transport in Silicon.

While the charge generated by an ionizing particle is being transported by the internal field in the detector, there is inevitably the process of diffusion which spreads out the original very fine column of charge during this transportation process. In the case of very highly ionizing particles (such as alphas) the original density of electrons and holes can be so high that space-charge effects are important. In the case of min-I particles, however, such effects are negligible and the time development of the electron and hole charge distributions may be treated by simple diffusion theory.

Consider a local region of charge (electrons or holes), for example a short section of the particle track length within the silicon. Under the influence of the internal field, this will be drifted through the material and at the same time will diffuse radially as shown in Figures 31 and 32. The RMS radius of the charge distribution increases as the square root of time (as (4.14)), with standard deviation $\sigma = \sqrt{2Dt}$. Thus 50% of the charge is contained within a radius of $0.95\sqrt{Dt}$. For electrons at room temperature this gives:

Time	Radius	Drift Distance
10 ns	6 μm	135 μm
1 μs	60 μm	14 mm
100 μs	0.6 mm	140 cm

The drift distances listed in the third column are obtained by assuming a 'typical' drift field in depleted silicon of 1 kV/cm, and using the fact that the drift velocity $V_d = -\mu_n E$.

Diffusive charge spreading is an attractive option for improving spatial precision beyond the limits of the detector granularity. For example, one might hope to achieve precision of one or two microns from a strip detector with 25 μm strip pitch, by centroid finding on the basis of measured pulse heights in adjacent strips. This depends on achieving a charge radius of $\gtrsim 30$ μm which (from the above table) implies large drift distances and/or very gentle drift fields. These constraints have different implications for different types of detector but in general, ideas for centroid finding lead to the need for processing high resistivity silicon.

528

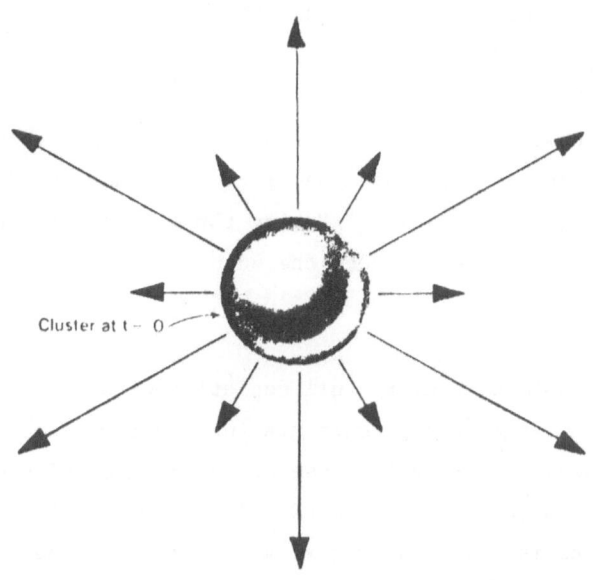

Fig. 31. From Reference 32.
  (a). depicts the random radial diffusion from a small
        central cluster of particles (electrons or holes).

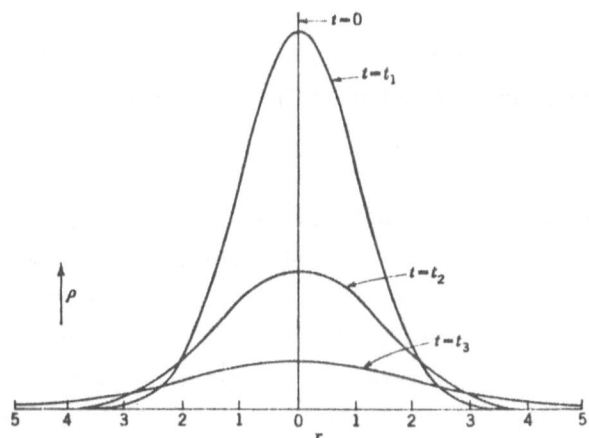

Fig. 31(b). Radial density distribution as a function of time over
        3 equal time intervals.

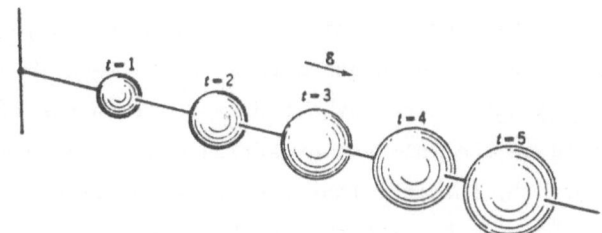

Fig. 32. From Reference 32. Combined drift and diffusion of an intially
        small cluster of particles (electrons or holes) as a function
        of time over equal time intervals. The drift distance is
        proportional to t but the radius grows only as $t^{\frac{1}{2}}$.

5. MICROSTRIP DETECTORS AND SILICON DRIFT CHAMBERS

5.1 Microstrip Detectors

These are based on fully depleted silicon slices with the general
diode structure described in Section 4.3, of thickness typically 300 μm.
The practical lower limit on thickness is set by the need for good
signal/noise for min-I particles. Making the detector thinner loses signal
charge directly and further reduces the output voltage due to the increased
capacitance of the detector.

It is desirable to achieve full depletion with bias voltage $\lesssim$ 100 V
in order to avoid large leakage currents (which produce noise or even
internal breakdown). Therefore, these detectors are built on high
resistivity silicon (typically $\gtrsim$ 10 KΩ cm). Even so, the mean collection
time for electrons is only ~ 4 ns giving a radius for the charge collected
at the surface of approximately 4 μm.

One surface is electrically subdivided into conducting strips for
charge collection on a pitch of typically 20 μm. Readout can be connected
to every strip or to every n^{th} strip, using capacitive charge division
between the floating and connected strips to provide the position
co-ordinate.

Figure 33 shows one option. Readout is connected to every 5th strip.
In this pioneering detector[33], the diode structure was made by the surface
barrier technique. More recently[34] detectors have been built with the
diode structure made by separate p$^+$ implants beneath each one of the
aluminium strips. Such detectors have proven to be very efficient. The
precision when reading out one strip in 3 (with 20 μm pitch) is found to be
σ = 4.5 μm. This precision degrades by about a factor of 2 if only one
strip in 6 is read out.

One limitation with the interpolation approach is the inevitable
degradation in 2-track resolution, which gives rise to a major loss in
precision of pairs of tracks which are merged at the level of the
granularity of the readout strips (even though their charge distributions
may be well separated on the full set of strips on the detector). This
problem is exacerbated by the fact that the number of tracks giving cluster
sizes > 2 goes well beyond the expectations based on δ-electron production.
In fact, with the 1 in 3 readout, 30% of tracks produce these fat clusters,
which can be understood on the basis of capacitances between the strip
holding the signal charge and non-neighbouring readout strips.

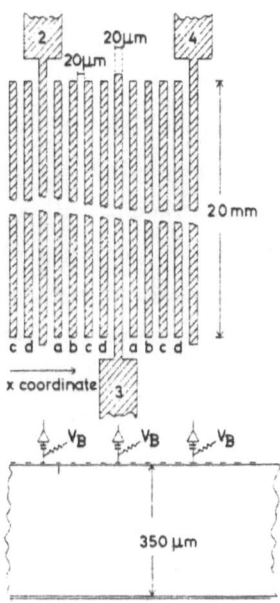

Fig. 33. From Reference 33. Layout of strips on one face of the detector, with one strip in 5 connected to the bias voltage and to the readout system via a locally connected preamplifier.

In order to take advantage of the precision of a detector in measuring impact parameters, it is important to have the first detector plane as close to the vertex as possible, for reasons of multiple scattering as well as geometry. Thus the 2-track resolution is in practice as important, in evaluating detector performance, as the precision. For this reason, the trend in microstrip detectors is to go to readout of every strip. The reason for not doing this in the first place is that with conventional electronics it is just impossible. As an example, consider the microstrip detectors of the NA32 experiment on the CERN SPS. To cover 24 x 36 mm with detectors equipped with 1 in 3 readout in the central $\frac{1}{3}$ of the detector, and 1 in 6 readout in the outer $\frac{2}{3}$ of the detector, involves fanouts, preamplifiers and connectors which extend outwards over a radius of 60 cm and a thickness of about 2 cm, which sets limits on the stacking density of the detectors. So the ratio of areas of the local readout to the detector is greater than 1000. Connecting readout to every strip would increase this ratio to over 5000. On top of this, one has to make room for

the tonnage of cabling which leads to the racks of remote readout
electronics.

In spite of these complications, the MPI Group within the ACCMOR
Collaboration have succeeded in connecting all strips to external readout
in some special detectors used as elements of an active target. These
detectors needed only to have 1 mm of active width (50 strips) so that the
total number of channels could be held down to a reasonable level. With
these detectors, the precision of measurement improved to $\sigma = 3.0$ μm and
the number of clusters more than 2 strips wide was reduced to a very low
level, confirming that the problems with the earlier detectors are a
peculiarity of the charge division readout. The main remaining problems
are:

(a) for angles $\gtrsim$ 100 mrad a rapid loss of precision for the reasons
      discussed in Section 3, which are common to all detectors of the
      thickness needed for microstrip readout.
(b) the density of off-chip (but necessarily local) electronics.

The solution to (b) is under development by several groups. The aims
in all cases are:

(a) to avoid fanouts from the detectors. The electronics is
      fabricated to provide preamp inputs on a pitch of about 50 μm so
      that by connecting alternate strips at each end of the detector
      one can deal with all strips with a pitch of 25 μm or greater.
(b) to provide analogue storage for typically 128 signals on a chip,
      and multiplexed analogue outputs, so that the number of output
      cables is reduced by a factor of 100.

The best results to-date have been made by physicists who have
developed and tested what they call the microplex chip[35]. The general
layout is shown in Figure 34. The chip butts against the detector at its
left-hand edge. The staggering of input pads (IP) is necessary because of
the lack of wire bonders capable of bonding at 50 μm pitch. The staggered
bonding is easily done with a modern computer controlled wedge bonder.

Figure 35 is a block diagram showing the main elements of the
microplex chip. The signals are fed through a charge sensitive preamlifier
of gain 500 onto a storage capacitor. In collider operation, the storage
capacitors would be reset between beam crossings, and in fixed target

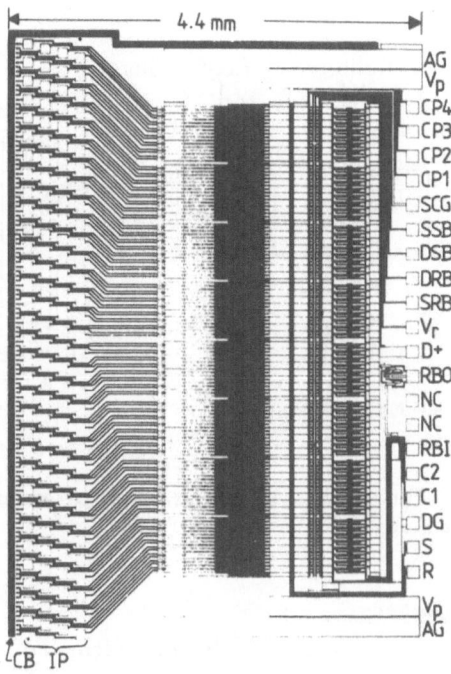

Fig. 34. From Reference 35. Microplex chip layout showing the input pads (IP) on a pitch of 50 μm.

applications they are reset frequently during the spill so that the data are adequately free of out-of-time track information. The stored information is read out at ~ 1 MHz (~ 150 μs for readout of the complete chip). The analogue signal from each capacitor in turn is connected to a line driver for remote digitization.

The first results from a microstrip detector read out with the microplex chips have been reported[36], using signals from a β source. Subsequently, a beam test has been performed, with excellent results[37]; signal/noise ≈ 17, where 'signal' means the mean pulse height from a min-I particle, and 'noise' means the measured noise on a single strip of the detector. (There is some arbitrariness in this, since signals are frequently shared between several strips.)

In the last few years, silicon microstrip detectors have moved from being special in-house production devices from a few privileged laboratories, to commercially available off-the-shelf items available to all groups. The consequent use in many experiments is proving a source of rich physics.

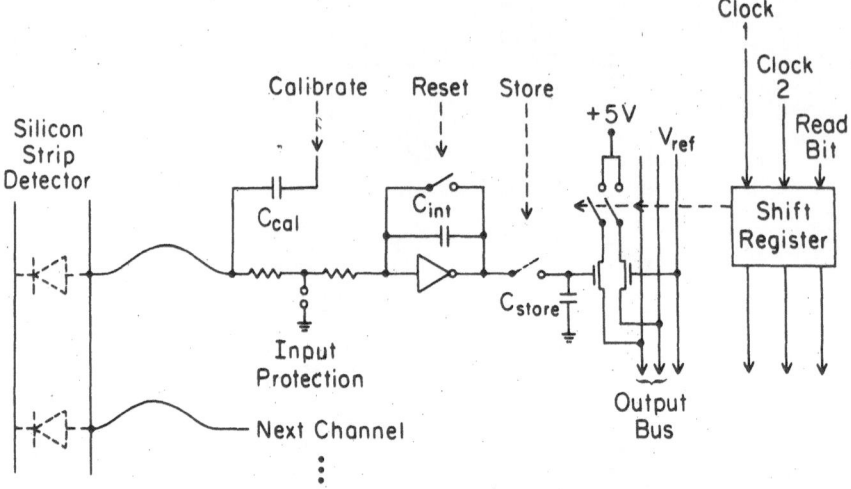

Fig. 35. From Reference 35. Block diagram of readout electronics contained on the microplex chip.

5.2   Silicon Drift Chambers

The silicon microstrip detector is analogous to a gaseous multiwire proportional chamber. Both types of detector are sometimes difficult to use due to the large number of outputs. Just as gaseous drift chambers were developed to greatly reduce this problem, so there has for years been speculation about the possibility of some form of 'silicon drift chamber'. A specific proposal was published by Gatti and Rehak[38], and built in Munich by Kemmer and co-workers. The basic idea is as follows. Starting from a wafer of high resistivity n-type silicon, p^+ implants are made on both surfaces (Figure 36) and covered with grounded metallic electrodes. Then, at a single electrode near the edge of the detector, a positive potential is applied which increasingly reverse-biases both junctions. The depletion regions expand from both sides; eventually they meet and produce a fully depleted (pinch-off) condition. The pinch-off occurs in the mid-plane of the detector. Figure 37 shows the field and potential distributions through the silicon under several different conditions. (a) shows the case of both surfaces being grounded. The electric field grows rapidly once one penetrates the depleted p^+ silicon, has a maximum at the np junction, then falls gradually through the lightly doped bulk material. It changes sign at the mid-plane of the detector. The corresponding potential distribution has the familiar form of a combination of quadratics, and leads to a shallow potential energy well for electron storage at the detector mid-plane. This is not an equilibrium condition, and thermal generation of electrons would cause an accumulation in this region leading to a loss of depletion. If the surface voltages are altered, the potential energy

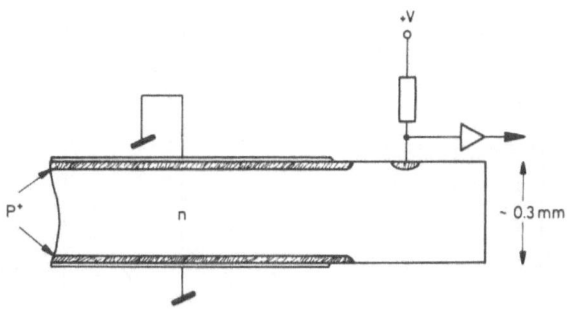

Fig. 36. Cross-section of basic structure used for the silicon drift chamber.

minimum can be shifted close to either face (Figure 37(b)) or even eliminated completely (Figure 37(c)). In the last case any generated electrons would be collected by the more positive surface electrode. The scheme used in the actual detector is indicated in Figure 38. The surfaces are subdivided into strips, whose potentials are graded, becoming steadily more positive from right to left, inducing a very gentle drift field within the detector. This field is small compared with the typical internal fields in silicon induced by the depletion process, as it needs to be if it is to lead to measurably long drift times. Typically, with an electrode pitch of 150 μm and ΔV = 2.3 volts/strip, we have a drift field of 150 V/cm.

From Section 4.3.3,

$$\text{drift velocity } V_d = -\mu_n E$$
$$= 1350 \times 150 = 2.0 \times 10^5 \text{ cm/s}$$
$$= 2.0 \text{ μm/ns}$$

As the electrons approach the anode, the potentials of strips on opposite faces of the detector are made more and more unequal, so that the electrons are slewed towards one face of the detector as shown in Figure 37 and 3-dimensionally in the plots of Figure 39. Given a narrow anode strip and good timing based on zero crosser discrimination, high precision may be achievable in the co-ordinate of the particle normal to the drift direction. As with the microstrip detector, we are discussing a 1-dimensional detector, but with the advantage of a great reduction in the number of output channels. In principle some measure of 2-dimensionality is also possible by using a subdivided anode, and work on this development is proceeding.

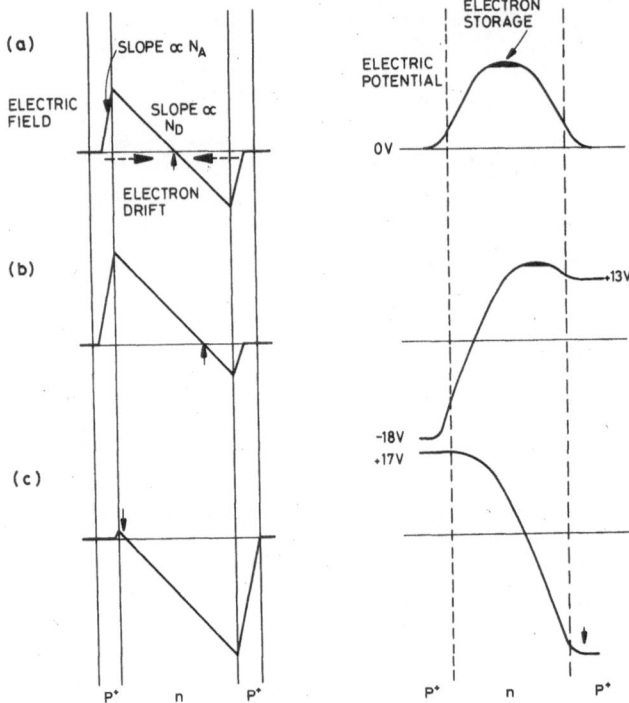

Fig. 37. Field and potential distributions in the silicon drift
chamber for different values of the surface potentials.

For this detector to work with high precision, it is essential to
have very uniformly doped material, in order to avoid inhomogeneities and
consequent smearing of the electron times of arrival at the anode.

Tests with min-I particles have resulted in a measured precision of
20 μm over a 4 mm drift length, compared with 5 μm when a small light spot
was used to generate a surface charge[39]. The difference may be due to some
smearing associated with the extended source of charge generated by the
traversing particle. One possible limitation of this type of detector may
be the 2-track resolution, but this would be helped by a subdivided anode
structure.

6. CHARGE-COUPLED DEVICES

The charge-coupled device (CCD) was invented in 1970 at Bell Labs[40]
and has been developed in two general directions. Linear CCD arrays are
used for optical imaging, for analogue signal storage, as delay lines, and
in a large variety of signal processing applications. Imaging
2-dimensional CCDs take advantage of the high performance of silicon for
photon detection which we discussed in Section 4. These CCDs are used
increasingly in TV cameras, for night vision systems (and general

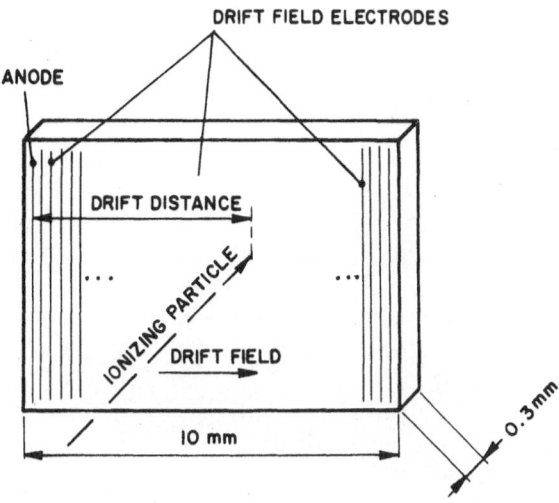

Fig. 38. From Reference 38. Silicon drift chamber. The surface is covered by a strip array of p^+ junction electrodes, which provides the depletion and the drift field. Electrons produced by the passage of a fast charged particle drift towards the anode, which is the only readout channel on the wafer.

surveillance), in astronomy (visible and X-ray, where they have many advantages over photographic plates) and in hybrid form as infra-red detectors. In this last application, the primary photon detector is a narrow band-gap material, and the charge collected is transferred to a CCD via an array of bump bonds. The possibility of using CCDs as high precision detectors of min-I particles was evaluated theoretically about 5 years ago[41].

It is with the 2-dimensional area arrays that we have to deal in the context of particle detection. These consist in general of a fine matrix of potential wells just below the surface of the silicon, typically 20 μm x 20 μm x 10 μm deep. Each well constitutes an element of a picture in normal imaging applications and is referred to as a pixel. The CCD has in addition some surface structure which allows charges from each pixel to be transported and deposited in turn onto the output node of the detector. These signals are sensed by on-chip circuitry in order to minimise noise. Let us examine in some detail, with the aid of the general discussion of Section 4, how such a detector can be built. For more detailed information, there are some excellent books on CCDs[42,43] as well as CCD Conference proceedings and hundreds of published papers.

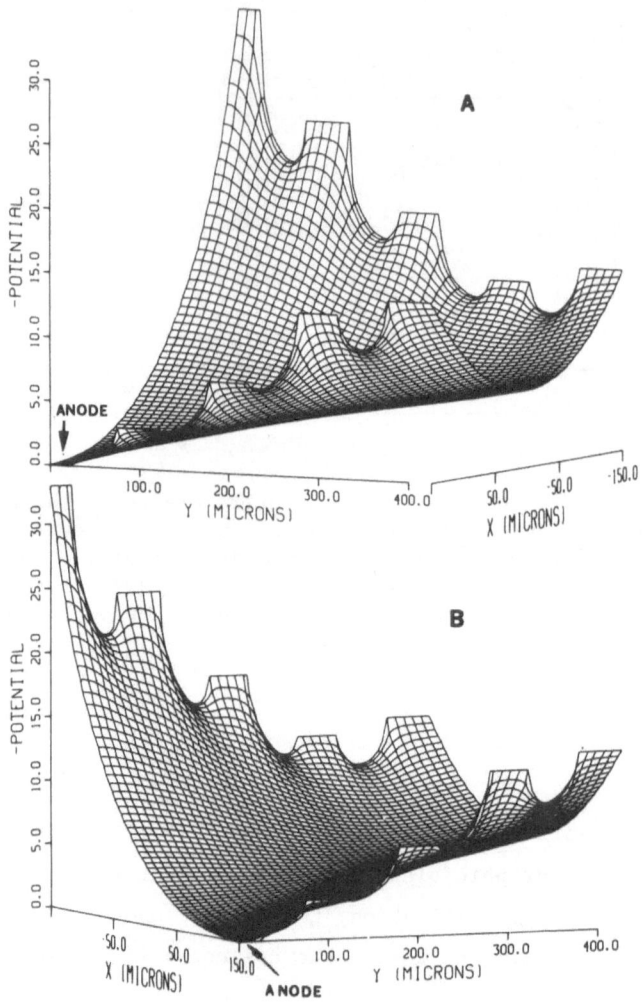

Fig. 39. From Reference 38. The potential within the semiconductor
drift chamber close to the readout anode. (Two views of
the same potential from different points are shown.)

6.1   Structure and Operation of 2-d CCDs

Let us first consider the steps in making a device which would have
some (but not yet all) of the features of a CCD. Starting with a
low-resistivity suitably inert substrate (see Figure 40(a) to (c)) we
proceed to grow an epitaxial layer of higher resistivity silicon with a
thickness adequate to contain all the necessary structures and associated
field penetration. We next make an np junction by the introduction of a
shallow (~ 1 μm) implant of n-type dopant. The surface is oxidised to make
an insulating layer and on top of this is deposited a thin conducting
layer. The simplest would be aluminium, but for light detection a high
degree of transparency is important, and about 0.3 μm low resistivity

'polysilicon' (amorphous silicon) would commonly be used. By analogy with FETs, the conducting surface layer is termed a gate.

Let us now put some bias voltage onto the structure, as shown in Figures 40(d) to (f). Grounding the substrate ($V_{ss} = 0$) we apply V_c to the n-channel and V_G to the gate. Initially assume $V_c = V_G$. Even with $V_c = 0$, as we learned in our discussion of the np junction, there will be a thin depletion layer around the interface between the two types of silicon. By increasing V_c, we are able to deplete more of the material as the junction becomes more and more strongly reverse biased. With the parameters chosen

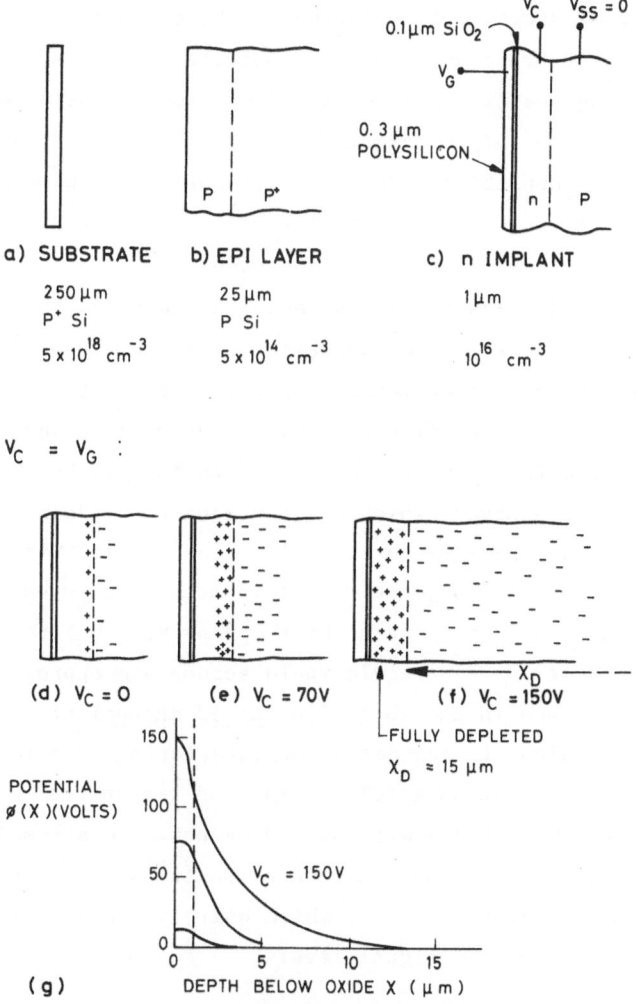

a) SUBSTRATE   b) EPI LAYER   c) n IMPLANT

250 μm   25 μm   1 μm
P⁺ Si   P Si
5×10^{18} cm⁻³ 5×10^{14} cm⁻³ 10^{16} cm⁻³

$V_C = V_G$:

(d) $V_C = 0$ (e) $V_C = 70V$ (f) $V_C = 150V$

FULLY DEPLETED
$X_D \approx 15$ μm

POTENTIAL
$\phi(X)$(VOLTS)

150
100
50
0

$V_C = 150V$

0   5   10   15
(g)   DEPTH BELOW OXIDE X (μm)

Fig. 40(a). to (c). show (with increasing magnification for each stage) the successive stages in making a CCD-like structure.
(d). to (f). The depletion process which would apply if V_C and V were increased together.
(g). The corresponding potential distributions as a function of G depth in the silicon.

539

in this example, a high voltage would be needed to achieve complete depletion of the n-channel, at which point we should have depleted about 20 μm of the p-type substrate. The potential distributions for increasing values of V_c are shown in Figure 40(g). For V_c = 150 V, such a device when traversed by particles would transport the generated electrons to the surface (Si/SiO$_2$ interface) and dump the holes into the undepleted substrate.

Now (Figure 41(a) and (b)) consider what happens if V_c is increased from 0 while V_G is held at 0 volts. Here the situation is entirely different; the large capacitance between the n-channel and the gate provides a further mechanism for depletion of the channel. The depletion around the np junction proceeds as before, but the voltage across the oxide induces an increasing positive charge, starting from the Si/SiO$_2$ surface and growing into the body of the n-channel. At a very low value of V_c (about 8 volts) these depletion regions meet, causing the phenomenon known as pinch-off. The corresponding value of V_c is called the pinch-off voltage and when it is reached further increases of V_c (which can be controlled say by an edge connection) have no influence on the potential over the area of the detector. The depletion depth in the p-type material is only about 6 μm in this case. What is particularly interesting is the potential distribution in the silicon. This is shown in Figure 41(c); look initially at the curve for V_G = 0. The quadratic form in both types of silicon is of course preserved (this is a consequence of Poisson's equation and uniform doping) but there is now a maximum in the electric potential just below the depth of the np junction. This acts as a potential energy minimum for electrons, so (in contrast to the case V_G = V_c) the electrons liberated by the passage of a particle would accumulate approximately 1 μm below the silicon surface in the so-called buried channel of the device. This is a vital ingredient in the design of CCDs for our application. Tiny charges (< 10 electrons) can be safely stored and transported as long as they are held in the bulk of the silicon. Once they are allowed to make contact with the surface they encounter numerous traps which cause serious loss of charge. Surface-channel CCDs, while quite commonly used, should be avoided for work with very low signal levels.

Notice that the situation depicted in Figure 41(c) represents a non-equilibrium condition. Thermally generated electrons would accumulate in the potential energy minimum and drive more and more of the n-channel out of depletion. CCD operation relies on some procedure for keeping the channel swept clean of electrons at an adequate rate.

Assuming that we avoid this accumulation of electrons, the effect of now varying the gate voltage V_G is to a first approximation simply to vary the depth (in volts) of the potential well but hardly at all to change its depth (in microns) below the silicon. There is in fact a slow variation in the depletion depth with V_G, as can be seen from the figure. The quantitative calculation follows easily from what we have done in Section 4; see for example Reference 42 for the details.

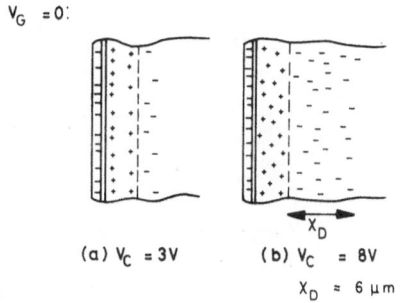

$V_G = 0$

(a) $V_C = 3V$ (b) $V_C = 8V$

$X_D \approx 6\,\mu m$

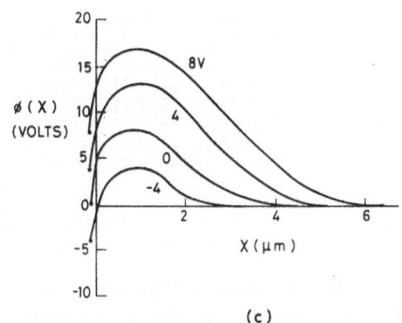

(c)

Fig. 41(a). and (b). The depletion process in normally biased CCD operation with V_G negative with respect to V_C.
   (c). The corresponding potential distributions after channel pinch-off for various values of V_G.

The device we have created has all the depth characteristics of an imaging CCD, but it still lacks two important features before it will have the necessary pixel structure over the surface. These are illustrated in Figure 42. Firstly, at the required pixel granularity (say 20 μm) p^+ implants are introduced of approximately 1 μm width and 1 μm depth. These become fully depleted as part of the overall biasing of the CCD, and so provide stripes of intense negative space charge which effectively repel electrons. Thus the electrons in the buried channel will now be confined to separate storage wells which run from top to bottom of the detector, in

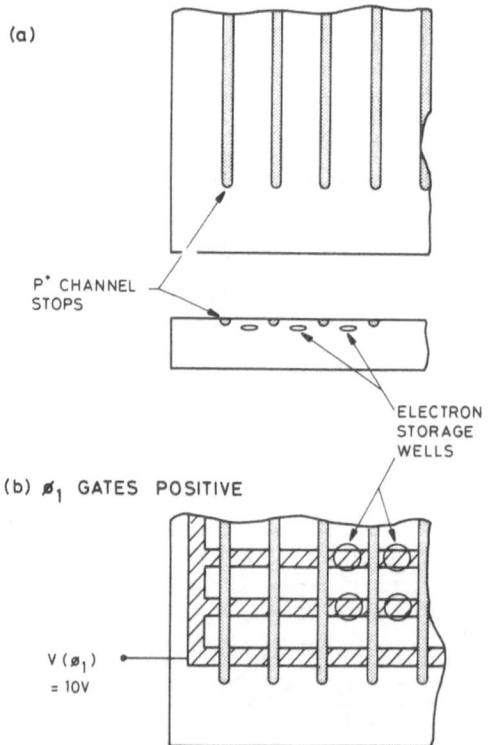

(a)

P⁺ CHANNEL STOPS

ELECTRON STORAGE WELLS

(b) ⌀₁ GATES POSITIVE

V (⌀₁) = 10V

Fig. 42. Establishing the potential well structure:
(a). Channel stops create potential barriers running vertically on the device.
(b). Gates create horizontal potential barriers. The combined result is a matrix of localised wells, each of which constitutes a pixel.

the view shown in Figure 42(a). The typical doping level of the channel stops is $N_a = 10^{18}$ cm^{-3}.

Secondly, the charges are confined in the vertical direction by making a polysilicon gate structure which is not uniform across the surface but which consists of a series of horizontal bars. By biasing these positively (see Figure 41(c) and Figure 42(b)) we can achieve potential wells under each of the intersections between these gate electrodes and the regions midway between the channel stops. We now have a matrix of discrete potential wells which may approach 10^6 in number on a typical CCD (400 channel stops x 600 gate electrodes).

But still we do not have a working CCD, since those potential wells are immobile. We can accumulate charge images but cannot read them out. To do this, we make a more complicated gate structure (Figure 43). We arrange these gates in triplets (ϕ_1, ϕ_2, ϕ_3) in this so-called 3 phase CCD structure. The static situation is for one phase (say ϕ_1) to be high, so that the electrons are stored under this phase. Then by manipulating the voltages between ϕ_1 and ϕ_2 as shown in the figure, the electons are moved to ϕ_2. Keeping ϕ_3 low throughout this operation ensures that the charges between adjacent pixels cannot be smeared together. The total physical width of ϕ_1 + ϕ_2 + ϕ_3 electrodes together constitutes one pixel, eg 3 x 7 μm = 21 μm.

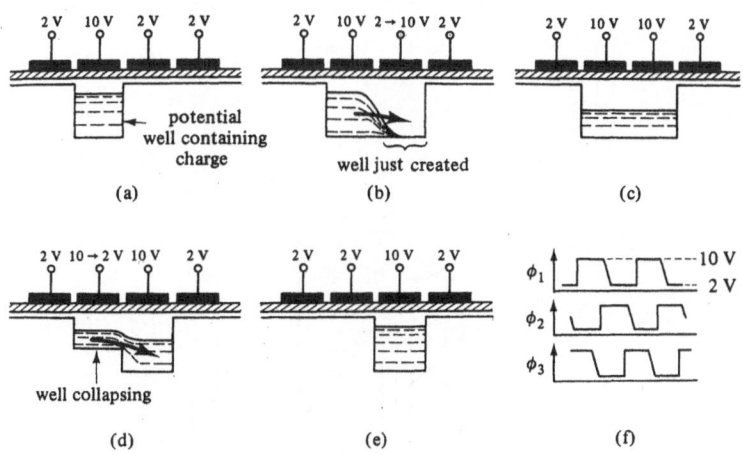

Fig. 43. From Reference 42.
        (a).to (e). Movement of potential well and associated charge packet by clocking of gate electrode voltages.
        (f). Clocking waveforms for a 3-phase CCD.

Now we have developed the capability to move all the stored charges down the device (for example) by one pixel at a time. Apart from 3 phase CCDs, there exist other varieties (4 phase, 2 phase, virtual phase, etc) but we can ignore these in an initial discussion of these detectors.

One further element in the system is needed. At the bottom of the area array called the imaging or I array is a linear CCD, the output register or R register into which the charges stored in the bottom row of the I array can be shifted. Once in this register, that row can be shifted sideways so that the charge contained in each pixel is sensed in turn by an on-chip circuit.

Figure 44 shows a diagram of one corner of a CCD detector, including the general charge collection operation for the passage of a min-I particle. One pixel is shown as a shaded area, covering the height of 3 I gates (Iϕ1 to 3) and bounded by two channel stops. The charge in the bottom row of pixels can be shifted first to the bottom gate of those pixels (Iϕ3) and then into the output register (Rϕ1). Once in the R register, these charges are transported sideways to the output node of the CCD.

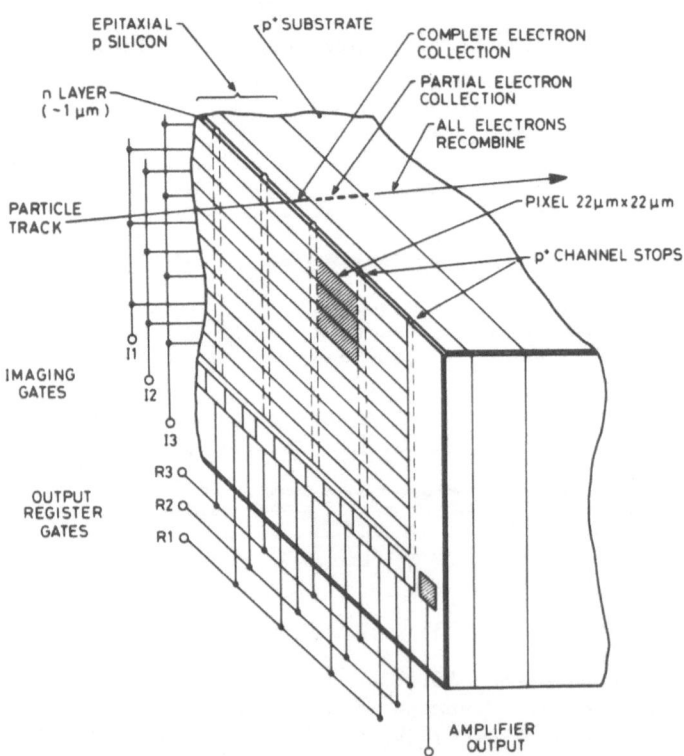

Fig. 44. One corner of a CCD enlarged to show details of the pixel (storage element) structure.

Figure 45 is a photograph of the corner of a 3 phase CCD. Apart from the features noted in the caption, the elbow-shaped aluminium track to the left of the R register carries the stored charge from the R register to the gate of an on-chip FET. This gate (dark structure on the figure) can be seen sandwiched between the source and drain of the small on-chip FET, whose connections (further aluminium electrodes) can be seen disappearing off the bottom of the figure.

The CCD structure shown in Figures 44 and 45 is sensitive to light or to particles over the full active area. It should be noted that this is not true of all imaging CCDs. Some, for example, have more complex channel stops; pnp structures which can be used for anti-blooming or for fast-clearing the CCDs. Such devices have dead bands between each pixel, a feature which makes them unacceptable for most applications as particle

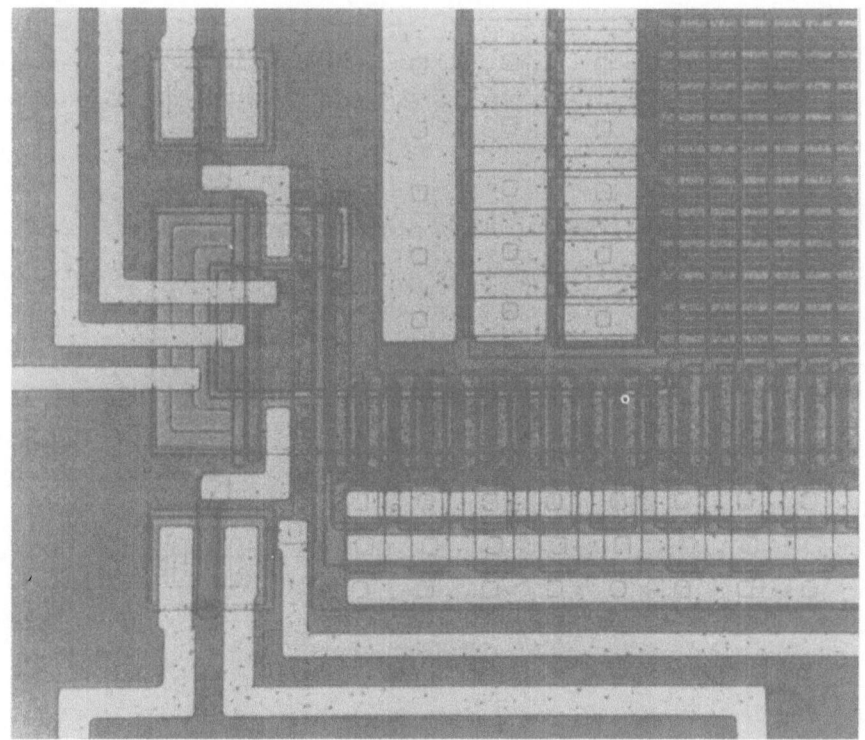

Fig. 45. Courtesy of GEC, England. Photograph of one corner of a CCD showing the pixels of the imaging area (upper right quadrant), readout register (running along the bottom of the imaging area and extending 10 pixels to the left of it) and output FET (below and to the left of the readout register). The light coloured structures are aluminium tracks which carry the drive pulses to the gates, connections to the FETs, etc. The 3 broad bus-lines running vertically carry the Iϕ voltages, and the 3 narrow lines running horizontally carry the Rϕ voltages.

detectors. As shown in Figure 44, the charge generated by a min-I particle along its track falls into 3 classes. There is a region of typically 10 µm below the surface for which the charge is within the depletion depth (see Figure 41) and is fully collected into the relevant pixel. Next, the charge from the 10 µm of undepleted epitaxial silicon (which generally has a long diffusion length) diffuses isotropically. About half of it diffuses into the depletion region and is caught in the relevant pixel or in neighbouring ones. Finally the electrons and holes generated in the p^+ substrate recombine very readily. The "effective thickness" of the CCD for particle detection is thus approximately 15 µm.

As has already been noted, the CCD potential wells represent a non-equilibrium condition. Thermal generation of electron-hole pairs in the material provides a source of electrons which accumulate. For TV imaging, these constitute a minor background, but for astronomy the long integration times and low signal levels necessitate cooling, typically to liquid nitrogen temperatures. For particle detection the requirements are less stringent and operating temperatures around 200° K may be entirely adequate, but this depends strongly on the timing of the clearing and readout of the detectors.

The rate capability of the driving electronics can be made quite high. It is (for example) no problem to shift the charges down the I array at 3 MHz or across the R register at 10 MHz with extremely high charge transfer efficiency. This quantity, often abbreviated to CTE, is the efficiency with which a bucket of charge is transferred from one pixel to the next. In buried channel CCDs, the charge transfer inefficiency may be as low as 10^{-5}. What has given CCDs their reputation of being very slow detectors has been the time required to sample the signal in order to achieve a sufficiently low noise level. This point will now be considered in some detail.

6.2   Readout Electronics

As shown in Figure 46, the charge from the pixel at the end of R register can be transferred onto the gate of an on-chip FET which is normally operated as a source follower. Its output is connected to a local preamplifier which senses the voltage change induced by the charge on the FET gate. This point can also be reset to some standard voltage V_R via an on-chip reset transistor, and this is conventionally done between each R transfer.

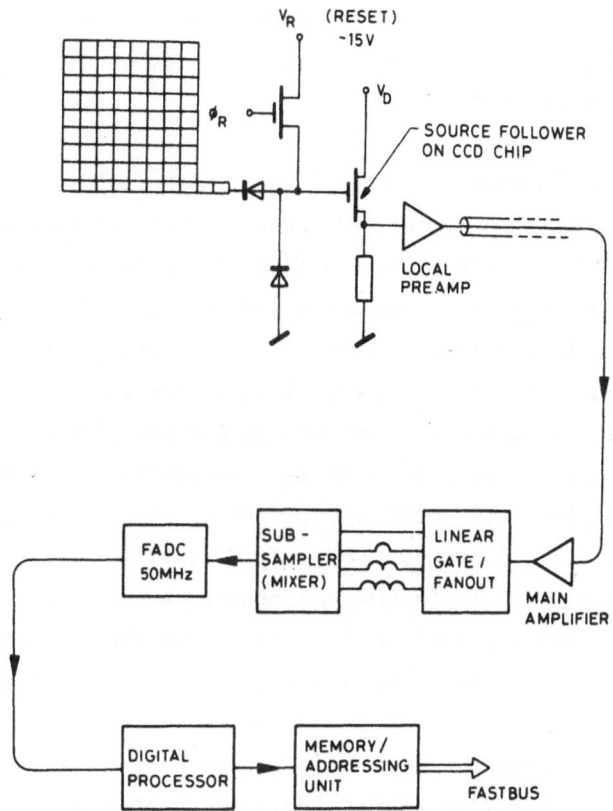

Fig. 46. Block diagram of low noise signal processing, which uses a flash ADC for digital sampling of the pulse height information.

For high signal levels (eg TV imaging) there is very little to add to this description. The outgoing voltage level provides a direct measurement of the stored charge (ie brightness) of that part of the image, which is built up line by line as the contents of successive rows are sequentially transferred from I to R register and along the R register to the output.

Typical TV imaging involves signals near to the well capacity of the devices (several x 10^5 electrons/pixel). The sensing of very small signal levels in CCDs has been pioneered by night vision specialists and by astronomers, where total readout noise of less than 10 electrons RMS is regularly achieved. In this way images with signals as small as 50 electrons/pixel can be reconstructed with good quality. The price paid by these special readout systems is speed. Given that in astronomical applications the images may be accumulated over periods of 20 minutes or more, a readout time of 10 seconds is completely acceptable. For particle detection in typical high energy physics experiments, a readout time of some milliseconds or at most tens of milliseconds is required. An important recent development has been the achievement of such fast readout without a serious degradation in noise.

Let us first consider the conventional method of low noise CCD readout, beyond the necessity already mentioned of adequately cooling the detector. The node is reset before each R transfer. Due to the inevitable existence of reset noise (the voltage actually set fluctuates uncontrollably by typically 200 electrons) it is next necessary to determine this precise voltage level. This is not entirely trivial for the following reason. In order to have good sensitivity (μV/pC of charge stored) the FET gate capacitance must be made small. It is typically held down to ~ 0.1 pF, but small FETs are inevitably noisy. Thus, in order to determine the output voltage to the required precision some signal averaging is necessary. Typically, astronomers use analogue integration of the output signal for a period of say 20μs. Next the R register is clocked, and the pixel charge is depositied onto the output node. A second integration is made and the voltage difference yields the charge contents of that pixel. This readout technique is generally known as correlated double sampling, and implies readout times of about 40 μs/pixel, ie 10 s for a complete CCD.

Such a readout system was used in the first measurement of the efficiency and precision of CCDs as detectors of min-I particles[29]. Subsequently, my group addressed the problem of speeding up the readout to the point where it could be used in a real experiment, and the block diagram of Figure 46 and pulse trains of Figure 47 show how this was done.

The first (and in some ways the most difficult) step was to develop a sufficiently clean CCD drive system. Any of the external drive pulses (Iϕ(1-3), Rϕ(1-3) or ϕ_R, the Reset) can couple capacitively to the CCD output, inducing various types of feedthrough. Given that these pulses are ~ 10 V and the min-I signal is ~ 1 mV, the feedthrough must be kept extremely small in order not to swamp the signal. This implies very clean drive pulses (sharp edges, no ringing) and careful layout of the external circuitry. The layout on the CCD itself has been very carefully designed to avoid problems in that area. Since the I gate capacitances are large (~ 10^4 pF) the drive system needs to provide these clean, relatively high voltage pulses also with high current. Fortunately, the overall readout time is dominated by the R drive system, since there are about 400 R transfers for every I transfer. The R gate capacitances are only about

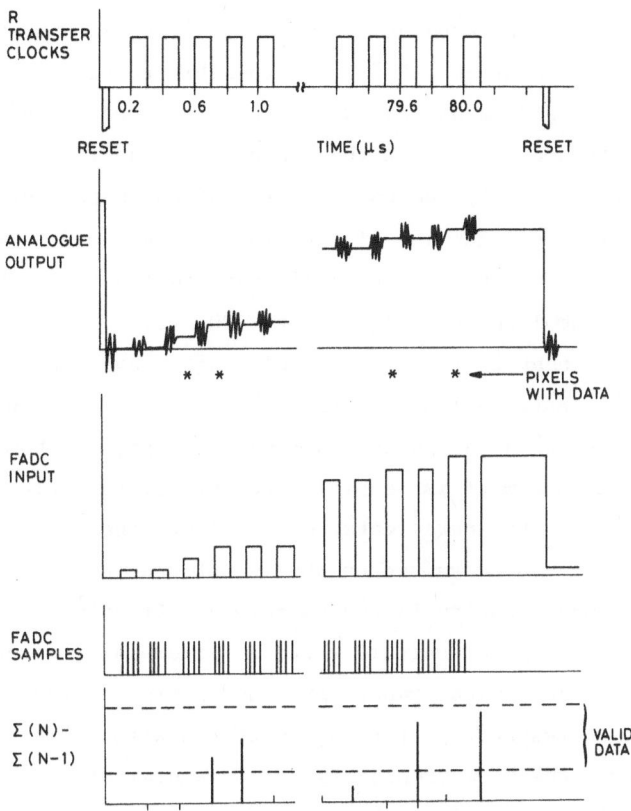

Fig. 47. Sequences of signals through one row of CCD data processing. After processing a row, the output ode is reset and the I gates are clocked to shift the next row into the now-empty R register. The process is repeated for each of the 600 rows of the detector.

1000 pF, so the drive currents are much lower than for the I gates. Eventually, a system has been developed which can drive the R register at 10 MHz with good charge transfer efficiency, as well as being sufficiently well decoupled from the analogue output to allow sampling to commence within 20 ns of the end of the Rϕ triplet.

The local preamplifier has $\sim$ 40 MHz bandwidth and sufficiently low noise ($\sim$ 3 nV/Hz$^{\frac{1}{2}}$) that it does not seriously increase the noise induced by the source follower ($\sim$ 10 nV/Hz$^{\frac{1}{2}}$). The signals are then fed via a remote main amplifier to a linear gate and fanout. The linear gate is opened only outside the duration of the Rϕ triplet, so protecting the later electronics from the large amplified spikes induced by the CCD clocking. Four outputs from the fanout are fed into an analogue mixer after relative delays of 0, 5, 10 and 15 ns. The mixer output is thus somewhat smoothed, as would be achieved by signal averaging at 200 MHz. This output is then sampled by an 8 bit flash ADC system operated at 50 MHz. By taking 4 samples over 80 ns (effectively 16 samples at 200 MHz) it is possible to achieve RMS noise levels on individual pixels of approximately 50 electrons equivalent pixel charge. While this does not match the measurement quality of the astronomers, as it obviously can not, in view of the much shorter sampling period, it is still less than 10% of the min-I signal, and so is entirely adequate for particle detection with good centroid finding. The transfer of charge and its sampling occupies in total 200 ns. The next feature of the fast readout is that instead of resetting after every pixel, as is customary in TV and astronomical imaging, we reset only once per row. Thus the analogue output builds up, with a series of steps, as shown in Figure 47. This arrangement is feasible because for particle detection we are always concerned with sparse data, which allows the contents of 400 pixels (mostly empty) to be summed on the output node without any problems of saturation or non-linearity in the on-chip or off-chip electronics. The flash ADC input (sub-sampler output) shown in Figure 47 thus consists of a series of levels, mostly equal (apart from noise fluctuations) but with occasional steps corresponding to pixels with non-zero stored charge. The digital output from the FADC sampler is fed to a digital processor which sums the samples in groups of 4, then makes a subtraction between these sums for successive pixels. The result may be positive or negative due to noise, but in the case of a genuine min-I signal it will be positive and within a range for valid data (Figure 47). Signals from pixels having valid data (in particular satisfying a threshold requirement) are sent to a memory unit (Figure 46) which includes synchronised clocking information so that it can store the pixel address along with the digitised pulse height.

6.3   Use of CCDs for Min-I Particle Detection

In the first tests of a telescope of CCDs in a beam line[29], it was established that they had high efficiency (98 ± 2%), good precision in x and y (~ 5 μm) and good 2-track resolution. The last feature is illustrated in Figures 48 and 49.

More recently, CCDs are being used in a charm production experiment (NA32 at the CERN SPS). By placing the detectors close to the target, the entire spectrometer aperture can be covered by $\frac{1}{3}$ of the CCD area, as shown in Figures 50 and 51. By using the fast readout system described in Section 6.2, this area can be read out within 12 ms, compatible with the readout time of the rest of the equipment in the spectrometer. However, for fixed target applications, this is not sufficient. The CCDs will be

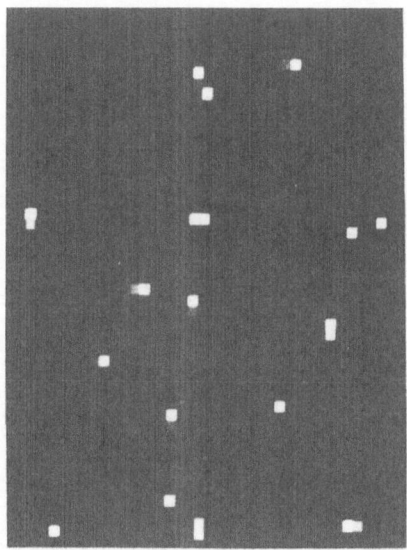

Fig. 48. Zoomed online display of a high density region of tracks traversing a CCD detector. The display shows 17 beam tracks in an area of 1 square millimetre.

traversed by beam particles and by the products of out-of-time interactions which would build up a large background of spurious hits, making pattern recognition impossible. In the NA32 experiment, for example, the CCDs are traversed by a beam of about 10^6 particles per second. How can this be tolerated?

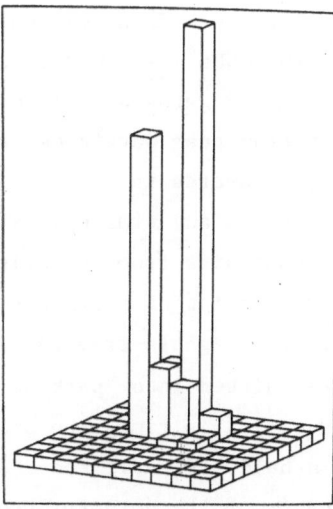

Fig. 49. Isometric offline reconstruction of two particles separated by 40 μm in a CCD detector. The height of each element represents the pulse height measured in the corresponding pixel.

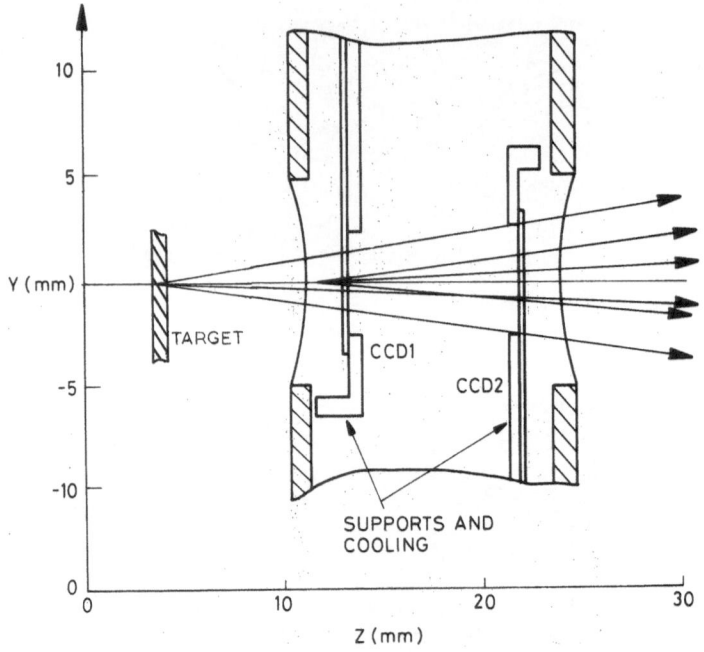

Fig. 50. Layout of CCDs in experiment NA32. A pair of detectors, in a low temperature cryostat with thin vacuum windows, is located ~ 10 mm downstream of a thin copper target. The CCDs are in metal packs which provide thermal coupling to the refrigeration (cold nitrogen gas) but these packs are cut out in the regions used for particle detection.

The method adopted in NA32 is illustrated in Figure 52. The beam profile is made thin vertically and wide horizontally, illuminating the regions of detector shown. The CCDs are run during the beam burst in a "drift chamber" mode. The I and R gates are clocked continuously together at about 2 MHz. The signals from beam particles traversing the CCDs are transported upwards in CCD1, downwards in CCD2, and eventually dumped at the output node. The detectors at any time have a low and perfectly acceptable density of background hits from recently arrived beam particles and interaction products. On receipt of a trigger the fast shifting continues for about 200 μs. The signals from the triggering event (shown as crosses in Figure 52) are shifted into "parking areas" at which stage the fast shifting stops. During this time, the beam is turned off with a simple kicker magnet, and is held off until the end of the readout period. The data in the parking area ($\frac{1}{3}$ of the detector nearest the R register) can now be read out in background-free conditions.

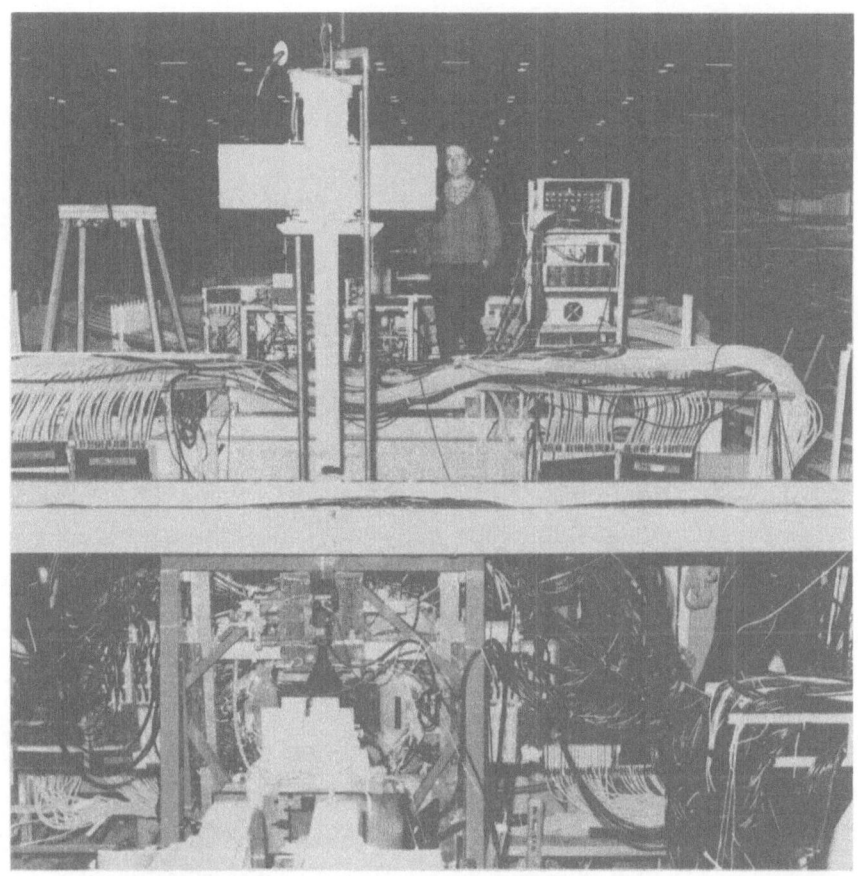

Fig. 51(a). General view of the NA32 CCD detector. The drive signals
are generated on the large drive cards at the top and fed
down to the CCDs on a system of low impedance strip-lines.
The detector is shown in the 'raised' position. For
running, it is lowered into the congested target area which
is visible below.

This method of fast-clearing is applicable in general. Coupled with
the high data storage capability of CCDs (10 spurious tracks per mm^2 would
occupy only 1% of the pixels) these detectors, even without a genuine
fast-clear capability, can easily be used in high rate environments.

6.4 CCD/MSD Comparison

So far, of the various ideas for silicon detectors, only MSDs and
CCDs have been used for high precision particle tracking in experiments.
Let us try to summarise their relative attributes.

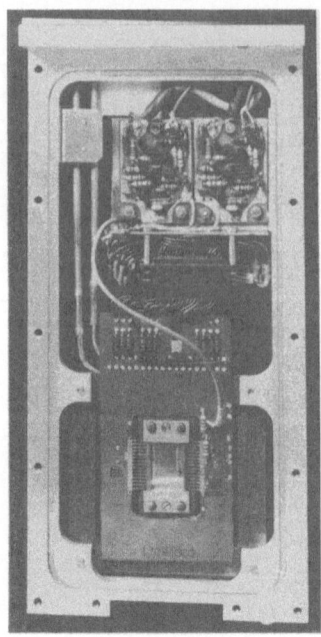

Fig. 51(b). Close-up of CCD1 at the bottom of the cryostat.

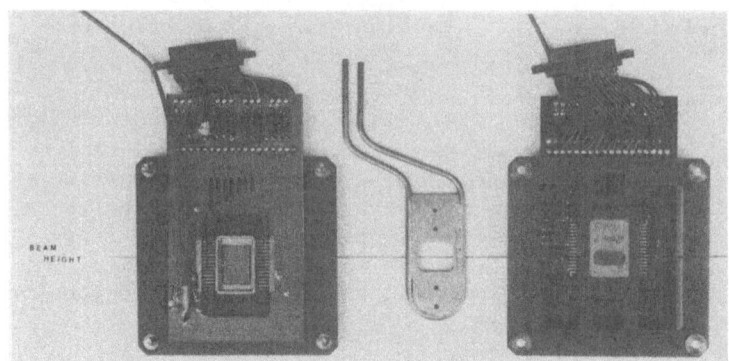

Fig. 51(c). Component elements displaced sideways; CCD1 seen from the front, cooling plate, CCD2 seen from the back.

Effective Detector Thickness

$$CCD \quad \sim 15 \ \mu m$$
$$MSD \quad \sim 300 \ \mu m$$

In both cases, these thicknesses are about at the lower limit for good signal/noise with current readout electronics. However, there is considerable scope for development. For example, CCD output circuits specially adapted for small signals are being designed with

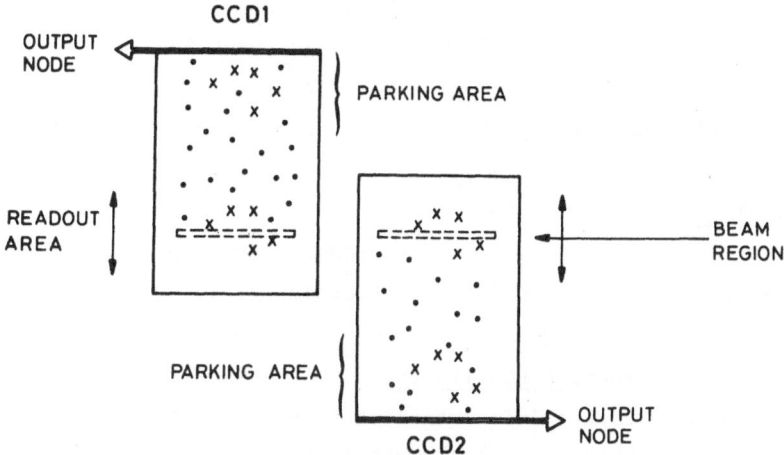

Fig. 52. Beam's eye view of the NA32 CCDs displaced sideways (as in Figure 43(c)) illustrating the "drift chamber" mode of readout.

1 electron RMS noise. CCDs are normally backed by inert silicon giving a <u>total</u> thickness similar to MSDs. If multiple scattering is important, they can conveniently be thinned to less than 100 μm.

<u>Most Probable Energy Loss</u>

$$15 \times 200 \text{ eV} = 3.0 \text{ keV for CCDs}$$
$$300 \times 280 \text{ eV} = 84 \text{ keV for MSDs}$$

<u>Typical Readout Noise</u>

$$0.21 \text{ keV for CCDs}$$
$$8.5 \text{ keV for MSDs}$$

<u>Measurements</u>

x, y points for CCDs

x or y co-ordinates for MSDs

Developments are under way to read x and y simultaneously from one MSD (orthogonal strips on front and back) but of course this does not resolve the pairing ambiguities.

<u>Precision</u> (0°)

5×5 μm for CCDs

3 μm for MSDs if every strip read

Development of current ideas could lead to 1 μm MSD precision, and 0.2×0.2 μm CCD precision.

Precision (Angled Tracks)

Due to the thickness, this degrades much faster for MSDs than for
CCDs. Energy loss fluctuations lead to non-Gaussian tails. At 45°:

> CCDs have 10% probability of error > 4 μm
>
> MSDs have 10% probability of error > 80 μm

2-Track Resolution

> 40 μm in space   for CCDs (reading every pixel)
>
> 40 μm in projection for MSDs (reading every strip)

Flux Limit Per Crossing (Colliders)

Due to the much higher information storage capacity of CCDs (2500 per
mm^2 compared with 50 per mm^2 for MSDs) they have a much higher
capability of absorbing extraneous hits from background interactions,
synchrotron radiation, etc. The actual limit depends on the quality
of the overall tracking system (ie how precisely a track can be
projected onto the CCD detectors in order to determine which hits are
signal and which background).

Beam Rate (Fixed Target Experiment)

> ~ 1 MHz for CCDs
>
> ~ 10 MHz for MSDs with fast readout

Use of multiplexing readout tends obviously to reduce the rate
capability of MSDs.

Readout Time

> ~ 40 ms for CCDs
>
> ~ 1 ms for MSDs via multiplexing

MSDs can give fast outputs (eg for triggering) in those cases where
one can tolerate independent readout electronics on each strip
(practicable over a small area only).

Area Coverage

> ~ 4 cm^2 for currently available CCDs
>
> ~ 20 cm^2 for currently available MSDs

For larger area coverage (eg colliders) in both cases there are plans
for mosaics of many detectors.

Overall, we may say that CCDs have a role when one wants to work as
close as possible to the primary vertex, where track merging and/or
background flux would prohibit MSDs. They then have the advantages of

minimal extrapolation length of tracks to the primary vertex (most important if multiple scattering is serious), unique space points and potentially higher precision.

MSDs provide less information but have advantages in the case of higher continuous rates and where larger area coverage is essential (eg photon beams or LEP).

Note also that both forms of detector are advancing rapidly. The limitations of 2 years ago (slowness of CCDs, very large volume of off-chip electronics for microstrips) are now much reduced. Furthermore there are some very interesting ideas for hybrid detectors (MSDs with linear CCD readout, deep depletion silicon detectors connected by bump bonding to 2-d CCDs, etc) which will not be discussed here since they are rather speculative. But they may be very important in future. Hybrid detectors have been widely developed for other fields, especially in infra-red imaging. For a review of the inter-connection possibilities, there is an excellent paper by Chan[44] on this subject.

Finally, it is perhaps interesting to comment on a major technical challenge in this field. The granularity of CCDs in most vertex detector applications is more than adequate. But (remembering Figure 4) one would welcome improved measurement precision below 5 μm in all cases where multiple scattering is not the limiting factor. This applies especially in fixed target experiments where secondary momenta are high and the limit from multiple scattering is typically much less than 1 μm. The aim for improved precision can in principle be met by some technique for charge spreading. CCDs are used in star guidance systems for space based applications where, by defocusing the star images on the CCD surface and using a centroid finding approach, precisions of $\sigma_x = \sigma_y = 0.1$ μm have been achieved[45], despite the pixel sizes of 20 x 20 μm. The principle of this technique is illustrated in Figure 53. How could this be used for particle detection? In view of the very unfavourable diffusion vs drift characteristics mentioned in Section 4.3.3, the solution probably does not lie in this direction. One idea, which has not been fully evaluated, is to build a very shallow depletion CCD (~ 1 μm depletion depth) on an epitaxial layer of ~ 10 μm thickness (see Figure 54). In this case most of the signal would be generated by the isotropically diffusing electrons in field-free undepleted material, which might give adequate charge spreading for measurement precision of 0.2 μm to be achieved. But this is a very new area and there is certainly room for imaginative thinking on the problem.

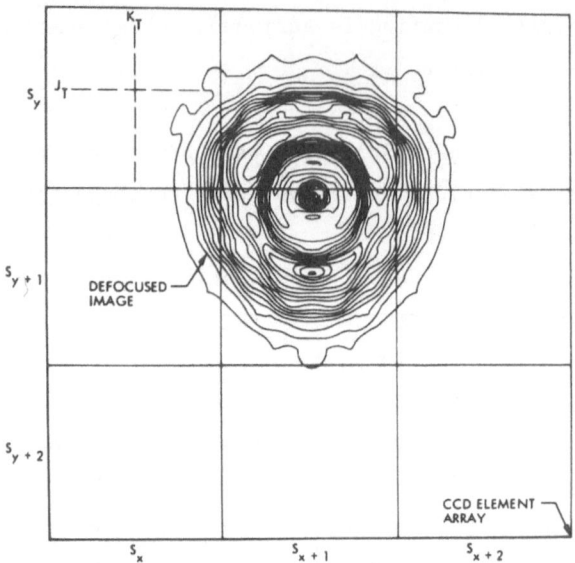

Fig. 53. From Reference 45. Each square represents a CCD pixel. The contours represent a defocused star image of optimal size for position determination by centroid finding.

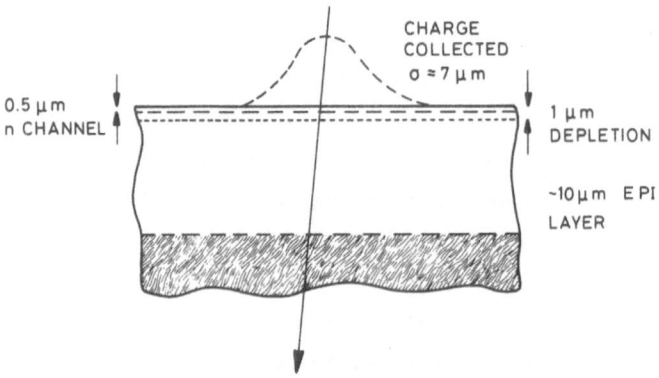

Fig. 54. An idea for a particle tracking CCD with precision of about 0.2 μm. It consists of a very shallow depletion device, and relies for most of the signal on diffusion from the undepleted epitaxial substrate.

7. APPLICATIONS

We shall here look at some examples of the use of silicon detectors as high precision tracking devices for secondary vertex detection. One should not forget the contrasting approach of the FRAMM Collaboration and others (eg NA1 experiment on the CERN SPS) who look for secondary vertices based on a change in pulse height using multiplicity counters downstream of the primary vertex. We consider one example from currently running experiments and one future application as an illustration of current trends.

7.1   Current Results

The first charm lifetime measurements using electronic tracking detectors came from the ACCMOR Collaboration in the NA11 experiment at the CERN SPS, using a telescope of 6 planes of microstrip detectors beyond a beryllium target. This setup is shown in Figure 55, and was followed by a large multiparticle spectrometer which included equipment for the detection of single electrons. This requirement in the trigger and offline analysis produced a charm enrichment factor of about 20 with respect to all inelastic interactions. The incident beam was 200 GeV/c π^-.

With the single electron selection, charm production events could be found, albeit on a substantial background. The addition of the microstrip detectors was particularly useful in rejecting most combinatorial background. With the help of these detectors, it was possible to select the particles emerging from the secondary vertex for making mass fits. The precision on the impact parameter was typically 15 μm, and the effect on the backgrounds was dramatic, due to the rejection of most events (for which all tracks came from the primary vertex) and the rejection of most tracks from the remaining events. Given mean multiplicities for tracked charged particles in these 200 GeV/c π Be interactions of about 10, the background rejection factors from the secondary vertex cuts were as follows:

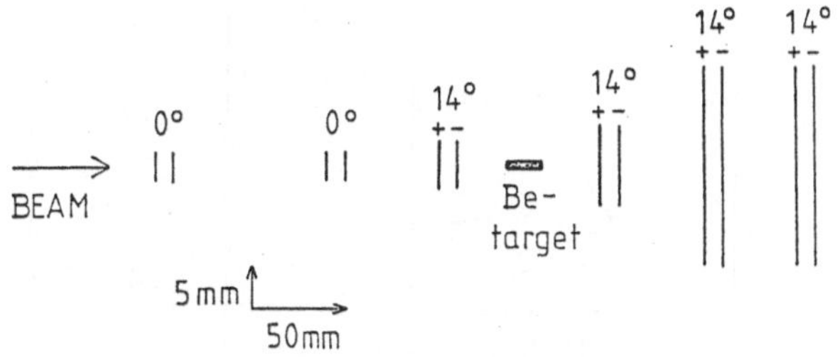

Fig. 55. Side view of NA11 target region showing the beryllium target and the silicon strip detectors. The inclination of the strips to the horizontal direction is indicated.

Decay Mode	Background Rejection
Kπ	300
Kππ	1500
K3π	7100

The result was twofold: clean charm signals on almost no background, and measurement of lifetimes[46].

The microstrip detectors have given some beautifully clean charm events (like the one shown in Figure 56) but in the case of high multiplicity events there are often problems due to merging of the tracks in one or more MSD planes, despite the fact that the first plane is about 4 cm from the vertex. Indeed, this illustrates a general restriction in the power of microstrip detectors. In order to obtain substantial cross-sections for charm, beauty, etc, one is pushed to higher energies. This results in the production being embedded in high multiplicity, tightly collimated jets of secondary particles. In order to make the best possible measurement of the impact parameters, one naturally wants to have the first detector plane as close as possible to the target. In the NA11/NA32 experiment, many events are confused in the reconstruction (assignment of MSD hits to tracks found in the drift chambers) due to the high density of hits (many of them merged together) in the MSDs. This problem could of course be reduced by moving the detectors further away from the interaction point, but at the expense of vertex precision. In order to clean up

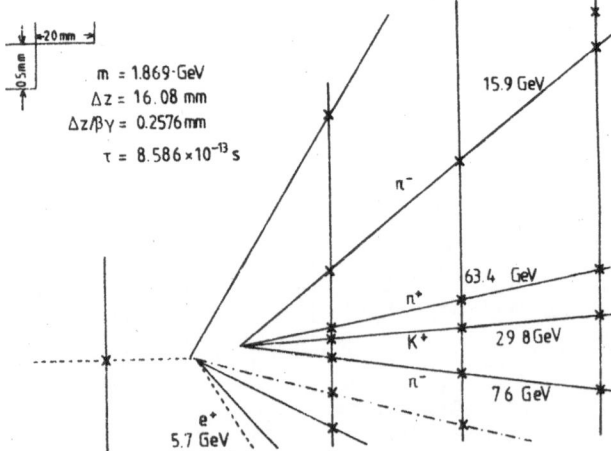

Fig. 56. Reconstruction in one MSD view of the primary and decay vertices in a charm-production event. The decay $\bar{D}^{\circ} \rightarrow K^{+}\pi^{+}\pi^{-}\pi^{-}$ is seen, with all final state particles identified by Cerenkov hodoscopes.

events, and in general to make precise position measurements as close to
the vertex as possible, it was decided to include CCD detectors in the
setup, using the slim cryostat already described in Section 6.3. These are
operated with the first detector as close as 10 mm from the target, and
have allowed unprecedented cleanliness and precision in the vertex
reconstruction. Figure 57 shows a typical event, where there is evident
track merging in the first MSD plane. This is a representation of the
event in one of the MSD viewing directions (14° to the horizontal). In
this view the CCD information looks very confused, but the correct way to
look at the event in the CCDs is face on to the detectors, ie along the
beam direction. This is shown in Figure 58. The upstream detector has 7
hits in 1 mm^2, and shows no problem at all of track merging in spite of the

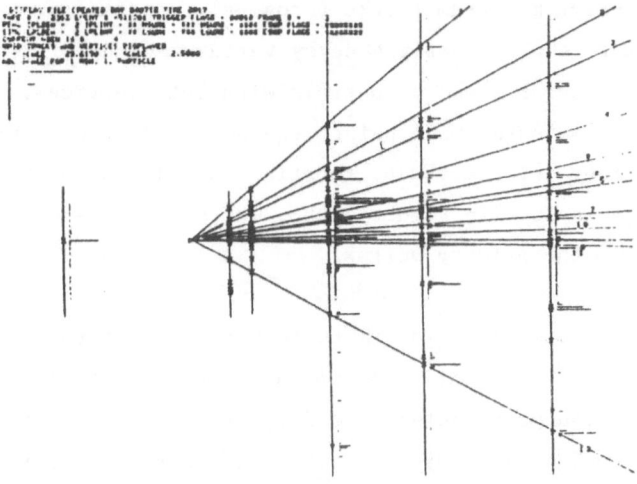

Fig. 57. A complex event in which there is evident track merging in the
first MSD plane. The data from two CCD planes are seen in
projection.

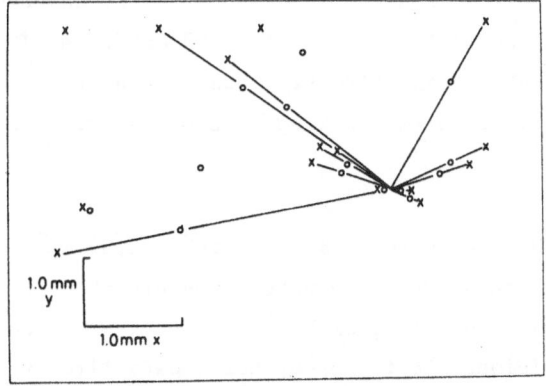

Fig. 58. The event of Figure 57 as seen by the CCD detectors looking
along the beam direction. The circles represent hits in CCD1
and the crosses represent hits in CCD2.

fact that the detector is only 15 mm from the vertex in a high multiplicity interaction at 200 GeV/c. The accidental background is completely negligible. During 1985, data have been taken in a CCD-optimised geometry, using a thin (1 mm) copper target placed 10 mm in front of the first CCD. Charm decays should then be cleanly seen with secondary vertices reconstructed in air (negligible secondary interaction backgrounds). The typical beam flux used in data taking was around 10^6 particles per second through the CCDs. The first charm decay found with this setup is shown in Figure 59. The reconstructed tracks of the outgoing particles are illustrated in Figure 59(a). Note the distorted scale used for visual clarity; the real tracks are all in a very small forward cone. Also shown are the CCD hits used in the track fitting (circles) and unused background hits due to out-of-time beam tracks (crosses). Apart from the production vertex, the figure shows clearly a decay vertex made up of tracks 1, 2 and 5 and possibly 3, which looks compatible with both vertices. The beauty of CCDs is that they measure space points (being pixel-based devices) and so one can get further information by rotating the viewing direction about the beam axis. This is done in Figure 59(b), where one sees that track 3 indeed comes from the primary vertex.

The clearest visualisation is to look face-on at the CCDs; this is seen in Figure 59(c), a beam's view of the event. Virtually all the background hits disappear (being beyond the range of the plot in X or Y) and the event topology is obvious to the eye. Note that this reveals two additional tracks (broken lines) from the primary vertex, which were not found by the reconstruction program, since they fell outside the acceptance of the rest of the spectrometer.

The decay tracks 1, 2 and 5 are identified by Cerenkov hodoscopes and the magnetic spectrometer to be a π^- of 18.80 GeV/c, a K^+ of 22.35 GeV/c and a π^- of 43.22 GeV/c respectively. These reconstruct to an effective mass of 1869 ± 6 MeV so we are looking at a D^- decay. The lifetime was 5.48×10^{-13} s.

Note that CCD1 reconstructs 8 hits within 0.04 mm^2, a density of 200 hits/mm^2. With these very accurate detectors which can be placed so close to the interaction point, we have a "vertex microscope" of unprecedented precision. It is hoped, using data already on tape, to determine precisely the lifetimes of the shorter lived charm particles F and Λ_c.

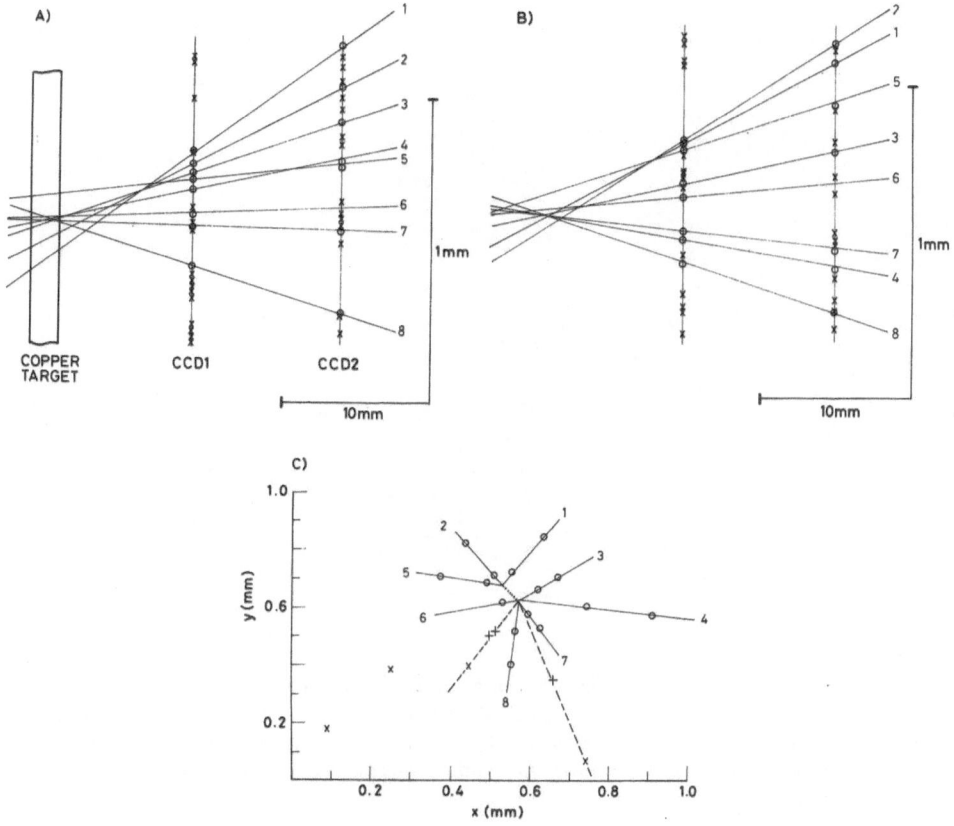

Fig. 59. The first charm decay seen with a CCD vertex detector, from experiment NA32. The crosses are unused CCD hits and the circles are used hits. In figure (c), the inner hit on each track is from CCD1, and the outer hit from CCD2.

7.2   Proposed Future Applications

As we noted in Section 1, events of the type $e^+e^- \rightarrow Z^\circ \rightarrow q\bar{q}$ will be a rich source of physics involving short-lived particles. SLC is a particularly promising environment since the beam pipe can be made small (less synchrotron radiation at large radii than in the circular collider LEP). Thus the detectors can be placed close to the primary vertex. Figure 60 shows the planned barrel vertex detector for SLD[47]. Based on a beam pipe radius of 1 cm, this envisages an inner barrel of CCDs at about

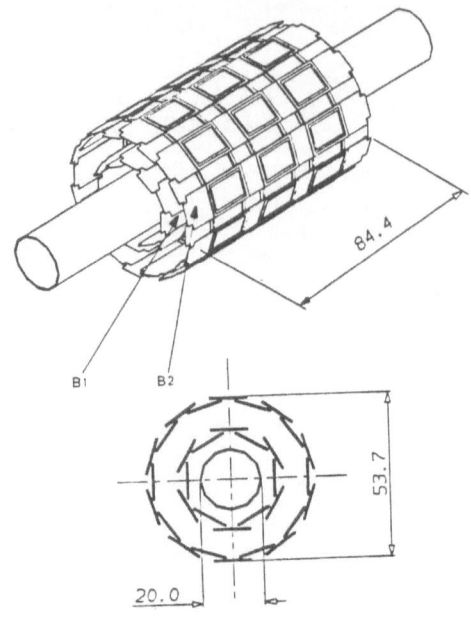

Fig. 60. Arrangement of CCDs to make up the vertex detector for SLD. There is an inner barrel which fits closely around the beam pipe and an outer barrel (partly cut away in the drawing to reveal the inner one) at about twice the radius of the beam pipe.

this radius, and an outer barrel at double this radius. Figure 61 shows the first prototype 'ladder' of CCDs, the basic unit from which the barrels are to be built up.

The calculated detector performance is summarised in Figure 62. This shows the precision of the projected impact parameter with respect to the

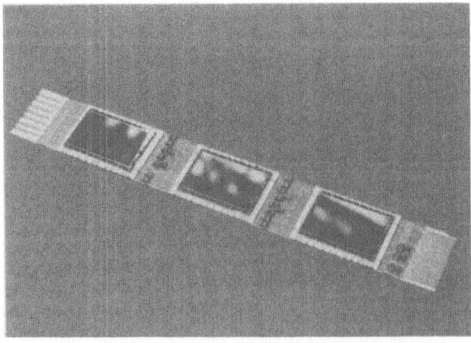

Fig. 61. Prototype CCD ladder for the SLD vertex detector.

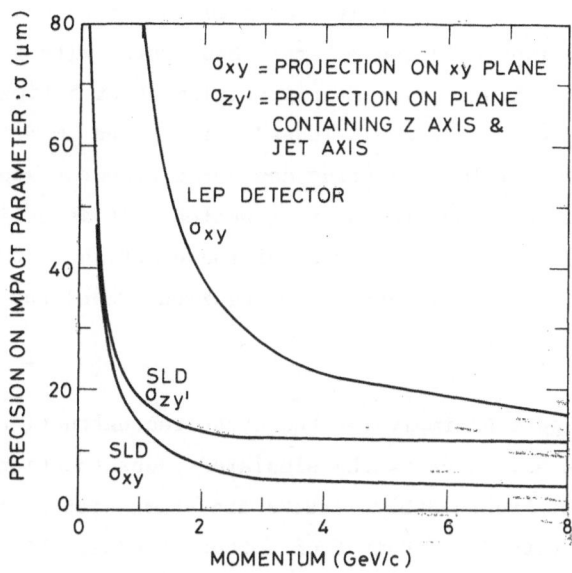

Fig. 62. Precision of track impact parameters with respect to the
primary vertex in two projections
(a) onto a plane (XY) normal to the beam direction
(b) onto a plane (ZY') containing the beam direction and
the jet axis.
The lower curves refer to the SLD design including the CCD
vertex detector and the upper curve refers to the LEP
geometry and assumes 3 barrels of microstrip
detectors.

primary vertex, assuming the combined reconstruction power of the CCD
vertex detector and the SLD central drift chamber. In the beam view of the
event (XY plane) the full precision of the drift chamber gives the curve
shown. The precision degrades badly below 1 GeV/c due to multiple
scattering in the beam pipe and vertex detector. In the orthogonal plane
(ZY') the precision in the vertex detector is unchanged, but the precision
of the impact parameter is worse since here one relies on the Z precision
of the central drift chamber in the overall fit, and this is considerably
worse than the precision in the azimuth. Again at low momenta multiple
scattering dominates, but in both projections there is a large part of the
useful momentum range over which the full precision of the vertex detector
is used.

In contrast, the situation at LEP will be rather less favourable. Operating on an inner detector radius of 8 cm (due to synchrotron radiation background) even a sophisticated 3 barrel MSD system with 3 μm measurement precision will yield impact parameter precision about 5 times worse than at SLD, due to the larger lever arm. Indeed, at LEP the impact parameter precision will be multiple scattering dominated even far above 10 GeV/c, where very few secondary particles are expected. It is unlikely that MSDs at LEP will provide an orthogonal view of the event, and CCDs are excluded because of the very large detector area required around the thick beam pipe.

What does Figure 62 imply for the event reconstruction? An example is shown in Figure 63, which is the simulated reconstruction of the b$\bar{\text{b}}$ production event of Figure 3(b). Tracks are drawn with solid lines if they can be uniquely assigned to their true vertex (whether this be the primary vertex, the B (or $\bar{\text{B}}$) decay vertex or the D (or $\bar{\text{D}}$) decay vertex. Figure 63(a) assumes the SLD precision. We see that all the primary tracks are correctly associated, that 2 of the 6 B decay tracks are unique, and that both D decay tracks are unique. For the LEP reconstruction (Figure 63(b)) all but one of the primary tracks and all of the B decay tracks are ambiguous, while the D decay tracks are unique.

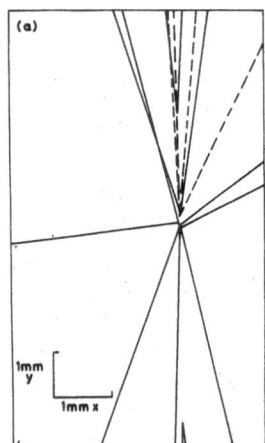

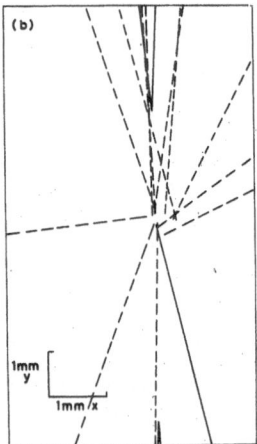

(a). SLD simulation   **(b).** LEP simulation

Fig. 63 Reconstruction of the event Z° → b$\bar{\text{b}}$ of Figure 3(b). Tracks which are ambiguous as to their vertex assignment are shown by broken lines.

By looking at a large number of events of various types, we can arrive at the efficiency figures with which the planned SLD detector is able to uniquely associate tracks with vertices; these are summarised in the table. For comparison, the corresponding figures for the assumed LEP detector are included.

Event Type	Unique Tracks SLD		Unique Tracks LEP	
	Primary Vertex	Decay Tracks	Primary Vertex	Decay Tracks
$Z^\circ \rightarrow c\bar{c}$	91%	83%	51%	41%
$Z^\circ \rightarrow b\bar{b}$	86%	76%	48%	30%
$Z^\circ \rightarrow t\bar{t}$	85%	65%	43%	16%

In the case of the SLD detector, the efficiency falls slowly with increasing quark mass due partly to the increasing topological complexity of the events (more vertices give a higher probability of confusion) and partly due to the falling momenta and so worse multiple scattering of the final state particles (remember Figure 5). These efficiencies are sufficiently high that one can be assured of a great deal of interesting physics with such detectors looking at Z° decays. In the LEP case, the effects of multiple scattering are correspondingly worse.

Microstrip detectors at LEP can cover the large areas needed, which CCDs could not. Microstrip detectors may also have an important role at SLC. Their use is <u>disfavoured</u> for the reasons that

(a) they provide only one view

(b) they need to be placed at a larger radius in order to have a tolerably low level of track merging. This means that they will have poorer precision in measuring the impact parameters of low momentum tracks.

However, their use is <u>favoured</u> for the reason that they may be simpler to implement, especially if the SLC synchrotron radiation background necessitates operation of the inner barrel at a radius of 2 cm or greater.

Fixed target experiments have been a very important testing ground for these high precision detectors. Much physics may still be possible in such experiments, specially at the higher energies available at the Tevatron. From the point of view of event reconstruction, the fixed target environment is preferred to colliders, due to the higher momenta. But the low cross-section for charm and beauty production in hadronic collisions means that a lifetime trigger is highly desirable. To date, there has been less success with trigger schemes than with tracking detectors, and there is an urgent need for new ideas in this area.

Another very challenging problem is that of efficient reconstruction of the topology of complex decays. Even given excellent detector performance, topologies which may be quite clear to the eye in event displays with 3-d rotation capability are not easy to sort out by software. There appears to be almost no published work on this topic, which so far is being handled in a restricted sense by different groups. As collision energies increase, so will the topological complexity of the events. Further constructive thinking in this area would be extremely useful.

8. RADIATION DAMAGE

Silicon detectors form a subset of all MOS devices. Radiation damage is relevant to many users of these devices and has been extensively studied. Much of the work is motivatd by industrial and military applications, but in the field of low level optical imaging the space-based astronomers have particular interest due to Van Allen radiation belts and on-board nuclear power plants.

Microstrip detectors, CCDs, silicon drift chambers, MOS multiplexing chips, all need individual consideration with regard to radiation damage. Rather than go through each in detail, we can get a general feeling for the different classes of effect, and consider CCDs as a typical example which embodies most of the effects relevant to the other devices.

8.1 General Discussion

There are essentially three classes of effect, namely ionization effects, atomic collisions with sufficient momentum transfer to disturb the atom in the crystal lattice, and nuclear interactions which may result in chemical changes (eg $Si \rightarrow P$) and large energy release by nuclear disintegration including α emission, etc.

568

8.1.1 <u>Bulk Damage</u>. For military applications, where intense bursts of
radiation are relevant, bulk effects due to ionization may be important,
causing latch-up and other short term effects. For astronomy and high
energy physics the bulk effects of ionization begin and end with the
generation of charge which may affect the operation of the device at the
time (background signal in detectors, injection of false signal on storage
capacitors etc in readout devices), but which has no long term
consequences. Once the ionizing radiation is switched off, the bulk
material returns to its previous state.

In contrast to this, atomic collisions with high momentum transfer
and nuclear interactions can permanently alter the properties of the bulk
material. Such processes are grouped together as the source of
<u>displacement damage</u>, in which silicon atoms are displaced from their normal
lattice locations. These effects may be local single-atom displacements,
in which case the damage is classified as a <u>point defect</u>; such defects
commonly result from high energy photon or electron irradiation.
Displacement damage may also give rise to <u>damage clusters</u> which consist of
relatively large disturbed regions within the crystal; such defects
commonly result from nuclear interactions of (for example) neutrons and
protons. The most probable events of this type are small angle elastic
Coulomb scattering of silicon nuclei by the incident high energy (charged)
particle. Thus a 50 keV recoil nucleus can create clusters of damage (with
knock-on and stopping of other nuclei) over a volume of several hundred
Ångstroms typical dimensions.

The bulk damage due to the passage of high energy particles can be
described by the number of atomic (silicon) displacements per cm of track
length. For protons traversing silicon, this rate falls from $\sim 10^4$/cm at
1 MeV to $\sim 10^2$/cm at 1 GeV. The displaced atoms give rise to vacancy-
related energy levels within the band gap of the silicon. Many of these
are double levels, and some can acquire charges of twice the electron
charge (see Figue 64).

The first effect of these additional energy levels is to reduce the
minority carrier lifetimes due to the high capture probability, the capture
being followed by recombination with an opposite-sign majority carrier from
the bulk material. Additional effects are:
    a decrease in carrier concentration (at higher irradiation)
    a reduction in carrier mobility (at still higher irradiation)

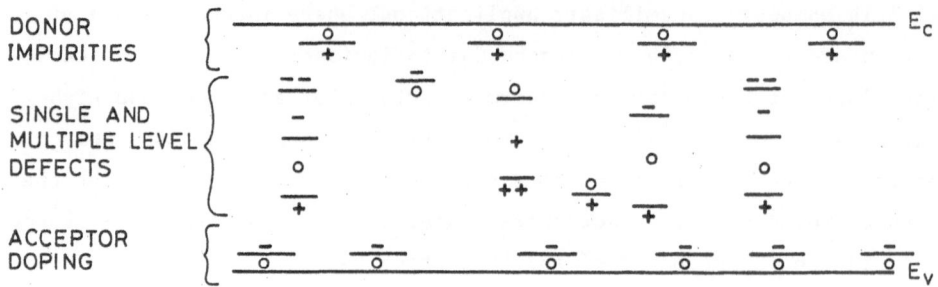

Fig. 64. Sketch indicating the energy levels and charge states of some vacancy-related simple defects in silicon. The charge states are those which apply when the Fermi level is on the corresponding side of the defect level.

In depleted material, specially in silicon close to the internal breadown fields, the damage centres can act as generators of large dark current spikes. (Virtual phase CCDs, for example, are particularly prone to this effect.) In general, the presence of numerous extra energy levels results in additional dark current due to the resultant carrier generation-recombination processes.

A further effect (specially noticeable in high resistivity material) is _defect doping_. Due to the low dopant density in high resistivity material (n type, for example), the Fermi level is initially well above the valence band edge. As the density of radiation-induced defect levels builds up, the inter-band gap becomes heavily populated with a complex system of (almost) randomly distributed energy states. In such a system, the Fermi level will shift so as to achieve charge neutrality in the most economical way, which generally means that it moves towards the middle of the band gap.

Bulk devices fabricated on high resistivity silicon, such as microstrip detectors, are prone to the above effects. Three important phenomena have been observed[48].

The detectors used in the NA32 experiment were 280 μm thick, fabricated on 3 K Ω cm n-type silicon. A high energy pion beam traversed the central region of each detector, resulting (after running times of over a year) in accumulated doses of up to 10^{14} particles/cm^2.

With the detectors operated at room temperature, the first effect
noticed is an incrase in dark current. This follows the form

$$I_D = I_0 + \alpha F \qquad \text{where F is the particle flux per cm}^2.$$

I_D is the dark current per cm^3.

I_0 is the value obtained with a new detector ($\sim$ 100 nA/cm^3).

The measured value of α (from several detectors) is

$$1.3 \pm 0.4 \times 10^{-8} \text{ nA/cm}$$

eg for 10^{13} pions per cm^2, we would have I_D = 130 μA/cm^3 or 3.6 μA/cm^2 in
detectors of the abovementioned thickness. Since 3.8×10^7 min-I
particles/cm^2 deposits 1 rad (100 ergs/gm) in silicon, this implies dark
current generation of 0.014 nA/rad cm^2 in detectors of this thickness.
This is in good agreement with figures from other groups.

The second effect observed in these microstrip detectors is a
decrease in the apparent donor concentration in the n-type material. The
irradiated parts of the detector require an ever decreasing voltage to
reach full depletion, as the Fermi level drifts more and more towards the
mid-band region. This has been explained above, and does not imply a real
reduction in the donor concentration. On the contrary, the increasing dark
current indicates a falling resistivity. Nevertheless, the material does
appear to become more nearly intrinsic when viewed from the standpoint of
the depletion voltage. The effective rate of donor reduction is given by

$$\Delta N_d = \beta F$$

where β is found to be -0.050 ± 0.015 cm^{-1}. This applies to the 3 K Ω cm
material. Eventually the rate of change would necessarily diminish and the
material would settle down with its Fermi level at the mid-band gap. But
in practice, the detectors would be unusable due to dark current before
this stage is reached.

Finally, the reduction in apparent donor concentration implies a
reduced space charge in the irradiated region. Given a rapid gradient in
the irradiation, this leads to distortion of the electric field within the
semiconductor. The column of charge generated by a particle is no longer
collected by drifting normal to the face of the detector, but at an angle
to it. This causes a distortion in the reconstructed track position which
can require quite large corrections.

We should note that any detector based on collecting a signal by
drifting within the bulk of high resistivity silicon will be subject to
such distortions. The silicon drift chamber may be particularly vulnerable
to radiation damage, for this reason.

8.1.2 <u>Damage due to Ionization.</u> While the small energy transfers associated with the ionization process cause no long term effects in bulk silicon, this is not true of the oxide. Electron-hole pairs are liberated in this layer (Figure 65). The band gap in SiO_2 is 8.8 eV but on average 18 eV of ionizing radiation is needed to release an electron-hole pair in silicon dioxide. Some of the electrons and holes recombine, but if there is an electric field across the oxide (as is normally the case in MOS devices) the electrons are rapidly swept out of the oxide layer. The remaining holes drift very slowly (time constant can be $\sim 10^3$ seconds at low temperature). They would eventually disappear at the metal/oxide or silicon/oxide interface depending on the bias polarity. In the case shown in the figure, most holes are transported into the bulk silicon but some are trapped in deep <u>interface traps</u>, where they remain indefinitely (unless the device can be heated to several hundred degrees Centigrade). The probability f_c of hole trapping is a strong function of the processing details, and can vary between 2% (from manufacturers who have the best capability for producing radiation hard devices) to as high as 40%.

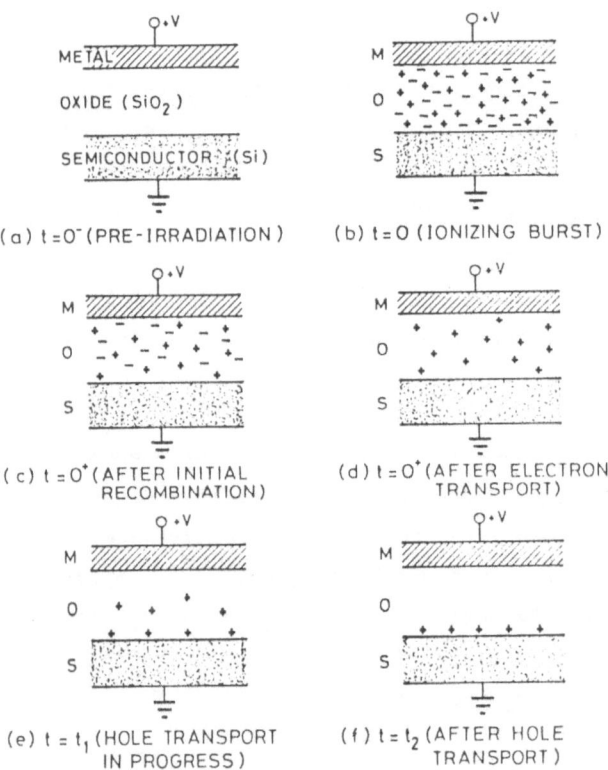

Fig. 65. (From Reference 49.) Time development of charge distributions following a burst of ionizing radiation on a positively biased MOS structure.

The effects of charge buildup can be described in terms of a flat-band voltage shift ΔV_{FB}, this being the gate voltage required to restore the valence and conduction band edges to the levels they have in the deep bulk silicon.

$$\Delta V_{FB} = \frac{Q_{tr}}{C_{ox}}$$

The trapped charge $Q_{tr} = f_c Q_g$

$$\propto t_{ox}$$

where Q_g is the generated charge, which is proportional to the oxide thickness t_{ox}.

$$C_{ox} \propto \frac{1}{t_{ox}}$$

$$\therefore \Delta V_{FB} \propto t_{ox}^2$$

For very thin oxides the dependence can be even faster, like t_{ox}^3.

Thus the effect of ionizing radiation on the oxide can be minimised by

- optimised processing (mainly a question of clean, strain-free original material, use of low temperature processes only, after oxide growth, and the avoidance of impurities which can create deep interface traps).
- thinnest possible gate oxides.

For many devices, the flat-band voltage shift is all that matters, but for surface channel devices the performance depends also on the density of fast surface states. For example, for MOSFETs, these states affect the transconductance and threshold voltage of the transistors. The interface state density is proportional to the dose of ionizing radiation and is not removable by annealing.

8.2 Radiation Damage in CCDs

Bulk damage mainly affects 'bulk-effect' devices such as bipolar transistors. Surface damage mainly affects MOS devices. CCDs are in fact sensitive to both types of damage. In general,

10^4 rads is acceptable (but there are exceptions!)

10^6 rads produces serious degradation or failure,
  unless special precautions are taken in manufacture
  and operation.

10^6 rads corresponds to:

$$10^{15} \quad 15 \text{ MeV neutrons/cm}^2$$
$$2 \times 10^{15} \quad 1 \text{ MeV } \gamma\text{s/cm}^2$$
$$4 \times 10^{13} \quad \text{Min-I particles/cm}^2$$

In e^+e^- collider experiments, the main concern comes from synchrotron radiation-induced soft X-rays. But the limit from synchrotron radiation will normally be set by the confusion from background hits in the detector. This occurs well below the level at which radiation effects become serious.

In fixed target experiments, the passage of an intense hadron beam through the detector can cause local radiation damage on a timescale shorter than the life of a typical experiment.

In hadron colliders such as SSC the radiation levels due to the beam-beam interactions may limit the use of silicon detectors to greater radii than one would like for optimal vertex detection.

There has been a great deal of excellent work on the problem of radiation damage in CCDs, much of it by J Killiany and co-workers at the Naval Research Laboratory in Washington. Reference 50 provides a very useful review of the subject.

8.2.1 <u>Bulk Damage in CCDs</u>. 10^4 rads of neutrons give a factor 100 increase in the room-temperature dark current. The dark current may still be tolerable in cooled CCDs, except for virtual phase devices which can develop such a high density of intense dark current spikes as to be unusable.

The same neutron flux induces trapping centres in the silicon bulk which result in a loss of charge transfer efficiency. The probability of an electron failing to move from one pixel to the next during the charge shifting process becomes very high (about 2×10^{-3}). This gives rise to serious loss of signal and variation in response across the detector.

Fortunately, the scale and character of the damage resulting from the same dose of γs or min-I particles is much reduced. More than 10^6 rads are needed to produce equivalent effects.

The relative seriousness of neutron induced radiation damage should not be overlooked in experimental situations, where measurements of neutron background may explain the degradation of all types of silicon detectors, not only CCDs. Improved neutron shielding may be important for increasing the lifetime of the detectors.

8.2.2 Ionization-Induced Damage. Consider first the problem of charge buildup in the oxide. This affects different sections of the device differently. The charge transfer section is mainly affected by the shift in the flat-band voltage, all gate voltages needing to be correspondingly displaced if the device performance (eg well capacity) is to remain constant. The problems are obviously much worse if the irradiation is non-uniform across the detector area, a phenomenon quite common in vertex detection systems, but not relevant to military or astronomical applications. In such cases, there is not much to be gained by feedback-controlled gate adjustment, a technique which is useful in the case of uniform irradiation. The only clean solution is to select a technology which minimises the charge buildup.

For operation in the temperature range 200-300° K the situation can be made not too serious. Firstly, one should use an n-buried channel structure since in this case the gate voltages are negative with respect to the channel potential and the charge accumulates near to the metal/oxide interface, not the oxide/silicon interface. Secondly, one should use a planar insulator (avoiding the stepped oxide used in some CCD manufacturing processes). Thirdly, one should use a "hard oxide technology"; this covers a whole art which may mean different things to different practitioners. By combining these elements, it is possible to retain good CCD operation after 10^6 rads of ionizing radiation, whereas by neglecting them the devices may not be usable beyond 10^4 rads.

Figure 66(a) shows that while the situation can be made tolerable at higher temperatures, there is a rapid degradation below 200° K. The reason for this is that at lower temperatures relatively shallow trapping centres are able to retain fixed charges, whereas at higher temperatures the thermal energy makes them ineffective. Figure 66(b) shows that even at low temperatures, the problems are very dependent on the applied gate voltage.

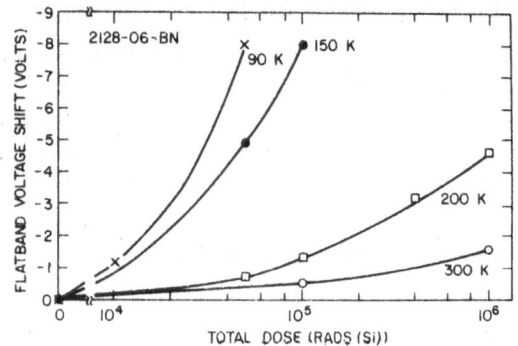

Fig. 66. From Reference 50. Various plots indicating the sensitivity of radiation damage to time, temperature and other parameters. **(a).** Flat-band voltage shift for the CCD radiation-hard oxide as a function of dose at several temperatures, illustrating the increased oxide charge trapping effects at low temperature.

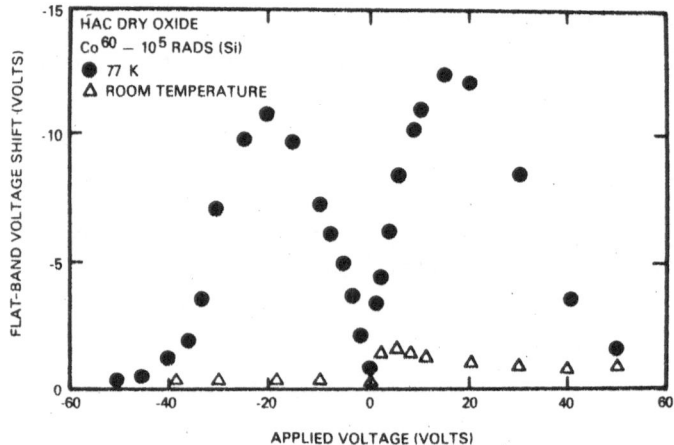

Fig. 66(b). Flat-band voltage shift at 10^5 rads (Si) as a function of applied voltage, illustrating the electric field dependence of the flat-band voltage shift at 77° K.

While it is not normally practicable to have zero bias during irradiation, it is important to be aware of this sharp voltage dependence in considering the optimisation of operating conditions.

These radiation effects are further complicated by the phenomenon of self-annealing. Figure 66(c) shows the time dependence of the flat-band voltage shift after a short burst of radiation. As expected, the self-annealing is most effective at higher temperatures, where the trapped holes have an increased probability of being released by fluctuations in the thermal energy. Even if a detector needs to be operated cold, it can be restored to health by periodically warming it to room temperature, as shown in Figure 66(d). This in fact refers to an input gate threshold voltage shift, but the principle applies generally. The cumulative radiation damage after each anneal is very slow, and corresponds exactly to that which would be obtained in continuous room temperature operation.

If it is absolutely necessary to run a CCD at liquid nitrogen temperature, then a change of insulator is needed. Killiany has manufactured CCDs using a thin oxide/thick nitride insulator (1000 Å of Si_3N_4 and 100 Å of SiO_2) and obtained flat-band voltage shifts as low as -1 V/10^6 rads at 80° K. For particle detection applications, however, one should virtually never need to use such low operating temperatures.

So far we have been discussing the effect of charge buildup in the charge transfer section of the CCD. Let us now turn to its effect on the output section of the device.

The CCD output section is in fact the least radiation sensitive part of the detector due to the following factors:

(a) The reset MOSFET is purely a switch, and so is insensitive to threshold shifts.

(b) The threshold voltage shift in the output MOSFET is reduced due to the fact that the electric field strength within its oxide is typically 4 times less than in the charge transfer section. (Remember Figure 66(b).)

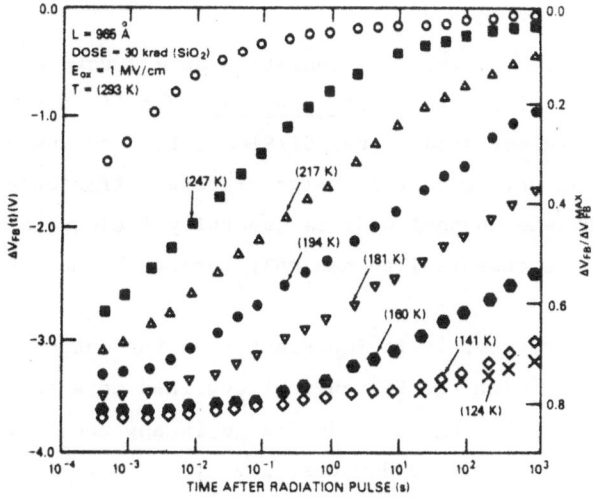

Fig. 66(c). Flat-band voltage shift as a function of time after 30 krad radiation pulse, illustrating annealing of the flat-band voltage shifts at various temperatures.

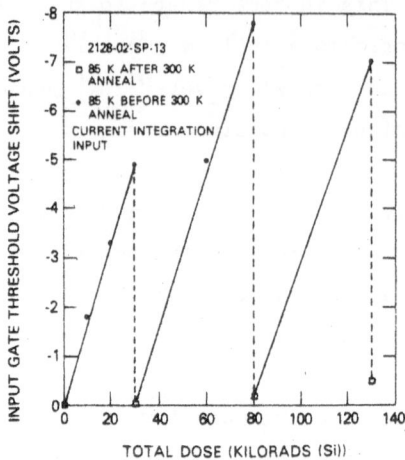

Fig. 66(d). Radiation-hard CCD input gate threshold voltage shift at 85° K as a function of dose and 300° K annealing.

(c) The source follower is AC coupled to the local preamp, so small DC shifts are unimportant. DC coupled systems are much more difficult to control.

(d) The output FET is used as a source follower, and so we are insensitive to changes in g_m. In contrast, high gain stages will inevitably be more radiation sensitive.

This completes the discussion of charge buildup in the oxide. We have still to consider the ionization-induced increase in density of interface states.

This phenomenon (at very low radiation doses) completely ruins the charge transfer efficiency in <u>surface channel</u> CCDs (ie devices in which the potential energy minimum sits at the Si/SiO_2 interface instead of about 1 μm below it). In any case, the charge transfer efficiency of non-irradiated surface channel CCDs is generally inadequate for very small signal levels; it is essential to use only buried channel CCDs.

For the same reason, it is important that the output MOSFET be run in buried channel mode (with the FET current confined to a channel displaced below the surface). This again is desirable in any case for noise optimisation in non-irradiated devices.

The interface states act also as generation centres for the surface component of the dark current. For example, at room temperature, after 10^6 rads the dark current density can increase by a factor of 500, from

2 nA/cm^2 to 1 µA/cm^2. The dark current may still be negligible at 200° K if reasonably fast device readout is employed.

While being applied specifically to CCDs, these comments on radiation damage illustrate most of the factors which apply to other silicon detectors and local preamplifier and multiplexing electronics. Each case is obviously different, and different parts of a device need to be considered separately. There are no simple numbers which one can quote as to the radiation hardness of silicon devices, simply because the variety of structures is so much greater than (for example) plastic scintillator. While this variety complicates the understanding of the problem, it also in many cases leads to a level of flexibility in which solutions can be found. The progress made in radiation hardening of CCDs is the best possible argument for a careful and systematic study of radiation damage, as a necessary prelude to minimising its effect in any type of detector.

9.   FUTURE PROSPECTS

As we have seen, many types of vertex detector have made important contributions to the study of short-lived particles. For the future, we have to think about higher energies which implies higher multiplicities and more tightly collimated jets. Given the clear advantages of having the first layer of vertex detection as close as possible to the production point, the trend will probably be towards pixel-based rather than 1-dimensional detector planes. (Remember that 10 × 1 mm^2 of detector with 20 µm granularity has 50 detecting elements if it is in strip form, and 25000 if it is in pixel form.) A significant reason for the promise of pixel-based detectors is their general utility as imaging devices. Their applicability is truly multi-disciplinary, which is not the case for 1-dimensional detectors. Indeed, the trend towards high resolution TV imaging is already pushing CCDs to granularity as small as 6 × 6 µm (see Reference 51). The requirements of radiation-hard processing are spreading from the defence-related origins of these studies. Many features (such as low temperature processing) are also demanded for other reasons, such as the minimization of implant diffusion in modern very high resolution MOS devices.

The noise optimisation of detectors with extremely small capacitance is a very exciting growth area. The node capacitance of CCDs for low level imaging (astronomy, X-ray imaging, min-I particle detection) can be reduced well below the present levels. Such signals would be optimally sensed with very small on-chip FETs. These FETs would have poor drive capability, and would best be linked to an on-chip drive stage based on a large FET with high tranconductance. Very interesting results have already been achieved[52], and one can look foward to 10 MHz readout rates with less than 10 electrons RMS readout noise. To take advantage of such fine quality in the output stage will require some very excellent design improvements in the imaging area of the device.

Many applications of CCDs involve low data occupancy (most pixels empty). This does not normally apply to optical imaging, but certainly does apply in most X-ray imaging applications, where one wants a low probability of two X-ray conversions in one pixel in order to achieve a reliable energy distribution as well as a 2-d image of the X-ray source. Low data occupancy is clearly a feature of most particle detection applications. This raises the question of a more sophisticated pixel-based device which could be read out selectively, either only the hit pixels, or small groups of pixels known to include one or more hit ones. There are at least two ideas for doing this. One of these[53] is intended to allow truly random access readout of individual pixels with local sensing circuits. The information as to which pixels to read is provided (with some level of ghosts due to ambiguities) by strip sensing in two orthogonal directions. This very interesting idea is likely to be applicable to a pixel-based tracking detector for very high multiplicity conditions where the ambiguity level from strips alone is too high. The pixel granularity may be limited to $\sim$ 50 μm, so the application to vertex detection may be rather limited. Another idea[54] is to add to a conventional CCD the capability of subdivided vertical clocking (in groups of say 10 columns wide) and charge sensing in the corresponding segments of the readout register. This would allow the register to be stacked with a high density of hits ($>$ 1 per 10 pixels) before each line readout. In this way, the readout time for a 1 cm^2 chip could be reduced from $\sim$ 50 ms to 1 ms. This idea can be applied to vertex detectors since it does not imply any coarsening of the granularity of the detector.

A word of caution. Those device experts who have lived through the development of CCDs to their present level of performance warn of many beautiful ideas which have been discarded along the way. Semiconductor devices are full of subtleties, to the point that experts frequently find that ideas which seem very simple, often prove highly complicated in practice, and vice versa. What is most important is that one should have access to the best facilities where these ideas can be tried out in the form of test devices.

An even more important point for high energy physics is to be clear about where we are trying to get to, in defining a long term R & D strategy. A number of points seem to be emerging, on which there is more or less agreement.

(a) To date, all topological reconstructions of vertices have been achieved in fixed target experiments (emulsions, bubble chambers, silicon microstrip detectors and CCDs). The Lorentz boost to the momenta and the corresponding angular compression makes this environment particularly favourable for detectors whose coverage and transparency are at best restricted.

Looking to the future of not-yet approved accelerators (SSC, LHC, colliding e^+e^- linacs) we must aim for vertex detectors operable in the centre of mass of the collisions. We should note in passing that non-CM colliders like HERA are really bad for vertex detection; the products of short-lived decays are thrown forward, but contained for a long way inside the beam pipe and so inaccessible to close-in detectors.

(b) A colliding e^+e^- linac of TeV CM energy presents an enormous challenge to the machine builders, but if one can be built, it should be a wonderful environment for vertex detectors. SLC is the prototype for such developments, and the synchrotron radiation backgrounds in a linear collider will be even less. The very highly collimated jets, rich in all possible quark flavours, will be best studied with very fine grained pixel devices. Given the possibility of switching off the machine (and so the backgrounds) during readout, as at SLC, CCDs (by this time with precision < 1 μm) should be very suitable for the vertex detection.

(c) SSC is possibly less of a challenge from the point of view of accelerator design, but it will be very difficult to use at the design luminosity of 10^{33} cm^{-2} s^{-1}. The beam-beam collisions throw ~ 800 watts of power into the detectors. A vertex detector sitting at a 1 cm radius would receive ~ 10^7 rads/year, and 3×10^6 rads/year at 2 cm. To date, CCDs are limited to ~ 10^6 rads even using the very best processing techniques. But the technology of growing highly reliable thin oxide is continually evolving, and 10 Mrad hard devices may well be available within the next 10 yars. There is fortunately a strong non-HEP impetus to make such developments. Given the intrinsically low cost of MOS fabrication, there is no reason to reject the idea of throwing away the detector chips and replacing them once or twice a year.

The desirability of pixel devices (as opposed to strips) is clear, but could they conceivably be used in the background conditions of SSC? One option is to have a fast clear/fast readout capability, as in the design being studied by the LBL group[53]. This could well be precisely what is needed for a compact tracking detector for SSC. To achieve the desired precision for vertex detection, however, an inner barrel of a more 'pure' CCD type would be needed. The suggested fast CCD[54] could certainly handle the general flux of particles traversing it during the pre-trigger period (by using the fast-clearing scheme which in NA32 succeeded in achieving negligible backgrounds with a flux of 10^6 particles cm^{-2} s^{-1} through the CCDs; see Section 6). However, even with a 1 ms readout time, this flux would be unacceptable from the point of view of pixel occupancy and pattern recognition. 1000 hits/mm^2 with 2500 pixels/mm^2 is at least a factor of 100 too high. One option which should be seriously considered is to design SSC so that it can be de-tuned (on receipt of a trigger) with a time constant of $\sim$ 1 μs, to 1% or less of its normal luminosity, and retuned on a similar timescale at the end of the detector readout. This feature could perhaps be implemented by kicker dipoles and/or pulsed quadrupoles to modify the beam focus conditions. Such an arrangement may be advantageous, if not essential, for other SSC detectors. The trend (seen already on the planned SLD detector) is towards on-board multiplexing electronics, so as to reduce almost to zero the quantity of cabling emerging from the detector. (This is essential if good hermeticity of the calorimetry is to be achieved, and allows unprecedented granularity of all detectors, needed in view of the finely collimated jets of high multiplicity.) All this on-board electronics is radiation sensitive, not only from the point of view of damage but also due to charge buildup on storage capacitors during the readout period. For the readout electronics at moderate radius, the abovementioned idea of turning down SSCs luminosity during readout may be very useful.

In conclusion, silicon vertex detectors seem capable of meeting some if not all of the challenges of topological reconstruction in the colliders which are now being considered for construction. Pixel based devices seem more promising than 1-dimensional detectors such as silicon microstrips and optical fibres. However, the technical challenges are considerable, and it may be that the problems will be solved in an entirely different manner.

582

What is clear is that while there is an increasing number of groups participating in these developments, the efforts are not yet matched to the scale of the problem, nor to the physics potential. What is needed are both new ideas and the backing of people and facilities to conduct a steady long term R & D programme in this area. The work going on at the Naval Research Laboratory in Washington for more than a decade to evolve ever more radiation resistant CCDs for military applications is an example of what could be done within the HEP laboratories. In any case, we shall contintue to profit from extremely close collaboration with other fields and with industry, especially if we are dealing with pixel-based devices which have a general imaging capability as well as detecting particles.

ACKNOWLEDGEMENTS

I owe a great debt to many experts on silicon detectors and signal processing who have provided me with much valuable information and advice. Among these are David Burt of GEC (England), Joe Killiany of the Naval Research Laboratory, Washington, Velkjo Radeka of BNL, Emilio Gatti of Milano, Rangy Kandiah of AERE (Harwell), Pier Francesco Manfredi of Pavia, and Erik Heijne of CERN. In addition, some of my fellow high energy physicists who became interested in silicon detectors around the same time as I did have provided a very exciting and stimulating exchange of ideas. I am especially grateful to Berhard Hyams of CERN, Robert Klanner and Gerhard Lutz of MPI, Munich and Sherwood Parker of Hawaii. My colleagues at RAL and Brunel University have formed a splendid collaboration, all of them making essential contributions to the success of the CCD detector work, particularly Gerry Agnew, Bob English, Tony Gillman, Laurie Lintern, Richard Stephenson, Colin Uden, Julia Walsh, Steve Watts and Fred Wickens.

Finally, I would like to thank Tom Ferbel for inviting me to give these lectures at the Advanced Study Institute.

REFERENCES

1.   B Stech, Heidelberg Preprint HD-THEP-84-8 (1984).

2.   A J Buras et al, MPI-PAE/PTh 7/84 (1984).

3.   V Barger and R J N Phillips, MAD/PH/155 (1984).

4.   H Lynch, SLD New Detector Note No 117 (1984).

5.   V Barger et al, MAD/PH/150 (1984).

6.   V Barger et al, MAD/PH/152 (1984).

7.   V Barger et al, MAD/PH/184 (1984).

8.   K Niu et al, Prog Theor Physics $\underline{46}$ (1971) 1644.

9.   M I Adamovich et al, Physics Lett $\underline{99B}$ (1981) 271.

10. K Niu, Proceedings of the International Symposium on Cosmic Rays and Particle Physics, Tokyo (1984) 642.

11. M I Adamovich et al, CERN-SPSC-85-60 (1985).

12. J P Albanese et al, Phys Lett $\underline{158B}$ (1985) 186.

13. M Aguilar-Benitez et al, CERN EP 85-103 (1985).

14. J A Jaros, SLAC Report 250 (1982) 29.

15. G Goldhaber, Berkeley Preprint, LBL-20445 (1985).

16. E Fernandez et al, SLAC-PUB-3390 (1984).

17. R Ruchti et al, Proc 1984 Summer Study on SSC, Snowmass Colarado (1984) 615.

18. M N Atkinson et al, Nucl Instr and Methods $\underline{225}$ (1984) 1.

19. M Atkinson et al, CERN/SPSC 85-81.

20. K G McKay, Phys Rev $\underline{84}$ (1951) 829.

21. B Rossi, 'High Energy Particles', Prentice-Hall (1952).

22. H A Bethe, Handb Physik $\underline{24/1}$ (1933) 491.

23. L Landau, J Phys USSR $\underline{8}$ (1944) 201.

24. O Blunck and S Leisegang, Z Physik $\underline{128}$ (1950) 500.

25. V A Chechin and V C Ermilova, Nucl Instr & Methods $\underline{136}$ (1976) 551.

26. V C Ermilova et al, Nucl Instr & Methods $\underline{145}$ (1977) 555.

27. W Allison and J Cobb, Ann Rev Nucl Sci $\underline{30}$ (1980) 253.

28. G Hall, Nucl Instr & Methods $\underline{220}$ (1984) 356.

29. R Bailey et al, Nucl Instr & Methods $\underline{213}$ (1983) 201.

30. R S Muller and T I Kamins, 'Device Electronics for Integrated Circuits', Wiley (1977).

31. A S Grove, 'Physics and Technology of Semiconductor Devices', Wiley (1967).

32. R A Greiner, 'Semiconductor Devices and Applications', McGraw-Hill (1961).

33. J B A England et al, Nucl Instr & Methods $\underline{185}$ (1981) 43.

34. B Hyams et al, Nucl Instr & Methods $\underline{205}$ (1983) 99.

35. J T Walker et al, Nucl Instr & Methods $\underline{226}$ (1984) 200.

36. G Anzivino et al, CERN-EP/85-148 (1985).

37. S Parker (Private Communication).

38. E Gatti and P Rehak, Nucl Instr & Methods $\underline{225}$ (1984) 608.

39. P Rehak et al, NIM $\underline{A235}$ (1985) 224.

40. W S Boyle and G E Smith, Bell System Technical Journal $\underline{49}$ (1970) 587.

41. C J S Damerell et al, Nucl Instr & Methods $\underline{185}$ (1981) 33.

42. J D E Beynon and D R Lamb, 'Charge-Coupled Devices and their Applications', McGraw-Hill (1980).

43. M J Howes and D V Morgan (Ed), 'Charge-Coupled Devices and Systems', Wiley (1979).

44. W S Chan, Journal of the Society of Photo-Optical Instrumentation Engineers (SPIE) 244 (1980) 81.

45. P M Salomon, Journal of the Society of Photo-Optical Instrumentation Engineers (SPIE) 203 (1979) 130.

46. Proc Int Europhysics Conference on HEP, Bari (1985) p223.

47. SLD Design Report SLAC-273 (1984).

48. R Klanner. Results presented at the 4th European Symposium on Semiconductor Detectors, Munich (1986).

49. V van Lint. Reported at the 4th European Symposium on Semiconductor Detectors, Munich (1986).

50. J M Killiany, Topics in Apply Physics 38 (1980) 147.

51. G J Declerck et al, IEEE Trans on Electron Devices ED-32 (1985) 1551.

52. I Akiyoma et al, Proc IEEE International Solid State Circuits Conference (1986) 96.

53. H Spieler of LBL Berkeley. Reported at RAL Meeting on Imaging Applications of Solid State Detectors (1986).

54. C Damerell, Possibility For a "Fast" CCD. Internal RAL Note.

Participants at the ASI From left to right: S. Mishra, J. Schukraft, C. Lirakis, D. Hitlin, B. Strongin, V. Mertens, S. Weseler, K. Bachmann, T. Yasuda, D. Gross, T. Yamanaka, S. Stapnes, M. Woods, H. Ziock, G. Wolf, M. Caccia, B. DeLotto, P. Campana, C. Damerell, P. Rowson, D. Dimitroyannis, N. Wermes, J. Dunlea, I. Kourbanis, E. Low, I. Brock, R. Kazimi, T. Behnke, C. Matthews, A. Farilla, P. Tomkins, R. Magahiz, C. Adolphsen, A. Satir, S. Tzamarias, K. Thorne, F. Bird, A. Sinanidis, K. Yasuoka, M. Franklin, F. Dittus, K. O'Shaughnessy, R. Yoshida, D. Nesic, D. Clarke, T. Mori, G. Siroli, A. Maio, C. Yosef, G. Ballocchi, R. Jones, T. Ferbel, R. Cirio, M. Panter, E. Milotti, C. Biino, A. Lanaro, S. Palestini, R. Breedon, M. Mattingly, A. Savoy-Navarro, C. Hawkes, T. Alvarez, I. Hinchliffe, L. DiLella, and J. Ryan. Missing: J. Rees

LECTURERS

C. Damerell	Rutherford-Appleton Laboratory
L. DiLella	CERN
D. Gross	Princeton University
I. Hinchliffe	Lawrence Berkeley Laboratory
D. Hitlin	California Inst. of Technology
J. Rees	Stanford University
A. Savoy-Navarro	CEN Saclay
G. Wolf	DESY

SCIENTIFIC ADVISORY COMMITTEE

M. Jacob	CERN
R. Palmer	Brookhaven National Laboratory
R. Peccei	DESY
C. Quigg	Fermilab
P. Soding	DESY
R. Taylor	Stanford Linear Accelerator
M. Tigner	Cornell University

SCIENTIFIC DIRECTOR

T. Ferbel	University of Rochester

INDEX